Student Solutions Manual and Study Guide

FOR

SERWAY AND JEWETT'S

PHYSICS

FOR SCIENTISTS AND ENGINEERS

VOLUME TWO
EIGHTH EDITION

John R. Gordon
Emeritus, James Madison University

Ralph V. McGrew
Broome Community College

Raymond A. Serway
Emeritus, James Madison University

Australia • Brazil • Japan • Korea • Mexico • Singapore • Spain • United Kingdom • United States

ISBN-13: 978-1-4390-4852-8
ISBN-10: 1-4390-4852-5

Brooks/Cole
20 Channel Center St.
Boston, MA 02210

Cengage Learning products are represented in Canada by Nelson Education, Ltd.

For your course and learning solutions, visit **www.cengage.com**

Purchase any of our products at your local college store or at our preferred online store **www.ichapters.com**

Printed in United States of America
1 2 3 4 5 6 7 14 13 12 11 10

PREFACE

This *Student Solutions Manual and Study Guide* has been written to accompany the textbook **Physics for Scientists and Engineers,** Eighth Edition, by Raymond A. Serway and John W. Jewett, Jr. The purpose of this *Student Solutions Manual and Study Guide* is to provide students with a convenient review of the basic concepts and applications presented in the textbook, together with solutions to selected end-of-chapter problems from the textbook. This is not an attempt to rewrite the textbook in a condensed fashion. Rather, emphasis is placed upon clarifying typical troublesome points and providing further practice in methods of problem solving.

Every textbook chapter has in this book a matching chapter, which is divided into several parts. Very often, reference is made to specific equations or figures in the textbook. Each feature of this Study Guide has been included to ensure that it serves as a useful supplement to the textbook. Most chapters contain the following components:

- **Equations and Concepts:** This represents a review of the chapter, with emphasis on highlighting important concepts, and describing important equations and formalisms.

- **Suggestions, Skills, and Strategies:** This offers hints and strategies for solving typical problems that the student will often encounter in the course. In some sections, suggestions are made concerning mathematical skills that are necessary in the analysis of problems.

- **Review Checklist:** This is a list of topics and techniques the student should master after reading the chapter and working the assigned problems.

- **Answers to Selected Questions:** Suggested answers are provided for approximately 15 percent of the objective and conceptual questions.

- **Solutions to Selected End-of-Chapter Problems:** Solutions are shown for approximately 20 percent of the problems from the text, chosen to illustrate the important concepts of the chapter. The solutions follow the *Conceptualize—Categorize—Analyze—Finalize* strategy presented in the text.

A note concerning significant figures: When the statement of a problem gives data to three significant figures, we state the answer to three significant figures. The last digit is uncertain; it can for example depend on the precision of the values assumed for physical constants and properties. When a calculation involves several steps, we carry out intermediate steps to many digits, but we write down only three. We "round off" only at the end of any chain of calculations, never anywhere in the middle.

We sincerely hope that this *Student Solutions Manual and Study Guide* will be useful to you in reviewing the material presented in the text and in improving your ability to solve problems and score well on exams. We welcome any comments or suggestions which could help improve the content of this study guide in future editions, and we wish you success in your study.

John R. Gordon
Harrisonburg, Virginia

Ralph V. McGrew
Binghamton, New York

Raymond A. Serway
Leesburg, Virginia

Acknowledgments

We are glad to acknowledge that John Jewett and Hal Falk suggested significant improvements in this manual. We are grateful to Charu Khanna and the staff at MPS Limited, A Macmillan Company for assembling and typesetting this manual and preparing diagrams and page layouts. Susan English of Durham Technical Community College checked the manual for accuracy and suggested many improvements. We thank Brandi Kirksey (Associate Developmental Editor), Mary Finch (Publisher), and Cathy Brooks (Senior Content Project Manager) of Cengage Learning, who coordinated this project and provided resources for it. Finally, we express our appreciation to our families for their inspiration, patience, and encouragement.

Suggestions for Study

We have seen a lot of successful physics students. The question, "How should I study this subject?" has no single answer, but we offer some suggestions that may be useful to you.

1. Work to understand the basic concepts and principles before attempting to solve assigned problems. Carefully read the textbook before attending your lecture on that material. Jot down points that are not clear to you, take careful notes in class, and ask questions. Reduce memorization of material to a minimum. Memorizing sections of a text or derivations would not necessarily mean you understand the material.

2. After reading a chapter, you should be able to define any new quantities that were introduced and discuss the first principles that were used to derive fundamental equations. A review is provided in each chapter of the Study Guide for this purpose, and the marginal notes in the textbook (or the index) will help you locate these topics. You should be able to correctly associate with each *physical quantity* the *symbol* used to represent that quantity (including vector notation if appropriate) and the SI *unit* in which the quantity is specified. Furthermore, you should be able to express each important formula or equation in a concise and accurate prose statement.

3. Try to solve plenty of the problems at the end of the chapter. The worked examples in the text will serve as a basis for your study. This Study Guide contains detailed solutions to about fifteen of the problems at the end of each chapter. You will be able to check the accuracy of your calculations for any odd-numbered problems, since the answers to these are given at the back of the text.

4. Besides what you might expect to learn about physics concepts, a very valuable skill you can take away from your physics course is the ability to solve complicated problems. The way physicists approach complex situations and break them down into manageable pieces is widely useful. At the end of Chapter 2, the textbook develops a general problem-solving strategy that guides you through the steps. To help you remember the steps of the strategy, they are called *Conceptualize, Categorize, Analyze,* and *Finalize.*

General Problem-Solving Strategy

Conceptualize

- The first thing to do when approaching a problem is to *think about* and *understand* the situation. Read the problem several times until you are confident you understand what is being asked. Study carefully any diagrams, graphs, tables, or photographs that accompany the problem. Imagine a movie, running in your mind, of what happens in the problem.

- If a diagram is not provided, you should almost always make a quick drawing of the situation. Indicate any known values, perhaps in a table or directly on your sketch.

- Now focus on what algebraic or numerical information is given in the problem. In the problem statement, look for key phrases such as "starts from rest" ($v_i = 0$), "stops" ($v_f = 0$), or "falls freely" ($a_y = -g = -9.80$ m/s^2). Key words can help simplify the problem.

- Next, focus on the expected result of solving the problem. Precisely what is the question asking? Will the final result be numerical or algebraic? If it is numerical, what units will it have? If it is algebraic, what symbols will appear in the expression?

- Incorporate information from your own experiences and common sense. What should a reasonable answer look like? What should its order of magnitude be? You wouldn't expect to calculate the speed of an automobile to be 5×10^6 m/s.

Categorize

- After you have a really good idea of what the problem is about, you need to *simplify* the problem. Remove the details that are not important to the solution. For example, you can often model a moving object as a particle. Key words should tell you whether you can ignore air resistance or friction between a sliding object and a surface.

- Once the problem is simplified, it is important to *categorize* the problem. How does it fit into a framework of ideas that you construct to understand the world? Is it a simple *plug-in problem*, such that numbers can be simply substituted into a definition? If so, the problem is likely to be finished when this substitution is done. If not, you face what we can call an *analysis problem*—the situation must be analyzed more deeply to reach a solution.

- If it is an analysis problem, it needs to be categorized further. Have you seen this type of problem before? Does it fall into the growing list of types of problems that you have solved previously? Being able to classify a problem can make it much easier to lay out a plan to solve it. For example, if your simplification shows that the problem can be treated as a particle moving under constant acceleration and you have already solved such a problem (such as the examples in Section 2.6), the solution to the new problem follows a similar pattern. From the textbook you can make an explicit list of the analysis models.

Analyze

- Now, you need to analyze the problem and strive for a mathematical solution. Because you have categorized the problem and identified an analysis model, you can select relevant equations that apply to the situation in the problem. For example, if your categorization shows that the problem involves a particle moving under constant acceleration, Equations 2.13 to 2.17 are relevant.

- Use algebra (and calculus, if necessary) to solve symbolically for the unknown variable in terms of what is given. Substitute in the appropriate numbers, calculate the result, and round it to the proper number of significant figures.

Finalize

- This final step is the most important part. Examine your numerical answer. Does it have the correct units? Does it meet your expectations from your conceptualization of the problem? What about the algebraic form of the result—before you substituted numerical values? Does it make sense? Try looking at the variables in it to see whether the answer would change in a physically meaningful way if they were drastically increased or decreased or even became zero. Looking at limiting cases to see whether they yield expected values is a very useful way to make sure that you are obtaining reasonable results.

- Think about how this problem compares with others you have done. How was it similar? In what critical ways did it differ? Why was this problem assigned? You should have learned something by doing it. Can you figure out what? Can you use your solution to expand, strengthen, or otherwise improve your framework of ideas? If it is a new category of problem, be sure you understand it so that you can use it as a model for solving future problems in the same category.

When solving complex problems, you may need to identify a series of subproblems and apply the problem-solving strategy to each. For very simple problems, you probably don't need this whole strategy. But when you are looking at a problem and you don't know what to do next, remember the steps in the strategy and use them as a guide.

Work on problems in this Study Guide yourself and compare your solutions to ours. Your solution does not have to look just like the one presented here. A problem can sometimes be solved in different ways, starting from different principles. If you wonder about the validity of an alternative approach, ask your instructor.

5. We suggest that you use this Study Guide to review the material covered in the text and as a guide in preparing for exams. You can use the sections Review Checklist, Equations and Concepts, and Suggestions, Skills, and Strategies to focus in on points that require further study. The main purpose of this Study Guide is to improve the efficiency and effectiveness of your study hours and your overall understanding of physical concepts. However, it should not be regarded as a substitute for your textbook or for individual study and practice in problem solving.

TABLE OF CONTENTS

23

Electric Fields

EQUATIONS AND CONCEPTS

Coulomb's law gives the magnitude of the electrostatic force between two stationary point charges, q_1 and q_2, separated by a distance r. The term *point charge* is used to indicate a particle of zero size that carries an electric charge.

$$F_e = k_e \frac{|q_1||q_2|}{r^2} \quad (23.1)$$

The **Coulomb constant,** k_e, in Equation 23.1 can be expressed in terms of the permittivity of free space, ϵ_0.

$$k_e = 8.987\,6 \times 10^9 \text{ N}\cdot\text{m}^2/\text{C}^2 \quad (23.2)$$

$$k_e = \frac{1}{4\pi \epsilon_0} \quad (23.3)$$

$$\epsilon_0 = 8.854\,2 \times 10^{-12} \text{ C}^2/\text{N}\cdot\text{m}^2 \quad (23.4)$$

An **approximate value** for k_e may be used in solving end-of-chapter problems.

$$k_e = 8.99 \times 10^9 \text{ N}\cdot\text{m}^2/\text{C}^2$$

The **elementary charge**, e, is the magnitude of charge on an electron $(-e)$ and a proton $(+e)$.

$$e = 1.602\,18 \times 10^{-19} \text{ C} \quad (23.5)$$

The **vector form of Coulomb's law** includes a unit vector, $\hat{\mathbf{r}}$. The vector $\vec{\mathbf{F}}_{12}$ is the force on q_2 due to q_1. The unit vector $\hat{\mathbf{r}}$ is directed from q_1 to q_2. *Coulomb's law applies exactly only to point charges. Regardless of the relative magnitudes of the two charges,* $\vec{\mathbf{F}}_{21} = -\vec{\mathbf{F}}_{12}$*; this follows from Newton's third law.*

$$\vec{\mathbf{F}}_{12} = k_e \frac{q_1 q_2}{r^2} \hat{\mathbf{r}}_{12} \quad (23.6)$$

The direction of the electrostatic force on a charge is determined from the experimental observation that like sign charges experience forces of mutual repulsion and unlike sign charges attract each other.

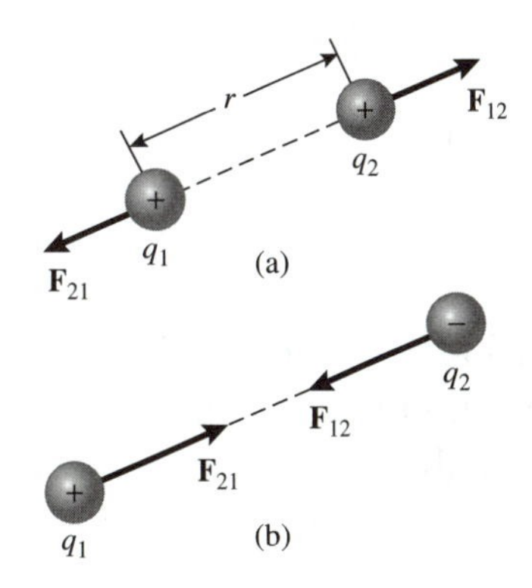

The principle of superposition applies when more than two point charges are present. The resultant electrostatic force exerted on any one charge equals the vector sum of the forces exerted on that charge by each of the other charges individually. The figure at right shows the resultant force (the open arrow) on q_3 due to q_1 and q_2.

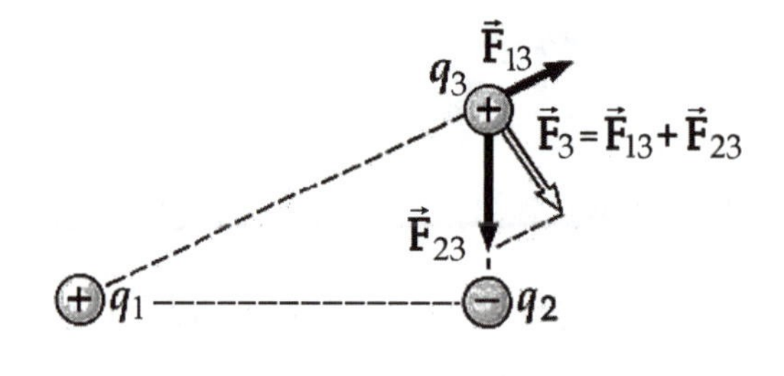

The **electric field vector at any point in space** is defined as the electric force per unit charge exerted on a small positive test charge placed at the point where the field is to be determined. In the figure at right, it is important to note that $\vec{\mathbf{E}}$ is the electric field produced by the source charge $+Q$, and is not the field due to the test charge q_0. *The direction of the electric field at any point is the direction of force on a* <u>*positive*</u> *test charge placed at the point.*

$$\vec{\mathbf{E}} \equiv \frac{\vec{\mathbf{F}}_e}{q_0} \qquad (23.7)$$

Q

q_0 $\vec{\mathbf{E}}$

Test charge

The **electric field a distance *r* from a point charge** is given by Equation 23.9. The unit vector $\hat{\mathbf{r}}$ is directed away from q and toward the point where the field is to be calculated. *The direction of the electric field is radially outward from a positive point charge and radially inward toward a negative point charge.*

$$\vec{\mathbf{E}} = k_e \frac{q}{r^2}\hat{\mathbf{r}} \qquad (23.9)$$

q_0 $\vec{\mathbf{E}}$ P r q (a)

q_0 P $\vec{\mathbf{E}}$ q (b)

The **electric field due to a group of charges** equals the vector sum of the electric fields due to each of the individual charges. *First use Equation 23.9 to calculate the electric field vector due to each individual charge and then find the vector sum.*

$$\vec{\mathbf{E}} = k_e \sum_i \frac{q_i}{r_i^2}\hat{\mathbf{r}}_i \text{ (vector sum)} \qquad (23.10)$$

Electric field lines are a convenient graphical representation of the electric field in the vicinity of a group of charges or a charge distribution. Lines are drawn so that:

- Lines begin on a positive charge and terminate on a negative charge (or at infinity).
- The number of lines leaving a positive charge (or approaching a negative charge) is proportional to the magnitude of the charge.
- No two field lines can cross.

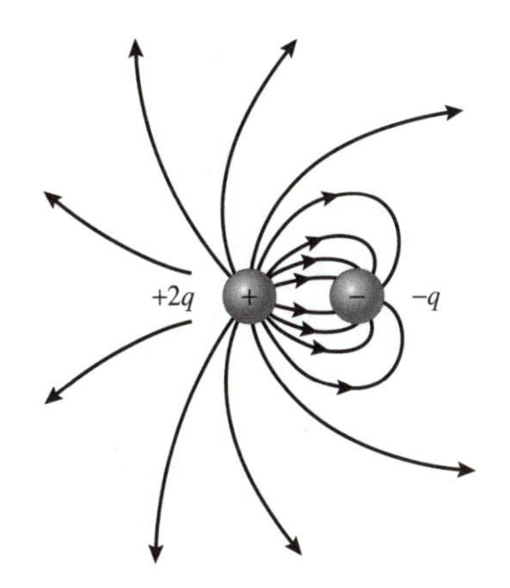

Electric field lines are related to the electric field in the following manner:

- The electric field vector is tangent to an electric field line at each point.
- The number of lines per unit area through a surface perpendicular to the field lines is proportional to the magnitude of the electric field in that region.

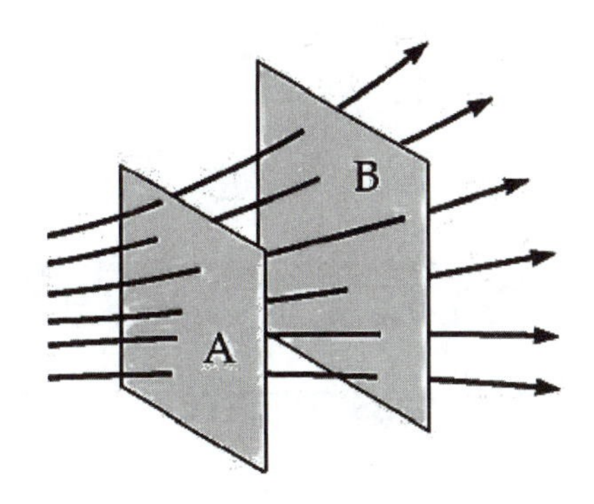

The **electric field of a continuous charge distribution** is found by integrating over the entire region that contains the charge. This is a vector operation and can usually be carried out easily when the charge is distributed along a line, over a surface, or throughout a volume and when there is a high degree of symmetry.

$$\vec{\mathbf{E}} = k_e \int \frac{dq}{r^2}\hat{\mathbf{r}} \qquad (23.11)$$

The concept of charge density is utilized to perform the integration shown in Equation 23.11 above. It is convenient to represent a charge increment dq as the product of an element of length, area, or volume and the charge density over that region.

For an element

of length dx, $\quad dq = \lambda\, dx$

of area dA, $\quad dq = \sigma\, dA$

of volume dV, $\quad dq = \rho\, dV$

For **uniform charge distributions** the volume charge density (ρ), the surface charge density (σ), and the linear charge density (λ) can be calculated based on the total charge and the geometric size (length, area, or volume) of the charge distribution.

$$\lambda \equiv \frac{Q}{\ell} \qquad \sigma \equiv \frac{Q}{A} \qquad \rho \equiv \frac{Q}{V}$$

For a non-uniform charge distribution, the densities λ, σ, and ρ must be stated as functions of position.

The **acceleration of a charged particle** in an electric field has a magnitude that is proportional to the magnitude of the field and has a direction that depends on the sign of the charge. *Positively charged particles accelerate along the direction of the field; and negatively charged particles accelerate opposite the direction of the field.*

$$\vec{\mathbf{a}} = \frac{q\vec{\mathbf{E}}}{m} \tag{23.12}$$

SUGGESTIONS, SKILLS, AND STRATEGIES

PROBLEM-SOLVING STRATEGY FOR ELECTRIC FORCES AND FIELDS

- **Units:** When performing calculations that involve the use of the Coulomb constant k_e that appears in Coulomb's law, charges must be in coulombs and distances in meters. If they are given in other units, you must convert them to SI.
- **Applying Coulomb's law to point charges:** It is necessary to use the *superposition principle* properly when dealing with a collection of interacting point charges. When several charges are present, the resultant force on any one charge is found by finding the individual force that every other charge exerts on that one, and then finding the vector sum of all these forces. The magnitude of the force that any charged object exerts on another is given by Coulomb's law, and the direction of the force is found by noting that the forces are repulsive between like charges and attractive between unlike charges.
- **Calculating the electric field of point charges:** Remember that the superposition principle can be applied to electric fields, which are vector quantities. To find the total electric field at a given point, first use Equation 23.9 to calculate the electric field at the point due to each individual charge. The resultant field at the point is the vector sum of the fields due to the individual charges. If charge q_2 happens to be located between charge q_1 and the point where the field is to be calculated ("blocking the line of sight" from q_1 to the field point), the contribution of q_1 to the field is not affected by the presence of q_2. Each charge contributes to the field according to Equation 23.9.

- **Evaluating the electric field of a continuous charge distribution:** To evaluate the electric field of a continuous charge distribution, it is convenient to employ the concept of charge density. Depending on the particular charge distribution, charge density can be written as: charge per unit volume (ρ), charge per unit area (σ), or charge per unit length (λ). The total charge distribution is then subdivided into small elements of volume dV, area dA, or length dx. Each element contains an increment of charge dq (equal to ρdV, σdA, or λdx). If the charge is non-uniformly distributed over the region, then the charge densities must be written as functions of position. For example, if the charge density along a line or long bar is proportional to the distance from one end of the bar, then the linear charge density could be written as $\lambda = bx$, where b is a constant of proportionality, and the charge increment dq becomes $dq = (bx)dx$.

- **Symmetry:** When dealing with either a distribution of point charges or a continuous charge distribution, take advantage of any symmetry in the system to simplify your calculations.

REVIEW CHECKLIST

You should be able to:

- Describe the fundamental properties of electric charge and the nature of electrostatic forces between charged bodies. (Section 23.1)

- Use Coulomb's law to determine the net electrostatic force (magnitude and direction) on a given point charge due to a known distribution of a finite number of other point charges. (Section 23.3)

- Calculate the electric field (magnitude and direction) at a specified location in the vicinity of a group of point charges. (Section 23.4)

- Calculate the electric field due to a continuous charge distribution. The charge may be distributed uniformly or non-uniformly along a line, over a surface, or throughout a volume. (Section 23.5)

- Correctly use the rules to draw and interpret electric field lines in the vicinity of point charges. (Section 23.6)

- Describe quantitatively the motion of a charged particle in a uniform electric field. (Section 23.7)

ANSWER TO AN OBJECTIVE QUESTION

15. A free electron and a free proton are released in identical electric fields. **(i)** How do the magnitudes of the electric force exerted on the two particles compare? (a) It is millions of times greater for the electron. (b) It is thousands of times greater for the electron. (c) They are equal. (d) It is thousands of times smaller for the electron. (e) It is millions of times smaller for the electron. **(ii)** Compare the magnitudes of their accelerations. Choose from the same possibilities as in part (i).

Answer **(i)** (c) The electric forces exerted by the field on the electron and on the proton are the same in magnitude, because the two particles have charges of the same magnitude. The two forces are opposite in direction, because the electron has a negative charge and the proton has a positive charge. These results imply that the net electric force on a neutral atom in a uniform electric field is precisely zero, and this fact is known to extraordinarily high precision.

(ii) (b) The proton's mass is 1836 times larger than that of the electron. When they feel forces of equal size, the electron moves with an acceleration larger in magnitude by 1836 times, compared to the proton. The accelerations, like the forces, are opposite in direction.

□ □ □ □

ANSWERS TO SELECTED CONCEPTUAL QUESTIONS

9. A balloon clings to a wall after it is negatively charged by rubbing. (a) Does this occur because the wall is positively charged? (b) Why does the balloon eventually fall?

Answer (a) The wall does not have a net positive charge. The balloon induces polarization of the molecules in the wall, so that a layer of positive charge exists near the balloon. This is just like the situation in Figure 23.4a, except that the signs of the charges are reversed. The attraction between these positive charges and the negative charges on the balloon is stronger than the repulsion between the negative charges on the balloon and the negative charges in the polarized molecules (because they are farther from the balloon), so that there is a net attractive force toward the wall.

(b) Ionization processes in the air surrounding the balloon provide ions to which excess electrons in the balloon can transfer, reducing the charge on the balloon and eventually causing the attractive force to be insufficient to support the weight of the balloon.

□ □ □ □

11. (a) Would life be different if the electron were positively charged and the proton were negatively charged? (b) Does the choice of signs have any bearing on physical and chemical interactions? Explain your answers.

Answer (a) No, life would not be different.

(b) The character and effect of electric forces is defined by (1) the fact that there are only two types of electric charge—positive and negative, and (2) the fact that opposite charges attract, while like charges repel. The choice of which sign is which is completely arbitrary.

As a related exercise, you might consider how the world would be very different if there were three types of electric charge, or if opposite charges repelled and like charges attracted.

□ □ □ □

SOLUTIONS TO SELECTED END-OF-CHAPTER PROBLEMS

4. Nobel laureate Richard Feynman (1918–1988) once said that if two persons stood at arm's length from each other and each person had 1% more electrons than protons, the force of repulsion between them would be enough to lift a "weight" equal to that of the entire Earth. Carry out an order-of-magnitude calculation to substantiate this assertion.

Solution

Conceptualize: We can think of this problem as showing that in a kind of controlled-experiment fair comparison, the electrical force exerted by one ordinary object on another is much stronger than the gravitational force. Stated differently, the problem will demonstrate that the imbalances between electron and proton numbers in macroscopic charged objects are much less than one-percent imbalances.

Categorize: We will estimate the charges on the two people in coulombs and then use Coulomb's law to calculate the magnitude of the force each exerts on the other.

Analyze: Suppose each person has mass 70 kg. In terms of elementary charges, each person consists of precisely equal numbers of protons and electrons and a nearly equal number of neutrons. The electrons comprise very little of the mass, so for each person we find the total number of protons and neutrons, taken together:

$$(70\text{ kg})\left(\frac{1\text{ u}}{1.66\times 10^{-27}\text{ kg}}\right) = 4\times 10^{28}\text{ u}$$

Of these, nearly one half, 2×10^{28}, are protons, and 1% of this is 2×10^{26}, constituting a charge of $(2 \times 10^{26})(1.60 \times 10^{-19}\text{ C}) = 3 \times 10^{7}\text{ C}$.

Thus, Feynman's force has magnitude

$$F = \frac{k_e q_1 q_1}{r^2} = \frac{(8.99\times 10^9\text{ N}\cdot\text{m}^2/\text{C}^2)(3\times 10^7\text{ C})^2}{(0.5\text{ m})^2} \sim 10^{26}\text{ N}$$

where we have used a half-meter arm's length. According to the particle in a gravitational field model, if the Earth were in an externally-produced uniform gravitational field of magnitude 9.80 m/s^2, it would weigh $F_g = mg = (6 \times 10^{24}\text{ kg})(10\text{ m/s}^2) \sim 10^{26}\text{ N}$.

Thus, the forces are of the same order of magnitude. ■

Finalize: In the interactions of electrons and protons, the electrical force is vastly larger than the gravitational force. For the gravitational force that the Earth exerts on you to be larger than an electrical force, the net charges on you and the Earth must be very close to zero.

13. Three charged particles are located at the corners of an equilateral triangle as shown in Figure P23.13. Calculate the total electric force on the 7.00-μC charge.

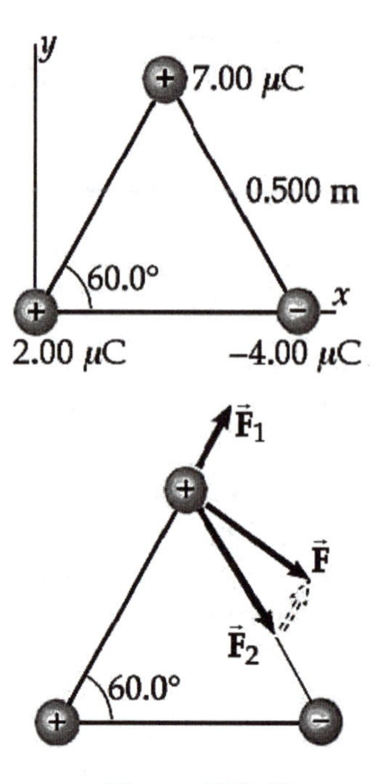

Figure P23.13

Solution

Conceptualize: The 7.00-μC charge experiences a repulsive force $\vec{\mathbf{F}}_1$ due to the 2.00-μC charge, and an attractive force $\vec{\mathbf{F}}_2$ due to the –4.00-μC charge, where $F_2 = 2F_1$. If we sketch vectors representing $\vec{\mathbf{F}}_1$ and $\vec{\mathbf{F}}_2$ and their sum, $\vec{\mathbf{F}}$ (see the second diagram), we find that the resultant appears to be about the same magnitude as F_2 and is directed to the right about 30.0° below the horizontal.

Categorize: We can find the net electric force by adding the two separate forces acting on the 7.00-μC charge. These individual forces can be found by applying Coulomb's law to each pair of charges.

Analyze: The force exerted on the 7.00-μC charge by the 2.00-μC charge is

$$\vec{\mathbf{F}}_1 = k_e \frac{q_1 q_2}{r^2}\hat{\mathbf{r}}$$

$$= \frac{(8.99\times10^9\ \text{N}\cdot\text{m}^2/\text{C}^2)(7.00\times10^{-6}\ \text{C})(2.00\times10^{-6}\ \text{C})}{(0.500\ \text{m})^2}(\cos 60°\hat{\mathbf{i}} + \sin 60°\hat{\mathbf{j}})$$

$$\vec{\mathbf{F}}_1 = (0.252\,\hat{\mathbf{i}} + 0.436\,\hat{\mathbf{j}})\ \text{N}$$

Similarly, the force on the 7.00-μC charge by the –4.00-μC charge is

$$\mathbf{F}_2 = k_e \frac{q_1 q_3}{r^2}\hat{\mathbf{r}}$$

$$= -\frac{(8.99\times10^9\ \text{N}\cdot\text{m}^2/\text{C}^2)(7.00\times10^{-6}\ \text{C})(-4.00\times10^{-6}\ \text{C})}{(0.500\ \text{m})^2}(\cos 60°\hat{\mathbf{i}} - \sin 60°\hat{\mathbf{j}})$$

$$\vec{\mathbf{F}}_2 = (0.503\,\hat{\mathbf{i}} - 0.872\,\hat{\mathbf{j}})\ \text{N}$$

Thus, the total force on the 7.00-μC charge, expressed as a set of components, is

$$\vec{\mathbf{F}} = \vec{\mathbf{F}}_1 + \vec{\mathbf{F}}_2 = (0.755\,\hat{\mathbf{i}} - 0.436\,\hat{\mathbf{j}})\ \text{N}$$ ■

We can also write the total force as:

$$\vec{\mathbf{F}} = \sqrt{(0.755\ \text{N})^2 + (0.436\ \text{N})^2}\ \text{at}\ \tan^{-1}\left(\frac{0.436\ \text{N}}{0.755\ \text{N}}\right)\ \text{below the } +x \text{ axis}$$

$\vec{\mathbf{F}} = 0.872$ N at 30.0° below the $+x$ axis ■

Finalize: Our calculated answer agrees with our initial estimate. An equivalent approach to this problem would be to find the net electric field at the location of the upper charge due to the two lower charges, and then apply $\vec{\mathbf{F}} = q\vec{\mathbf{E}}$ to find the force on the upper charge in this electric field.

---

21. In Figure P23.21, determine the point (other than infinity) at which the electric field is zero.

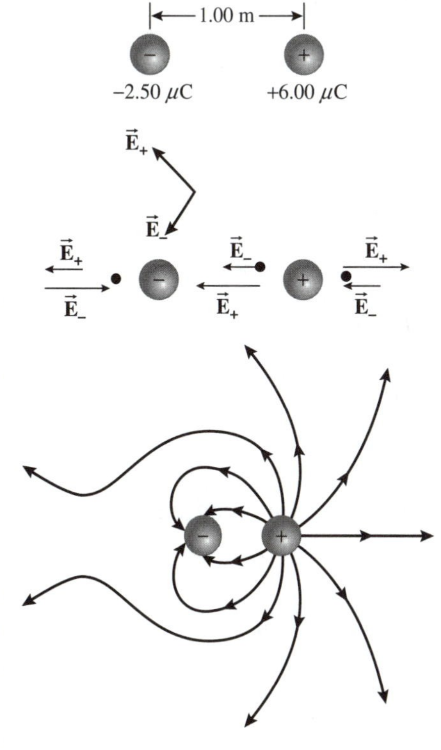

Figure P23.21

Solution

Conceptualize: Each charged particle produces a field that gets weaker farther away, so the net field due to both does indeed approach zero as the distance goes to infinity in any direction. But the problem asks about a point where the nonzero fields of the two particles add to zero…

Categorize: …as oppositely-directed vectors of equal magnitude. The field of the positively-charged object is everywhere pointing radially away from its location. The object with negative charge creates everywhere a field pointing toward its different location. These two fields are directed along different lines at any point in the plane except for points along the extended line joining the particles; so the two fields cannot be oppositely-directed to add to zero except at some location along this line, which we take as the x axis. Observing the middle diagram, we see that at points to the left of the negatively-charged object, this particle creates field pointing to the right and the positive object creates field to the left. At some point along this segment the fields will add to zero. At locations in between the objects, both create fields pointing toward the left, so the total field is not zero. At points to the right of the positive 6-μC object, its field is directed to the right and is stronger than the leftward field of the −2.5-μC object, so the two fields cannot be equal in magnitude to add to zero. We have argued that only at a certain point straight to the left of both charges can the fields they separately produce be opposite in direction and equal in strength to add to zero. We will find the location of this point by describing the magnitude of each field with the expression, derived from Coulomb's force law, for the field of a charged particle.

Analyze: The geometrical reasoning we did in the Categorize step could qualify as analysis, but here we proceed with algebraic reasoning. Let x represent the distance from the negatively-charged particle (charge q_-) to the zero-field point to its left. Then 1.00 m + x is the distance from the positive particle (of charge q_+) to this point. Each field is separately described by

$$\vec{\mathbf{E}} = k_e q \hat{\mathbf{r}} / r^2$$

so the equality in magnitude required for the two oppositely-directed vector fields to add to zero is described by

$$\frac{k_e|q_-|}{r^2} = \frac{k_e|q_+|}{(1\text{ m}+r)^2}$$

It is convenient to solve by taking the square root of both sides and cross-multiplying to clear of fractions:

$$|q_-|^{1/2}(1\text{ m}+r) = q_+^{1/2}r$$

$$1\text{ m}+r = \left(\frac{6.00}{2.50}\right)^{1/2} r = 1.55r$$

$$1\text{ m} = 0.549r$$

and $r = 1.82$ m to the left of the negatively-charged object ■

Finalize: The third panel in the diagram suggests the pattern of the net field created by the two charges. If we had taken the negative square root, the equation $1\text{ m} + r = -1.55r$ would have given us a location where the two charged particles create equal fields in the same direction, not adding to zero. Then the steps of our logic reveal that this problem has a unique solution. Other charge configurations could have several points, or none, where the field takes on a particular value. To visualize a positively-charged particle creating electric field throughout the surrounding empty space, it may aid your memory to think of a porcupine with quills sticking out radially in all directions, extending in principle to infinity. A negatively-charged particle is like a pincushion. The positive charge in this problem creates field on the far side of the negatively-charged ball just as if the ball were not there—there is no "shadowing," "blocking," or "screening" effect.

---

31. A uniformly charged ring of radius 10.0 cm has a total charge of 75.0 μC. Find the electric field on the axis of the ring at (a) 1.00 cm, (b) 5.00 cm, (c) 30.0 cm, and (d) 100 cm from the center of the ring.

Solution

Conceptualize: At the very center ($x = 0$) the field is zero because it has no direction to point in. At an infinite distance the field is zero again. We should see evidence for a maximum field somewhere between answers (b) and (c). The field points away from the ring along its axis, because the ring has uniformly distributed positive charge. Different elements of charge in the ring contribute to the total field components that sum to zero parallel to the face of the ring. The components perpendicular to the face of the ring add together to form the total field along the axis.

Categorize: We repeatedly evaluate the result of Example 23.7.

Analyze: We may particularize the result of Example 23.7 to

$$E = \frac{k_e x Q}{\left(x^2+a^2\right)^{3/2}} = \frac{(8.99\times10^9\text{ N}\cdot\text{m}^2/\text{C}^2)(75.0\times10^{-6}\text{ C})x}{\left(x^2+(0.100\text{ m})^2\right)^{3/2}} = \frac{\left(6.74\times10^5\text{ N}\cdot\text{m}^2/\text{C}\right)x}{\left(x^2+(0.100\text{ m})^2\right)^{3/2}}$$

Now just use your calculator to substitute different values for x.

(a) At $x = 0.010\,0$ m, $\vec{\mathbf{E}} = 6.64 \times 10^6 \hat{\mathbf{i}}$ N/C ■

(b) At $x = 0.050\,0$ m, $\vec{\mathbf{E}} = 2.41 \times 10^7 \hat{\mathbf{i}}$ N/C ■

(c) At $x = 0.300$ m, $\vec{\mathbf{E}} = 6.40 \times 10^6 \hat{\mathbf{i}}$ N/C ■

(d) At $x = 1.00$ m, $\vec{\mathbf{E}} = 6.64 \times 10^5 \hat{\mathbf{i}}$ N/C ■

Finalize: We have specified the field as being in the positive x direction at points on the positive x axis. The field points away from the ring on both sides. The 24 meganewtons per coulomb at 5 cm is indeed larger than the 6 or 0.6 MN/C at other points. Make sure you have gotten enough practice in evaluating a quantity and raising it to the 3/2 power.

---

35. A uniformly charged insulating rod of length 14.0 cm is bent into the shape of a semicircle, as shown in Figure P23.35. The rod has a total charge of $-7.50\ \mu$C. Find (a) the magnitude and (b) the direction of the electric field at O, the center of the semicircle.

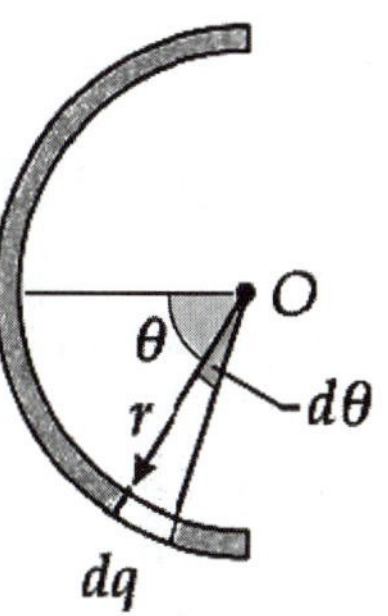

Figure P23.35

Solution

Conceptualize: The bottom and top halves of the shape will create downward and upward fields at O that will add to zero. We expect a net field of some millions of newtons per coulomb straight to the left, toward the midpoint of the negative charge.

Categorize: We will think of an incremental bit of charge as a charged particle, identify its contribution to the field at O, and then integrate to find the total field.

Analyze: Let λ be the charge per unit length. The diagram shows the bit of charge $dq = \lambda ds = \lambda r d\theta$.

The field it creates has magnitude $dE = \dfrac{k_e dq}{r^2} = \dfrac{k_e \lambda r d\theta}{r^2}$.

In component form, $E_y = 0$ (from symmetry) and $dE_x = dE\cos\theta$.

Integrating, $$E_x = \int_{\text{all charge}} dE_x = \int \frac{k_e \lambda r \cos\theta}{r^2} d\theta$$

$$E_x = \frac{k_e \lambda}{r}\int_{-\pi/2}^{\pi/2} \cos\theta\, d\theta = \frac{k_e\lambda}{r}\sin\theta\Big|_{-\pi/2}^{\pi/2} = \frac{k_e\lambda}{r}[1-(-1)] = \frac{2k_e\lambda}{r}$$

But $Q_{\text{total}} = \lambda\ell$, where $\ell = 0.140$ m and $r = \ell/\pi$

Thus, $E_x = \dfrac{2\pi k_e Q}{\ell^2} = \dfrac{2\pi\left(8.99\times10^9\ \text{N}\cdot\text{m}^2/\text{C}^2\right)\left(-7.50\times10^{-6}\ \text{C}\right)}{(0.140\ \text{m})^2}$

$\vec{\mathbf{E}} = \left(-2.16\times10^7\ \text{N/C}\right)\hat{\mathbf{i}}$ = (a) 21.6 MN/C (b) to the left in the diagram ■

Finalize: We were correct about several million newtons per coulomb straight to the left. It is important to think of the field as a vector—that is where the cosine factor in the integrand comes from.

---

37. A thin rod of length ℓ and uniform charge per unit length λ lies along the x axis, as shown in Figure P23.37. (a) Show that the electric field at P, a distance d from the rod along its perpendicular bisector, has no x component and is given by $E = 2k_e\lambda\sin\theta_0/d$. (b) **What if?** Using your result to part (a), show that the field of a rod of infinite length is $E = 2k_e\lambda/d$.

Solution

Conceptualize: The proportionalities of the field to the charge and to the Coulomb's-law constant k_e are reasonable. The proportionality to $1/d$ instead of $1/d^2$ is a bit remarkable.

Categorize: We calculate the field at P due to an element of the rod of length dx, which has a charge λdx. Then we do an integral to include the contributions of all bits of the charge to the field. We could say it is obvious that the x component of the field above the rod's midpoint is zero, but we will also prove it explicitly.

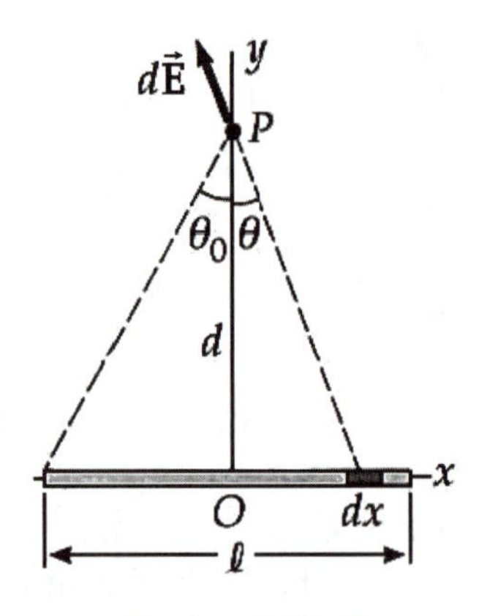

Figure P23.37

Analyze:

(a) The segment of rod from x to $(x + dx)$ has a charge of λdx, and creates an electric field upward along the line from dq to P:

$$d\vec{\mathbf{E}} = \frac{k_e dq}{r^2}\hat{\mathbf{r}} = \frac{k_e\lambda dx}{d^2+x^2}\hat{\mathbf{r}}$$

This bit of field has an x component

$$dE_x = \frac{k_e\lambda dx}{d^2+x^2}(-\sin\theta) = \frac{-k_e\lambda x\,dx}{\left(d^2+x^2\right)^{3/2}}$$

and a y component $\quad dE_y = \dfrac{k_e\lambda\,dx}{d^2+x^2}(\cos\theta) = \dfrac{k_e\lambda d\,dx}{\left(d^2+x^2\right)^{3/2}}$

The total field has an x component $E_x = \displaystyle\int_{\text{All } q} dE_x = \int_{-\ell/2}^{\ell/2} -\frac{k_e\lambda x\,dx}{\left(d^2+x^2\right)^{3/2}}$

To integrate, we make the change of variables to θ, such that $x = d\tan\theta$.
When $x = -\ell/2$, $\theta = -\theta_0$ and when $x = \ell/2$, $\theta = \theta_0$.
Further, $(d^2 + x^2)^{3/2} = (d^2 + d^2\tan^2\theta)^{3/2} = d^3\sec^3\theta$ and $dx = d\sec^2\theta\, d\theta$.

$$\text{Thus, } E_x = \int_{-\theta_0}^{\theta_0} \frac{-k_e\lambda d(\tan\theta)d\left(\sec^2\theta\right)d\theta}{d^3\sec^3\theta} = -\frac{k_e\lambda}{d}\int_{-\theta_0}^{\theta_0}\sin\theta\, d\theta$$

$$E_x = +\frac{k_e\lambda}{d}\cos\theta\Big]_{-\theta_0}^{+\theta_0} = \frac{k_e\lambda}{d}\left(\cos\theta_0 - \cos(-\theta_0)\right) = \frac{k_e\lambda}{d}\left(\cos\theta_0 - \cos\theta_0\right) = 0$$

This answer has to be zero because each segment of rod on the left produces a field whose contribution cancels out that of the corresponding segment of rod on the right. But every incremental bit of charge produces at P a contribution to the field with upward y component:

$$E_y = \int_{\text{All } q} dE_y = \int_{-\ell/2}^{\ell/2} \frac{k_e\lambda d\, dx}{\left(d^2 + x^2\right)^{3/2}}$$

Think of k_e, λ, and d as known constants. Now E_y is the unknown and x is the variable of integration, which we again change to θ, with $x = d\tan\theta$:

$$E_y = \int_{-\theta_0}^{\theta_0} \frac{k_e\lambda d\left(d\sec^2\theta\right)d\theta}{d^3\sec^3\theta} = \frac{k_e\lambda}{d}\int_{-\theta_0}^{\theta_0}\cos\theta\, d\theta = \frac{k_e\lambda}{d}\sin\theta\Big]_{-\theta_0}^{\theta_0}$$

$$E_y = \frac{k_e\lambda}{d}\left(\sin\theta_0 - \sin(-\theta_0)\right) = \frac{k_e\lambda}{d}\left(\sin\theta_0 + \sin\theta_0\right) = \frac{2k_e\lambda\sin\theta_0}{d}$$

■

(b) As ℓ goes to infinity, θ_0 goes to 90° and $\sin\theta_0$ becomes 1. Then the infinite amount of charge produces a finite field at P:

$$\vec{\mathbf{E}} = 0\hat{\mathbf{i}} + \frac{2k_e\lambda}{d}\hat{\mathbf{j}}$$

■

Finalize: The equations we were asked to prove are in fact true. In the next chapter we will see a different derivation of the $2k_e\lambda/d = \lambda/2\pi d\epsilon_0$ result.

---

39. A negatively charged rod of finite length carries charge with a uniform charge per unit length. Sketch the electric field lines in a plane containing the rod.

Solution

Conceptualize: Calculating the field at an arbitrary point near the rod would be a difficult integration. But drawing the field lines will be made remarkably easy by…

Categorize: …following the rules that field lines point toward negative charge, meeting the rod perpendicularly and ending there.

Analyze: Since the rod has negative charge, field lines point inwards. Any field line points nearly toward the center of the rod at large distances, where the rod would look like just a charged particle.

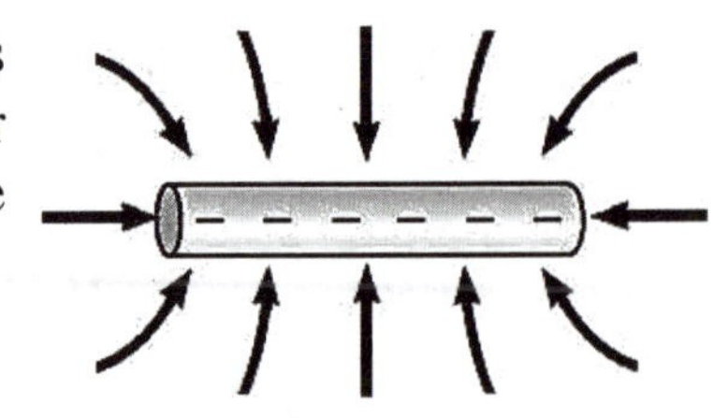

The lines curve to reach the rod perpendicular to its surface, where they end at equally-spaced points. ■

Finalize: The spacing of the field lines suggests that the field is strongest above the middle of the rod.

---

42. Three equal positive charges q are at the corners of an equilateral triangle of side a as shown in Figure P23.42. Assume that the three charges together create an electric field. (a) Sketch the field lines in the plane of the charges. (b) Find the location of one point (other than ∞) where the electric field is zero. What are (c) the magnitude and (d) the direction of the electric field at P due to the two charges at the base?

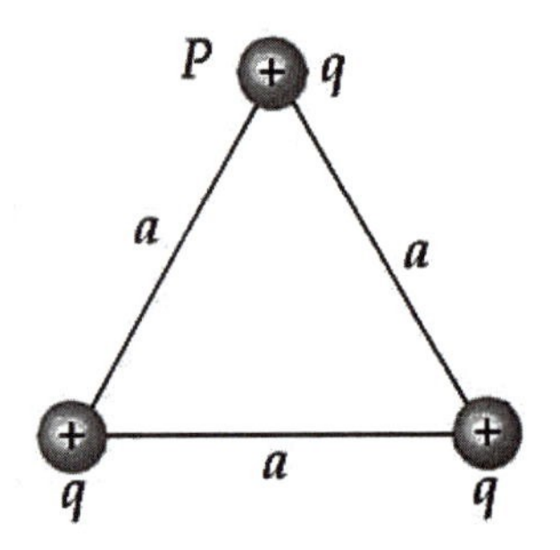

Figure P23.42

Solution

Conceptualize:

(a) The electric field has the general appearance shown by the curving black arrows in the figure. (b) This drawing indicates that $\vec{\mathbf{E}} = 0$ at the center of the triangle, since a small positive charge placed at the center of this triangle will be pushed away from each corner equally strongly. This fact could be verified by vector addition as in parts (c) and (d) below. ■

(c) and (d) The electric field at point P should be directed upwards and about twice the magnitude of the electric field due to just one of the lower charges as shown in the diagram at the right. For these parts of the problem, we must ignore the effect of the charge at point P, because a charge cannot exert a force on itself.

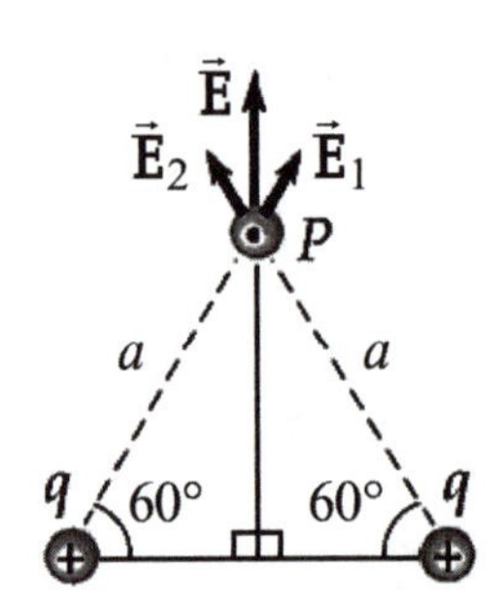

Categorize: The electric field at point P can be found by adding the electric field vectors due to each of the two lower charged particles: $\vec{\mathbf{E}} = \vec{\mathbf{E}}_1 + \vec{\mathbf{E}}_2$

(c) and (d) **Analyze:** The electric field from a charged particle is $\vec{\mathbf{E}} = k_e \dfrac{q}{r^2}\hat{\mathbf{r}}$

As shown in the figure, $\vec{\mathbf{E}}_1$ and $\vec{\mathbf{E}}_2$ both point 60° above the horizontal,

$$\vec{\mathbf{E}}_1 = k_e\frac{q}{a^2} \text{ to the right and upward, } \vec{\mathbf{E}}_2 = k_e\frac{q}{a^2} \text{ to the left and upward}$$

$$\vec{\mathbf{E}} = \vec{\mathbf{E}}_1 + \vec{\mathbf{E}}_2 = k_e\frac{q}{a^2}\left[\left(\cos 60°\,\hat{\mathbf{i}} + \sin 60°\,\hat{\mathbf{j}}\right) + \left(-\cos 60°\,\hat{\mathbf{i}} + \sin 60°\,\hat{\mathbf{j}}\right)\right]$$

$$\vec{\mathbf{E}} = k_e\frac{q}{a^2}\left[2\left(\sin 60°\,\hat{\mathbf{j}}\right)\right] = 1.73k_e\frac{q}{a^2}\hat{\mathbf{j}}$$ ■

Finalize: The net electric field at point *P* is indeed nearly twice the magnitude due to a single charge and is entirely vertical as expected from the symmetry of the configuration. In addition to the center of the triangle, the curving gray electric field lines in the figure above indicate three other points near the middle of each leg of the triangle where $\vec{\mathbf{E}} = 0$, but they are more difficult to find mathematically.

---

45. A proton accelerates from rest in a uniform electric field of 640 N/C. At one later moment, its speed is 1.20 Mm/s (nonrelativistic, because v is much less than the speed of light). (a) Find the acceleration of the proton. (b) Over what time interval does the proton reach this speed? (c) How far does it move in this time interval? (d) What is its kinetic energy at the end of this interval?

Solution

Conceptualize: A proton has so little mass that we expect some huge acceleration compared to any macroscopic object. It might accelerate over centimeters in nanoseconds to attain some tiny fraction of a joule.

Categorize: We use the particle in an electric field model and the particle under net force model. Then we use the particle under constant acceleration model and the definition of kinetic energy.

Analyze:

(a) $F = qE$ represents the definition of electric field, and tells us the one force in $\Sigma F = ma$. Then the acceleration is

$$a = \frac{F}{m} = \frac{qE}{m} = \frac{\left(1.602\times10^{-19}\text{ C}\right)\left(640\text{ N/C}\right)}{1.67\times10^{-27}\text{ kg}} = 6.14\times10^{10}\text{ m/s}^2$$ ■

(b) The particle under constant acceleration model gives us the equations $v_f = v_i + at$

$$t = \frac{v_f - 0}{a} = \frac{1.20\times10^{6}\text{ m/s}}{6.14\times10^{10}\text{ m/s}^2} = 19.5\ \mu\text{s} \quad \text{and}$$ ■

(c) $\Delta x = v_i t + \frac{1}{2}at^2 = 0 + \frac{1}{2}(6.14\times10^{10}\text{ m/s}^2)(19.5\times10^{-6}\text{ s})^2 = 11.7\text{ m}$ ■

(d) $K = \frac{1}{2}mv^2 = \frac{1}{2}(1.67\times10^{-27}\text{ kg})(1.20\times10^6\text{ m/s})^2 = 1.20\times10^{-15}\text{ J}$ ■

Finalize: The little review of Newton's second law and the equations about motion with constant acceleration belongs logically with studying a new kind of force. An electron would have started moving more abruptly, but the more massive proton sped up over meters in microseconds, not over centimeters in nanoseconds. The charge of the proton is $+e$. The field does on it work $F\Delta x = (640\text{ N/C})\,e\,(11.7\text{ m}) = 7490\,e$ J/C. Since it started from rest, its final kinetic energy is $7490\,e$ J/C. In a couple of chapters we will see that this is expressed most naturally as 7.49 keV = 7 490 electronvolts.

---

47. The electrons in a particle beam each have a kinetic energy K. What are (a) the magnitude and (b) the direction of the electric field that will stop these electrons in a distance d?

Solution

Conceptualize: We should expect that a larger electric field would be required to stop electrons with greater kinetic energy. Likewise, $\vec{\mathbf{E}}$ must be greater for a shorter stopping distance d. The electric field should be in the same direction as the motion of the negatively charged electrons in order to exert an opposing force that will slow them down.

Categorize: The electrons will experience an electrostatic force $\vec{\mathbf{F}} = q\vec{\mathbf{E}}$. Therefore, the work done by the electric field on each electron can be related to its initial kinetic energy according to the work-kinetic energy theorem.

Analyze: The work done on the charge is $W = \vec{\mathbf{F}}\cdot\vec{\mathbf{d}} = q\vec{\mathbf{E}}\cdot\vec{\mathbf{d}}$ and the kinetic energy changes according to $W = K_f - K_i = 0 - K$.

Assuming $\vec{\mathbf{v}}$ is in the $+x$ direction, we have $(-e)\vec{\mathbf{E}}\cdot d\hat{\mathbf{i}} = -K$.

Then $e\vec{\mathbf{E}}\cdot\left(d\hat{\mathbf{i}}\right) = K$.

Thus $\vec{\mathbf{E}}$ is in the direction of the electron beam. Both its (a) magnitude and (b) direction are specified by the vector expression $\vec{\mathbf{E}} = \frac{K}{ed}\hat{\mathbf{i}}$. ■

Finalize: As expected, the electric field is proportional to K, and inversely proportional to d. The direction of the electric field is important; if it were otherwise the electron would speed up instead of slowing down! If the particles were protons instead of electrons, the electric field would need to be directed opposite to the velocity in order for the particles to slow down.

---

49. A proton moves at 4.50×10^5 m/s in the horizontal direction. It enters a uniform vertical electric field with a magnitude of 9.60×10^3 N/C. Ignoring any gravitational effects, find (a) the time interval required for the proton to travel 5.00 cm horizontally, (b) its vertical displacement during the time interval in which it travels 5.00 cm horizontally, and (c) the horizontal and vertical components of its velocity after it has traveled 5.00 cm horizontally.

Solution

Conceptualize: The proton will maintain a constant horizontal velocity component. In less than a microsecond it could traverse a vacuum tube while moving with a huge vertical acceleration to get some measurable deflection.

Categorize: This is a projectile-motion problem. One component of the motion has zero acceleration and the other constant nonzero acceleration.

Analyze: $\vec{\mathbf{E}}$ is directed along the y direction; therefore, $a_x = 0$ and $x = v_{xi}t$.

(a) $t = \dfrac{x}{v_{xi}} = \dfrac{0.050\,0 \text{ m}}{4.50 \times 10^5 \text{ m/s}} = 1.11 \times 10^{-7} \text{ s}$ ■

(b) $a_y = \dfrac{qE_y}{m} = \dfrac{\left(1.60 \times 10^{-19} \text{ C}\right)\left(9.60 \times 10^3 \text{ N/C}\right)}{1.67 \times 10^{-27} \text{ kg}} = 9.20 \times 10^{11} \text{ m/s}^2$

$y = v_{yi}t + \frac{1}{2}a_y t^2 = \left(\frac{1}{2}\right)\left(9.20 \times 10^{11} \text{ m/s}^2\right)(1.11 \times 10^{-7} \text{ s})^2 = 5.68 \text{ mm}$ ■

(c) $v_{xf} = v_{xi} = 4.50 \times 10^5 \text{ m/s}$ ■

$v_{yf} = v_{yi} + a_y t = 0 + (9.20 \times 10^{11} \text{ m/s}^2)(1.11 \times 10^{-7} \text{ s}) = 1.02 \times 10^5 \text{ m/s}$ ■

Finalize: Beyond the problem statement, we have direct evidence that gravity is negligible, from the huge size of the vertical acceleration caused by the electrical force. In laboratory it is much easier to work with electron beams than with proton beams, but the equations are the same.

---

59. A charged cork ball of mass 1.00 g is suspended on a light string in the presence of a uniform electric field as shown in Figure P23.59. When $\vec{\mathbf{E}} = (3.00\hat{\mathbf{i}} + 5.00\hat{\mathbf{j}}) \times 10^5$ N/C, the ball is in equilibrium at $\theta = 37.0°$. Find (a) the charge on the ball and (b) the tension in the string.

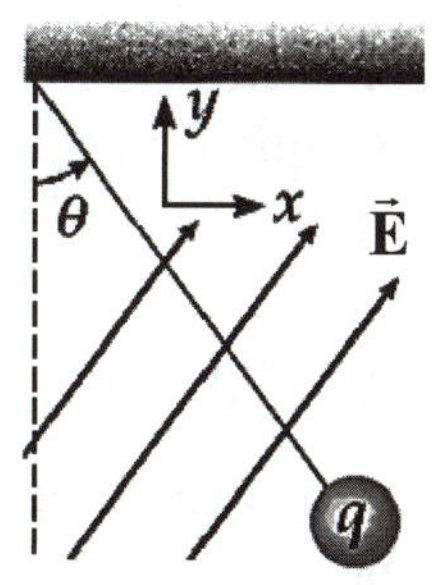

Figure P23.59

Solution

Conceptualize: Since the electric force must be in the same direction as $\vec{\mathbf{E}}$, the ball must be positively charged. If we examine the free body diagram that shows the three forces acting on the ball, the sum of which must be zero, we can see that the tension is about half the magnitude of the weight.

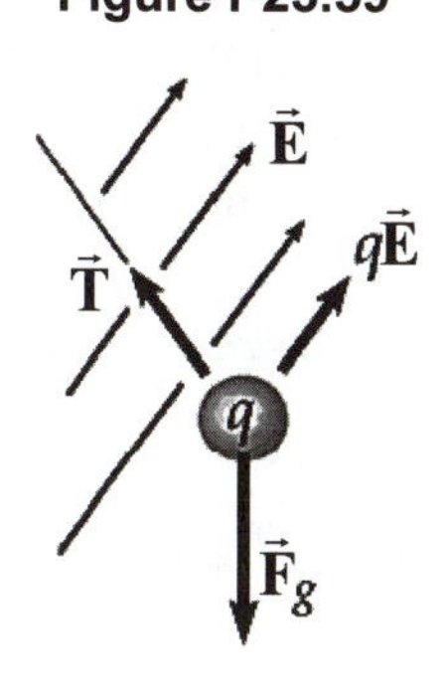

Categorize: The tension can be found from applying Newton's second law to this statics problem (electrostatics, in this case!). Since the force vectors are in two dimensions, we must apply $\Sigma\vec{\mathbf{F}} = m\vec{\mathbf{a}}$ to both the x and y components.

Analyze: $\Sigma\vec{\mathbf{F}} = \vec{\mathbf{T}} + q\vec{\mathbf{E}} + \vec{\mathbf{F}}_g = 0$

We are given $E_x = 3.00 \times 10^5$ N/C

and $E_y = 5.00 \times 10^5$ N/C

Applying Newton's second law or the first condition for equilibrium in the x and y directions,

$$\sum F_x = qE_x - T\sin 37.0° = 0 \qquad [1]$$

$$\sum F_y = qE_y + T\cos 37.0° - mg = 0 \qquad [2]$$

(a) Substitute T from Eq. [1] into Eq. [2] to obtain

$$q = \frac{mg}{E_y + \dfrac{E_x}{\tan 37.0°}} \qquad q = \frac{(1.00 \times 10^{-3}\text{ kg})(9.80\text{ m/s}^2)}{5.00 \times 10^5\text{ N/C} + \dfrac{3.00 \times 10^5\text{ N/C}}{\tan 37.0°}}$$

$$q = 1.09 \times 10^{-8}\text{ C}$$ ■

(b) Using this result for q in Equation [1], we find that the tension is

$$T = \frac{qE_x}{\sin 37.0°} = 5.44 \times 10^{-3}\text{ N}$$ ■

Finalize: The tension is slightly more than half the weight of the ball ($F_g = 9.80 \times 10^{-3}$ N), so our result seems reasonable based on our initial prediction.

---

60. A charged cork ball of mass m is suspended on a light string in the presence of a uniform electric field as shown in Figure P23.59. When $\vec{\mathbf{E}} = A\hat{\mathbf{i}} + B\hat{\mathbf{j}}$, where A and B are positive quantities, the ball is in equilibrium at the angle θ. Find (a) the charge on the ball and (b) the tension in the string.

Solution

Conceptualize: This is the symbolic or general version of the preceding problem. The known quantities are A, B, m, g, and θ. The unknowns are q and T.

Categorize: The approach to this problem will be the same as for the last problem, but without numbers to substitute for the variables. We use the free body diagram given in the solution to Problem 59. Compare the solutions step by step.

Analyze: The particle in equilibrium model gives us

$$-T\sin\theta + qA = 0 \qquad [1]$$

and $+T\cos\theta + qB - mg = 0$ **[2]**

(a) Substituting $T = qA/\sin\theta$ into Eq. [2] gives $\dfrac{qA\cos\theta}{\sin\theta} + qB = mg$

Isolating q on the left, $q = \dfrac{mg}{(A\cot\theta + B)}$ ■

(b) Substituting this value into Eq. [1], $T = \dfrac{mgA}{(A\cos\theta + B\sin\theta)}$ ■

Finalize: The tension is proportional to the weight of the particle. If we had solved this general problem first, we would only need to substitute the appropriate values in the equations for q and T to find the numerical results needed for Problem 59. If you find this problem more difficult than Problem 59, the little list at the Conceptualize step is useful. It shows what symbols to think of as known data, and what to consider unknown. The list is a guide for deciding what to solve for in the Analyze step, and for recognizing when we have an answer.

---

71. Two small spheres of mass m are suspended from strings of length ℓ that are connected at a common point. One sphere has charge Q and the other charge $2Q$. The strings make angles θ_1 and θ_2 with the vertical. (a) Explain how θ_1 and θ_2 are related. (b) Assume θ_1 and θ_2 are small. Show that the distance r between the spheres is approximately

$$r \approx \left(\frac{4k_eQ^2\ell}{mg}\right)^{1/3}$$

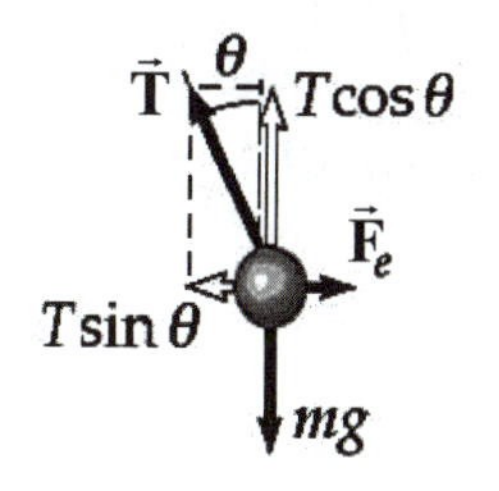

Solution

Conceptualize: The formula we are to prove has these reasonable features: The distance between the particles will increase if their charges increase and decrease in a stronger gravitational field. We may be surprised about where the one-third power comes from.

Categorize: Part (a) is about how different the idea of force is from the idea of charge. In part (b) we use the particle in equilibrium model.

Analyze:

(a) The spheres have different charges, but each exerts an equal force on the other, given by $F_e = k_e(Q)(2Q)/r^2$, where r is the distance between them. Because their masses are also equal, $\theta_2 = \theta_1$. ■

(b) For equilibrium the equation $\Sigma F_y = 0$ becomes $T\cos\theta - mg = 0$ and similarly $\Sigma F_x = 0$ becomes $F_e - T\sin\theta = 0$.

To solve simultaneously we substitute $T = \dfrac{mg}{\cos\theta}$

to obtain $F_e = \dfrac{mg\sin\theta}{\cos\theta} = mg\tan\theta$

For small angles, $\tan\theta \approx \sin\theta = \dfrac{r}{2\ell}$

Therefore, $F_e \approx mg\dfrac{r}{2\ell}$

The force F_e is $\dfrac{k_e Q(2Q)}{r^2} \approx mg\dfrac{r}{2\ell}$

so that $4k_e Q^2\ell \approx mgr^3$ and $r \approx \left(\dfrac{4k_e Q^2\ell}{mg}\right)^{1/3}$ ■

Finalize: Let us check for dimensional correctness: we have $\left(\dfrac{\mathrm{N\cdot m^2\cdot C^2\cdot m}}{\mathrm{C^2\cdot N}}\right)^{1/3} = \mathrm{m}$ as required. The one-third power means physically that since the electrical force depends so strongly on the distance between the charges, this distance will not change by very much if we change the charge or the mass or the string length or the free-fall acceleration by some factor.

---

74. Review. A negatively charged particle $-q$ is placed at the center of a uniformly charged ring, where the ring has a total positive charge Q as shown in Figure P23.74. The particle, confined to move along the x axis, is moved a small distance x along the axis (where $x << a$) and released. Show that the particle oscillates in simple harmonic motion with a frequency given by

$$f = \frac{1}{2\pi}\left(\frac{k_e qQ}{ma^3}\right)^{1/2}$$

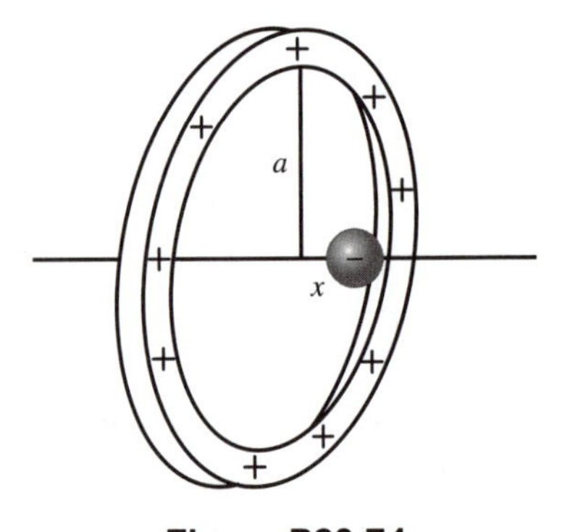

Figure P23.74

Solution

Conceptualize: The negative charge is attracted toward the center of the ring if it is located on the axis on either side, so the restoring force can be expected to cause vibration.

Categorize: To find the frequency we must examine the force law and compare it with $\dfrac{d^2\vec{\mathbf{x}}}{dt^2} = -\omega^2\vec{\mathbf{x}}$, the defining equation for simple harmonic motion.

Analyze: From Example 23.7, the electric field at points along the x axis is

$$\vec{\mathbf{E}} = \frac{k_e xQ\hat{\mathbf{i}}}{\left(x^2+a^2\right)^{3/2}}$$

The field is zero at $x = 0$, so the negatively-charged particle is in equilibrium at this point. When it is displaced by an amount x that is small compared to a,

$$\sum \vec{\mathbf{F}} = m\frac{d^2\vec{\mathbf{x}}}{dt^2} \quad \text{becomes} \quad (-q)\frac{k_e x Q\hat{\mathbf{i}}}{a^3} = m\frac{d^2\vec{\mathbf{x}}}{dt^2}$$

The particle's acceleration is proportional to its displacement (x) from the equilibrium position and is oppositely directed, so it moves in simple harmonic motion. ■

Since the angular frequency of any simple harmonic motion is set by $\dfrac{d^2\vec{\mathbf{x}}}{dt^2} = -\omega^2\vec{\mathbf{x}}$, we have by comparison $\omega = \left(\dfrac{k_e qQ}{ma^3}\right)^{1/2} = 2\pi f$

and $f = \dfrac{1}{2\pi}\left(\dfrac{k_e qQ}{ma^3}\right)^{1/2}$ ■

Finalize: A larger charge on either the particle or the ring would make the restoring force larger and increase the vibration frequency. Larger mass would make the frequency smaller. Larger ring size a would have a bigger effect in reducing the frequency.

24

Gauss's Law

EQUATIONS AND CONCEPTS

Electric flux through a plane surface is equal to the product of the magnitude of the field and the projection of the area onto a plane perpendicular to the direction of the field. Equation 24.2 gives the value of the flux through a plane area when a *uniform field* makes an angle θ with the normal to the surface. Electric flux is a scalar quantity and has SI units of $N \cdot m^2/C$.

$$\Phi_E = EA\cos\theta \quad (24.2)$$

(uniform surface)

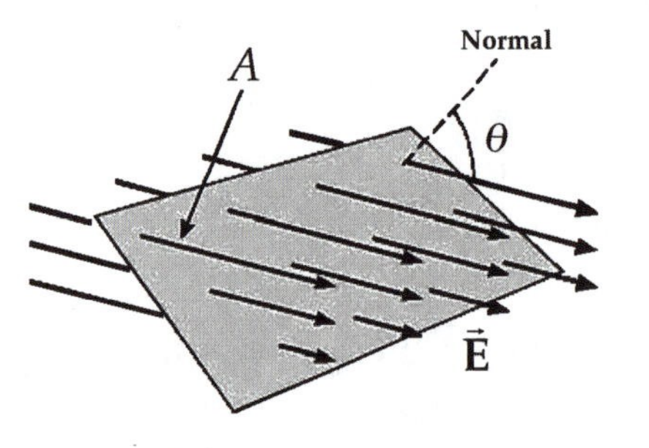

For a general surface in a non-uniform field, the flux is calculated by integrating the normal component of the field over the surface in question. *The integrand is the scalar or "dot" product of two vectors and the integral must be evaluated over the entire surface in question.*

$$\Phi_E \equiv \int_{\text{surface}} \vec{\mathbf{E}} \cdot d\vec{\mathbf{A}} \quad (24.3)$$

(non-uniform surface)

The **net flux through a closed surface** is proportional to the net number of lines leaving the surface (the number of lines leaving the surface minus the number entering the surface). E_n *represents the component of the electric field normal to the surface, and the integral must be evaluated over a closed surface.*

$$\Phi_E = \oint E_n dA \quad (24.4)$$

(closed surface)

Gauss's law states that the net flux through any closed surface surrounding a net charge q equals the net charge enclosed by the surface divided by the constant ϵ_0. The net flux is independent of the shape and size of the surface. *The surface is called a "gaussian" surface.*

$$\Phi_E = \oint \vec{\mathbf{E}} \cdot d\vec{\mathbf{A}} = \frac{q_{in}}{\epsilon_0} \quad (24.6)$$

The **magnitude of the electric field due to symmetric and uniform charge distributions** (e.g., solid insulating or conducting sphere, spherical shell, long line, plane) can be calculated by evaluating Equation 24.6 over a gaussian surface appropriate for each case.

The electric field at a point P, located:

a **distance r from a point charge q.**

$$E = k_e \frac{q}{r^2}$$

exterior to a uniformly charged insulating sphere of radius a and total charge Q.

$$E = k_e \frac{Q}{r^2} \qquad \text{for } r > a$$

interior to a uniformly charged insulating sphere of radius a and total charge Q.

$$E = k_e \left(\frac{Q}{a^3}\right) r \qquad \text{for } r < a$$

interior to a uniformly charged conducting sphere of radius a and total charge Q.

$$E = 0 \qquad \text{for } r < a$$

outside a thin uniformly charged spherical shell of radius a and charge Q.

$$E = k_e \frac{Q}{r^2} \qquad \text{for } r > a$$

a **distance r from an infinitely long uniform line of charge** with linear charge density λ.

$$E = 2k_e \frac{\lambda}{r} \qquad (24.7)$$

any distance from **an infinite nonconducting plane of charge** with surface charge density σ.

$$E = \frac{\sigma}{2\epsilon_0} \qquad (24.8)$$

just outside the surface of a charged conductor in equilibrium with surface charge density σ.

$$E = \frac{\sigma}{\epsilon_0} \qquad (24.9)$$

SUGGESTIONS, SKILLS, AND STRATEGIES

Gauss's law is a very powerful theorem that relates any charge distribution to the resulting electric field at any point in the vicinity of the charge. In this chapter you should learn how to apply Gauss's law to those cases in which the charge distribution has a sufficiently high degree of symmetry. As you review the examples presented in Section 24.3 of the text, observe how each of the following steps has been included in the application of the equation $\oint \vec{\mathbf{E}} \cdot d\vec{\mathbf{A}} = q/\epsilon_0$ to that particular situation.

- The gaussian surface must be a closed surface and should be chosen to have the **same symmetry as the charge distribution.**
- The dimensions of the surface should be such that **the surface includes the point where the electric field is to be calculated.**

- From the symmetry of the charge distribution, you should be able to correctly determine the direction of the electric field vector relative to the direction of the area vector for each element of the gaussian surface.

- Subdivide the gaussian surface into regions of area such that over each region the electric field has a constant value. Each of the separate regions should satisfy one or more one of the following:

  $\vec{\mathbf{E}} \perp d\vec{\mathbf{A}}$ so that $E\,dA\cos\theta = 0$ (as is the case over each end of the cylindrical gaussian surface in the figure).

  $\vec{\mathbf{E}} \parallel d\vec{\mathbf{A}}$ so that $E\,dA\cos\theta = E\,dA$ (as is the case over the curved portion of the cylindrical gaussian surface in the figure).

  $\vec{\mathbf{E}}$ and $d\vec{\mathbf{A}}$ are oppositely directed so that $E\,dA\cos\theta = -E\,dA$.

  $\vec{\mathbf{E}} = 0$ (as is the case over a surface inside a conductor).

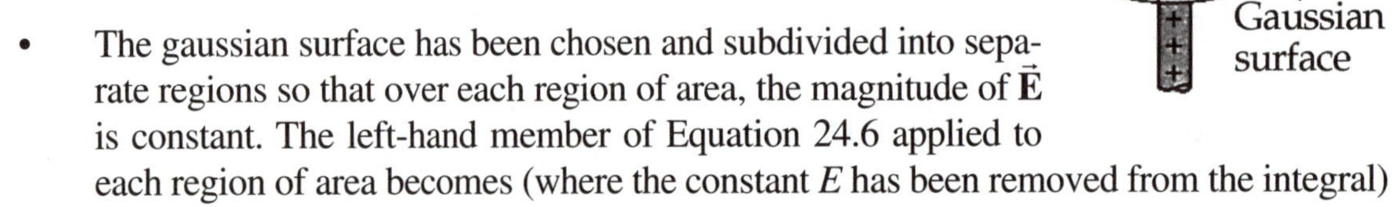

- The gaussian surface has been chosen and subdivided into separate regions so that over each region of area, the magnitude of $\vec{\mathbf{E}}$ is constant. The left-hand member of Equation 24.6 applied to each region of area becomes (where the constant E has been removed from the integral)

$$\int \vec{\mathbf{E}} \cdot d\vec{\mathbf{A}} = E\int dA = EA$$

- Evaluation of the right-hand side of Equation 24.6 requires a calculation of the total charge enclosed by the gaussian surface, $q = \int dq$. It is often convenient to represent the charge distribution in terms of the charge density ($dq = \lambda dx$ for a line of charge, $dq = \sigma\,dA$ for a surface of charge, or $dq = \rho\,dV$ for a volume of charge). The integral of dq is then evaluated only over that length, area, or volume which **includes that portion of the charge inside the gaussian surface.**

- In the two previous steps, the components of Equation 24.6 (the mathematical statement of Gauss's law) have been evaluated; you can now calculate the electric field on the gaussian surface if the charge distribution is given. Conversely, if the electric field is known, you can determine the charge distribution that produces the field.

REVIEW CHECKLIST

You should be able to:

- Calculate the electric flux through a surface; in particular, find the net electric flux through a closed surface. (Section 24.1)

- Construct a gaussian surface to match charge distributions which have a high degree of symmetry (spherical, cylindrical, or planar); and use Gauss's law to evaluate the electric field at points interior or exterior to the charge distributions. (Sections 24.2 and 24.3)

- Describe the properties which characterize an electrical conductor in electrostatic equilibrium. (Section 24.4)

ANSWER TO AN OBJECTIVE QUESTION

4. A particle with charge q is located inside a cubical gaussian surface. No other charges are nearby. **(i)** If the particle is at the center of the cube, what is the flux through each one of the faces of the cube? (a) 0 (b) $q/2\epsilon_0$ (c) $q/6\epsilon_0$ (d) $q/8\epsilon_0$ (e) depends on the size of the cube **(ii)** If the particle can be moved to any point within the cube, what maximum value can the flux through one face approach? Choose from the same possibilities as in part (i).

Answer **(i)** The cube has six faces (front, back, left, right, top, and bottom). The particle at the center of the cube sends equal amounts of electric flux through each face. Gauss's law says that the total flux is q/ϵ_0. Then the flux through each separate face is $q/6\epsilon_0$, answer (c).

(ii) To maximize the flux through the top face of the cube, move the charge to a position just below the center of the top face, still barely inside the cube. Just half of the flux created by the charge is associated with electric field with an upward vertical component, and the other half is flux of field that has a downward component. As the distance from the charge up to the top of the cube approaches zero, the top of the cube will intercept essentially all of the flux that goes through a hemisphere above the charge. The amount of this flux is one half of q/ϵ_0, so the answer is (b).

□ □ □ □

ANSWERS TO SELECTED CONCEPTUAL QUESTIONS

4. If the total charge inside a closed surface is known but the distribution of the charge is unspecified, can you use Gauss's law to find the electric field? Explain.

Answer No. If we wish to use Gauss's law to find the electric field, we must be able to bring the electric field $\vec{\mathbf{E}}$ out of the integral. This can be done in some cases—when the field is constant, for example. However, since we do not know the charge distribution, we cannot claim that the field is constant, and thus cannot find the electric field.

To illustrate this point, consider a sphere that contains a net charge of 100 μC. The charges could be located near the center, or they could all be grouped at the northernmost point within the sphere. Between the two cases, the net electric flux would be the same, but the electric field would vary greatly.

□ □ □ □

7. A person is placed in a large hollow metallic sphere that is insulated from ground. (a) If a large charge is placed on the sphere, will the person be harmed upon touching the inside of the sphere? (b) Explain what will happen if the person also has an initial charge whose sign is opposite that of the charge on the sphere.

Answer (a) The metallic sphere is a good conductor, so any excess charge on the sphere will reside on the outside of the sphere. From Gauss's law, we know that the field inside the sphere will then be zero. As a result, when the uncharged person touches the inside of the sphere, no charge will be exchanged between the person and the sphere, and the person will not be harmed.

(b) What happens, then, if the person has an initial charge? Regardless of the sign of the person's initial charge, the charges in the conducting surface will redistribute themselves to maintain a net zero charge within the **conducting metal.** Thus, if the person has a 5.00-μC charge on his skin, exactly –5.00 μC will gather on the inner surface of the sphere, so that the electric field inside the metal will be zero. When the person touches the metallic sphere then, he will receive a shock due only to the charge on his own skin.

□ □ □ □

9. A common demonstration involves charging a rubber balloon, which is an insulator, by rubbing it on your hair, and touching the balloon to a ceiling or wall, which is also an insulator. Because of the electrical attraction between the charged balloon and the neutral wall, the balloon sticks to the wall. Imagine now that we have two infinitely large flat sheets of insulating material. One is charged and the other is neutral. If these sheets are brought into contact, does an attractive force exist between them, as there was for the balloon and the wall?

Answer There will not be an attractive force. There are two factors to consider in the attractive force between a balloon and a wall, or between any pair of charged and neutral objects. The first factor is that the molecules in the wall will orient themselves with their positive ends toward the balloon, and their negative ends pointing away from the balloon. The second factor to consider is that the balloon is of finite size, and thus the molecules in the wall are in a nonuniform electric field. Therefore the nearby "positive ends" of the molecules in the wall will experience an attractive electrostatic force that will be greater in magnitude than the repulsive force exerted on the more distant "negative ends" of the molecules. The net result is an overall force of attraction.

Now consider the infinite sheets brought into contact. The polarization of the molecules in the neutral sheet will indeed occur, as in the wall. But the electric field from the charged sheet is **uniform,** and therefore is independent of the distance from the sheet. Thus, both the negative and positive charges in the neutral sheet will experience the same electric field and the same magnitude of electric force. The attractive force on the positive charges will cancel with the repulsive force on the negative charges, and there will be no net force.

□ □ □ □

SOLUTIONS TO SELECTED END-OF-CHAPTER PROBLEMS

1. A 40.0-cm-diameter loop is rotated in a uniform electric field until the position of maximum electric flux is found. The flux in this position is measured to be $5.20 \times 10^5\ \mathrm{N \cdot m^2/C}$. What is the magnitude of the electric field?

Solution

Conceptualize: Visualize orienting a flat-plate solar energy collector so that it intercepts maximum flux of sunlight.

Categorize: We use the definition of electric flux.

Analyze: For a uniform field the flux is $\Phi = \vec{\mathbf{E}} \cdot \vec{\mathbf{A}} = EA\cos\theta$.

The maximum value of the flux occurs when $\theta = 0$. This means, when the field is in the same direction as the area vector, which is defined as having the direction of the perpendicular to the area.

Therefore, we can calculate the field strength at this point as

$$E = \frac{\Phi_{\max}}{A} = \frac{\Phi_{\max}}{\pi r^2}$$

$$E = \frac{5.20 \times 10^5 \text{ N} \cdot \text{m}^2/\text{C}}{\pi(0.200 \text{ m})^2} = 4.14 \times 10^6 \text{ N/C}$$ ■

Finalize: This is a straightforward problem emphasizing the point that the flux depends on the strength of the field and the size of the area, but also on the orientation of the field to the area.

---

9. The following charges are located inside a submarine: 5.00 μC, −9.00 μC, 27.0 μC, and −84.0 μC. (a) Calculate the net electric flux through the hull of the submarine. (b) Is the number of electric field lines leaving the submarine greater than, equal to, or less than the number entering it?

Solution

Conceptualize: The hull of a submarine is definitely a closed surface, so this problem is a natural for . . .

Categorize: . . . applying Gauss's law, which should tell us the answers directly.

Analyze: The total charge within the closed surface is

$$5.00\ \mu\text{C} - 9.00\ \mu\text{C} + 27.0\ \mu\text{C} - 84.0\ \mu\text{C} = -61.0\ \mu\text{C}$$

(a) so the total electric flux is

$$\Phi_E = \frac{q}{\epsilon_0} = \frac{-61.0 \times 10^{-6} \text{ C}}{\left(8.85 \times 10^{-12} \text{ C}^2/\text{N} \cdot \text{m}^2\right)} = -6.89 \times 10^6 \text{ N} \cdot \text{m}^2/\text{C}$$ ■

(b) The minus sign means that fewer lines leave the surface than enter it. ■

Finalize: We need not know about the size of the submarine, or the strengths of the field at different points—that is, the concentration of the field lines. There might be a large outward electric flux near a positive charge, but the total enclosed charge tells us unerringly the net flux of electric field.

---

17. A particle with charge Q is located a small distance δ immediately above the center of the flat face of a hemisphere of radius R as shown in Figure P24.17. What is the electric flux (a) through the curved surface and (b) through the flat face as $\delta \to 0$?

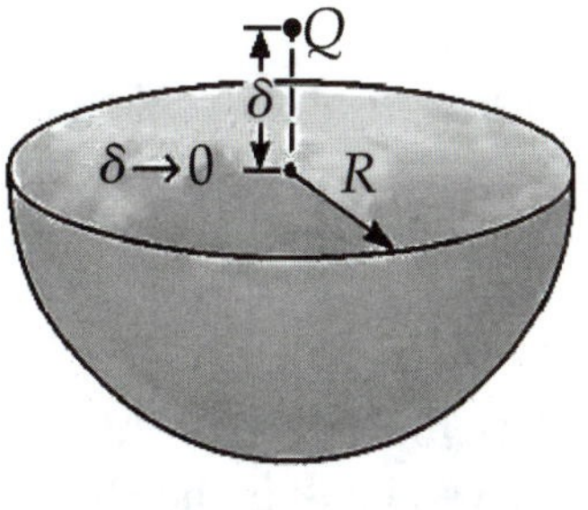

Figure P24.17

Solution

Conceptualize: From Gauss's law, the flux through a sphere with a charged particle in it should be Q/ϵ_0, so we should expect the electric flux through a hemisphere to be half this value:

$$\Phi_{\text{curved}} = Q/2\epsilon_0$$

The flat section appears like an infinite plane to a point just above its surface. Then half of all the field lines from the charged particle are intercepted by the flat surface, so the flux through this section should also equal $Q/2\epsilon_0$.

Categorize: We can apply the definition of electric flux directly for part (a) and then use Gauss's law to find the flux for part (b).

Analyze:

(a) With δ very small, all points on the hemisphere are nearly at distance R from the charge, so the field everywhere on the curved surface is $k_e Q/R^2$ radially outward (normal to the surface). Therefore, the flux is this field strength times the area of half a sphere:

$$\Phi_{\text{curved}} = \int \vec{\mathbf{E}} \cdot d\vec{\mathbf{A}} = E_{\text{local}} A_{\text{hemisphere}}$$

$$\Phi_{\text{curved}} = \left(k_e \frac{Q}{R^2}\right)\left(\tfrac{1}{2}\right)\left(4\pi R^2\right) = \frac{1}{4\pi \epsilon_0} Q(2\pi) = \frac{Q}{2\epsilon_0} \quad \blacksquare$$

(b) The closed surface encloses zero charge, so Gauss's law gives

$$\Phi_{\text{curved}} + \Phi_{\text{flat}} = 0 \qquad \text{or} \qquad \Phi_{\text{flat}} = -\Phi_{\text{curved}} = \frac{-Q}{2\epsilon_0} \quad \blacksquare$$

Finalize: The direct calculations of the electric flux agree with our predictions, except for the negative sign in part (b), which comes from the fact that the area unit vector is defined as pointing outward from an enclosed surface, and in this case, the electric field has a component in the opposite direction (down).

---

23. A large flat horizontal sheet of charge has a charge per unit area of 9.00 μC/m^2. Find the electric field just above the middle of the sheet.

Solution

Conceptualize: Visualize rain falling vertically from a deck of stratus clouds covering the whole sky. Grumpy tourists get equally wet at the top or at the bottom of the Grand Canyon. The sheet of charge creates uniform field pointing up away from it, the same at all distances.

Categorize: We think, to make sure it applies, and then use the appropriate equation derived back in the chapter text.

Analyze: For a large uniformly charged sheet, $\vec{\mathbf{E}}$ will be perpendicular to the sheet, and will have a magnitude of

$$E = \frac{\sigma}{2\epsilon_0} = 2\pi k_e \sigma = (2\pi)\left(8.99\times10^9\ \text{N}\cdot\text{m}^2/\text{C}^2\right)\left(9.00\times10^{-6}\ \text{C/m}^2\right)$$

so $$\vec{\mathbf{E}} = 5.08\times10^5\ \text{N/C}\ \hat{\mathbf{j}}$$ ■

Finalize: Doubling the distance from a charged particle reduces the field by a factor of four. Doubling the distance from a line of charge reduces the magnitude of $\vec{\mathbf{E}}$ by a factor of two. In this problem doubling the distance from the sheet of charge leaves the field unchanged. Gauss's law is not just about geometry, but some of its implications are directly about geometry.

---

27. Consider a thin spherical shell of radius 14.0 cm with a total charge of 32.0 μC distributed uniformly on its surface. Find the electric field (a) 10.0 cm and (b) 20.0 cm from the center of the charge distribution.

Solution

Conceptualize: The field inside the shell should be zero. The field outside the shell should be just like that created by a charged particle at its center.

Categorize: The field must be radially outward and uniform in size over any sphere centered at the center of the shell. Knowing this lets us use spherical gaussian surfaces through the field points to find the field magnitudes there.

Analyze:

(a) A gaussian sphere, radius 10.0 cm, encloses 0 charge. Then at the surface of this gaussian sphere, inside the charged shell, we have $\vec{\mathbf{E}} = 0$. ■

(b) For a gaussian sphere of radius 20.0 cm, we apply $\oint \vec{\mathbf{E}}\cdot d\vec{\mathbf{A}} = \frac{q_{\text{in}}}{\epsilon_0}$

The field is radially outward, and $4\pi r^2 E = q/\epsilon_0$

$$E = \frac{k_e q}{r^2} = \frac{\left(8.99\times10^9\ \text{N}\cdot\text{m}^2/\text{C}^2\right)\left(32.0\times10^{-6}\ \text{C}\right)}{\left(0.200\ \text{m}\right)^2} = 7.19\times10^6\ \text{N/C}$$

so $$\vec{\mathbf{E}} = (7.19\times10^6\ \text{N/C})\ \hat{\mathbf{r}}$$ ■

Finalize: Our results agree with our predictions. The chapter-opener photograph in the textbook suggests the situation of the field lines starting on the surface of the sphere.

---

29. A uniformly charged, straight filament 7.00 m in length has a total positive charge of 2.00 μC. An uncharged cardboard cylinder 2.00 cm in length and 10.0 cm in radius surrounds the filament at its center, with the filament as the axis of the cylinder. Using reasonable approximations, find (a) the electric field at the surface of the cylinder and (b) the total electric flux through the cylinder.

Solution

Conceptualize: The filament creates electric field pointing radially away from it everywhere. The field lines diverge, so the field gets weaker farther away. The field "shines" perpendicularly through the cardboard, to produce flux through it.

Categorize: An equation back in the chapter gives the field near a very long straight charged filament. The flux will be all the flux that is produced by the charge that resides on 2 cm of the filament.

Analyze: The approximation in this case is that the filament length is so large when compared to the cylinder length that the "infinite line" of charge can be assumed.

(a) We have $E = \dfrac{2k_e\lambda}{r}$ where the linear charge density is

$$\lambda = \frac{2.00 \times 10^{-6}\ \text{C}}{7.00\ \text{m}} = 2.86 \times 10^{-7}\ \text{C/m}$$

so $$E = \frac{(2)(8.99 \times 10^{9}\ \text{N}\cdot\text{m}^2/\text{C})(2.86 \times 10^{-7}\ \text{C/m})}{0.100\ \text{m}} = 5.14 \times 10^{4}\ \text{N/C} \quad \text{outward}$$ ■

(b) We can find the flux by multiplying the field and the lateral surface area of the cylinder:

$$\Phi_E = 2\pi rLE = 2\pi rL\left(\frac{2k_e\lambda}{r}\right) = 4\pi k_e\lambda L$$

so $$\Phi_E = 4\pi(8.99 \times 10^{9}\ \text{N}\cdot\text{m}^2/\text{C}^2)(2.86 \times 10^{-7}\ \text{C/m})(0.0200\ \text{m})$$
$$= 6.46 \times 10^{2}\ \text{N}\cdot\text{m}^2/\text{C}$$ ■

Finalize: We could also find the flux directly from Gauss's law. The charge enclosed by the cardboard is $\lambda L = (2.86 \times 10^{-7}\ \text{C/m})(0.0200\ \text{m})$. The net flux it creates is

$$q/\epsilon_0 = 5.72 \times 10^{-9}\ \text{C}/(8.85 \times 10^{-12}\ \text{C}^2/\text{N}\cdot\text{m}^2) = 6.46 \times 10^{2}\ \text{N}\cdot\text{m}^2/\text{C}$$

If the length of the cylinder, here 2 cm, were a considerable fraction of 700 cm, the field would start pointing out through both of the ends of the cylinder to reduce the answer to part (a). But the answer to part (b) would still be the same.

---

33. Consider a long cylindrical charge distribution of radius R with a uniform charge density ρ. Find the electric field at distance r from the axis where $r < R$.

Solution

Conceptualize: At $r = 0$, the field should be zero because it could not point in one direction more than in any other direction. It would make sense for the field to be radially outward at all nonzero values of r, and to grow larger as r increases.

Categorize: We use our knowledge of the direction of the field to choose a gaussian surface. We evaluate both sides of Gauss's law to obtain an equation involving the unknown field magnitude, which we can then solve.

Analyze: If ρ is positive, the field must everywhere be radially outward. Choose as the gaussian surface a cylinder of length L and radius r, contained inside the charged rod. Its volume is $\pi r^2 L$ and it encloses charge $\rho \pi r^2 L$. The circular end caps have no electric flux through them; there $\vec{\mathbf{E}} \cdot d\vec{\mathbf{A}} = E\,dA \cos 90.0° = 0$. The curved surface has $\vec{\mathbf{E}} \cdot d\vec{\mathbf{A}} = E\,dA \cos 0°$, and E must be the same strength everywhere over the curved surface.

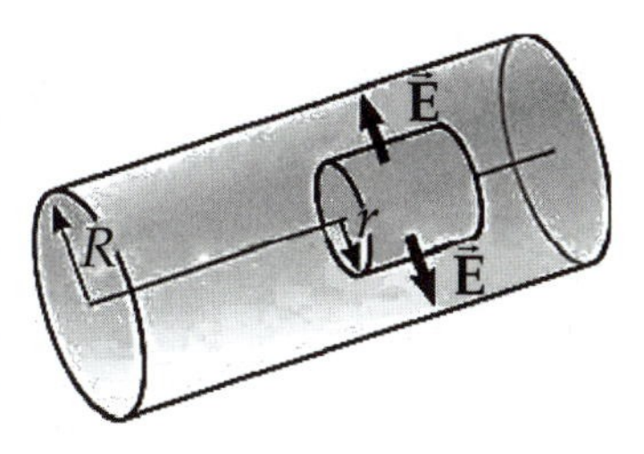

Then $\oint \vec{\mathbf{E}} \cdot d\vec{\mathbf{A}} = \dfrac{q_{inside}}{\epsilon_0}$ becomes $E \int_{\text{Curved Surface}} dA = \dfrac{\rho \pi r^2 L}{\epsilon_0}$

Noting that $2\pi r L$ is the lateral surface area of the cylinder, we have

$$E(2\pi r)L = \frac{\rho \pi r^2 L}{\epsilon_0}$$

Thus, $\vec{\mathbf{E}} = \dfrac{\rho r}{2\epsilon_0}$ radially away from the axis. ■

Finalize: We do indeed have zero field on the axis of the rod. The field is strongest at the surface of the rod. Inside it grows proportionately to the radius. Outside it decreases in proportion to $1/r$.

---

35. A long, straight metal rod has a radius of 5.00 cm and a charge per unit length of 30.0 nC/m. Find the electric field (a) 3.00 cm, (b) 10.0 cm, and (c) 100 cm from the axis of the rod, where distances are measured perpendicular to the rod's axis.

Solution

Conceptualize: The field will be (ten times?) weaker at ten times farther away from the axis of the rod. But we must look out for the field inside and the fact that the rod is metallic.

Categorize: We assume the rod is many meters long, so we can model it as a long filament for the exterior points. But at 3 cm from the axis we are inside the metal, and …

Analyze:

(a) … inside the conductor, $E = 0$. ■

The metal rod carries charge only on its surface. Charge deposited anywhere within the metal runs away from itself. The bits of charge repel one another and get as far apart as they can, by moving out to the surface. This charge produces no field under the surface.

(b) Outside the conductor, $E = 2k_e\lambda/r$. At $r = 0.100$ m,

$$E = 2\frac{k_e\lambda}{r} = 2\frac{(8.99 \times 10^9\ \text{N}\cdot\text{m}^2/\text{C}^2)(30.0 \times 10^{-9}\ \text{C/m})}{0.100\ \text{m}}$$

$$= 5.39 \times 10^3\ \text{N/C directed radially away from the axis}$$ ■

(c) At $r = 1.00$ m,

$$E = 2\frac{k_e\lambda}{r} = 2\frac{(8.99 \times 10^9\ \text{N}\cdot\text{m}^2/\text{C}^2)(30.0 \times 10^{-9}\ \text{C/m})}{1.00\ \text{m}}$$

$$= 539\ \text{N/C radially outward from the positive charge}$$ ■

Finalize: The answer to (c) is indeed one-tenth of the answer to (b). Another way of stating the solution to part (a): If there were any nonzero electric field, free charges in the metal would be moving and the electrostatic situation would not yet be established. The charge would still be in the process of running away to the surface.

---

43. A long, straight wire is surrounded by a hollow metal cylinder whose axis coincides with that of the wire. The wire has a charge per unit length of λ, and the cylinder has a net charge per unit length of 2λ. From this information, use Gauss's law to find (a) the charge per unit length on the inner surface of the cylinder, (b) the charge per unit length on the outer surface of the cylinder and (c) the electric field outside the cylinder, a distance r from the axis.

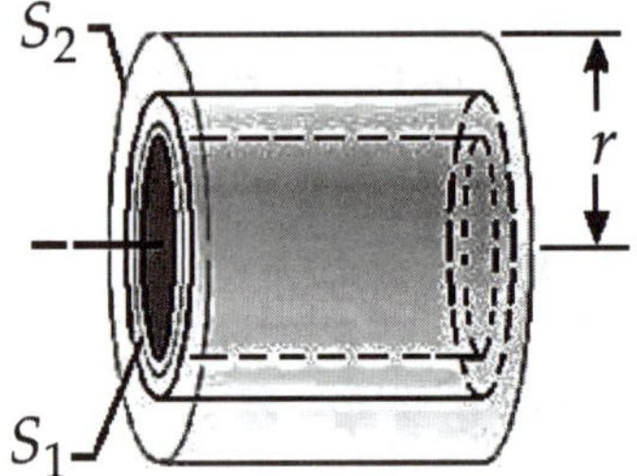

Solution

Conceptualize: Assume λ represents a positive number. The positively charged wire induces a negative charge on the inner surface of the metal shell. The shell has polarization contributing to the charge on its outer surface, as well as a net charge.

Categorize: By considering surface after nesting gaussian surface we will build up our answers.

Analyze:

(a) Use a cylindrical gaussian surface S_1 within the metal of the conducting cylinder. Here $E = 0$.

Thus, $\oint E_n dA = \left(\frac{1}{\epsilon_0}\right) q_{\text{in}} = 0$ and $\lambda_{\text{inner}} = -\lambda$ ■

(b) Next, the net charge on the metal pipe is described by

$$\lambda_{\text{inner}} + \lambda_{\text{outer}} = 2\lambda \quad \text{so} \quad \lambda_{\text{outer}} = 3\lambda$$ ■

(c) For a gaussian surface S_2 outside the conducting cylinder,

$$\oint E_n dA = \left(\frac{1}{\epsilon_0}\right) q_{\text{in}} \quad \text{or} \quad E(2\pi rL) = \frac{1}{\epsilon_0}(\lambda - \lambda + 3\lambda)L:$$

$$E = \frac{3\lambda}{2\pi \epsilon_0 r} \text{ radially away from the axis}$$ ■

Finalize: In the space, empty of matter, between the wire and the inner wall of the cylinder, the field is $E = \dfrac{\lambda}{2\pi \epsilon_0 r}$ away from the axis. It is unaffected by the metal cylinder. The field must be zero within the metal material, but it can be nonzero in a hollow space enclosed by metal.

---

44. A thin square conducting plate 50.0 cm on a side lies in the xy plane. A total charge of 4.00×10^{-8} C is placed on the plate. Find (a) the charge density on each face of the plate, (b) the electric field just above the plate, and (c) the electric field just below the plate. You may assume the charge density is uniform.

Solution

Conceptualize: The sheet of charge creates uniform away-pointing field, upward in the space above the sheet and downward in the space below.

Categorize: We compute the charge density from its definition. We compute the field from the equation that the chapter proves for a charged sheet.

Analyze: We ignore "edge" effects and assume that the total charge distributes itself uniformly over each side of the plate, with one half the total charge on each side.

(a) $$\sigma = \frac{q}{A} = \frac{4.00 \times 10^{-8}\ \text{C}}{2(0.500\ \text{m})^2} = 8.00 \times 10^{-8}\ \text{C/m}^2$$ ■

(b) Just above the plate,

$$E = \frac{\sigma}{\epsilon_0} = \frac{8.00 \times 10^{-8}\ \text{C/m}^2}{8.85 \times 10^{-12}\ \text{C}^2/\text{N}\cdot\text{m}^2} = 9.04 \times 10^3\ \text{N/C} \quad \text{upward}$$ ■

(c) Just below the plate, $E = \dfrac{\sigma}{\epsilon_0} = 9.04 \times 10^3\ \text{N/C}$ downward ■

Finalize: This is not a trick question. The answers to (b) and (c) have the same numerical value and follow the same rule about direction (that positive charge creates away-pointing electric field throughout the surrounding space, like a crabby porcupine) but they are fair questions. Drawing a picture may help you, but the reasoning required is really three-dimensional reasoning. Think about the quills of a big group of porcupines who are crowded close together on a flat mesh hammock, and are taking care not to poke one another. Their quills point straight up and straight down.

---

48. A sphere of radius R surrounds a particle with charge Q, located at its center as shown in Figure P24.47. Find the electric flux through a circular cap of half-angle θ.

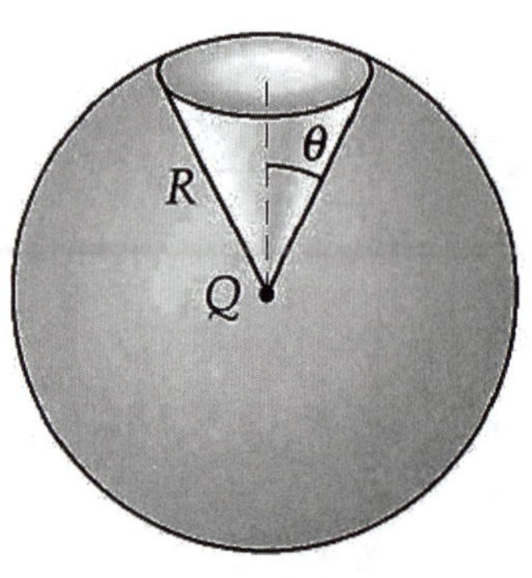

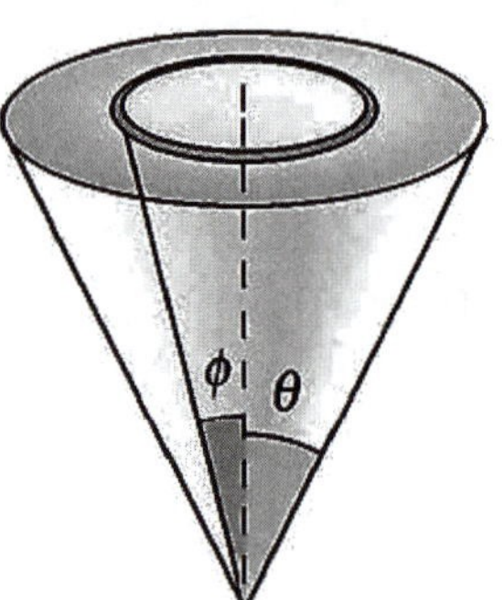

Figure P24.47

Solution

Conceptualize: The charged particle creates field pointing away in all radial directions and uniform over the surface of the sphere. The cap covers a fraction of the sphere's surface and intercepts that fraction of the total flux.

Categorize: The field we can write down directly. The surface area of the cap is curved. We will do an integral to find the area.

Analyze:

The electric field of the charged particle has constant strength k_eQ/R^2 over the cap and points radially outward. To find the area of the curved cap, we think of it as formed of rings, each of radius $r = R \sin \phi$, where ϕ ranges from 0 to θ. The width of each ring is $ds = Rd\phi$, so its area is the product of its two perpendicular dimensions,

$$dA = (2\pi r)ds = 2\pi(R\sin\phi)(R\,d\phi)$$

The whole cap has an area of

$$A = \int dA = \int_0^\theta 2\pi R^2 \sin\phi\; d\phi = 2\pi R^2(-\cos\phi)\big|_0^\theta = 2\pi R^2(-\cos\theta + 1)$$

The flux through it is

$$\Phi_E = \int \vec{\mathbf{E}}\cdot d\vec{\mathbf{A}} = \int E\,dA\cos 0° = E\int dA = EA$$

$$\Phi_E = \frac{k_eQ}{R^2}2\pi R^2(1-\cos\theta) = \left(\frac{1}{4\pi\epsilon_0}\right)(2\pi Q)(1-\cos\theta) = \frac{Q}{2\epsilon_0}(1-\cos\theta) \quad \blacksquare$$

Finalize: We can check the derived expression for three particular angles:

For $\theta = 90°$, the cap is a hemisphere and intercepts half the flux from the charge:

$$\Phi_E = \frac{Q}{2\epsilon_0}(1-\cos 90°) = \frac{Q}{2\epsilon_0}$$

For $\theta = 180°$, the cap is a full sphere and all the field lines go through it:

$$\Phi_E = \frac{Q}{2\epsilon_0}(1-\cos 180°) = \frac{Q}{\epsilon_0}$$

The flux must go to zero when the cap area goes to zero for $\theta = 0$; and $(1 - \cos 0) = (1 - 1)$ does have this property.

52. A solid, insulating sphere of radius a has a uniform charge density throughout its volume and a total charge Q. Concentric with this sphere is an uncharged, conducting hollow sphere whose inner and outer radii are b and c, as shown in Figure P24.51. We wish to understand completely the charges and electric fields at all locations. (a) Find the charge contained within a sphere of radius $r < a$. (b) From this value, find the magnitude of the electric field for $r < a$. (c) What charge is contained within a sphere of radius r where $a < r < b$? (d) From this value, find the magnitude of the electric field for r where $a < r < b$. (e) Now consider r where $b < r < c$. What is the magnitude of the electric field for this range of values of r? (f) From this value, what must be the charge on the inner surface of the hollow sphere? (g) From part (f), what must be the charge on the outer surface of the hollow sphere? (h) Consider the three spherical surfaces of radii a, b, and c. Which of these surfaces has the largest magnitude of surface charge density?

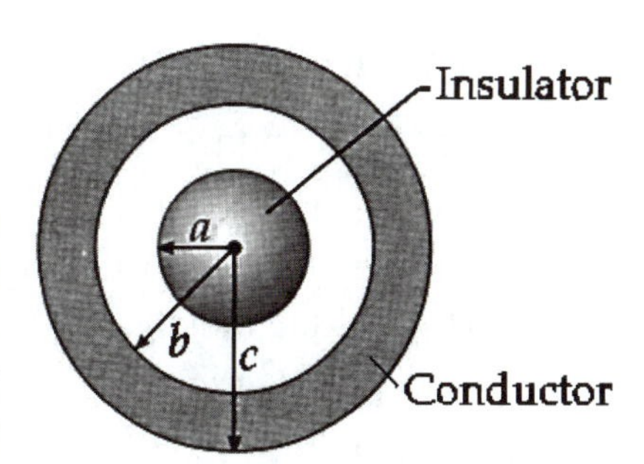

Figure P24.51

Solution

Conceptualize: Suppose Q is positive. The charge on the inner solid sphere will create outward field to polarize the metal shell, inducing negative charge on its inner surface. The net charge of the shell is zero, so it must have an equal amount (Q) of positive charge on its outer surface. The charge density will be predictably lower on the larger outer surface.

Categorize: We consider nesting gaussian surfaces from the inside out. We will use the identification of the field as zero between b and c. We will use the property that the metal has zero total charge.

Analyze: Choose as each gaussian surface a concentric sphere of radius r. The electric field will be perpendicular to its surface, and will be uniform in strength over its surface. The density of charge in the insulating sphere is $\rho = Q/\left(\frac{4}{3}\pi a^3\right)$.

(a) The sphere of radius $r < a$ encloses charge

$$\rho\left(\tfrac{4}{3}\pi r^3\right) = Q\left(\tfrac{4}{3}\pi r^3\right)/\left(\tfrac{4}{3}\pi a^3\right) = Q(r/a)^3 \qquad \blacksquare$$

(b) Applying Gauss's law to this sphere reveals the magnitude of the field at its surface.

Here $\Phi = q/\epsilon_0$ becomes $E\left(4\pi r^2\right) = \dfrac{Q\left(r/a\right)^3}{\epsilon_0}$ so $E = \dfrac{Qr}{4\pi\epsilon_0 a^3} = \dfrac{k_e Qr}{a^3}$ ■

(c) For a sphere of radius r with $a < r < b$, the whole insulating sphere is enclosed, so the charge within is Q. ■

(d) Gauss's law for this sphere becomes $E\left(4\pi r^2\right) = \dfrac{Q}{\epsilon_0}$ and $E = \dfrac{Q}{4\pi\epsilon_0 r^2} = k_e Q/r^2$ ■

(e) For $b < r < c$, we must have $E = 0$ ■

because any nonzero field would be moving charges in the metal.

(f) A spherical gaussian surface within the metal must enclose zero total charge. Part of it is charge Q on the insulating sphere, and the rest of it must be charge $-Q$ on the inner surface of the metal, at radius b. ■

This induced charge was deposited on the inner wall of the shell when charges were moving in the metal, when the shell was first set up to enclose the charged sphere.

(g) The shell is uncharged as a whole. With charge $-Q$ on its inner surface and no charge in the bulk of the metal, charge $+Q$ is left on its outer surface, at radius c. ■

This charge was deposited here in the polarization process, when charges were moving in the metal to reduce the field within it to zero.

(h) The field is strongest at radius a, but the insulating sphere carries charge in the volume of its material. With zero thickness, the surface of radius a has zero volume and zero surface charge density. Then the surface of radius b, smaller in area than the surface of radius c, carries charge with the greatest surface density magnitude. ■

Finalize: The answers are specified in terms of some of the symbols we take as given information, namely Q, r, a, b, c, and the universal constant $k_e = 1/4\pi\epsilon_0$. When you pay careful attention to where the zeros are and to the reasons for them, you will be thinking carefully about charge and field, just as you should.

It would be natural to continue the logic and find the field outside the shell, for $r > c$. Here $\Phi = q/\epsilon_0$ becomes $E(4\pi r^2) = \dfrac{Q+Q-Q}{\epsilon_0}$ and $E = \dfrac{Q}{4\pi\epsilon_0 r^2} = k_e Q / r^2$.

---

54. Two infinite, nonconducting sheets of charge are parallel to each other as shown in Figure P24.54. The sheet on the left has a uniform surface charge density σ, and the one on the right has a uniform charge density $-\sigma$. Calculate the electric field at points (a) to the left of, (b) in between, and (c) to the right of the two sheets. (d) **What if?** Find the electric fields in all three regions if both sheets have *positive* uniform surface charge densities of value σ.

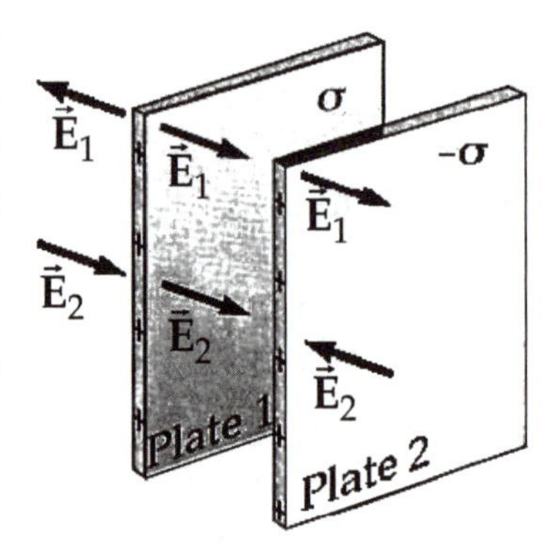

Figure P24.54

Solution

Conceptualize: When the sheets have opposite-sign charges with the same charge density, a positive test charge at a point midway between them will experience the same force in the same direction, as a force of attraction to one sheet and a force of repulsion from the other. Therefore, the electric field here will add up to $E = \sigma/\epsilon_0$, twice as large as a single sheet of charge would produce. Because each infinite sheet produces a field that is uniform in space, the test charge will experience an equally large force everywhere in the region between the plates. Outside the oppositely-charged sheets, the electric fields they separately produce will point in opposite directions, so they add to zero.

Categorize: The principle of superposition can be applied to add the electric field vectors due to each sheet of charge.

Analyze: Each sheet separately creates electric field at all points given by $\vec{\mathbf{E}} = \sigma/2\epsilon_0$ directed away from the positive sheet and toward the negative sheet.

(a) At any point to the left of the two parallel sheets

$$\vec{\mathbf{E}} = E_1(-\hat{\mathbf{i}}) + E_2(+\hat{\mathbf{i}}) = -\frac{\sigma}{2\epsilon_0}\hat{\mathbf{i}} + \frac{\sigma}{2\epsilon_0}\hat{\mathbf{i}} = 0 \quad \blacksquare$$

(b) At any point between the two sheets $\vec{\mathbf{E}} = E_1\hat{\mathbf{i}} + E_2\hat{\mathbf{i}} = 2E\hat{\mathbf{i}} = \frac{\sigma}{\epsilon_0}\hat{\mathbf{i}}$ ■

(c) At any point to the right of the two sheets, $\vec{\mathbf{E}} = E_1\hat{\mathbf{i}} + E_2(-\hat{\mathbf{i}}) = 0$ ■

(d) If both sheets are positive, we must reverse all three arrows labeled $\vec{\mathbf{E}}_2$ in the diagram. Now at a point to the left the addition is

$$\vec{\mathbf{E}} = E_1(-\hat{\mathbf{i}}) + E_2(-\hat{\mathbf{i}}) = -\frac{\sigma}{2\epsilon_0}\hat{\mathbf{i}} - \frac{\sigma}{2\epsilon_0}\hat{\mathbf{i}} = -\frac{\sigma}{\epsilon_0}\hat{\mathbf{i}} \quad \blacksquare$$

In between the total field is $\vec{\mathbf{E}} = E_1\hat{\mathbf{i}} + E_2(-\hat{\mathbf{i}}) = 0$ ■

At a point to the right of both sheets, $\vec{\mathbf{E}} = E_1\hat{\mathbf{i}} + E_2\hat{\mathbf{i}} = 2E\hat{\mathbf{i}} = \frac{\sigma}{\epsilon_0}\hat{\mathbf{i}}$ ■

Finalize: We essentially solved parts (a) through (c) in the Conceptualize step, so it is no surprise that the results are what we expected. In Chapter 26 we will identify a pair of oppositely-charged plates as a capacitor. The results of part (d) are complementary to the case where the plates are oppositely charged. Note that the field-creating influence of one sheet can be thought of as reaching right through the other sheet and its charge, unimpeded.

---

62. A solid insulating sphere of radius R has a nonuniform charge density that varies with r according to the expression $\rho = Ar^2$, where A is a constant and $r < R$ is measured from the center of the sphere. (a) Show that the magnitude of the electric field outside ($r > R$) the sphere is $E = AR^5/5\epsilon_0 r^2$. (b) Show that the magnitude of the electric field inside ($r < R$) the sphere is $E = Ar^3/5\epsilon_0$. *Note:* The volume element dV for a spherical shell of radius r and thickness dr is equal to $4\pi r^2 dr$.

Solution

Conceptualize: The equations we prove will show that the field is zero at the center, where it has no direction to point in. The field grows as we move away from the center, has its maximum magnitude at the surface, and then shows familiar inverse-square law decrease outside $r = R$.

Categorize: We will apply Gauss's law to surfaces both smaller and larger than R. Gauss's law is about surface integrals. We will also have to do volume integrals to find the enclosed charges creating the fields.

Analyze:

(a) We call the constant A', reserving the symbol A to denote area. The whole charge of the ball is

$$Q = \int_{\text{ball}} dQ = \int_{\text{ball}} \rho dV = \int_{r=0}^{R} A'r^2 4\pi r^2 dr = 4\pi A' \frac{r^5}{5}\bigg|_0^R = \frac{4\pi A' R^5}{5}$$

To find the electric field, consider as gaussian surface a concentric sphere of radius r outside the ball of charge:

In this case, $\oint \vec{\mathbf{E}} \cdot d\vec{\mathbf{A}} = \frac{Q}{\epsilon_0}$ reads $EA \cos 0° = \frac{Q}{\epsilon_0}$

Solving, $E(4\pi r^2) = \frac{4\pi A' R^5}{5\epsilon_0}$

and the electric field is $E = \frac{A' R^5}{5\epsilon_0 r^2}$ ■

(b) Let the gaussian sphere lie inside the ball of charge:

$$\oint_{\substack{\text{spherical surface,} \\ \text{radius } r}} \vec{\mathbf{E}} \cdot d\vec{\mathbf{A}} = \int_{\substack{\text{spherical volume,} \\ \text{radius } r}} dQ / \epsilon_0$$

Now the integrals become $E(\cos 0)\oint dA = \int \frac{\rho dV}{\epsilon_0}$ or $EA = \int_0^r \frac{A' r^2 (4\pi r^2) dr}{\epsilon_0}$

Performing the integration, $E(4\pi r^2) = \left(\frac{A' 4\pi}{\epsilon_0}\right)\left(\frac{r^5}{5}\right)\bigg|_0^r = \frac{A' 4\pi r^5}{5\epsilon_0}$

And the field is $E = \frac{A' r^3}{5\epsilon_0}$ ■

Finalize: We drew no two-dimensional pictures. You must think three-dimensionally about volume and surface area of spheres within and without the ball of charge. A check: the two expressions for field must agree at $r = R$. We have from (a) $E = \frac{A' R^5}{5\epsilon_0 r^2} = \frac{A' R^5}{5\epsilon_0 R^2} = \frac{A' R^3}{5\epsilon_0}$ and evaluating the expression from (b) gives the same result.

---

66. Review. A slab of insulating material (infinite in the y and z directions) has a thickness d and a uniform positive charge density ρ. An edge view of the slab is shown in Figure P24.59. (a) Show that the magnitude of the electric field a distance x from its center and inside the slab is $E = \rho x/\epsilon_0$. (b) **What if?** Suppose an electron of charge $-e$ and mass m_e can move freely within the slab. It is released from rest at a distance x from the center. Show that the electron exhibits simple harmonic motion with a frequency

$$f = \frac{1}{2\pi}\sqrt{\frac{\rho e}{m_e \epsilon_0}}$$

Solution

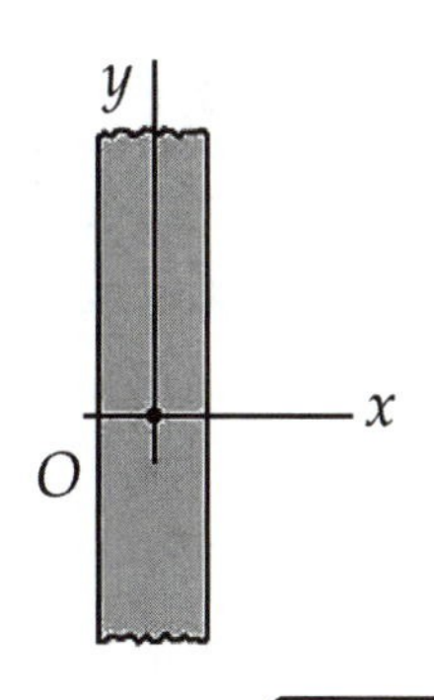

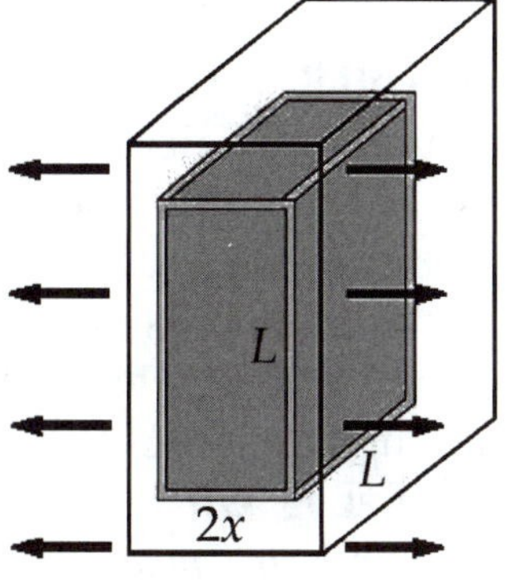

Figure P24.59

Conceptualize: At the midplane $x = 0$ the field is zero so the electron has an equilibrium position. When it is off-center the negative electron is attracted to the positive charge of the sheet, so it feels a restoring force. To prove that its motion is simple harmonic we must show…

Categorize: … that the restoring force is proportional to the excursion from equilibrium. Gauss's law will let us evaluate the field to do this. Then comparison of Newton's second law here to $a = -\omega^2 x$ will let us identify the frequency $f = \omega/2\pi$.

Analyze:

(a) The slab has left-to-right symmetry, so its field must be equal in strength at x and at $-x$. The field points everywhere away from the central plane. Take as gaussian surface a rectangular box of thickness $2x$ and height and width L, centered on the $x = 0$ plane. The gaussian surface, shown shaded in the second drawing, lies inside the slab. The charge the surface contains is $\rho V = \rho(2xL^2)$. The total flux leaving it is EL^2 through the right face, EL^2 through the left face, and zero through each of the other four sides.

Thus Gauss's law $\oint \vec{\mathbf{E}} \cdot d\vec{\mathbf{A}} = \dfrac{q}{\epsilon_0}$ becomes $2EL^2 = \dfrac{\rho 2xL^2}{\epsilon_0}$

so the field is $$E = \frac{\rho x}{\epsilon_0}$$ ■

(b) The electron experiences a force opposite to $\vec{\mathbf{E}}$. When displaced to $x > 0$, it experiences a restoring force to the left.

For it, $\sum \vec{\mathbf{F}} = m_e \vec{\mathbf{a}}$ reads $q\vec{\mathbf{E}} = m_e \vec{\mathbf{a}}$ or $\dfrac{-e\rho x\hat{\mathbf{i}}}{\epsilon_0} = m_e \vec{\mathbf{a}}$

Solving for the acceleration, $\vec{\mathbf{a}} = -\left(\dfrac{e\rho}{m_e \epsilon_0}\right) x\hat{\mathbf{i}}$ or $\vec{\mathbf{a}} = -\omega^2 x\hat{\mathbf{i}}$

That is, its acceleration is proportional to its displacement and oppositely directed, as is required for simple harmonic motion.

Solving for the frequency, $\omega^2 = \dfrac{e\rho}{m_e \epsilon_0}$ and $f = \dfrac{\omega}{2\pi} = \dfrac{1}{2\pi}\sqrt{\dfrac{e\rho}{m_e \epsilon_0}}$ ■

Finalize: The charge-to-mass ratio of the electron controls its frequency of oscillation. And a higher charge density in the slab will create more field to increase the electron's frequency.

25

Electric Potential

EQUATIONS AND CONCEPTS

Potential difference between points Ⓐ and Ⓑ in an electric field ($\Delta V = V_{Ⓑ} - V_{Ⓐ}$) is defined as the *change* in the potential energy of the charge-field system when a test charge (q_0) is moved from point Ⓐ to point Ⓑ divided by the test charge. Potential difference (ΔV) can be evaluated by integrating $\vec{\mathbf{E}} \cdot d\vec{\mathbf{s}}$ along any path from Ⓐ to Ⓑ. *Electric potential is a scalar quantity characteristic of an electric field. Potential energy is characteristic of a charge-field system.*

$$\Delta V = \frac{\Delta U}{q_0} = -\int_{Ⓐ}^{Ⓑ} \vec{\mathbf{E}} \cdot d\vec{\mathbf{s}} \quad (25.3)$$

The **work required by an external force** to move a charge q from point Ⓐ to point Ⓑ in an electric field is independent of the path. *Zero work is required to move a charge between two points that are at the same potential.* The SI unit of electric potential is the volt (V).

$$W = q\Delta V \quad (25.4)$$

$$1 \text{ V} \equiv 1 \text{ J/C}$$

$$1 \text{ N/C} = 1 \text{ V/m}$$

The **electron volt** (eV) is a unit of energy which is defined as the energy that a charge-field system gains or loses when a charge of magnitude e is moved through a potential difference of 1 V.

$$1 \text{ eV} = 1.60 \times 10^{-19} \text{ J} \quad (25.5)$$

The **potential difference between two points in a uniform electric field** depends on the displacement $\vec{\mathbf{d}}$ along the direction parallel to $\vec{\mathbf{E}}$ and on the magnitude of $\vec{\mathbf{E}}$. *Electric field lines always point in the direction of decreasing electric potential. In the figure, the potential at point Ⓑ is lower than that at point Ⓐ.*

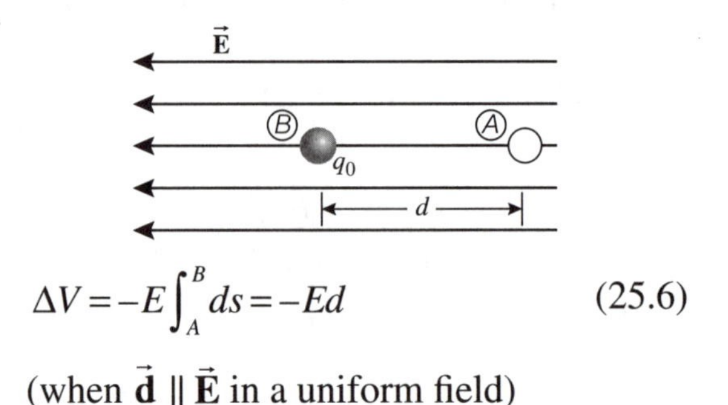

$$\Delta V = -E\int_{A}^{B} ds = -Ed \quad (25.6)$$

(when $\vec{\mathbf{d}} \parallel \vec{\mathbf{E}}$ in a uniform field)

The **change in potential energy**, ΔU, of a charge-field system as a charged particle moves from point Ⓐ to point Ⓑ in an electric field depends on the sign and magnitude of the charge as well as on the change in potential, ΔV. *The electric potential energy of a charge-field system decreases when a positive charge moves in the direction of the field. See figure accompanying Equation 25.6.*

$$\Delta U = q_0 \Delta V = -q_0 Ed \tag{25.7}$$

(for $d \parallel \vec{\mathbf{E}}$ in a uniform field)

The **electric potential due to a single point charge**, at a distance r from the charge, depends inversely on the distance from the charge. The potential is given by Equation 25.11 when the zero reference level for potential is taken to be at infinity. *Electric potential is a scalar quantity.*

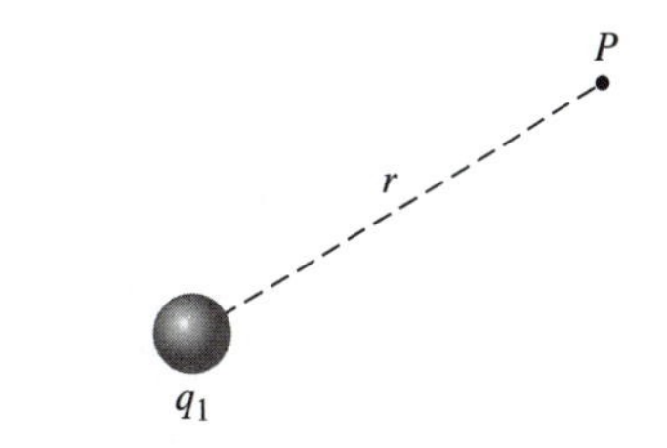

$$V = k_e \frac{q}{r} \tag{25.11}$$

The **electric potential at a point in the vicinity of several charged particles** is the scalar sum of the potentials due to the individual charges.

$$V = k_e \sum_i \frac{q_i}{r_i} \tag{25.12}$$

The **potential energy of a system of two charged particles** separated by a distance r_{12} represents the work required to assemble the charges from an infinite separation. Hence, the negative of the potential energy equals the minimum work required to separate them by an infinite distance with no final kinetic energy. *The electric potential energy associated with a system of two charged particles is positive if the two charges have the same sign, and negative if they are of opposite sign.*

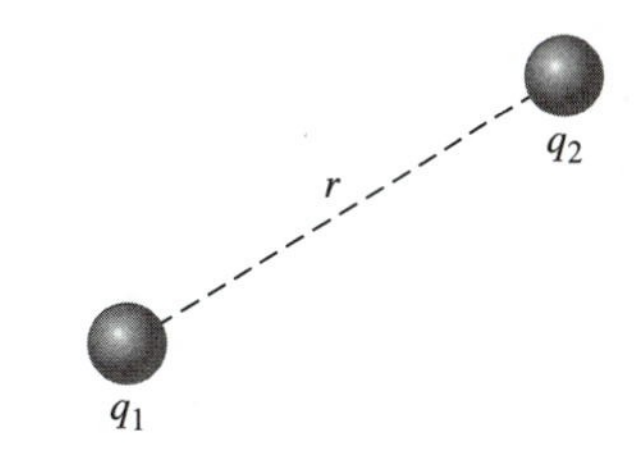

$$U = k_e \frac{q_1 q_2}{r_{12}} \tag{25.13}$$

The **total potential energy of a system of three charged particles** is found by calculating U for each pair of charges and summing the terms algebraically. *An expression similar to Equation 25.14, including additional terms, can be used to calculate the potential energy of a system consisting of more than three charged particles.*

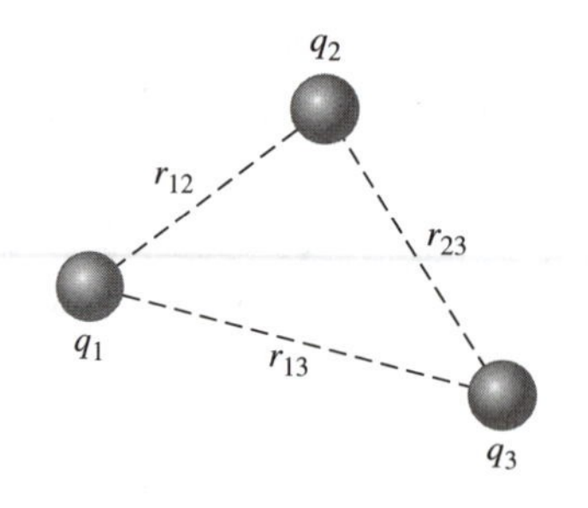

$$U = k_e\left(\frac{q_1q_2}{r_{12}} + \frac{q_1q_3}{r_{13}} + \frac{q_2q_3}{r_{23}}\right) \tag{25.14}$$

The **components of the electric field** are the negative partial derivatives of the electric potential. The electric field components in Cartesian coordinates are given in Equation 25.18.

$$E_x = -\frac{\partial V}{\partial x} \tag{25.18}$$

$$E_y = -\frac{\partial V}{\partial y}$$

$$E_z = -\frac{\partial V}{\partial z}$$

The **vector expression for the electric field** can be evaluated at any point $P(x, y, z)$ within a region if the electric potential over the region is known.

$$\vec{\mathbf{E}} = -\left(\frac{\partial}{\partial x}\hat{\mathbf{i}} + \frac{\partial}{\partial y}\hat{\mathbf{j}} + \frac{\partial}{\partial z}\hat{\mathbf{k}}\right)V$$

The **potential** (relative to zero at infinity) **due to a continuous charge distribution** can be calculated by integrating the contribution due to each charge element dq over the line, surface, or volume which contains all the charge. *Recall, if the electric field is known (e.g., from Gauss's law) the potential can be calculated by using Equation 25.3.*

$$V = k_e\int\frac{dq}{r} \tag{25.20}$$

SUGGESTIONS, SKILLS, AND STRATEGIES

The vector expression giving the electric field over a region can be obtained from the scalar function which describes the electric potential, V, over the region by using a vector differential operator called the gradient operator, ∇:

$$\nabla = \hat{\mathbf{i}}\frac{\partial}{\partial x} + \hat{\mathbf{j}}\frac{\partial}{\partial y} + \hat{\mathbf{k}}\frac{\partial}{\partial z}$$

$$\vec{\mathbf{E}} = -\nabla V$$

This is equivalent to $$\vec{\mathbf{E}} = -\left(\hat{\mathbf{i}}\frac{\partial V}{\partial x} + \hat{\mathbf{j}}\frac{\partial V}{\partial y} + \hat{\mathbf{k}}\frac{\partial V}{\partial z}\right)$$

The derivatives in the above expression are called *partial derivatives*. This means that when the derivative is taken with respect to any one coordinate, any other coordinates which appear in the expression for the potential function are treated as constants.

Since the electrostatic force is a conservative force, the work done by the electrostatic force in moving a charge q from an initial point Ⓐ to a final point Ⓑ depends only on the location of the two points and is independent of the path taken between Ⓐ and Ⓑ. When calculating potential differences using the equation

$$\Delta V = -\int_{Ⓐ}^{Ⓑ} \vec{\mathbf{E}} \cdot d\vec{\mathbf{s}} \tag{25.3}$$

any path between Ⓐ and Ⓑ may be chosen to evaluate the integral; therefore you should select a path for which the evaluation of the "line integral" in Equation 25.3 will be as convenient as possible.

For example, if $\vec{\mathbf{E}}$ is in the form $\vec{\mathbf{E}} = Lx\hat{\mathbf{i}} + My\hat{\mathbf{j}} + Nz\hat{\mathbf{k}}$, where L, M, and N are constants, the potential is integrated as

$$\int_{Ⓐ}^{Ⓑ} \vec{\mathbf{E}} \cdot d\vec{\mathbf{s}} = \int_{Ⓐ}^{Ⓑ} (Lx\,dx + My\,dy + Nz\,dz)$$

This integral can be most easily evaluated over a path which moves first parallel to the x-axis from $x_{Ⓐ}$ to $x_{Ⓑ}$, then parallel to the y-axis, then parallel to the z-axis. Along the first leg of this path, for instance, $dy = dz = 0$.

Thus Equation 25.3 becomes

$$V_{Ⓑ} - V_{Ⓐ} = -\left[L\int_{x_{Ⓐ}}^{x_{Ⓑ}} x\,dx + M\int_{y_{Ⓐ}}^{y_{Ⓑ}} y\,dy + N\int_{z_{Ⓐ}}^{z_{Ⓑ}} z\,dz\right]$$

and $$V_{Ⓑ} - V_{Ⓐ} = \frac{L}{2}\left(x_{Ⓐ}^2 - x_{Ⓑ}^2\right) + \frac{M}{2}\left(y_{Ⓐ}^2 - y_{Ⓑ}^2\right) + \frac{N}{2}\left(z_{Ⓐ}^2 - z_{Ⓑ}^2\right)$$

PROBLEM-SOLVING STRATEGY

- When working problems involving electric potential, remember that potential is a *scalar quantity* (rather than a vector quantity like the electric field), so there are no components to worry about. Therefore, when using the superposition principle to evaluate the electric potential at a point due to a system of point charges, you simply take the algebraic sum of the potentials due to each charge. However, you must keep track of signs. The potential ($V = k_e q/r$) for each positive charge is positive, while the potential for each negative charge is negative.
- Usually, only *changes* in electric potential are significant; hence the point where you choose the potential to be zero is arbitrary. When dealing with point charges or a finite-sized charge distribution, we usually define $V = 0$ to be at a point infinitely far from the charges. However, if the charge distribution itself extends to infinity, some other nearby point must be selected as the reference point.

- The electric potential at some point P due to a continuous distribution of charge can be evaluated by dividing the charge distribution into infinitesimal elements of charge dq located at a distance r from the point P. You then treat this element as a point charge, so that the potential at P due to the element is $dV = k_e\, dq/r$. The total potential at P is obtained by integrating dV over the entire charge distribution. In performing the integration for most problems, it is necessary to express dq and r in terms of a single variable. In order to simplify the integration, it is important to give careful consideration to the geometry involved in the problem.
- Another method that can be used to obtain the potential due to a finite continuous charge distribution is to start with the definition of the potential difference given by Equation 25.3. If $\vec{\mathbf{E}}$ is known or can be obtained easily (e.g., from Gauss's law), then the line integral of $\vec{\mathbf{E}}\cdot d\vec{\mathbf{s}}$ can be evaluated.
- When you know the electric potential in a region around a point, it is possible to obtain the electric field at that point by remembering that

$$E_x = -\frac{\partial V}{\partial x}, \quad E_y = -\frac{\partial V}{\partial y}, \quad \text{and} \quad E_z = -\frac{\partial V}{\partial z}$$

- In this chapter the symbol V is used to represent the electric potential at some point and ΔV is used to represent the potential difference between two points. For example, the expression $\Delta V = V_{Ⓑ} - V_{Ⓐ}$ is read "*the potential difference between points* Ⓐ *and* Ⓑ *is the potential at point* Ⓑ *minus the potential at point* Ⓐ."

In practice, a variety of phrases are used to describe the potential difference between two points, the most common being "voltage." A voltage applied to a device or across a device has the same meaning as the potential difference across the device. For example, if we say that the voltage across a certain capacitor is 12 volts, we mean that the potential difference between the plates of the capacitor is 12 volts.

REVIEW CHECKLIST

You should be able to:

- Calculate the change in electric potential energy (or work done by an external force) when a charge q moves between any two points in an electric field. (Section 25.1)
- Calculate the electric potential difference between any two points in a uniform electric field, and the electric potential difference between any two points in the vicinity of a group of point charges. (Sections 25.1 and 25.2)
- Calculate the electric potential energy associated with a group of point charges. (Section 25.3)
- Obtain an expression for the electric field (a vector quantity) over a region of space if the scalar electric potential function for the region is known. (Section 25.4)
- Calculate the electric potential due to continuous charge distributions of reasonable symmetry—such as a charged ring, sphere, line, disk, or infinite plane. (Section 25.5)

ANSWER TO AN OBJECTIVE QUESTION

12. A particle with charge −40.0 nC is on the x axis at the point with coordinate $x = 0$. A second particle, with charge −20.0 nC, is on the x axis at $x = 0.500$ m. **(i)** Is the point at a finite distance where the electric field is zero (a) to the left of $x = 0$, (b) between $x = 0$ and $x = 0.500$ m, (c) to the right of $x = 0.500$ m? **(ii)** Is the electric potential zero at this point? (a) No; it is positive. (b) Yes. (c) No; it is negative. **(iii)** Is there a point at a finite distance where the electric potential is zero? (a) Yes; it is to the left of $x = 0$. (b) Yes; it is between $x = 0$ and $x = 0.500$ m. (c) Yes; it is to the right of $x = 0.500$ m. (d) No.

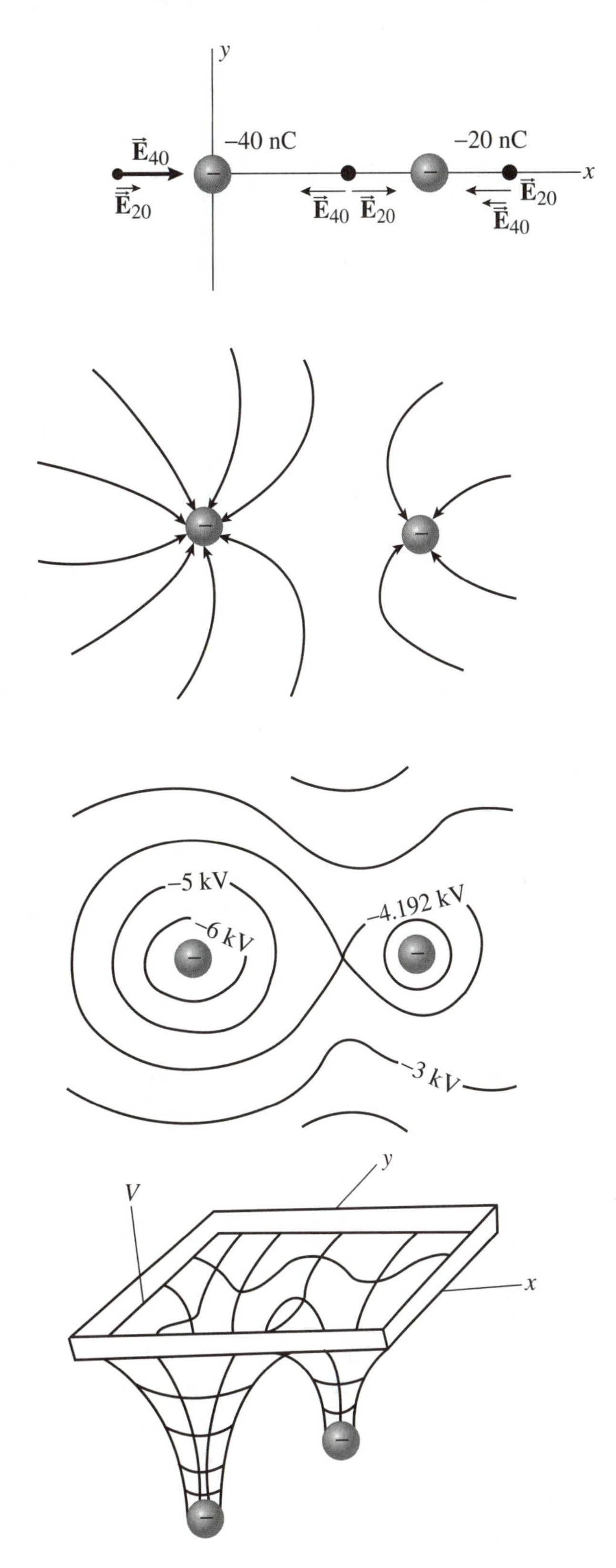

Figure OQ25.12

Answer **(i)** Refer to Figure OQ25.12. In it, the first panel represents how each negative charge creates its own electric field pointing toward itself. The separate fields of the two charges are both toward the right in the $x < 0$ part of the axis, and both toward the left in the $x > 0.5$ m section. In between the two charged particles, their fields are in opposite directions and add to zero at one particular point, which is somewhat closer to the smaller-magnitude charge. The answer is (b). The second panel in the diagram is a sketch of electric field lines.

The two negative charges create negative potential throughout the whole surrounding space, separately and even more working together. The potential is not zero at any point a finite distance away. So the answer to **(ii)** is (c) and the answer to **(iii)** is (d). On a map of potential, the zero-field point shows up as a special saddle point. The third panel in the diagram is a contour map of the potential, and the zero-field point is where contour lines cross. The fourth panel in the diagram represents a three-dimensional graph of V versus position (x, y). Here the special point is the one point where the tangent plane to the surface representing V is horizontal.

□ □ □ □

ANSWER TO A CONCEPTUAL QUESTION

3. When charged particles are separated by an infinite distance, the electric potential energy of the pair is zero. When the particles are brought close, the electric potential energy of a pair with the same sign is positive, whereas the electric potential energy of a pair with opposite signs is negative. Give a physical explanation of this statement.

Answer You may remember from the chapter on gravitational potential energy that potential energy of a system is defined to be positive when positive work must have been performed by an external agent to change the system from an initial configuration, to which we assign a zero value of potential energy, to a final configuration. For example, the system of a flag and the Earth has positive potential energy when the flag is raised if we define zero potential energy as the configuration with the flag at the ground, since positive work must be done by an external force in order to lift it from the ground to the top of the pole.

When assembling particles with like charges from an infinite separation, a configuration for which we have defined the potential energy as having a value of zero, it takes work to move them closer together to some distance r; therefore energy is being stored in the system of charges, and the potential energy is positive.

When assembling two particles with opposite charges from an infinite separation, the particles tend to accelerate toward each other, and thus energy is released as they approach a separation at distance r. Therefore, the potential energy of a pair of unlike charges is negative.

□ □ □ □

SOLUTIONS TO SELECTED END-OF-CHAPTER PROBLEMS

1. (a) Calculate the speed of a proton that is accelerated from rest through an electric potential difference of 120 V. (b) Calculate the speed of an electron that is accelerated through the same potential difference.

Solution

Conceptualize: Since 120 V is only a modest potential difference, we might expect that the final speed of the particles will be substantially less than the speed of light. We should also expect the speed of the electron to be significantly greater than the proton because, with $m_e << m_p$, an equal force on both particles will result in a much greater acceleration for the electron.

Categorize: Conservation of energy of the proton-field system can be applied to this problem to find the final speed from the kinetic energy of the particles. (Review this work-energy theory of motion from Chapter 8 if necessary.)

Analyze:

(a) For the particle-field system, energy is conserved as the proton moves from high to low potential, which can be defined for this problem as moving from 120 V down to 0 V:

$$K_i + U_i = K_f + U_f \qquad\qquad 0 + qV_i = \tfrac{1}{2}mv_p^2 + 0$$

The initial energy is $qV_i = (1.60 \times 10^{-19}\ \text{C})(120\ \text{V})\left(\dfrac{1\ \text{J/C}}{1\ \text{V}}\right) = 1.92 \times 10^{-17}\ \text{J}$

The final speed is given by

$$v_p = \left(\frac{2qV_i}{m}\right)^{1/2} = \left(\frac{2\ (1.60 \times 10^{-19}\ \text{C})\ 120\ \text{J/C}}{1.67 \times 10^{-27}\ \text{kg}}\right)^{1/2} = 1.52 \times 10^5\ \text{m/s} \qquad \blacksquare$$

(b) The electron will gain speed in moving the other way, from $V_i = 0$ to $V_f = 120$ V. Here $K_i + U_i = K_f + U_f$ becomes

$$0 + 0 = \tfrac{1}{2}mv_e^2 + qV_f = \tfrac{1}{2}mv_e^2 - eV_f$$

so $$v_e = \left(\frac{-2qV_f}{m}\right)^{1/2} = \left(\frac{-2\ (-1.60 \times 10^{-19}\ \text{C})\ 120\ \text{J/C}}{9.11 \times 10^{-31}\ \text{kg}}\right)^{1/2} = 6.49 \times 10^6\ \text{m/s} \qquad \blacksquare$$

Finalize: Both of these speeds are significantly less than the speed of light, as expected, which also means that we were justified in not using the relativistic kinetic energy formula. (For precision to three significant digits, the relativistic formula is only needed if v is greater than about 0.1 c.)

Both the electron and the proton here are described as having energy 120 electron volts. Look back at the problem solution to see that each really does have kinetic energy (e) (120 V) = 120 eV. One eV is just a unit, so the energy of any object can be quoted as a number of eV's. But for a beam of charged particles, this specification tells just how high to turn the power supply knob for the accelerating voltage.

---

5. An electron moving parallel to the x axis has an initial speed of 3.70×10^6 m/s at the origin. Its speed is reduced to 1.40×10^5 m/s at the point $x = 2.00$ cm. (a) Calculate the electric potential difference between the origin and that point. (b) Which point is at the higher potential?

Solution

Conceptualize: A coasting skateboarder slows when moving uphill. We assume the electron is moving in vacuum. It slows because it is coasting up to higher electric potential energy. With negative charge, it must be moving to a location with a more negative potential. The starting point is at higher voltage.

Categorize: We use the energy version of the isolated system model to equate the energy of the electron-field system when the electron is at $x = 0$ to the energy when the electron is at $x = 2$ cm. The unknown will be the difference in potential $V_f - V_i$.

Analyze: Thus, $K_i + U_i = K_f + U_f$

becomes $\frac{1}{2}mv_i^2 + qV_i = \frac{1}{2}mv_f^2 + qV_f$

or $\frac{1}{2}m(v_i^2 - v_f^2) = q(V_f - V_i)$ so $(V_f - V_i) = \dfrac{m(v_i^2 - v_f^2)}{2q}$

(a) Noting that the electron's charge is negative, and evaluating the potential difference, we have

$$V_f - V_i = \frac{(9.11 \times 10^{-31}\ \text{kg})\left[(3.70 \times 10^6\ \text{m/s})^2 - (1.40 \times 10^5\ \text{m/s})^2\right]}{2(-1.60 \times 10^{-19}\ \text{C})} = -38.9\ \text{V}$$ ■

(b) The negative sign means that the 2.00-cm location is lower in potential than the origin. ■

Finalize: Our calculation agrees with our qualitative argument that the origin is the point of higher potential. A particle with positive charge would slow in free flight toward higher voltage, but the negative electron slows as it moves into lower potential. The 2.00-cm distance was unnecessary information for this problem. **If** the field were uniform, we could use the 2-cm distance to find the x component of the field from $\Delta V = -E_x d$.

---

18. At a certain distance from a charged particle, the magnitude of the electric field is 500 V/m and the electric potential is –3.00 kV. (a) What is the distance to the particle? (b) What is the magnitude of the charge?

Solution

Conceptualize: The particle possesses its charge just at its own location. It creates both vector field and scalar potential throughout the surrounding empty space. The charge must be negative to create negative voltage. We expect less than a microcoulomb and a distance of several centimeters for a tabletop or vacuum-tube situation.

Categorize: We will use these equations: At a distance r from a charged particle, the voltage is $V = \dfrac{k_e q}{r}$ and the field magnitude is $E = \dfrac{k_e|q|}{r^2}$

Analyze: Combining them, $E = \dfrac{r|V|}{r^2} = \dfrac{|V|}{r}$

(a) so $r = \dfrac{|V|}{E} = \dfrac{3\,000 \text{ V}}{500 \text{ N/C}} = 6.00 \dfrac{\text{N}\cdot\text{m/C}}{\text{N/C}} = 6.00 \text{ m}$ ■

(b) and $q = \dfrac{rV}{k_e} = \dfrac{(6.00 \text{ m})(-3\,000 \text{ V})}{8.99 \times 10^9 \text{ N}\cdot\text{m}^2/\text{C}^2} = -2.00\ \mu\text{C}$ ■

Finalize: Charge and distance are a bit larger than our guesses. Recall that a particle need not be small. A big ball can carry this rather large charge. We know that the vector field is pointing toward the negative charge.

---

20. The three charged particles in Figure P25.20 are at the vertices of an isosceles triangle (where $d = 2.00$ cm). Taking $q = 7.00\ \mu$C, calculate the electric potential at point A, at the midpoint of the base.

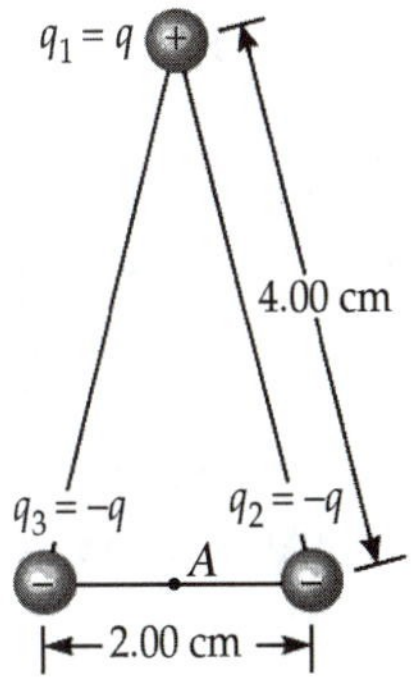

$q_1 = q$

4.00 cm

$q_3 = -q$ $q_2 = -q$

A

2.00 cm

Figure P25.20

Solution

Conceptualize: Being close to two negative charges, the point A probably has a negative potential of some kilovolts.

Categorize: Each charge creates potential at A on its own. We will add up the contributions.

Analyze: Define $q_1 = q$ and $q_2 = q_3 = -q$.

The charged particles are at distances from A of

$$r_1 = \sqrt{(0.040\,0 \text{ m})^2 - (0.010\,0 \text{ m})^2} = 3.87 \times 10^{-2} \text{ m}$$

and $r_2 = r_3 = 0.010\,0$ m

The potential at point A is

$$V = \frac{k_e q_1}{r_1} + \frac{k_e q_2}{r_2} + \frac{k_e q_3}{r_3} = k_e(q)\left(\frac{1}{r_1} + \frac{-1}{r_2} + \frac{-1}{r_3}\right)$$

Substituting,

$$V = \left(8.99 \times 10^9 \frac{\text{N}\cdot\text{m}^2}{\text{C}^2}\right)(7.00 \times 10^{-6} \text{ C})\left(\frac{1}{0.038\,7 \text{ m}} - \frac{1}{0.010\,0 \text{ m}} - \frac{1}{0.010\,0 \text{ m}}\right)$$

so $V = -11.0 \times 10^6$ V ■

Finalize: A microcoulomb is a large charge for a tabletop object. The net potential is in megavolts rather than in kilovolts. The two negative charges dominate the more distant positive charge in creating the net potential. For review and contrast, let us do this related calculation:

Calculate the electric field vector at the same point due to the three charged particles. The separate fields of the two negative charges are in opposite directions and add to zero. The net field is

$$\vec{\mathbf{E}} = \frac{k_e q_1}{r_1^2}\hat{\mathbf{r}}_1 = \frac{(8.99 \times 10^9 \text{ N}\cdot\text{m}^2/\text{C}^2)(7.00 \times 10^{-6}\text{ C})}{(0.040\ 0\text{ m})^2 - (0.010\ 0\text{ m})^2} \text{ down}$$

$$\vec{\mathbf{E}} = (42.0 \times 10^6 \text{ N/C})(-\hat{\mathbf{j}})$$

---

22. Show that the amount of work required to assemble four identical charged particles of magnitude Q at the corners of a square of side s is $5.41k_eQ^2/s$.

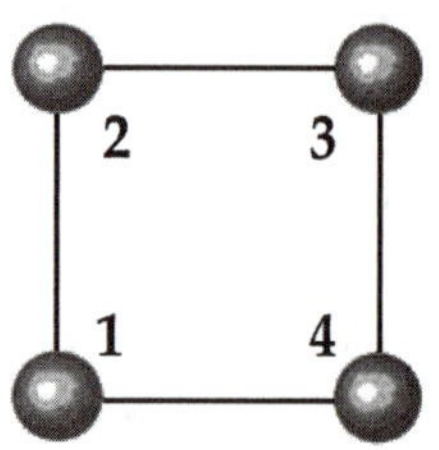

Solution

Conceptualize: Work is required to push the charges together against their mutual repulsion. It is reasonable that its magnitude increases if Q increases and if s decreases—if the charges are squeezed closer together.

Categorize: The work required equals the sum of the potential energies for all pairs of charges. No energy is involved in placing q_4 at a given position in empty space. When q_3 is brought from far away and placed close to q_4, the system potential energy can be expressed as q_3V_4, where V_4 is the potential at the position of q_3 established by charge q_4. When q_2 is brought into the system, it interacts with two other charges, so we have two additional terms q_2V_3 and q_2V_4 in the total potential energy. Finally, when we bring the fourth charge q_1 into the system, it interacts with three other charges, giving us three more energy terms.

Analyze: Thus, the complete expression for the energy is:

$$U = q_1V_2 + q_1V_3 + q_1V_4 + q_2V_3 + q_2V_4 + q_3V_4$$

$$U = \frac{q_1k_eq_2}{r_{12}} + \frac{q_1k_eq_3}{r_{13}} + \frac{q_1k_eq_4}{r_{14}} + \frac{q_2k_eq_3}{r_{23}} + \frac{q_2k_eq_4}{r_{24}} + \frac{q_3k_eq_4}{r_{34}}$$

$$U = \frac{Qk_eQ}{s} + \frac{Qk_eQ}{s\sqrt{2}} + \frac{Qk_eQ}{s} + \frac{Qk_eQ}{s} + \frac{Qk_eQ}{s\sqrt{2}} + \frac{Qk_eQ}{s}$$

Evaluating,

$$U = \frac{k_eQ^2}{s}\left(4 + \frac{2}{\sqrt{2}}\right) = 5.41k_e\frac{Q^2}{s} \quad \blacksquare$$

Finalize: In describing the four charged particles, we never multiply all four or any three together. We think only of pairs, and we think of all of the pairs. That is where the 5.41 factor comes from. The most generally useful things you learn about electricity may be what a volt is (a JOULE PER COULOMB) and the associated definition about electric potential energy being $U = qV$.

---

35. Over a certain region of space, the electric potential is $V = 5x - 3x^2y + 2yz^2$. (a) Find the expressions for the x, y, and z components of the electric field over this region. (b) What is the magnitude of the field at the point P that has coordinates (1.00, 0, –2.00) m?

Solution

Conceptualize: The given equation tells a voltage value for every point in three-dimensional space. The downward slope of the potential from point to point is the electric field we will compute.

Categorize: First, we find the x, y, and z components of the field from the partial derivatives of the potential function with respect to x, y, and z. Then, we evaluate them at point P. We assume that V is given in volts, as a function of distances in meters.

Analyze:

(a) The components of the field are

$$E_x = -\frac{\partial V}{\partial x} = -\frac{\partial}{\partial x}\left(5x - 3x^2y + 2yz^2\right) = -5 + 6xy + 0$$

and

$$E_y = -\frac{\partial V}{\partial y} = 3x^2 - 2z^2 \quad \text{and} \quad E_z = -\frac{\partial V}{\partial z} = -4yz \qquad \blacksquare$$

(b) At point P, $E_x = -5 + 6(1.00\text{ m})(0\text{ m}) = -5.00\text{ N/C}$

At point P, $E_y = 3(1.00\text{ m})^2 - 2(-2.00\text{ m})^2 = -5.00\text{ N/C}$

At point P, $E_z = -4(0\text{ m})(-2.00\text{ m}) = 0\text{ N/C}$

At P, the field's magnitude is

$$E = \sqrt{\left(-5.00\text{ N/C}\right)^2 + \left(-5.00\text{ N/C}\right)^2 + 0^2} = 7.07\text{ N/C} \qquad \blacksquare$$

Finalize: It is not hard to take a partial derivative after you learn the rule to think of the other variables as constants. It would be cumbersome to draw a graph of V as a function of the three spatial variables. Picture the field at the particular point in question as a weak field in the xy plane, at 225° to the x axis.

---

39. It is shown in Example 25.7 that the potential at a point P a distance a above one end of a uniformly charged rod of length ℓ lying along the x axis is

$$V = k_e \frac{Q}{\ell} \ln\left(\frac{\ell + \sqrt{a^2 + \ell^2}}{a}\right)$$

Use this result to derive an expression for the y component of the electric field at P.

Solution

Conceptualize: The y component of the field will point away from the rod if Q is positive. It will be proportional to Q and will decrease as y increases.

Categorize: In the given expression for the potential, we replace a with y, because this distance is the y coordinate of point P. Then we can take the partial derivative of the potential with respect to y to have the field component. The partial derivative reduces to an ordinary derivative because the expression for potential does not specify how V depends on the x coordinate.

Analyze: The y component of the electric field is

$$E_y = -\frac{\partial V}{\partial y} = -\frac{k_e Q}{\ell}\frac{d}{dy}\left[\ln\left(\ell + \sqrt{\ell^2 + y^2}\right) - \ln y\right]$$

$$= -\frac{k_e Q}{\ell}\left[\frac{2y/2\sqrt{\ell^2 + y^2}}{\ell + \sqrt{\ell^2 + y^2}} - \frac{1}{y}\right] = \frac{k_e Q}{y\sqrt{\ell^2 + y^2}} \qquad \blacksquare$$

Finalize: The field is indeed proportional to Q and decreases as y increases. It points away from the positively-charged rod because in this case E_y is positive when y is positive and negative when y is negative.

---

42. A rod of length L (Fig. P25.42) lies along the x axis with its left end at the origin. It has a nonuniform charge density $\lambda = \alpha x$, where α is a positive constant. (a) What are the units of α? (b) Calculate the electric potential at A.

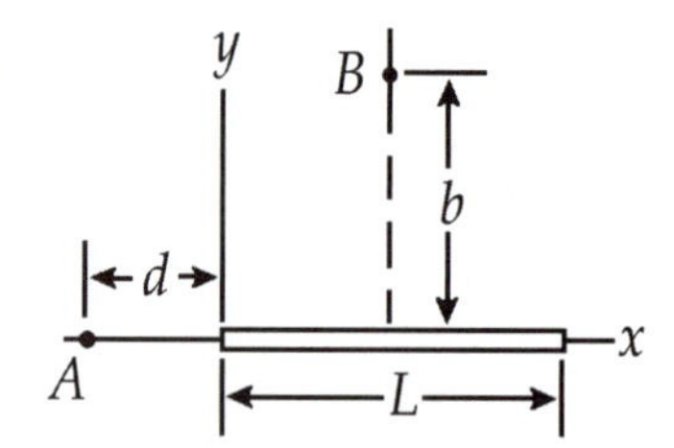

Figure P25.42

Solution

Conceptualize: The potential will be proportional to α. It will decrease as d increases.

Categorize: We think just of units in part (a). In part (b) we must do an integral, adding up infinitely many infinitesimal contributions to the total potential at A from all the elements of the rod.

Analyze:

(a) As a linear charge density, λ has units of C/m. So $\alpha = \lambda/x$ must have units of C/m^2. $\blacksquare$

(b) Consider a small segment of the rod at location x and of length dx. The amount of charge on it is $\lambda\, dx = (\alpha x)\, dx$. Its distance from A is $d + x$, so its contribution to the electric potential at A is

$$dV = k_e \frac{dq}{r} = k_e \frac{\alpha x\, dx}{d + x}$$

Relative to $V = 0$ infinitely far away, to find the potential at A we must integrate these contributions for the whole rod, from $x = 0$ to $x = L$. Then $V = \int_{\text{all } q} dV = \int_0^L \frac{k_e \alpha x}{d + x} dx$

To perform the integral, make a change of variables to

$u = d + x$ $\quad du = dx,$ $\quad u(\text{at } x = 0) = d,$ $\quad$ and $\quad u(\text{at } x = L) = d + L:$

$$V = \int_d^{d+L} \frac{k_e \alpha(u - d)}{u} du = k_e \alpha \int_d^{d+L} du - k_e \alpha d \int_d^{d+L} \left(\frac{1}{u}\right) du$$

$$V = k_e \alpha u \Big|_d^{d+L} - k_e \alpha d \ln u \Big|_d^{d+L} = k_e \alpha(d + L - d) - k_e \alpha d(\ln(d + L) - \ln d)$$

$$V = k_e \alpha L - k_e \alpha d \ln\left(\frac{d + L}{d}\right) \quad ■$$

Finalize: At the start or in the middle of the calculation, it is good to keep track of symbols: the unknown is V. The values k_e, α, d, and L are known and constant. While we are doing the integral x and u are variables, and will not appear in the answer. We have the answer when the unknown is expressed in terms of the d, L, and α mentioned in the problem and the universal constant k_e.

---

48. A spherical conductor has a radius of 14.0 cm and charge of 26.0 μC. Calculate the electric field and the electric potential at (a) $r = 10.0$ cm, (b) $r = 20.0$ cm, and (c) $r = 14.0$ cm from the center.

Solution

Conceptualize: A metal sphere carries a charge. We review Chapters 23 and 24 to identify or calculate the field that the charge creates inside the sphere, at its surface, and outside. The potential follows related rules that are different in detail.

Categorize: For points on the surface and outside, the sphere of charge behaves like a charged particle at its center, both for creating field and potential. But…

Analyze:

(a) …inside a conductor when charges are not moving, the electric field is zero and the potential is uniform, the same as on the surface. We have $\vec{\mathbf{E}} = 0$ ■

$$V = \frac{k_e q}{R} = \frac{(8.99 \times 10^9 \text{ N}\cdot\text{m}^2/\text{C}^2)(26.0 \times 10^{-6} \text{ C})}{0.140 \text{ m}} = 1.67 \times 10^6 \text{ V} \quad ■$$

(b) The sphere behaves like a charged particle at its center when you stand outside.

$$\vec{\mathbf{E}} = \frac{k_e q}{r^2}\hat{\mathbf{r}} = \frac{(8.99 \times 10^9 \text{ N}\cdot\text{m}^2/\text{C}^2)(26.0 \times 10^{-6} \text{ C})}{(0.200 \text{ m})^2}\hat{\mathbf{r}} = (5.84 \times 10^6 \text{ N/C})\hat{\mathbf{r}} \quad ■$$

$$V = \frac{k_e q}{r} = \frac{\left(8.99\times10^9\,\text{N}\cdot\text{m}^2/\text{C}^2\right)\left(26.0\times10^{-6}\,\text{C}\right)}{0.200\text{ m}} = 1.17\times10^6\text{ V}$$ ■

(c) $$\vec{\mathbf{E}} = \frac{k_e q}{r^2}\hat{\mathbf{r}} = \frac{\left(8.99\times10^9\,\text{N}\cdot\text{m}^2/\text{C}^2\right)\left(26.0\times10^{-6}\text{ C}\right)}{(0.140\text{ m})^2}\hat{\mathbf{r}} = \left(11.9\times10^6\text{ N/C}\right)\hat{\mathbf{r}}$$ ■

$V = 1.67\times10^6$ V as in part (a) ■

Finalize: The potential is a continuous function. It has the same value just below the surface and just above the surface. The field is discontinuous here. We have given in part (c) the field just outside the metal surface. It is zero just inside the surface.

---

49. Lightning can be studied with a Van de Graaff generator, which consists of a spherical dome on which charge is continuously deposited by a moving belt. Charge can be added until the electric field at the surface of the dome becomes equal to the dielectric strength of air. Any more charge leaks off in sparks as shown in Figure P25.49. Assume the dome has a diameter of 30.0 cm and is surrounded by dry air with a "breakdown" electric field of 3.00×10^6 V/m. (a) What is the maximum potential of the dome? (b) What is the maximum charge on the dome?

Solution

Conceptualize: Van de Graaff generators produce voltages that can make your hair stand on end, somewhere on the order of about 100 kV. With these high voltages, the maximum charge on the dome is probably more than typical tabletop charged-particle values of about 1 μC.

The maximum potential and charge will be limited by the electric field strength at which the air surrounding the dome will ionize. This critical value is determined by the **dielectric strength** of air which, from Table 26.1, is $E_{\text{critical}} = 3\times10^6$ V/m. An electric field stronger than this will cause the air to act like a conductor instead of an insulator. This process is called dielectric breakdown and may be seen as a spark.

Categorize: From the maximum allowed electric field, we can find the charge and potential that would create this situation. Since we are only given the diameter of the dome, we will assume that the conductor is spherical, which allows us to use the electric field and potential equations for a spherical conductor. With these equations, it will be easier to do part (b) first and use the result for part (a).

Analyze: At the surface of a spherical conductor with total charge Q,

(b) $$\left|\vec{\mathbf{E}}\right| = \frac{k_e Q}{r^2}:$$

$$Q = \frac{Er^2}{k_e} = \frac{\left(3.00\times10^6\text{ V/m}\right)(0.150\text{ m})^2}{8.99\times10^9\text{ N}\cdot\text{m}^2/\text{C}^2}\left(1\text{ N}\cdot\text{m/V}\cdot\text{C}\right) = 7.51\,\mu\text{C}$$ ■

(a) $$V = \frac{k_e Q}{r} = \frac{\left(8.99 \times 10^9\ \text{N} \cdot \text{m}^2/\text{C}^2\right)\left(7.51 \times 10^{-6}\ \text{C}\right)}{0.150\ \text{m}} = 450\ \text{kV}$$ ■

Finalize: These calculated results seem reasonable based on our predictions. The voltage is about 4 000 times larger than the 120 V of common electrical outlets, but the charge is similar in magnitude to many of the static charge problems we have solved earlier. This implies that many of the charge configurations mentioned in other problems would have to be in a vacuum because the electric field near these charged particles would be strong enough to cause sparking in air. For example, a charged ball with $Q = 1\ \mu\text{C}$ and $r = 1$ mm would have an electric field near its surface of

$$E = \frac{k_e Q}{r^2} = \frac{\left(9 \times 10^9\ \text{N} \cdot \text{m}^2/\text{C}^2\right)\left(1 \times 10^{-6}\ \text{C}\right)}{\left(0.001\ \text{m}\right)^2} = 9 \times 10^9\ \text{V/m}$$

which is well beyond the dielectric breakdown strength of air!

---

53. The liquid-drop model of the atomic nucleus suggests high-energy oscillations of certain nuclei can split the nucleus into two unequal fragments plus a few neutrons. The fission products acquire kinetic energy from their mutual Coulomb repulsion. Assume the charge is distributed uniformly throughout the volume of each spherical fragment and, immediately before separating, each fragment is at rest and their surfaces are in contact. The electrons surrounding the nucleus can be ignored. Calculate the electric potential energy (in electron volts) of two spherical fragments from a uranium nucleus having the following charges and radii: $38e$ and 5.50×10^{-15} m, $54e$ and 6.20×10^{-15} m.

Solution

Conceptualize: The answer should be on the order of a hundred million electron volts. The nuclear force ordinarily holds a nucleus together. No one knows how to write down an $F =$ formula for the nuclear force. It is a little remarkable that we can compute a good estimate for the energy released by nuclear fission just from knowing about the electric force…

Categorize: …or really about electric potential energy. The problem is equivalent to finding the potential energy of a charged particle with $q = 38e$ at a distance 11.7×10^{-15} m (the distance between their centers when in contact) from a charged particle of $54e$.

Analyze: The two spheres of charge have together electric potential energy

$$U = qV = k_e \frac{q_1 q_2}{r_{12}}$$

$$= (8.99 \times 10^9\ \text{J} \cdot \text{m/C}^2) \frac{(38)(1.60 \times 10^{-19}\ \text{C})(54)(1.60 \times 10^{-19}\ \text{C})}{\left(5.50 \times 10^{-15}\ \text{m} + 6.20 \times 10^{-15}\ \text{m}\right)}$$

$$U = 4.04 \times 10^{-11}\ \text{J} = 253\ \text{MeV}$$ ■

Finalize: The analysis gives a good account of the energy released by uranium fission. There are no electrons involved here, but the proton charge is the same magnitude as the electron charge, and it is remarkably convenient to use the electronvolt as a unit of energy in atoms, nuclei, and many laboratory processes.

---

60. Calculate the work that must be done on charges brought from infinity to charge a spherical shell of radius R to a total charge Q.

Solution

Conceptualize: Compare this problem with problem 22 above. Work must be done to squeeze the charge together against its mutual repulsion. The amount of work will increase with increasing Q and with decreasing R.

Categorize: No work is required to place the first infinitesimal bit of charge on the sphere. But thereafter the sphere is at a positive potential and more and more work is required to add more bits of charge. We will do an integral to add up all the work: $W = \int dW$.

Analyze: When the potential of the shell is V due to a charge q, the work required to add an additional increment of charge dq is $dW = V\,dq$ where $V = \frac{k_e q}{R}$. Then $dW = \left(\frac{k_e q}{R}\right) dq$ and the whole amount of work is $W = \frac{k_e}{R}\int_0^Q q\,dq$. We perform the integral to find

$$W = \left(\frac{k_e}{R}\right)\left(\frac{Q^2}{2}\right) \quad ■$$

Finalize: Compare again to problem 22. The required work shows the same proportionality to the square of the charge, and the same inverse proportionality to the size of the assemblage. The proportionality constant turns out to be simpler here than in problem 22.

---

61. From Gauss's law, the electric field set up by a uniform line of charge is

$$\vec{\mathbf{E}} = \left(\frac{\lambda}{2\pi \epsilon_0 r}\right)\hat{\mathbf{r}}$$

where $\hat{\mathbf{r}}$ is a unit vector pointing radially away from the line and λ is the linear charge density along the line. Derive an expression for the potential difference between $r = r_1$ and $r = r_2$.

Solution

Conceptualize: The bits of charge in the line create potential as well as field. We expect a larger ΔV for $(r_1 - r_2)$ larger and for λ larger.

Categorize: It would be possible but more difficult to calculate the answer from integrating over the whole infinite line the contributions to the total potential difference of the bits of charge comprising the filament. Instead we use the relationship between the potential and the known field.

Analyze: In Equation 25.3, $V_2 - V_1 = \Delta V = -\int_1^2 \vec{\mathbf{E}} \cdot d\vec{\mathbf{s}}$, think about stepping from distance r_1 out to the larger distance r_2 away from the charged line. Then $d\vec{\mathbf{s}} = dr\,\hat{\mathbf{r}}$, and we can make r the variable of integration:

$$V_2 - V_1 = -\int_{r_1}^{r_2} \frac{\lambda}{2\pi \epsilon_0 r} \hat{\mathbf{r}} \cdot dr\, \hat{\mathbf{r}} \qquad \text{with} \qquad \hat{\mathbf{r}} \cdot \hat{\mathbf{r}} = 1 \cdot 1 \cos 0^\circ = 1$$

The potential difference is $V_2 - V_1 = -\dfrac{\lambda}{2\pi \epsilon_0} \displaystyle\int_{r_1}^{r_2} \frac{dr}{r} = -\dfrac{\lambda}{2\pi \epsilon_0} \ln r \Big|_{r_1}^{r_2}$

and $$V_2 - V_1 = -\frac{\lambda}{2\pi \epsilon_0} \left(\ln r_2 - \ln r_1\right) = -\frac{\lambda}{2\pi \epsilon_0} \ln \frac{r_2}{r_1} \quad \blacksquare$$

Finalize: If $r_2 > r_1$, then $V_2 - V_1$ is negative. This means the potential decreases as we move away from a positively charged filament. As r_2 goes to infinity, we might hope that the potential would go to zero, as for a charged particle. But really it diverges to $-\infty$. The line of charge has an infinite total charge and a somewhat inconvenient infinity appears in the potential it creates. But the potential difference is finite between any two points at finite distances away.

---

71. An electric dipole is located along the y axis as shown in Figure P25.71. The magnitude of its electric dipole moment is defined as $p = 2aq$. (a) At a point P, which is far from the dipole ($r >> a$), show that the electric potential is

$$V = \frac{k_e p \cos\theta}{r^2}$$

(b) Calculate the radial component E_r and the perpendicular component E_θ of the associated electric field. Note that $E_\theta = -(1/r)(\partial V/\partial \theta)$. Do these results seem reasonable for (c) $\theta = 90°$ and $0°$? (d) For $r = 0$? (e) For the dipole arrangement shown, express V in terms of Cartesian coordinates using $r = (x^2 + y^2)^{1/2}$ and

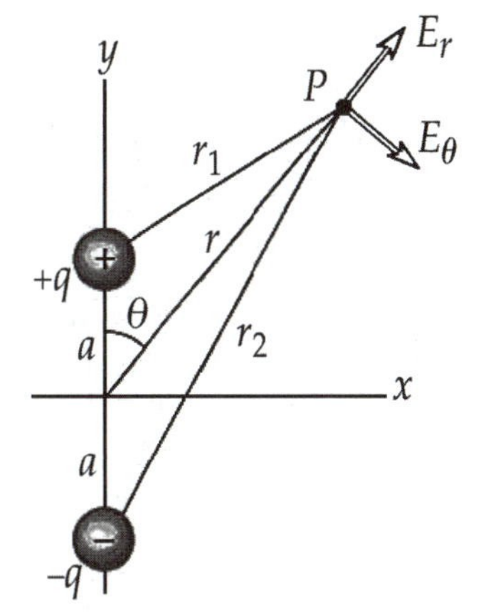

Figure P25.71

$$\cos\theta = \frac{y}{\left(x^2 + y^2\right)^{1/2}}$$

(f) Using these results and again taking $r >> a$, calculate the field components E_x and E_y.

Solution

Conceptualize: We know how a charged particle (a monopole) creates potential around it. This is about how a dipole creates potential.

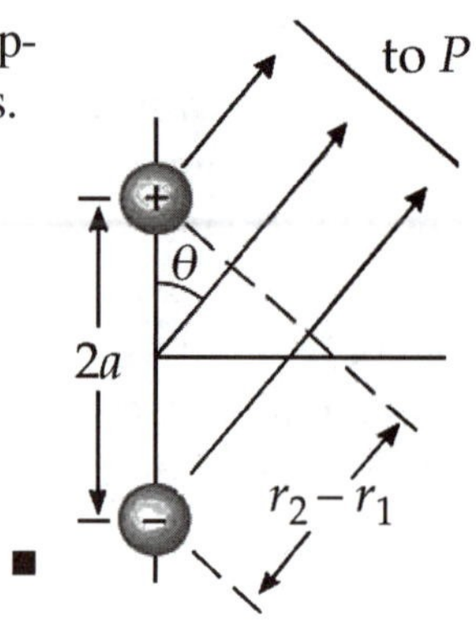

Categorize: We add up the potentials that the two charged particles separately create. Then we take partial derivatives to find field components.

Analyze:

(a) The total potential is $V = \dfrac{k_e q}{r_1} - \dfrac{k_e q}{r_2} = \dfrac{k_e q}{r_1 r_2}(r_2 - r_1)$

From the diagram, for $r >> a$ we have $r_2 - r_1 \approx 2a\cos\theta$ and $r_1 \approx r \approx r_2$, so

$$V \approx \frac{k_e q}{(r)(r)}(2a\cos\theta) = \frac{k_e p\cos\theta}{r^2}$$ ■

(b) $E_r = -\dfrac{\partial V}{\partial r} = -\dfrac{\partial}{\partial r}\left(\dfrac{k_e p\cos\theta}{r^2}\right) = \dfrac{2k_e p\cos\theta}{r^3} = E_r$ ■

In spherical coordinates,

$$E_\theta = -\frac{1}{r}\left(\frac{\partial V}{\partial \theta}\right) = -\frac{1}{r}\frac{\partial}{\partial \theta}\left(\frac{k_e p\cos\theta}{r^2}\right) = \frac{k_e p\sin\theta}{r^3} = E_\theta$$ ■

(c) The values at $\theta = 0$, $\theta = 90°$, and $\theta = 180°$ correspond to an electric field that points outward at $\theta = 0$ from the positive charge on top, loops around to point downward at $\theta = 90°$, and continues around to point upward toward the negative charge from below at $\theta = 180°$. This is reasonable. ■

(d) Taking $r = 0$ makes no sense, because we assumed that r is much larger than a in the derivation. The field is not infinite at $r = 0$, at the point halfway between the charges, although the distant-field equations suggest that it goes to infinity. ■

(e) For $r = \sqrt{x^2 + y^2}$, $\cos\theta = \dfrac{y}{\sqrt{x^2 + y^2}}$ and $V = \dfrac{k_e p y}{\left(x^2 + y^2\right)^{3/2}}$ ■

(f) Then $E_x = -\dfrac{\partial V}{\partial x} = -\dfrac{\partial}{\partial x}\left(\dfrac{k_e p y}{\left(x^2 + y^2\right)^{3/2}}\right) = \dfrac{3k_e p x y}{\left(x^2 + y^2\right)^{5/2}} = E_x$ ■

$$E_y = -\frac{\partial V}{\partial y} = -\frac{\partial}{\partial y}\left(\frac{k_e p y}{\left(x^2 + y^2\right)^{3/2}}\right) = \frac{k_e p\left(2y^2 - x^2\right)}{\left(x^2 + y^2\right)^{5/2}} = E_y$$ ■

Finalize: In part (a) we can argue that the proportionality to $\cos\theta$ is reasonable because for $\theta = 90°$ we are on the x axis, equidistant from the positive and negative charges. Here the total potential is zero, as the cosine function describes. If we take the special case $y = 0$ in part (f) of this problem and interchange the labels x and y, we should get the result derived in part C of Example 23.5. With these assignments we get $E_y = 0$ and $|E_x| = |k_e p(-y^2)/y^5| = k_e 2qa/y^3$, which does agree with the result of the Example.

If we take the special case $x = 0$ in part (f) and interchange x and y, we get the results $E_y = 0$ and $|E_x| = 4\,k_e\,q\,a/x^3$, agreeing with Example 25.4.

26

Capacitance and Dielectrics

EQUATIONS AND CONCEPTS

The **capacitance of a capacitor** is defined as the ratio of the magnitude of the charge on either conductor (or plate) to the magnitude of the potential difference between the conductors. *Capacitance is always a positive quantity and has SI units of farads (F).*

$$C \equiv \frac{Q}{\Delta V} \qquad (26.1)$$

$$1\ \text{F} = 1\ \text{C/V}$$

An **air-filled parallel-plate capacitor** has a capacitance that is proportional to the *area of overlap* between the plates and inversely proportional to the separation of the plates.

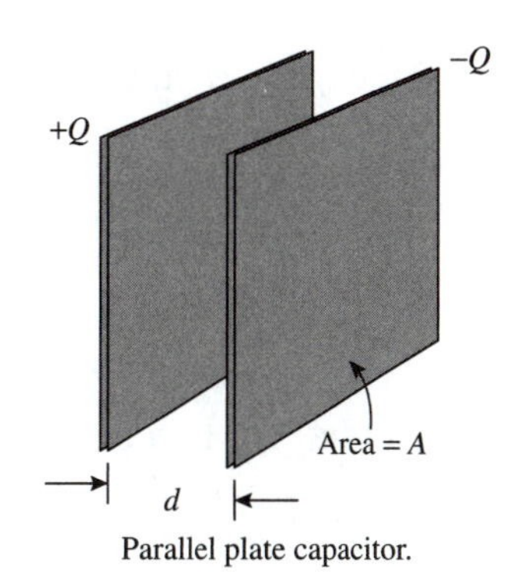

Parallel plate capacitor.

$$C = \frac{\epsilon_0 A}{d} \qquad (26.3)$$

When a **dielectric material** (insulator) completely fills the region between the plates of a capacitor, the capacitance increases by a factor κ. *Kappa, called the dielectric constant, is dimensionless and is characteristic of a particular material.*

$$C = \kappa C_0 \qquad (26.14)$$

It is necessary to recognize parallel and series combinations of capacitors, and to be able to calculate the effective capacitance of each combination.

When capacitors connected in parallel, each capacitor has two circuit points in common with each of the other capacitors in the group.

For **parallel combination of capacitors**:

- Equivalent capacitance equals the sum of the individual capacitances.

$$C_{eq} = C_1 + C_2 + C_3 + \ldots \quad (26.8)$$

- The total charge is the sum of the charges on the individual capacitors.

$$Q_{total} = Q_1 + Q_2 + Q_3 + \ldots$$

- The potential difference is the same for each capacitor.

$$\Delta V_{total} = \Delta V_1 = \Delta V_2 = \Delta V_3 = \ldots$$

When capacitors are connected in series, adjacent capacitors have only one circuit point in common.

For **series combination of capacitors**:

- The reciprocal of the equivalent capacitance is equal to the sum of the reciprocals of the individual capacitances.

$$\frac{1}{C_{eq}} = \frac{1}{C_1} + \frac{1}{C_2} + \frac{1}{C_3} + \ldots \quad (26.10)$$

- The potential difference across the group equals the sum of the potential differences across each capacitor.

$$\Delta V = \Delta V_1 + \Delta V_2 + \Delta V_3 + \ldots$$

- The total charge on the series combination is equal to the charge on each capacitor, each capacitor having the same charge.

$$Q_{total} = Q_1 = Q_2 = Q_3 = \ldots$$

For the **special case of two capacitors in series**, Equation 26.10 yields a convenient form for finding the equivalent capacitance.

$$C_{eq} = \frac{C_1 C_2}{C_1 + C_2}$$

A **series-parallel combination** of capacitors can be arranged using three or more capacitors. Using Equations 26.8 and 26.10, a series-parallel combination of capacitors can be reduced to a single equivalent capacitor.

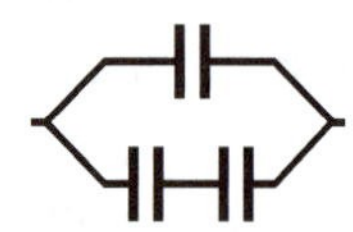

Two capacitors in series combined in parallel with a third.

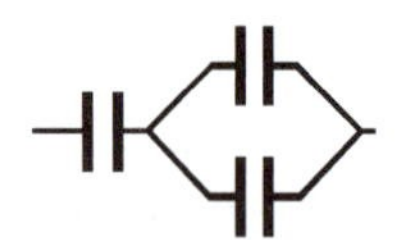

Two capacitors in parallel combined in series with a third.

The **potential energy** stored in the electrostatic field of a charged capacitor equals the work done (by a battery or other source) in charging the capacitor from $q = 0$ to $q = Q$. The potential energy increases as the charge and potential increase. *Equation 26.11 applies to capacitors of any geometry.*

$$U = \frac{Q^2}{2C} = \tfrac{1}{2}Q\Delta V = \tfrac{1}{2}C(\Delta V)^2 \quad (26.11)$$

The **energy density** (energy per unit volume) at any point in the electrostatic field of a charged capacitor is proportional to the square of the electric field intensity at that point. *This is true for any electric field, not just that of a parallel plate capacitor.*

$$u_E = \tfrac{1}{2}\epsilon_0 E^2 \qquad (26.13)$$

An **electric dipole** consists of two charges of equal magnitude and opposite sign separated by a distance $2a$. Vector $\vec{\mathbf{p}}$ is the **electric dipole moment** and is directed from the negative charge to the positive charge.

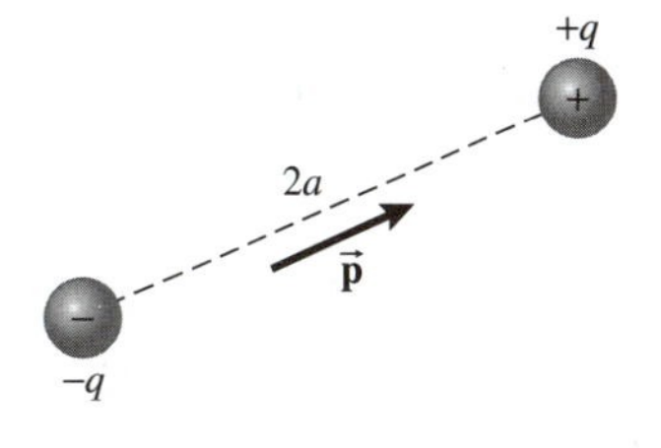

The **electric dipole moment** has a magnitude that depends on the value of q and the distance of separation of the two charges.

$$p \equiv 2aq \qquad (26.16)$$

The **net torque on an electric dipole** in a uniform electric field can be expressed as a vector cross product. *The torque will be maximum when the direction of the dipole moment is perpendicular to the direction of the electric field.*

$$\vec{\boldsymbol{\tau}} = \vec{\mathbf{p}} \times \vec{\mathbf{E}} \qquad (26.18)$$

The **potential energy** associated with a dipole-electric field system depends on the angle between the direction of the dipole moment and the direction of the magnetic field. The equation for potential energy can also be expressed as a dot product.

$$U = -pE\cos\theta \qquad (26.19)$$

$$U = -\vec{\mathbf{p}}\cdot\vec{\mathbf{E}} \qquad (26.20)$$

SUGGESTIONS, SKILLS, AND STRATEGIES

PROBLEM-SOLVING HINTS FOR CAPACITANCE

- When analyzing a series-parallel combination of capacitors to determine the equivalent capacitance, you should make a sequence of circuit diagrams which show the successive steps in the simplification of the circuit. At each step, combine those capacitors which are in simple-parallel or simple-series relationship to each other, and use appropriate equations for series or parallel capacitors at each step of the simplification. At each step, you know two of the three quantities: Q, ΔV, and C. You will be able to determine the remaining quantity using the relation $Q = C\Delta V$.

- When calculating capacitance, be careful with your choice of units. To calculate capacitance in farads, make sure that distances are in meters and use the SI value of ϵ_0. When checking consistency of units, remember that the electric field has units of newtons per coulomb (N/C) or volts per meter (V/m).

- When two or more unequal capacitors are connected in series, they carry the same charge, but their potential differences are not the same. The capacitances add as reciprocals, and the equivalent capacitance of the combination is always less than that of the smallest individual capacitor.

- When two or more capacitors are connected in parallel, the potential difference is the same for each capacitor. The charge on each capacitor is proportional to its capacitance; the capacitances add directly to give the equivalent capacitance of the parallel combination.

- A dielectric increases capacitance by the factor κ (the dielectric constant). This occurs because induced surface charges on the dielectric reduce the electric field inside the material from E to (E/κ).

- Be careful about problems in which you may be connecting or disconnecting a battery to a capacitor. It is important to note whether modifications to the capacitor are being made while the capacitor is connected to the battery or after it is disconnected. If the capacitor remains connected to the battery, the potential difference across the capacitor necessarily remains the same (equal to that of the battery), and the charge is proportional to the capacitance. However, it may be modified, for example, by insertion of a dielectric. On the other hand, if you disconnect the capacitor from the battery before making any modifications to the capacitor, then its charge remains the same. In this case, as you vary the capacitance, the potential difference across the plates changes in inverse proportion to capacitance, according to $\Delta V = Q/C$.

REVIEW CHECKLIST

You should be able to:

- Use the basic definition of capacitance and the equation for finding the potential difference between two points in an electric field in order to calculate the capacitance of a capacitor for cases of relatively simple geometry—parallel plates, cylindrical, spherical. (Sections 26.1 and 26.2)

- Determine the equivalent capacitance of a network of capacitors in series-parallel combination, calculate the final charge on each capacitor, and find the potential difference across each when a known potential is applied across the combination. (Section 26.3)

- Make calculations involving the relationships among potential difference, charge, capacitance, stored energy, and energy density for capacitors, and apply these results to the particular case of a parallel plate capacitor. (Section 26.4)
- Calculate the capacitance, potential difference, and stored energy of a capacitor which is partially or completely filled with a dielectric. (Section 26.5)
- Calculate the dipole moment, torque, and potential energy associated with an electric dipole in an external electric field. (Section 26.6).

ANSWER TO AN OBJECTIVE QUESTION

7. **(i)** What happens to the magnitude of the charge on each plate of a capacitor if the potential difference between the conductors is doubled? (a) It becomes four times larger. (b) It becomes two times larger. (c) It is unchanged. (d) It becomes one-half as large. (e) It becomes one-fourth as large. **(ii)** If the potential difference across a capacitor is doubled, what happens to the energy stored? Choose from the same possibilities as in part (i).

Answer **(i)** (b) The equation $Q = C\Delta V$ describes how the charge is directly proportional to the applied potential difference. **(ii)** (a) Since $U = C(\Delta V)^2/2$, doubling ΔV will quadruple the stored energy.

□ □ □ □

ANSWER TO A CONCEPTUAL QUESTION

3. If you were asked to design a capacitor in which small size and large capacitance were required, what would be the two most important factors in your design?

Answer You should use a dielectric filled capacitor whose dielectric constant is very large. Furthermore, you should make the dielectric as thin as you can, keeping in mind that the maximum allowed voltage should not produce dielectric breakdown of the insulator.

□ □ □ □

SOLUTIONS TO SELECTED END-OF-CHAPTER PROBLEMS

5. A 50.0-m length of coaxial cable has an inner conductor that has a diameter of 2.58 mm and carries a charge of 8.10 μC. The surrounding conductor has an inner diameter of 7.27 mm and a charge of –8.10 μC. Assume the region between the conductors is air.

(a) What is the capacitance of this cable? (b) What is the potential difference between the two conductors?

Solution

Conceptualize: We study parallel-plate and spherical capacitors. This problem is about a cylindrical capacitor, also briefly covered in the chapter text.

Categorize: We apply the definition of capacitance and the result proved in the chapter for this geometry.

Analyze:

(a) The capacitance of a cylindrical capacitor is

$$C = \frac{\ell}{2k_e \ln(b/a)} = \frac{50.0\text{ m}}{2(8.99\times10^9\text{ N}\cdot\text{m}^2/\text{C}^2)\ln(7.27/2.58)} = 2.68\times10^{-9}\text{ F} \quad \blacksquare$$

(b) From the definition of capacitance the potential difference between the cylindrical plates is

$$\Delta V = \frac{Q}{C} = \frac{8.10\times10^{-6}\text{ C}}{2.68\times10^{-9}\text{ F}} = 3.02\text{ kV} \quad \blacksquare$$

Finalize: A person ordinarily buys a coaxial cable to carry electric current shielded from noise due to environmental electric fields. We have shown here that the cable itself possesses capacitance, which may be large enough in some applications to be important to the circuit of which it is part.

---

7. When a potential difference of 150 V is applied to the plates of a parallel-plate capacitor, the plates carry a surface charge density of 30.0 nC/cm^2. What is the spacing between the plates?

Solution

Conceptualize: A practical capacitor often has the plates quite close together. We could guess a fraction of a millimeter for the spacing.

Categorize: We will be able to solve this problem with a simple equation obtained by combining the definition of capacitance and the capacitance of a parallel-plate capacitor.

Analyze: We have $Q = C\Delta V$ with $C = \epsilon_0 A/d$.

Thus, $Q = \epsilon_0 A \Delta V/d$

The surface charge density on each plate has the same magnitude, given by $\sigma = \frac{Q}{A} = \frac{\epsilon_0 \Delta V}{d}$

Thus, $d = \frac{\epsilon_0 \Delta V}{Q/A} = \frac{\left(8.85\times10^{-12}\ \text{C}^2/\text{N}\cdot\text{m}^2\right)(150\ \text{V})}{\left(30.0\times10^{-9}\ \text{C/cm}^2\right)}$

$$d = \left(4.42\times10^{-2}\frac{\text{V}\cdot\text{C}\cdot\text{cm}^2}{\text{N}\cdot\text{m}^2}\right)\frac{\left(1\ \text{m}^2\right)}{\left(10^4\ \text{cm}^2\right)}\frac{\text{J}}{\text{V}\cdot\text{C}}\frac{\text{N}\cdot\text{m}}{\text{J}} = 4.42\ \mu\text{m}$$ ■

Finalize: Depending on the plate area, we could have a device with small or large capacitance. The charge density and the plate spacing, as well as the voltage and internal field, are intensive rather than extensive properties of the capacitor. By this we mean that they do not depend on the physical size as measured by the plate area. The capacitance, charge, area, and stored energy are extensive properties.

---

8. An air-filled spherical capacitor is constructed with inner and outer shell radii of 7.00 cm and 14.0 cm, respectively. (a) Calculate the capacitance of the device. (b) What potential difference between the spheres results in a charge of 4.00 μC on the capacitor?

Solution

Conceptualize: Since the separation between the inner and outer shells is much larger than a typical electronic capacitor with $d \sim 0.1$ mm and capacitance in the microfarad range, we might expect the capacitance of this spherical configuration to be on the order of picofarads (based on a factor of about 700 times larger spacing between the conductors). The potential difference should be sufficiently low to prevent sparking through the air that separates the shells.

Categorize: The capacitance can be found from the equation for spherical shells, and the voltage can be found from $Q = C\Delta V$.

Analyze:

(a) For a spherical capacitor with inner radius a and outer radius b,

$$C = \frac{ab}{k_e(b-a)} = \frac{(0.070\,0\ \text{m})(0.140\ \text{m})}{\left(8.99\times10^9\ \text{N}\cdot\text{m}^2/\text{C}^2\right)(0.140\ \text{m} - 0.070\,0\ \text{m})} = 1.56\times10^{-11}\ \text{F}$$

$$= 15.6\ \text{pF}$$ ■

(b) $\Delta V = \frac{Q}{C} = \frac{4.00\times10^{-6}\ \text{C}}{1.56\times10^{-11}\ \text{F}} = 2.57\times10^5\ \text{V} = 257\ \text{kV}$ ■

Finalize: The capacitance agrees with our prediction, but the voltage seems rather high. We can check the reasonableness of this voltage by approximating the configuration as two charged parallel plates separated by $d = 7.00$ cm. The electric field between them is

$$E \sim \frac{\Delta V}{d} = \frac{2.57 \times 10^5 \text{ V}}{0.070\,0 \text{ m}} = 3.67 \times 10^6 \text{ V/m}$$

This electric field somewhat exceeds the dielectric breakdown strength of air (3×10^6 V/m), so it may not even be possible to place 4.00 μC of charge on this capacitor!

---

9. An isolated charged conducting sphere of radius 12.0 cm creates an electric field of 4.90×10^4 N/C at a distance 21.0 cm from its center. (a) What is its surface charge density? (b) What is its capacitance?

Solution

Conceptualize: We can think of the sphere as a charge that creates a potential at exterior points just as a charged particle at its center would. We can also think of using a battery to put the sphere's surface at some potential, relative to zero volts at infinite distance. This action deposits some charge on the surface, which we can regard as stored on a capacitor plate.

Categorize: We use equations about a charge with spherical symmetry (an example is $V = Q/4\pi\epsilon_0 R$) to relate the charge to its exterior effects.

Analyze:

(a) The electric field outside a spherical charge distribution of radius R is $E = k_e Q/r^2$. Therefore, $Q = Er^2/k_e$. Since the surface charge density is $\sigma = Q/A$,

$$\sigma = \frac{Er^2}{k_e 4\pi R^2} = \frac{\left(4.90 \times 10^4 \text{ N/C}\right)(0.210 \text{ m})^2}{\left(8.99 \times 10^9 \text{ N} \cdot \text{m}^2/\text{C}^2\right)\left(4\pi\right)(0.120 \text{ m})^2} = 1.33 \ \mu\text{C/m}^2 \quad \blacksquare$$

(b) For an isolated charged sphere of radius R,

$$C = 4\pi\epsilon_0 R = (4\pi)(8.85 \times 10^{-12} \text{ C}^2/\text{N}\cdot\text{m}^2)(0.120 \text{ m}) = 13.3 \text{ pF} \quad \blacksquare$$

Finalize: This is a straightforward problem. The sphere is as big as a large classroom world globe. Its capacitance is small, in micromicrofarads.

---

11. An air-filled capacitor consists of two parallel plates, each with an area of 7.60 cm^2, separated by a distance of 1.80 mm. A 20.0-V potential difference is applied to these plates. Calculate (a) the electric field between the plates, (b) the surface charge density, (c) the capacitance, and (d) the charge on each plate.

Solution

Conceptualize: This is a tabletop model capacitor. Its capacitance is small compared to the capacitances of inexpensive electronic components.

Categorize: We can substitute into simple equations to solve the parts in the order given. We have enough equations to do the parts in any order.

Analyze:

(a) The potential difference between two points in a uniform electric field is $\Delta V = Ed$, so

$$E = \frac{\Delta V}{d} = \frac{20.0 \text{ V}}{1.80 \times 10^{-3} \text{ m}} = 1.11 \times 10^{4} \text{ V/m} \qquad \blacksquare$$

(b) The electric field between capacitor plates is $E = \dfrac{\sigma}{\epsilon_0}$, so $\sigma = \epsilon_0 E$:

$$\sigma = \left(8.85 \times 10^{-12} \text{ C}^2/\text{N}\cdot\text{m}^2\right)\left(1.11 \times 10^{4} \text{ V/m}\right) = 9.83 \times 10^{-8} \text{ C/m}^2$$

$$= 98.3 \text{ nC/m}^2 \qquad \blacksquare$$

(c) For a parallel-plate capacitor, $C = \dfrac{\epsilon_0 A}{d}$:

$$C = \frac{\left(8.85 \times 10^{-12} \text{ C}^2/\text{N}\cdot\text{m}^2\right)\left(7.60 \times 10^{-4} \text{ m}^2\right)}{1.80 \times 10^{-3} \text{ m}} = 3.74 \times 10^{-12} \text{ F} = 3.74 \text{ pF} \qquad \blacksquare$$

(d) The charge on each plate is $Q = C\Delta V$

$$Q = (3.74 \times 10^{-12} \text{ F})(20.0 \text{ V}) = 7.47 \times 10^{-11} \text{ C} = 74.7 \text{ pC} \qquad \blacksquare$$

Finalize: This capacitor of a few picofarads is physically large compared to those typically used on electronic circuit boards, but its capacitance is small.

---

23. Four capacitors are connected as shown in Figure P26.23.

(a) Find the equivalent capacitance between points a and b.

(b) Calculate the charge on each capacitor, taking $\Delta V_{ab} = 15.0$ V.

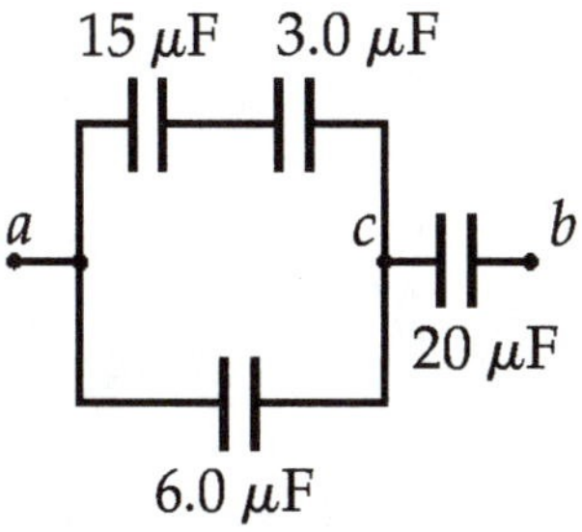

15 μF 3.0 μF a c b 20 μF 6.0 μF

Figure P26.23

Solution

Conceptualize: To exterior devices, the section of electric circuit between a and b would behave like a single capacitor. When 15 volts is applied between a and b, each one of the actual capacitors will have some smaller voltage across it. We will see that the 20 microfarads stores the whole charge that passes points a and b, but the other actual capacitors store less.

Categorize: We find a set of capacitors in series or in parallel, replace them with a single equivalent capacitor, draw the new circuit, and repeat. The circuit diagrams will guide our analysis of how the charge divides up.

Analyze:

(a) We simplify the circuit of Figure P26.23 in three steps as shown in Figures (a), (b), and (c).

First, the 15.0 μF and 3.00 μF in series are equivalent to

$$\frac{1}{(1/15.0\ \mu\text{F}) + (1/3.00\ \mu\text{F})} = 2.50\ \mu\text{F}$$

Next, 2.50 μF combines in parallel with 6.00 μF, creating an equivalent capacitance of 8.50 μF.

At last, 8.50 μF and 20.0 μF are in series, equivalent to

$$\frac{1}{(1/8.50\ \mu\text{F}) + (1/20.00\ \mu\text{F})} = 5.96\ \mu\text{F} \quad \blacksquare$$

(b) We find the charge on each capacitor and the voltage across each by working backwards through solution figures (c)–(a), alternately applying $Q = C\Delta V$ and $\Delta V = Q/C$ to every capacitor, real or equivalent. For the 5.96-μF capacitor, we have

$$Q = C\Delta V = (5.96\ \mu\text{F})(15.0\ \text{V}) = 89.5\ \mu\text{C}$$

Thus, if a is higher in potential than b, just 89.5 μC flows between the wires and the plates to charge the capacitors in each picture. In (b) we have, for the 8.5-μF capacitor,

$$\Delta V_{ac} = \frac{Q}{C} = \frac{89.5\ \mu\text{C}}{8.50\ \mu\text{F}} = 10.5\ \text{V}$$

and for the 20.0 μF in (b), (a), and the original circuit, we have $Q_{20} = 89.5\ \mu$C $\blacksquare$

$$\Delta V_{cb} = \frac{Q}{C} = \frac{89.5\ \mu\text{C}}{20.0\ \mu\text{F}} = 4.47\ \text{V}$$

Next, the circuit in diagram (a) is equivalent to that in (b), so $\Delta V_{cb} = 4.47$ V and $\Delta V_{ac} = 10.5$ V

For the 2.50 μF, $\Delta V = 10.5$ V and $Q = C\Delta V = (2.50\ \mu\text{F})(10.5\ \text{V}) = 26.3\ \mu\text{C}$

For the 6.00 μF, $\Delta V = 10.5$ V and $Q_6 = C\Delta V = (6.00\ \mu\text{F})(10.5\ \text{V}) = 63.2\ \mu\text{C}$ $\blacksquare$

Now, 26.3 μC having flowed in the upper parallel branch in (a), back in the original circuit we have $Q_{15} = 26.3\ \mu$C and $Q_3 = 26.3\ \mu$C $\blacksquare$

Finalize: Some students think this kind of problem is just about using the formulas for series and parallel capacitor combinations. Really to analyze circuits you need to develop some essential skills beyond using formulas. One skill is to *recognize* parallel and series connections. In the original diagram the 6 μF and the 15 μF are drawn on parallel lines, but they are not in parallel or in series. To be in parallel, capacitors must be connected "top end to top end and bottom end to bottom end," so that the same voltage must appear on both. To be in series, capacitors must have no junction between them, so that the same charge will necessarily appear on both. Another essential skill, after calculating the capacitance of an equivalent capacitor, is to *draw* it in a new circuit diagram. All the diagrams are necessary for letting you practice the next skill, applying $Q = C\,\Delta V$ and $\Delta V = Q/C$ to *find the charges and potential differences* for all the capacitors, real and equivalent, including those in the original circuit. To bring the problem to a logical conclusion we do this **Related Calculation**:

A problem on your next exam may also ask for the voltage across each of the original capacitors. The remaining answers are:

$$\Delta V_{15} = \frac{Q}{C} = \frac{26.3\ \mu\text{C}}{15.0\ \mu\text{F}} = 1.75\text{V} \quad \text{and} \quad \Delta V_3 = \frac{Q}{C} = \frac{26.3\ \mu\text{C}}{3.0\ \mu\text{F}} = 8.77\ \text{V}$$

---

24. Consider the circuit shown in Figure P26.24, where $C_1 = 6.00\ \mu\text{F}$, $C_2 = 3.00\ \mu\text{F}$, and $\Delta V = 20.0$ V. Capacitor C_1 is first charged by closing switch S_1. Switch S_1 is then opened, and the charged capacitor is connected to the uncharged capacitor by closing S_2. Calculate (a) the initial charge acquired by C_1 and (b) the final charge on each capacitor.

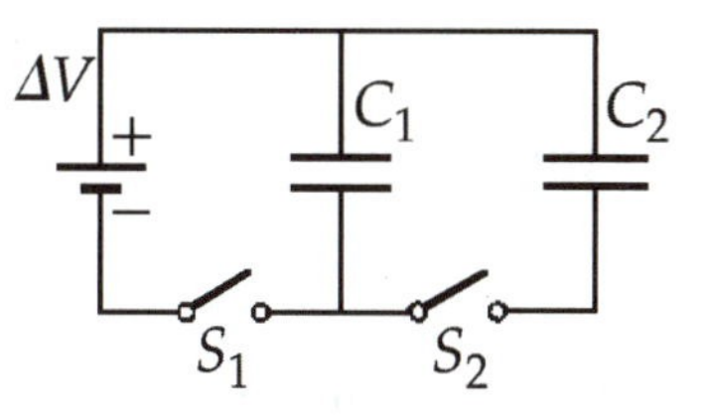

Figure P26.24

Solution

Conceptualize: The two switches are never closed at the same time. Capacitor 2 is never connected directly across the battery. The voltage across it will never be 20 V, but some smaller value after…

Categorize: …charge originally on capacitor 1 is shared with capacitor 2. We must visualize the sequence of operations. Then just the definition of capacitance, applied again and again, will give us the answers.

Analyze:

(a) When S_1 is closed, the charge on C_1 will be

$$Q_1 = C_1\Delta V_1 = (6.00\ \mu\text{F})(20.0\ \text{V}) = 120\ \mu\text{C} \qquad \blacksquare$$

(b) When S_1 is opened and S_2 is closed, the total charge will remain constant and be shared by the two capacitors. We let primed symbols represent the new charges on the capacitors, in $Q'_1 = 120\ \mu\text{C} - Q'_2$. The potential differences across the two capacitors will be equal.

$$\Delta V' = \frac{Q'_1}{C_1} = \frac{Q'_2}{C_2} \quad \text{or} \quad \frac{120\ \mu\text{C} - Q'_2}{6.00\ \mu\text{F}} = \frac{Q'_2}{3.00\ \mu\text{F}} \qquad \blacksquare$$

Then we do the algebra to find $Q'_2 = 40.0\ \mu\text{C}$ and ■

$$Q'_1 = 120\ \mu\text{C} - 40.0\ \mu\text{C} = 80.0\ \mu\text{C} \qquad \blacksquare$$

Finalize: It might help you to draw the circuit diagrams with switch 1 closed, then back to open, and then with just switch 2 closed. Note that the symbol for an open switch looks like an open door on a floor plan. An open door allows things to go through but an open switch does not allow charge to pass through it.

This problem may be the first you have seen to directly apply a universal law of nature, conservation of charge. This is the fundamental reason that the sum of the charges on the two capacitors in the final circuit must be 120 μC. Electrical potential energy is not conserved when the connection is made. Perhaps remarkably, the stored energy drops from 1200 μJ to 800 μJ. The missing energy becomes internal energy in the wires joining the capacitors or leaves the circuit as electromagnetic radiation. If it were not so, most of the charge would bounce between one capacitor and the other, oscillating forever.

---

27. A group of identical capacitors is connected first in series and then in parallel. The combined capacitance in parallel is 100 times larger than for the series connection. How many capacitors are in the group?

Solution

Conceptualize: Since capacitors in parallel add and ones in series add as inverses, 2 capacitors in parallel would have a capacitance 4 times greater than if they were in series, and 3 capacitors would give a ratio $C_p/C_s = 9$, so maybe $n = \sqrt{C_p/C_s} = \sqrt{100} = 10$.

Categorize: The ratio reasoning above seems like an efficient way to solve this problem, but we should check the answer with a more careful analysis based on the general relationships for series and parallel combinations of capacitors.

Analyze: Call C the capacitance of one capacitor and n the number of capacitors. The equivalent capacitance for n capacitors in parallel is

$$C_p = C_1 + C_2 + \cdots + C_n = nC$$

The relationship for n capacitors in series is

$$\frac{1}{C_s} = \frac{1}{C_1} + \frac{1}{C_2} + \cdots + \frac{1}{C_n} = \frac{n}{C}$$

Therefore $\dfrac{C_p}{C_s} = \dfrac{nC}{C/n} = n^2$ or $n = \sqrt{C_p/C_s} = \sqrt{100} = 10$ ■

Finalize: Our prediction was correct. A qualitative reason that $C_p/C_s = n^2$ is that the amount of charge that can be stored on the capacitors increases according to the area of the plates for a parallel combination, but it is the total voltage that increases for a series combination.

---

36. A parallel-plate capacitor has a charge Q and plates of area A. What force acts on one plate to attract it toward the other plate? Because the electric field between the plates is $E = Q/A\epsilon_0$, you might think that the force is $F = QE = Q^2/A\epsilon_0$. This conclusion is wrong, because the field E includes contributions from both plates, and the field created by the positive plate cannot exert any force on the positive plate. Show that the force exerted on each plate is actually $F = Q^2/2\epsilon_0 A$. *Suggestion:* Let $C = \epsilon_0 A/x$ for an arbitrary plate separation x and note that the work done in separating the two charged plates is $W = \int F dx$.

Solution

Conceptualize: The force *on* one plate is exerted *by* the other, through the electric field created by that single other plate:

$$E = \frac{\sigma}{2\epsilon_0} = \frac{Q}{2A\epsilon_0}$$

The force on each plate is: $F = (Q_{\text{self}})(E_{\text{other}}) = \dfrac{Q^2}{2A\epsilon_0}$

Categorize: To prove this, we follow the hint, and calculate the work done in separating the plates, which equals the potential energy stored in the charged capacitor:

$$U = \tfrac{1}{2}\frac{Q^2}{C} = \int F dx$$

Analyze: Now from the fundamental theorem of calculus, $dU = F dx$

and
$$F = \frac{d}{dx}U = \frac{d}{dx}\left(\frac{Q^2}{2C}\right) = \tfrac{1}{2}\frac{d}{dx}\left(\frac{Q^2}{A\epsilon_0/x}\right)$$

Performing the differentiation, $F = \tfrac{1}{2}\dfrac{d}{dx}\left(\dfrac{Q^2 x}{A\epsilon_0}\right) = \tfrac{1}{2}\left(\dfrac{Q^2}{A\epsilon_0}\right)$ ■

Finalize: The force exerted on one charged plate by another is sometimes used in a machine shop to hold a workpiece stationary. Larger charge on both plates will imply a larger field exerting on a larger charge an extra-large force, so the proportionality to charge squared is reasonable.

---

46. A commercial capacitor is to be constructed as shown in Figure 26.46. This particular capacitor is made from two strips of aluminum separated by a strip of paraffin-coated paper. Each strip of foil and paper is 7.00 cm wide. The foil is 0.004 00 mm thick, and the paper is 0.025 0 mm thick and has a dielectric constant of 3.70. What length should the strips have, if a capacitance of 9.50×10^{-8} F is desired before the capacitor is rolled up? (Adding a second strip of paper and rolling the capacitor effectively doubles its capacitance, by allowing charge storage on both sides of each strip of foil.)

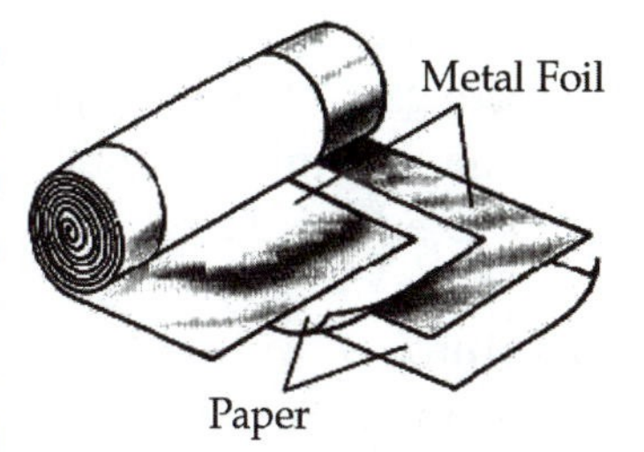

Figure P26.46

Solution

Conceptualize: Ninety-five nanofarads is a reasonable capacitance for a physically compact component from the electronics store. The thinness of the dielectric—the closeness of the plates—is a primary reason for how large its capacitance is, so the long dimension of the foil plates may be well under a meter.

Categorize: We use just the equation giving the capacitance of a parallel-plate capacitor with a dielectric between its plates.

Analyze: Our diagram exaggerates how the strips can be offset to avoid contact between the two foils. It shows how a second paper strip can be used to roll the capacitor into a convenient cylindrical shape with electrical contacts at the two ends. We suppose that the overlapping width of the two metallic strips is still $w = 7.00$ cm. Then for the area of the plates we have $A = \ell w$ in $C = \kappa\epsilon_0 A/d = \kappa\epsilon_0 \ell w/d$. Solving the equation gives

$$\ell = \frac{Cd}{\kappa \epsilon_0 w} = \frac{(9.50\times10^{-8}\text{ F})(2.50\times10^{-5}\text{ m})}{3.70(8.85\times10^{-12}\text{ C}^2/\text{N}\cdot\text{m}^2)(0.070\ 0\text{ m})} = 1.04\text{ m} \quad \blacksquare$$

Finalize: Oops! It's a bit over a meter, but the capacitor can still be smaller than a roll of coins. The distance between positive and negative charges is the thickness of one sheet of paper because these charges sit on adjacent surfaces of the metal foils. The electric field between the top and bottom surfaces of one piece of foil is zero. The thickness of the foil was given in the problem but it does not affect the capacitance.

---

51. A small, rigid object carries positive and negative 3.50-nC charges. It is oriented so that the positive charge has coordinates (–1.20 mm, 1.10 mm) and the negative charge is at the point (1.40 mm, –1.30 mm). (a) Find the electric dipole moment of the object. The object is placed in an electric field $\vec{\mathbf{E}} = (7.80\times10^3\,\hat{\mathbf{i}} - 4.90\times10^3\,\hat{\mathbf{j}})$ N/C. (b) Find the torque acting on the object. (c) Find the potential energy of the object–field system when the object is in this orientation. (d) Assuming the orientation of the object can change, find the difference between the maximum and minimum potential energies of the system.

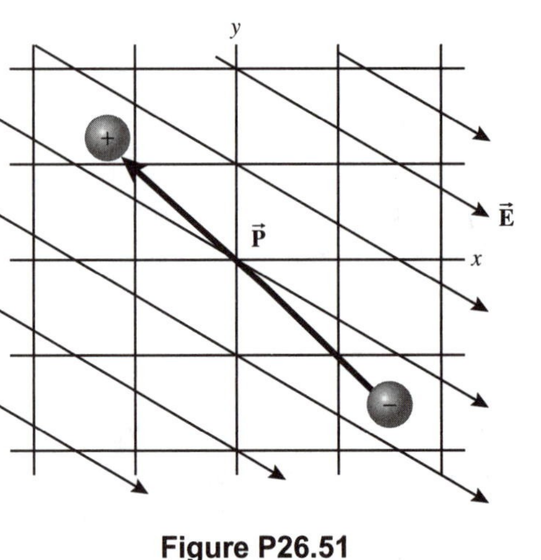

Figure P26.51

Solution

Conceptualize: An electric dipole in an external electric field behaves like a horizontal needle mounted on a vertical axle, with a light spiral spring causing it to have one preferred orientation. You can hold the needle in any orientation, exerting more torque to hold it farther from its lowest-energy orientation. If you release it, it will oscillate as a torsional pendulum, described in Section 15.5. The electric dipole in this problem is a small size that might be used for classroom observation. With nanocoulomb charges separated by millimeters, its dipole moment will be some picocoulomb-meters. The torque exerted on the dipole will be some nanonewton-meters, tending to turn it clockwise to align it with the field. The potential energies will be on the order of nanojoules. In the orientation described, the dipole is close to its maximum potential energy, since the dipole moment is close to being opposite the field direction.

Categorize: We will use directly the definition of dipole moment and the theorems about torque and potential energy that are proved in the chapter text. The problem contains review of vector dot and cross products.

Analyze:

(a) The displacement from negative to positive charge is

$$2\vec{\mathbf{a}} = \left(-1.20\hat{\mathbf{i}} + 1.10\hat{\mathbf{j}}\right)\text{ mm} - \left(1.40\hat{\mathbf{i}} - 1.30\hat{\mathbf{j}}\right)\text{ mm} = \left(-2.60\hat{\mathbf{i}} + 2.40\hat{\mathbf{j}}\right)\times 10^{-3}\text{ m}$$

The electric dipole moment is $\vec{\mathbf{p}} = 2\vec{\mathbf{a}}q$

$$\vec{\mathbf{p}} = \left(3.50\times 10^{-9}\text{ C}\right)\left(-2.60\hat{\mathbf{i}} + 2.40\hat{\mathbf{j}}\right)\times 10^{-3}\text{ m} = \left(-9.10\hat{\mathbf{i}} + 8.40\hat{\mathbf{j}}\right)\times 10^{-12}\text{C}\cdot\text{m} \quad \blacksquare$$

(b) The torque exerted by the field on the dipole is

$$\vec{\boldsymbol{\tau}} = \vec{\mathbf{p}}\times\vec{\mathbf{E}} = \left[\left(-9.10\hat{\mathbf{i}} + 8.40\hat{\mathbf{j}}\right)\times 10^{-12}\text{ C}\cdot\text{m}\right]\times\left[\left(7.80\hat{\mathbf{i}} - 4.90\hat{\mathbf{j}}\right)\times 10^{3}\text{ N/C}\right]$$

$$\vec{\boldsymbol{\tau}} = \left(+44.6\hat{\mathbf{k}} - 65.5\hat{\mathbf{k}}\right)\times 10^{-9}\text{ N}\cdot\text{m} = -2.09\times 10^{-8}\text{ N}\cdot\text{m }\hat{\mathbf{k}} \quad \blacksquare$$

(c) Relative to zero energy when it is perpendicular to the field, the dipole has potential energy

$$U = -\vec{\mathbf{p}}\cdot\vec{\mathbf{E}} = -\left[\left(-9.10\hat{\mathbf{i}} + 8.40\hat{\mathbf{j}}\right)\times 10^{-12}\text{ C}\cdot\text{m}\right]\cdot\left[\left(7.80\hat{\mathbf{i}} - 4.90\hat{\mathbf{j}}\right)\times 10^{3}\text{ N/C}\right]$$

$$U = \left(71.0 + 41.2\right)\times 10^{-9}\text{ J} = 112\text{ nJ} \quad \blacksquare$$

(d) For convenience we compute the magnitudes

$$\left|\vec{\mathbf{p}}\right| = \sqrt{(9.10)^2 + (8.40)^2}\times 10^{-12}\text{ C}\cdot\text{m} = 12.4\times 10^{-12}\text{ C}\cdot\text{m}$$

and $\left|\vec{\mathbf{E}}\right| = \sqrt{(7.80)^2 + (4.90)^2}\times 10^{3}\text{ N/C} = 9.21\times 10^{3}\text{ N/C}$

The maximum potential energy occurs when the dipole moment is opposite in direction to the field, and is

$$U_{\text{max}} = -\vec{\mathbf{p}}\cdot\vec{\mathbf{E}} = -\left|\vec{\mathbf{p}}\right|\left|\vec{\mathbf{E}}\right|(-1) = \left|\vec{\mathbf{p}}\right|\left|\vec{\mathbf{E}}\right| = 114\text{ nJ}$$

The minimum potential energy configuration is the stable equilibrium position with the dipole aligned with the field. The value is $U_{\text{min}} = -114\text{ nJ}$

Then the difference, representing the range of potential energies available to the dipole, is $U_{\text{max}} - U_{\text{min}} = 228\text{ nJ}$ $\blacksquare$

Finalize: Our estimates were all good. We actually reviewed multiplying a vector by a scalar, taking the dot product of vectors, and taking the cross product of vectors. Getting a physical feel for how electric dipoles behave is a good step toward understanding magnetic dipoles, and in particular compass needles and how atoms and nuclei behave in magnetic fields.

---

59. A parallel-plate capacitor is constructed using a dielectric material whose dielectric constant is 3.00 and whose dielectric strength is 2.00×10^8 V/m. The desired capacitance is 0.250 μF, and the capacitor must withstand a maximum potential difference of 4.00 kV. Find the minimum area of the capacitor plates.

Solution

Conceptualize: This is a good design problem. We must combine the requirement that the electric field not be too large with the requirement that the capacitance be large enough.

Categorize: Algebra of simultaneous equations is the natural way to combine the requirements. Here the equations will be simple.

Analyze: The dielectric strength is $E_{max} = 2.00 \times 10^8 \text{ V/m} = \dfrac{\Delta V_{max}}{d}$

so we have for the distance between plates $d = \dfrac{\Delta V_{max}}{E_{max}}$

Now to also satisfy $C = \dfrac{\kappa \epsilon_0 A}{d} = 0.250 \times 10^{-6}$ F with $\kappa = 3.00$, we combine by substitution to solve for the plate area

$$A = \frac{Cd}{\kappa \epsilon_0} = \frac{C\Delta V_{max}}{\kappa \epsilon_0 E_{max}} = \frac{(0.250 \times 10^{-6}\text{ F})(4\,000\text{ V})}{(3.00)(8.85 \times 10^{-12}\text{ F/m})(2.00 \times 10^8\text{ V/m})} = 0.188\text{ m}^2 \quad \blacksquare$$

Finalize: Along the way we have shown that the requirements are not mutually exclusive, and that there is only one possibility for the plate area. Neither of these facts might be obvious.

---

65. A capacitor of unknown capacitance has been charged to a potential difference of 100 V and then disconnected from the battery. When the charged capacitor is then connected in parallel to an uncharged 10.0-μF capacitor, the potential difference across the combination is 30.0 V. Calculate the unknown capacitance.

Solution

Conceptualize: The voltage across the combination ΔV_1 will be reduced according to the size of the added capacitance. (Example: If the unknown capacitance were $C = 10.0\ \mu$F, then ΔV_1 would be 50.0 V because the charge is now distributed evenly between the two capacitors.) Since the final voltage is less than half the original, we might guess that the unknown capacitor is about 5.00 μF.

Categorize: We can use the relationships for capacitors in parallel to find the unknown capacitance, along with the requirement that the charge on the unknown capacitor must be the same as the total charge on the two capacitors in parallel.

Analyze: We name our ignorance and call the unknown capacitance C_u. The charge originally deposited on **each** plate, + on one and – on the other, is

$$Q = C_u \Delta V = C_u (100 \text{ V})$$

Now in the new connection this same conserved charge redistributes itself between the two capacitors according to $Q = Q_1 + Q_2$.

$$Q_1 = C_u(30.0 \text{ V}) \quad \text{and} \quad Q_2 = (10.0\ \mu\text{F})(30.0 \text{ V}) = 300\ \mu\text{C}$$

We can eliminate Q and Q_1 by substitution:

$$C_u(100 \text{ V}) = C_u (30.0 \text{ V}) + 300\ \mu\text{C} \quad \text{so} \quad C_u = \frac{300\ \mu\text{C}}{70.0 \text{ V}} = 4.29\ \mu\text{F}$$ ■

Finalize: The calculated capacitance is close to what we expected, so our result seems reasonable. In this and other capacitance combination problems, it is important not to confuse the charge and voltage across the system with those across individual components. Careful attention can be given to the subscripts to avoid this confusion, but perhaps a diagram is better for you—draw it! Whatever you do, don't confuse the symbol "*C*" for capacitance with the unit of charge "C" for coulombs.

---

68. A parallel-plate capacitor of plate separation d is charged to a potential difference ΔV_0. A dielectric slab of thickness d and dielectric constant κ is introduced between the plates while the battery remains connected to the plates. (a) Show that the ratio of energy stored after the dielectric is introduced to the energy stored in the empty capacitor is $U/U_0 = \kappa$. (b) Give a physical explanation for this increase in stored energy. (c) What happens to the charge on the capacitor? *Note:* This situation is not the same as in Example 26.5, in which the battery was removed from the circuit before the dielectric was introduced.

Solution

Conceptualize: Sliding in the dielectric will increase the capacitance, so it seems natural that the stored energy should increase.

Categorize: If it helps your visualization, draw the original capacitor with vacuum between its plates and the final capacitor with the dielectric, both connected to the same battery. The equation for stored energy in these capacitors will yield our solution.

Analyze:

(a) The capacitance changes from, say, C_0 to κC_0. The battery will maintain constant voltage across it by pumping out extra charge.

The original energy is $U_0 = \frac{1}{2} C_0(\Delta V)^2$

and the final energy is $U = \frac{1}{2} \kappa C_0(\Delta V)^2$ so $U/U_0 = \kappa$ ■

(b) The extra energy comes from (part of the) electrical work done by the battery in separating extra charge. ■

(c) The original charge is $Q_0 = C_0\Delta V$

and the final value is $Q = \kappa C_0\Delta V$

so the charge increases by the factor κ. ■

Finalize: Observe how here, and also in problem 65, we freely introduced symbols for quantities we did not know. The symbols divided out in the particular answers required, but the solutions could still be valid if the symbols did not divide out.

---

70. The inner conductor of a coaxial cable has a radius of 0.800 mm, and the outer conductor's inside radius is 3.00 mm. The space between the conductors is filled with polyethylene, which has a dielectric constant of 2.30 and a dielectric strength of 18.0×10^6 V/m. What is the maximum potential difference that this cable can withstand?

Solution

Conceptualize: We can increase the potential difference between the core and the sheath until the electric field in the polyethylene starts to punch a hole through it, passing a spark.

Categorize: This will first happen at the surface of the inner conductor, where the electric field is strongest.

Analyze: From Example 26.1, when there is a vacuum between the conductors, the voltage between them is

$$\Delta V = |V_b - V_a| = 2k_e\lambda \ln\left(\frac{b}{a}\right) = \frac{\lambda}{2\pi \epsilon_0}\ln\left(\frac{b}{a}\right)$$

With a dielectric, a factor $1/\kappa$ must be included, and the equation becomes

$$\Delta V = \frac{\lambda}{2\pi\kappa \epsilon_0}\ln\left(\frac{b}{a}\right)$$

The electric field is $E = \dfrac{\lambda}{2\pi\kappa \epsilon_0 r}$

So when $E = E_{max}$ at $r = a$,

$$\frac{\lambda_{max}}{2\pi\kappa\epsilon_0} = E_{max}a \quad \text{and} \quad \Delta V_{max} = \frac{\lambda_{max}}{2\pi\kappa\epsilon_0}\ln\left(\frac{b}{a}\right) = E_{max}a\ \ln\left(\frac{b}{a}\right)$$

Thus, $\Delta V_{max} = (18.0 \times 10^6\ \text{V/m})(0.800 \times 10^{-3}\ \text{m})\ \ln\left(\frac{3.00\ \text{mm}}{0.800\ \text{mm}}\right) = 19.0\ \text{kV}$ ■

Finalize: We can think of the combination $\kappa\epsilon_0$ as the permittivity ϵ of the dielectric material, associated with how hard it is for charges to establish electric fields in that material, as opposed to in vacuum.

27

Current and Resistance

EQUATIONS AND CONCEPTS

Electric current (I) is defined as the rate at which charge flows through the cross-sectional area of a conductor. *Under the influence of an electric field, electric charges will move through gases, liquids, and solid conductors.*

$$I_{\text{ave}} = \frac{\Delta Q}{\Delta t} \qquad (27.1)$$

$$I \equiv \frac{dQ}{dt} \qquad (27.2)$$

The **SI unit of current is the ampere** (A). One coulomb of charge passing through a cross-sectional area of conductor in one second constitutes a current of one ampere.

$$1\ \text{A} = 1\ \text{C/s} \qquad (27.3)$$

The **direction of conventional current** is assigned to be the direction of flow of positive charge (as shown in the figure, from + to – outside the battery). *This is opposite the direction of electron flow in conductors.*

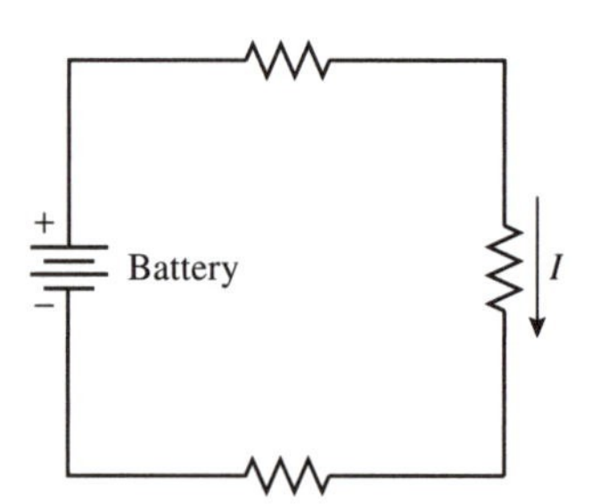

The **average current** in a conductor can be related to microscopic quantities in the conductor: the number of charge carriers (positive or negative) per unit volume (n), the quantity of charge (q) associated with each carrier, and the drift velocity (v_d) of the carriers.

$$I_{avg} = \frac{\Delta Q}{\Delta t} = nqv_d A \qquad (27.4)$$

Current density in a conductor is defined as current per unit area. Equation 27.5 *is valid when the surface of cross-sectional area A is perpendicular to the direction of the current and the current density is uniform.*

$$J \equiv \frac{I}{A} = nqv_d \qquad (27.5)$$

In **ohmic materials** (e.g., most metals), current density is proportional to the electric field in the conductor. *The conductivity of a conductor (σ) is characteristic of the conducting material and, in most cases, independent of the electric field.*

$$J = \sigma E \qquad (27.6)$$

σ = conductivity

Resistance is defined as the ratio of the potential difference across a conductor to the current in the conductor. The **SI unit of resistance** is the ohm (Ω). If a potential difference of 1 V across a conductor results in a current of 1 A, the resistance of the conductor is 1 Ω.

$$R \equiv \frac{\Delta V}{I} \qquad (27.7)$$

$$1\ \Omega = 1\ \text{V/A} \qquad (27.8)$$

Resistivity (ρ) is the inverse of conductivity and is an *intrinsic property of the material* of which a conductor is made. The SI units of resistivity are ohm meters. *Resistivity is not a property of a specific resistive component in a circuit.*

$$\rho = \frac{1}{\sigma} \qquad (27.9)$$

The **resistance of a given conductor** of uniform cross section depends on the length, cross-sectional area, and resistivity of the conducting material. Resistivity is characteristic of a particular *type of material* and depends on the electronic structure and varies with temperature.

$$R = \rho\frac{\ell}{A} \qquad (27.10)$$

R = resistance
ρ = resistivity

The **drift velocity** of an electron in a conductor is due to the electric field produced by a potential difference across the conductor. In Equation 27.13, τ is the average time between electron collisions in the conductor.

$$\vec{\mathbf{v}}_d = \frac{q\vec{\mathbf{E}}}{m_e}\tau \qquad (27.13)$$

In an **ohmic conductor** (one that obeys Ohm's law), conductivity and resistivity do not depend on the magnitude of the electric field. *Equations 27.15 and 27.16 express σ and τ in terms of microscopic properties of the conductor.*

$$\sigma = \frac{nq^2\tau}{m_e} \qquad (27.15)$$

$$\rho = \frac{1}{\sigma} = \frac{m_e}{nq^2\tau} \qquad (27.16)$$

The **temperature coefficient of resistivity** (α) determines the rate at which resistivity (and resistance) increases or decreases with a change in the tempeature of a conductor. T_0 is a stated reference temperature (usually 20 °C).

$$\rho = \rho_0\left[1+\alpha\left(T-T_0\right)\right] \qquad (27.17)$$

$$R = R_0\left[1+\alpha\left(T-T_0\right)\right] \qquad (27.19)$$

$$\alpha = \frac{1}{\rho_0}\frac{\Delta\rho}{\Delta T} \qquad (27.18)$$

Power is the rate at which energy is delivered to a resistor or other circuit element when a potential difference is maintained

$$P = I\Delta V \qquad (27.20)$$

$$P = I^2R = \frac{(\Delta V)^2}{R} \qquad (27.21)$$

between the terminals of the circuit element. The SI unit of power is the watt (W). For a device that obeys Ohm's law, the power converted can be expressed in alternative forms.

SUGGESTIONS, SKILLS, AND STRATEGIES

Equation 27.10, $R = \rho\ell/A$, can be used directly to calculate the resistance of a conductor of uniform cross-sectional area and constant resistivity. For those cases in which the area, resistivity, or both vary along the length of the conductor, the resistance must be determined as an integral of dR.

The conductor is subdivided into elements of length dx over which ρ and A may be considered constant in value and the total resistance is

$$R = \int \frac{\rho}{A} dx$$

Consider, for example, the case of a truncated cone of constant resistivity, radii a and b, and height h. The conductor should be subdivided into disks of thickness dx and radius r, and oriented parallel to the faces of the cone as shown in the figure below. Note from the geometry of the figure that

$$x = \left(\frac{r-a}{b-a}\right)h$$

so that $$r = \frac{x}{h}(b-a) + a$$

and $$R = \int_0^h \frac{\rho}{\pi r^2} dx$$

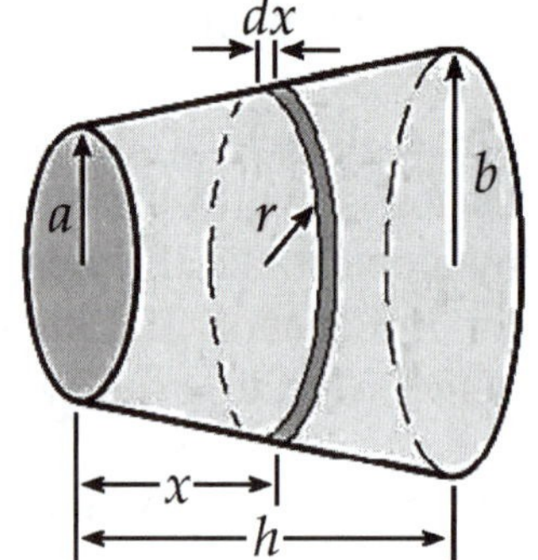

The remainder of this calculation is left as a problem for you to work out (see Problem 27.56 of the text).

REVIEW CHECKLIST

You should be able to:

- Calculate electron drift speed, and quantity of charge passing a point in a given time interval in a specified current-carrying conductor. (Section 27.1)
- Determine the resistance of a conductor using Ohm's law. Also calculate the resistance based on the physical characteristics of a conductor. (Section 27.2)
- Make calculations of the variation of resistance with temperature. (Section 27.4)
- Calculate the power, current, resistance, or potential difference when energy is delivered to a circuit element. (Section 27.6)

ANSWER TO AN OBJECTIVE QUESTION

6. Two wires A and B with circular cross sections are made of the same metal and have equal lengths, but the resistance of wire A is three times greater than that of wire B. **(i)** What is the ratio of the cross-sectional area of A to that of B? (a) 3 (b) $\sqrt{3}$ (c) 1 (d) $1/\sqrt{3}$ (e) 1/3. **(ii)** What is the ratio of the radius of A to that of B? Choose from the same possibilities as in part (i).

Answer **(i)** Since $R = \dfrac{\rho\ell}{\text{Area}}$, the ratio of resistances is given by $\dfrac{R_B}{R_A} = \dfrac{\text{Area}_A}{\text{Area}_B} = \dfrac{1}{3}$. The answer is (e).

(ii) From the ratio of the areas, $\dfrac{\pi r_A^2}{\pi r_B^2} = \dfrac{1}{3}$, we can calculate that the radius of wire A is $1/\sqrt{3}$ times the radius of wire B. The answer is (d).

□ □ □ □

ANSWERS TO SELECTED CONCEPTUAL QUESTIONS

4. Use the atomic theory of matter to explain why the resistance of a material should increase as its temperature increases.

Answer As the temperature increases, the amplitude of atomic vibrations increases. This makes it more likely that the drifting electrons will be scattered by atomic vibrations, and makes it more difficult for charged particles to participate in organized motion inside the conductor.

This answer applies to the situation in metallic conduction, and generally to situations where the number of charge carriers is constant. But note that in a semiconductor the number of charge carriers may change with temperature, to make the resistance behave differently.

□ □ □ □

7. If charges flow very slowly through a metal, why does it not require several hours for a light to come on when you throw a switch?

Answer Individual electrons move with a small average velocity through the conductor, but as soon as the voltage is applied, electrons all along the conductor start to move. Actually the current does not exist "immediately" when the switch is turned on, but rather the establishment of the electric field throughout the conductor is limited by the speed of light.

SOLUTIONS TO SELECTED END-OF-CHAPTER PROBLEMS

5. Suppose the current in a conductor decreases exponentially with time according to the equation $I(t) = I_0 e^{-t/\tau}$, where I_0 is the initial current (at $t = 0$) and τ is a constant having dimensions of time. Consider a fixed observation point within the conductor. (a) How much charge passes this point between $t = 0$ and $t = \tau$? (b) How much charge passes this point between $t = 0$ and $t = 10\tau$? (c) **What If?** How much charge passes this point between $t = 0$ and $t = \infty$?

Solution

Conceptualize: The exponential function of Euler's number to the negative power t/τ describes a current starting with its maximum value and decreasing to 36.9% of its previous value in every time interval τ thereafter. The answer to (b) must be larger than the answer to (a), but less than ten times larger. It is not obvious that the answer to (c) is finite.

Categorize: If the current were constant, we would just multiply its value by a time interval to find the charge passing in that time. Here the current is always changing, but we can use its definition to find the charge by integration.

Analyze: From $I = \dfrac{dQ}{dt}$, we have $dQ = I\,dt$.

From this, we derive the general integral $\quad Q = \int dQ = \int I\,dt$

In all three cases, define an end-time, T. $\quad Q = \int_0^T I_0 e^{-t/\tau}dt$

Integrating from time $t = 0$ to time $t = T$, $\quad Q = \int_0^T \left(-I_0\tau\right)e^{-t/\tau}\left(-\dfrac{dt}{\tau}\right)$

We perform the integral and set $Q = 0$ at $t = 0$

to obtain $\quad Q = -I_0\tau\left(e^{-T/\tau} - e^0\right) = I_0\tau\left(1 - e^{-T/\tau}\right)$

(a) If $T = \tau$, $\quad Q = I_0\tau(1 - e^{-1}) = 0.632\,1\,I_0\tau$ ■

(b) If $T = 10\tau$, $\quad Q = I_0\tau(1 - e^{-10}) = 0.999\,95\,I_0\tau$ ■

(c) If $T = \infty$, $\quad Q = I_0\tau(1 - e^{-\infty}) = I_0\tau$ ■

Finalize: The current decreases steeply enough in time that the total charge it transports is finite. In fact, a large majority of the charge is transported in the first "time constant" τ and nearly all of the charge in a time of 10τ. We will see in the next chapter that a capacitor discharging through a resistor produces a current with just this pattern of change in time.

---

11. The electron beam emerging from a certain high-energy electron accelerator has a circular cross section of radius 1.00 mm. (a) The beam current is 8.00 μA. Find the current density in the beam assuming it is uniform throughout. (b) The speed of the electrons is so close to the speed of light that their speed can be taken as 300 Mm/s with negligible error. Find the electron density in the beam. (c) Over what time interval does Avogadro's number of electrons emerge from the accelerator?

Solution

Conceptualize: The current density can be several amps per square meter, as for a small current in a metal wire. In a metal wire the electron density is high and the speed is low, but here the speed is high and the electron density is very low. Avogadro's number is so large that a long time will be required for that number of electrons to go past.

Categorize: This current is not in a wire, but we can use the same ideas of electron density and current density.

Analyze:

(a) The current density is

$$J = \frac{I}{A} = \frac{8.00 \times 10^{-6}\text{ A}}{\pi\left(1.00 \times 10^{-3}\text{ m}\right)^2} = 2.55\text{ A/m}^2 \qquad \blacksquare$$

(b) From $J = nev_d$, we have

$$n = \frac{J}{ev_d} = \frac{2.55\text{ A/m}^2}{\left(1.60 \times 10^{-19}\text{ C}\right)\left(3.00 \times 10^{8}\text{ m/s}\right)}$$

$$n = 5.31 \times 10^{10}\text{ m}^{-3} \qquad \blacksquare$$

(c) From $I = \Delta Q/\Delta t$, we have

$$\Delta t = \frac{\Delta Q}{I} = \frac{N_A e}{I} = \frac{\left(6.02 \times 10^{23}\right)\left(1.60 \times 10^{-19}\text{ C}\right)}{8.00 \times 10^{-6}\text{ A}}$$

$$\Delta t = 1.20 \times 10^{10}\text{ s (or about 382 years!)} \qquad \blacksquare$$

Finalize: In answer (b), tens of billions of electrons per cubic meter counts as a low density, compared to the density in matter. The problem mentions a high-energy accelerator, but it could be about the electron beam in a television picture tube. An electron-beam tube, a chemical battery, a fuel cell, metal wires, and an electrochemical plating tank can all be in the same electric circuit, carrying the same current, described everywhere by nqv_dA, with wildly different values for the factors in this expression.

---

14. A 0.900-V potential difference is maintained across a 1.50-m length of tungsten wire that has a cross-sectional area of 0.600 mm^2. What is the current in the wire?

Solution

Conceptualize: We expect a current of many milliamps, to noticeably warm the wire.

Categorize: We will use just the definition of electric resistance and the identification of how resistance depends on the length and cross-sectional area of a conductor.

Analyze: The definition of resistance is in $I = \dfrac{\Delta V}{R}$ where $R = \dfrac{\rho\ell}{A}$.

Therefore, $I = \dfrac{\Delta VA}{\rho\ell} = \dfrac{(0.900\ \text{V})(6.00 \times 10^{-7}\ \text{m}^2)}{(5.6 \times 10^{-8}\ \Omega\cdot\text{m})(1.50\ \text{m})} = 6.43\text{A}.$ ■

Finalize: These numbers could apply to a simple tabletop experiment or a practical warming appliance. It was just this sort of experiment that Georg Simon Ohm performed, treating the applied voltage and the conductor's dimensions and material as the independent variables. We can refer to the wire equally well as a resistor or a conductor.

---

15. Suppose you wish to fabricate a uniform wire from 1.00 g of copper. If the wire is to have a resistance of $R = 0.500\ \Omega$ and all the copper is to be used, what must be (a) the length and (b) the diameter of this wire?

Solution

Conceptualize: Copper is such a good conductor (its resistivity is so low) that we expect a very long thin wire.

Categorize: We will just use the identification $R = \rho\ell/A$ of how resistance depends on dimensions, together with the definition of density.

Analyze: Don't mix up symbols! Call the density ρ_d and the resistivity ρ_r. From $\rho_d = m/V$, the volume is $V = A\ell = m/\rho_d$. The resistance is $R = \rho_r\ell/A$.

(a) We can solve for ℓ by eliminating A:

$$A = \frac{m}{\ell\rho_d} \qquad R = \frac{\rho_r\ell}{m/\ell\rho_d} \qquad R = \frac{\rho_r\rho_d\ell^2}{m}$$

$$\ell = \sqrt{\frac{mR}{\rho_r\rho_d}} = \sqrt{\frac{(1.00 \times 10^{-3}\ \text{kg})(0.500\ \Omega)}{(1.70 \times 10^{-8}\ \Omega\cdot\text{m})(8.92 \times 10^3\ \text{kg/m}^3)}} = 1.82\ \text{m}$$ ■

(b) To have a single diameter, the wire has a circular cross section:

$$A = \pi r^2 = \pi\left(\frac{d}{2}\right)^2 = \frac{m}{\ell\rho_d}$$

$$d = \sqrt{\frac{4m}{\pi\ell\rho_d}} = \left[\frac{4(1.00 \times 10^{-3}\ \text{kg})}{\pi(1.82\ \text{m})(8.92 \times 10^3\ \text{kg/m}^3)}\right]^{1/2} = 2.80 \times 10^{-4}\ \text{m} = 0.280\ \text{mm}$$ ■

Finalize: The wires you use in lab are short and thick compared to this one, so their resistance is generally quite negligible—do not blame odd experimental results on resistance in the wires. Observe in Table 27.2 the fantastically wide range of resistivity values from good conductors to good insulators.

---

19. If the magnitude of the drift velocity of free electrons in a copper wire is 7.84×10^{-4} m/s, what is the electric field in the conductor?

Solution

Conceptualize: For electrostatic cases, we learned that the electric field inside a conductor is always zero. On the other hand, if there is a current, a non-zero electric field must be maintained by a battery or other source to make the charges flow. Therefore, we might expect the electric field to be small, but definitely **not** zero.

Categorize: The drift velocity of the electrons can be used to find the current density, which can be used with the definition of conductivity to find the electric field inside the conductor.

Analyze: We choose to find first the electron density in copper. Its mass density is 8.92 g/cm^3 and from a periodic table its molar mass is 63.5 g/mol, so the number density of atoms is

$$n = 8.92 \frac{\text{g}}{\text{cm}^3}\left(\frac{1\text{ mol}}{63.5\text{ g}}\right)\left(\frac{6.02\times 10^{23}\text{ atoms}}{1\text{ mol}}\right)\left(\frac{1\times 10^{6}\text{ cm}^3}{1\text{ m}^3}\right) = 8.46\times 10^{28}/\text{m}^3$$

We assume that each atom contributes one conduction electron. Then the density of conduction electrons is this same number.

The current density in this wire is then

$$J = nqv_d = (8.46 \times 10^{28}\text{ e}^-/\text{m}^3)(1.60 \times 10^{-19}\text{ C/e}^-)(7.84 \times 10^{-4}\text{ m/s})$$

$$J = 1.06 \times 10^{7}\text{ A/m}^2$$

The definition of conductivity can be stated as $J = \sigma E = E/\rho$

where $\rho = 1.7 \times 10^{-8}\ \Omega \cdot \text{m}$ for copper,

so then $E = \rho J = (1.70 \times 10^{-8}\ \Omega \cdot \text{m})(1.06 \times 10^{7}\text{ A/m}^2) = 0.180\text{ V/m}$ ■

Finalize: This electric field is certainly smaller than typical static values outside charged objects. The direction of the electric field should be along the length of the conductor, for otherwise the electrons would be forced to leave the wire! In fact, excess charges arrange themselves on the surface of the wire, notably at bends in it, to create an electric field that "steers" the free electrons to flow along the length of the wire from low to high potential (opposite the direction of a positive test charge). It is also interesting to note that when the electric field is being established it travels at the speed of light; but the drift velocity of the electrons is literally a "snail's pace"!

---

23. What is the fractional change in the resistance of an iron filament when its temperature changes from 25.0°C to 50.0°C?

Solution

Conceptualize: We are warming the material from 298 to 323 K, so we might guess a resistance increase of eight percent.

Categorize: We apply directly the definition of temperature coefficient of resistivity.

Analyze: If we ignore thermal expansion, the change in the material's resistivity with temperature $\rho = \rho_0 [1 + \alpha\Delta T]$ implies that the change in resistance is $R - R_0 = R_0\alpha\Delta T$. The fractional change in resistance is defined by $f = (R - R_0)/R_0$.

Therefore, $f = \dfrac{R_0\alpha\Delta T}{R_0} = \alpha\Delta T = (5.00 \times 10^{-3}\,°\text{C}^{-1})(50.0\,°\text{C} - 25.0°\text{C}) = 0.125$ ■

Finalize: It is close to 0.08, but not quite equal. The resistivity is not just proportional to the absolute temperature. The metal hangs together over such a wide range of temperatures and resistance is so easy to measure that an electric-resistance thermometer is a very useful device for measuring temperatures.

---

25. An aluminum wire with a diameter of 0.100 mm has a uniform electric field of 0.200 V/m imposed along its entire length. The temperature of the wire is 50.0°C. Assume one free electron per atom. (a) Use the information in Table 27.2 and determine the resistivity of aluminum at this temperature. (b) What is the current density in the wire? (c) What is the total current in the wire? (d) What is the drift speed of the conduction electrons? (e) What potential difference must exist between the ends of a 2.00-m length of the wire to produce the stated electric field?

Solution

Conceptualize: Aluminum is a very good conductor but this wire is quite thin. We expect a current of some milliamps driven by the (0.2 V/m)(2 m) = 0.4 V potential difference. We expect a drift speed in meters per hour.

Categorize: We will use the definition of the temperature coefficient of resistivity. Knowing that conductivity = 1/resistivity, we can find the current density directly from the applied field.

Analyze:

(a) The resistivity is computed from $\rho = \rho_0\left[1 + \alpha\left(T - T_0\right)\right]$

$$\rho = \left(2.82 \times 10^{-8}\ \Omega\cdot\text{m}\right)\left[1 + \left(3.90 \times 10^{-3}\,°\text{C}^{-1}\right)\left(30.0°\text{C}\right)\right] = 3.15 \times 10^{-8}\ \Omega\cdot\text{m}$$ ■

(b) The current density is

$$J = \sigma E = \frac{E}{\rho} = \left(\frac{0.200\ \text{V/m}}{3.15 \times 10^{-8}\ \Omega\cdot\text{m}}\right)\frac{1\ \Omega\cdot\text{A}}{\text{V}} = 6.35 \times 10^{6}\ \text{A/m}^2$$ ■

(c) The current density is related to the current by $J = \frac{I}{A} = \frac{I}{\pi r^2}$

$$I = J\pi r^2 = (6.35 \times 10^6 \text{ A/m}^2)\,[\pi(5.00 \times 10^{-5} \text{ m})^2] = 49.9 \text{ mA}$$ ■

(d) The mass density gives the number-density of free electrons; we assume that each atom donates one conduction electron:

$$n = \left(\frac{2.70 \times 10^3 \text{ kg}}{\text{m}^3}\right)\left(\frac{1 \text{ mol}}{26.98 \text{ g}}\right)\left(\frac{10^3 \text{ g}}{\text{kg}}\right)\left(\frac{6.02 \times 10^{23} \text{ free e}^-}{1 \text{ mol}}\right) = 6.02 \times 10^{28} \text{ e}^-/\text{m}^3$$

Now $J = nqv_d$ gives the drift speed as

$$v_d = \frac{J}{nq} = \frac{6.35 \times 10^6 \text{ A/m}^2}{(6.02 \times 10^{28} \text{ e}^-/\text{m}^3)(-1.60 \times 10^{-19} \text{ C/e}^-)} = -6.59 \times 10^{-4} \text{ m/s}$$ ■

The sign indicates that the electrons drift opposite to the field and current.

(e) The applied voltage is $\Delta V = E\ell = (0.200 \text{ V/m})(2.00 \text{ m}) = 0.400 \text{ V}$. ■

Finalize: It would be natural to compute the resistance of the two-meter strand as 8.02 Ω, but our set of equations is overcomplete and makes that identification optional. Observe that a "typical tabletop" current density in a metal wire is in megamperes per square meter. Aluminum and copper, quite different in mass density, have very similar densities of conduction electrons, on the order of 10^{29} per cubic meter.

---

33. A 100-W lightbulb connected to a 120-V source experiences a voltage surge that produces 140 V for a moment. By what percentage does its power output increase? Assume its resistance does not change.

Solution

Conceptualize: The voltage increases by about 20%, but since $P = (\Delta V)^2/R$, the power will increase as the square of the voltage:

$$\frac{P_f}{P_i} = \frac{(\Delta V_f)^2/R}{(\Delta V_i)^2/R} = \frac{(140 \text{ V})^2}{(120 \text{ V})^2} = 1.361 \text{ or a } 36.1\% \text{ increase}$$

Categorize: We have already found an answer to this problem by reasoning in terms of ratios, but we can also calculate the power explicitly for the bulb and compare with the original power by using the definition of resistance and the equation for electrical power. To find the power, we must first find the resistance of the bulb, which should remain roughly constant during the power surge (we can check the validity of this assumption later).

Analyze: From $P = (\Delta V)^2/R$, we find that $R = \frac{(\Delta V_i)^2}{P} = \frac{(120 \text{ V})^2}{100 \text{ W}} = 144\ \Omega$

The final current is $I_f = \dfrac{\Delta V_f}{R} = \dfrac{140\ \text{V}}{144\ \Omega} = 0.972\ \text{A}$

The power during surge is $P = \dfrac{\left(\Delta V_f\right)^2}{R} = \dfrac{(140\ \text{V})^2}{144\ \Omega} = 136\ \text{V}$

So the percentage increase is $\dfrac{136\ \text{W} - 100\ \text{W}}{100\ \text{W}} = 0.361 = 36.1\%$ ■

Finalize: Our result tells us that this 100-W light bulb momentarily acts like a 136-W light bulb, which explains why it would suddenly get brighter. Some electronic devices (like computers) are sensitive to voltage surges like this, which is the reason that **surge protectors** are recommended to protect these devices from being damaged.

In solving this problem, we assumed that the resistance of the bulb did not change during the voltage surge. We can check this assumption. Let us assume that the filament is made of tungsten and that its resistance will change linearly with temperature according to Equation 27.19. Let us further assume that the increased voltage lasts for a time interval long enough so that the filament comes to a new steady-state temperature. The temperature change can be estimated from the power surge according to Stefan's law (Equation 20.19), assuming that all the power loss is due to radiation. By this law, the absolute temperature is proportional to the fourth root of the power, so that a 36% change in power should correspond to only about an 8% increase in temperature. A typical operating temperature of a white light bulb is about 3 000°C, so $\Delta T \approx 0.08(3\ 273\ \text{K}) = 260°\text{C}$. Then the increased resistance would be roughly

$$R = R_0[1 + \alpha(T - T_0)] = (144\ \Omega)[1 + (4.5 \times 10^{-3})(260)] \approx 310\ \Omega$$

It appears that the resistance could double from 144 Ω. The assumption of constant resistance would be a poor model. On the other hand, if the voltage surge lasts only a very short time, the 136 W we calculated originally accurately describes the conversion of electrically transmitted energy into internal energy in the filament.

---

34. Review. A well-insulated electric water heater warms 109 kg of water from 20.0°C to 49.0°C in 25.0 min. Find the resistance of its heating element, which is connected across a 240-V potential difference.

Solution

Conceptualize: Everyone wants to take a shower as soon as you all get home from a camping trip. The water heater should have high power compared to other household appliances. For the power to be some kilowatts, the current should be several amperes and the resistance only a few ohms.

Categorize: We will identify the energy output required for the resistor, then its power, and then its resistance.

Analyze: If the tank has good insulation, essentially all of the energy electrically transmitted to the heating element becomes internal energy in the water: $\Delta E_{(\text{internal})} = E_{(\text{electrical})}$ Our symbol $E_{(\text{electrical})}$ represents the same thing as the textbook's T_{ET}, namely electrically transmitted energy.

Since $\Delta E_{(\text{internal})} = mc\Delta T$ and $E_{(\text{electrical})} = P\Delta t = (\Delta V)^2 \Delta t/R$

where $c = 4\ 186$ J/kg·°C

the resistance is

$$R = \frac{(\Delta V)^2 \Delta t}{cm\Delta T} = \frac{(240\ \text{V})^2(1\ 500\ \text{s})}{(4\ 186\ \text{J/kg}\cdot°\text{C})(109\ \text{kg})(29.0°\text{C})} = 6.53\ \Omega$$ ■

Finalize: Any resistor can be called a conductor and can be called a heating element. The one in the water heater has 100% efficiency in converting electrically transmitted energy into internal energy. The current going in comes out again unchanged in amount, but the electrically transmitted energy stays there in the tank of hot water. The current is quite large, 36.8 A, requiring a heavy cable running to the appliance.

If the heating element had lower resistance it would have power still higher than the 8.82 kW of this heater. This behavior may seem strange if you think of lower mechanical resistance as a smaller friction force stopping a certain sliding crate. With less friction, the kinetic energy of the crate would be entirely converted into internal energy over a longer time interval, with lower power. On the other hand, a device with less electrical resistance plugged into the same 240 V carries more current and warms up faster.

---

39. Assuming the cost of energy from the electric company is \$0.110/kWh, compute the cost per day of operating a lamp that draws a current of 1.70 A from a 110-V line.

Solution

Conceptualize: We estimate cheaper than a fancy cup of coffee.

Categorize: This problem is about identifying what we buy from the electric company. It is not charge, current, field, potential, potential difference, resistance, or power, but…

Analyze: …energy. The power of the lamp is $P = I\,\Delta V = U/\Delta t$, where U is the energy transformed. Then the energy you buy, in standard units, is

$$U = \Delta V\, I\, \Delta t = (110\ \text{V})(1.70\ \text{A})(1\ \text{day})\left(\frac{24\ \text{h}}{1\ \text{day}}\right)\left(\frac{3\ 600\ \text{s}}{\text{h}}\right)\left(\frac{1\ \text{J}}{\text{V}\cdot\text{C}}\right)\left(\frac{1\ \text{C}}{\text{A}\cdot\text{s}}\right) = 16.2\ \text{MJ}$$

In kilowatt hours, the energy is

$$U = \Delta V\, I\, \Delta t = (110\ \text{V})(1.70\ \text{A})(1\ \text{day})\left(\frac{24\ \text{h}}{1\text{day}}\right)\left(\frac{\text{J}}{\text{V}\cdot\text{C}}\right)\left(\frac{\text{C}}{\text{A}\cdot\text{s}}\right)\left(\frac{\text{W}\cdot\text{s}}{\text{J}}\right) = 4.49\ \text{kWh}$$

So operating the lamp costs (4.49 kWh)(\$0.110/kWh) = 49.4 cents. ■

Finalize: This is a 187-W bulb at 110 V, so it might be sold in the hardware store as a 200-W bulb at 120 V. The good news is that it is not expensive to operate. If your goal is saving energy and money, the bad news is that it does not help a lot to avoid using an electric pencil sharpener or corn popper. The heavy hitters in household energy use may be the furnace, air conditioner, water heater, and refrigerator; but your car can dwarf them all.

---

45. A certain toaster has a heating element made of Nichrome wire. When the toaster is first connected to a 120-V source (and the wire is at a temperature of 20.0°C), the initial current is 1.80 A. The current decreases as the heating element warms up. When the toaster reaches its final operating temperature, the current is 1.53 A. (a) Find the power delivered to the toaster when it is at its operating temperature. (b) What is the final temperature of the heating element?

Solution

Conceptualize: Most toasters are rated at about 1 000 W (usually stamped on the bottom of the unit), so we might expect this one to have a similar power rating. The temperature of the heating element should be hot enough to toast bread but low enough that the nickel-chromium alloy element does not melt. (The melting point of nickel is 1 455°C, and chromium melts at 1 907°C.)

Categorize: The power can be calculated directly by multiplying the current and the voltage. The temperature can be found from the linear resistivity equation for Nichrome, with $\alpha = 0.4 \times 10^{-3}\ ^\circ\text{C}^{-1}$ from Table 27.2.

Analyze:

(a) The power at the operating point is

$$P = \Delta V I = (120\text{ V})(1.53\text{ A}) = 184\text{ W} \qquad \blacksquare$$

(b) The resistance at 20.0°C is $R_0 = \dfrac{\Delta V}{I} = \dfrac{120\text{ V}}{1.80\text{ A}} = 66.7\ \Omega$

At the operating temperature, $R = \dfrac{120\text{ V}}{1.53\text{ A}} = 78.4\ \Omega$

Neglecting thermal expansion,

we have
$$R = \frac{\rho\ell}{A} = \frac{\rho_0\left[1+\alpha\left(T - T_0\right)\right]\ell}{A} = R_0\left[1+\alpha\left(T - T_0\right)\right]$$

$$T = T_0 + \frac{R/R_0 - 1}{\alpha} = 20.0^\circ\text{C} + \frac{78.4\ \Omega/66.7\ \Omega - 1}{0.4 \times 10^{-3}\,^\circ\text{C}^{-1}} = 461^\circ\text{C} \qquad \blacksquare$$

Finalize: Although this toaster appears to use significantly less power than most, the temperature seems high enough to toast a piece of bread in a reasonable amount of time. The absolute temperature of the filament in a typical 1 000-W toaster would be much less than five times higher because Stefan's radiation law (Equation 20.19) tells us that (assuming all power is lost through radiation) $T \propto \sqrt[4]{P}$, so that the temperature might be about 850°C. In either case, the operating temperature is well below the melting point of the heating element.

---

53. A charge Q is placed on a capacitor of capacitance C. The capacitor is connected into the circuit shown in Figure P27.53, with an open switch, a resistor, and an initially uncharged capacitor of capacitance $3C$. The switch is then closed, and the circuit comes to equilibrium. In terms of Q and C, find (a) the final potential difference between the plates of each capacitor, (b) the charge on each capacitor, and (c) the final energy stored in each capacitor. (d) Find the internal energy appearing in the resistor.

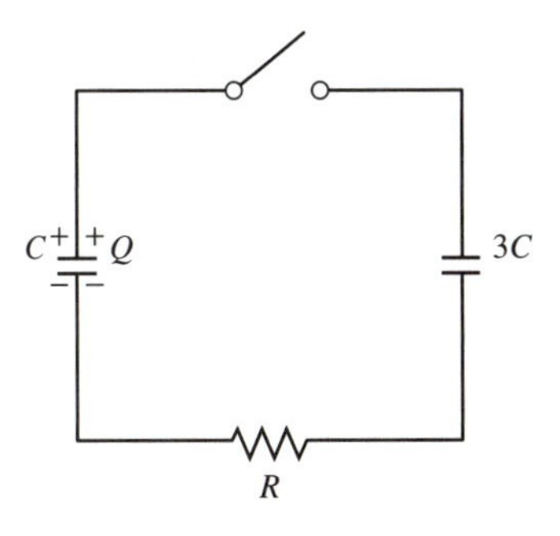

Figure P27.53

Solution

Conceptualize: When the charged capacitor is connected to the uncharged capacitor, charge moves in an electric current. The current in the resistor converts energy from electric potential energy into internal energy; we expect the amount to be proportional to the square of Q.

Categorize: We will use conservation of charge for the circuit as a system, with the definition of capacitance, to find the final potential difference. Then we can identify the initial and final electric potential energies of the capacitors. The internal energy appearing in the resistor must account for the difference. We assume that negligible energy is carried away by electromagnetic radiation.

Analyze: The original stored energy is $U_i = \frac{1}{2}Q\Delta V_i = \frac{1}{2}\frac{Q^2}{C}$.

(a) After the switch has been closed and the charge has come to equilibrium, electric current stops. The potential difference across the resistor is zero. The lower ends of the two capacitors are at the same potential. The upper ends of the two capacitors, connected directly through the switch, are also at the same potential. Then the two capacitors are in parallel. Immediately after the switch is closed, charge Q distributes itself over the plates of C and $3C$ in parallel, presenting equivalent capacitance $4C$.

Then the final potential difference is $\Delta V_f = \dfrac{Q}{4C}$ for both capacitors. ■

(b) The smaller capacitor then carries charge $C\Delta V_f = \dfrac{Q}{4C}C = \dfrac{Q}{4}$. ■

The larger capacitor carries charge $3C\dfrac{Q}{4C} = \dfrac{3Q}{4}$. ■

(c) The smaller capacitor stores final energy

$$\frac{1}{2}C\left(\Delta V_f\right)^2 = \frac{1}{2}C\left(\frac{Q}{4C}\right)^2 = \frac{Q^2}{32C}$$ ■

The larger capacitor possesses energy $\frac{1}{2}3C\left(\frac{Q}{4C}\right)^2 = \frac{3Q^2}{32C}$. ■

(d) The total final energy is $\frac{Q^2}{32C} + \frac{3Q^2}{32C} = \frac{Q^2}{8C}$. The loss of potential energy is the energy appearing as internal energy in the resistor:

$$\frac{Q^2}{2C} = \frac{Q^2}{8C} + \Delta E_{\text{int}} \qquad \Delta E_{\text{int}} = \frac{3Q^2}{8C}$$ ■

Finalize: The parts of the problem about the charge redistribution could have been assigned in Chapter 26. In Chapter 28 we will study the time variation of the current. Solving this problem in this chapter emphasizes the contrasting things that capacitors and resistors do with energy: a capacitor stores electric potential energy and a resistor converts it into internal energy.

---

54. An experiment is conducted to measure the electrical resistivity of Nichrome in the form of wires with different lengths and cross-sectional areas. For one set of measurements, a student uses 30-gauge wire, which has a cross-sectional area of 7.30×10^{-8} m^2. The student measures the potential difference across the wire and the current in the wire with a voltmeter and an ammeter, respectively. (a) For each set of measurements given in the table taken on wires of three different lengths, calculate the resistance of the wires and the corresponding values of the resistivity. (b) What is the average value of the resistivity? (c) Explain how this value compares with the value given in Table 27.2.

L (m)	ΔV (V)	I (A)	R (Ω)	ρ (Ω · m)
0.540	5.22	0.72		
1.028	5.82	0.414		
1.543	5.94	0.281		

Solution

Conceptualize: With material and cross-sectional area constant, we expect resistance to be proportional to length. It should be about twice as large for the second sample compared to the first, and three times as large for the third. The resistivity should be about the same for all three. The amount of scatter will suggest the experimental uncertainty.

Categorize: For each row in the table, we will first find the resistance from $R = \Delta V/I$, and then find the resistivity from $R = \rho\,\ell/A$ or $\rho = RA/\ell$.

Analyze:

(a) In the first row,

$$R = \frac{\Delta V}{I} = \frac{5.22\text{ V}}{0.72\text{ A}} = 7.25\ \Omega$$

and $\rho = \frac{RA}{\ell} = \frac{(7.25\ \Omega)(7.30 \times 10^{-8}\text{ m}^2)}{0.540\text{ m}} = 9.80 \times 10^{-7}\ \Omega \cdot \text{m}$

Applying this method to each entry, we obtain:

L(m)	$R(\Omega)$	$\rho(\Omega \cdot \text{m})$
0.540	7.25	9.80×10^{-7}
1.028	14.1	9.98×10^{-7}
1.543	21.1	1.00×10^{-6}

■

(b) Thus the average resistivity is $\rho = 9.93 \times 10^{-7}\ \Omega \cdot \text{m}$. ■

(c) This differs from the tabulated $1.00 \times 10^{-6}\ \Omega \cdot \text{m}$ by 0.7%. The difference is accounted for by the experimental uncertainty, which we may estimate as $(1.00 - 0.98)/1.00 = 2\%$. ■

Finalize: We have estimated the *amount* of the experimental uncertainty. If we looked at the apparatus and procedure, we might form hypotheses about the *causes* of the uncertainty, and these could suggest ways to reduce the uncertainty.

---

56. An all-electric car (not a hybrid) is designed to run from a bank of 12.0-V batteries with total energy storage of 2.00×10^7 J. If the electric motor draws 8.00 kW as the car moves at a steady speed of 20.0 m/s, (a) what is the current delivered to the motor? (b) How far can the car travel before it is "out of juice"?

Solution

Conceptualize: We guess a current on the order of a hundred amperes. A practical car should travel on the order of a hundred km on one charge.

Categorize: We need not know details about the car's construction to follow the electrically transmitted input and the mechanical output of the motor. (We will study motors in Chapters 29 and 31.)

Analyze:

(a) Since $P = I\Delta V$ we have $I = \frac{P}{\Delta V} = \frac{8.00 \times 10^3\text{ W}}{12.0\text{ V}} = 667\text{ A}$ ■

(b) From $P = U/\Delta t$, the time the car runs is

$$\Delta t = \frac{U}{P} = \left(\frac{2.00 \times 10^7\ \text{J}}{8.00 \times 10^3\ \text{W}}\right)\left(\frac{1\ \text{W}\cdot\text{s}}{\text{J}}\right) = 2.50 \times 10^3\ \text{s}$$

So it moves a distance of

$$\Delta x = v\Delta t = (20.0\ \text{m/s})(2.50 \times 10^3\ \text{s}) = 50.0\ \text{km} \qquad \blacksquare$$

Finalize: A fifty-kilometer range can be sufficient for commuting and local deliveries, but this car does not sound very practical. A current of hundreds of amperes is hard to handle. Connecting some of the batteries in series to run the motor at higher voltage could let it run on less current.

---

72. Material with uniform resistivity ρ is formed into a wedge as shown in Figure P27.72. Show that the resistance between face A and face B of this wedge is

$$R = \rho\frac{L}{w(y_2 - y_1)}\ln\left(\frac{y_2}{y_1}\right)$$

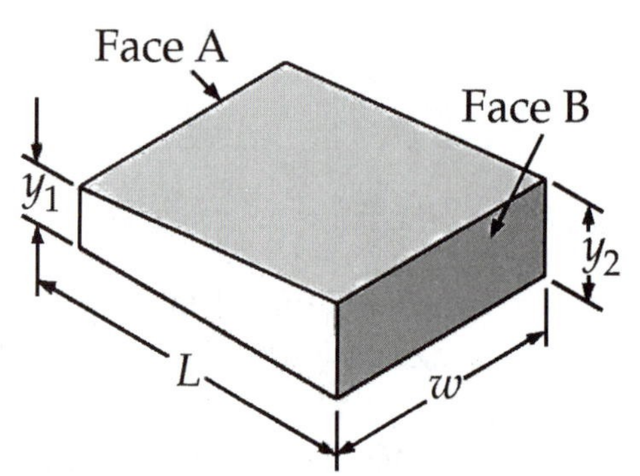

Figure P27.72

Solution

Conceptualize: The proportionalities of resistance to resistivity and to length L are reasonable. The inverse proportionality to width w is reasonable, because a smaller width crowds current into a smaller cross-sectional area. As for the y's, see below.

Categorize: We will apply an incremental version of $R = \rho L/A$ to one vertical slice of the wedge, and then add up the whole resistance of all the slices.

Analyze: The current flows generally parallel to L. Consider a slice of the material perpendicular to this current, of thickness dx, and at distance x from face A. Then the other dimensions of the slice are w and y,

where by proportion $\dfrac{y - y_1}{x} = \dfrac{y_2 - y_1}{L}$ so $y = y_1 + (y_2 - y_1)\dfrac{x}{L}$

The bit of resistance which this slice contributes is

$$dR = \frac{\rho dx}{A} = \frac{\rho dx}{wy} = \frac{\rho dx}{w\left(y_1 + \left(y_2 - y_1\right)\left(x/L\right)\right)}$$

The whole resistance is that of all the slices:

$$R = \int_{x=0}^{L} dR = \int_0^L \frac{\rho dx}{w\left(y_1 + \left(y_2 - y_1\right)\left(x/L\right)\right)}$$

$$= \frac{\rho}{w}\frac{L}{y_2 - y_1}\int_{x=0}^{L}\frac{\left((y_2 - y_1)/L\right)dx}{y_1 + (y_2 - y_1)(x/L)}$$

With $u = y_1 + (y_2 - y_1)\frac{x}{L}$ this is of the form $\int \frac{du}{u}$,

so $R = \frac{\rho L}{w(y_2 - y_1)}\ln\left[y_1 + (y_2 - y_1)(x/L)\right]_{x=0}^{L}$

$$R = \frac{\rho L}{w(y_2 - y_1)}(\ln y_2 - \ln y_1) = \frac{\rho L}{w(y_2 - y_1)}\ln\frac{y_2}{y_1}$$ ■

Finalize: We have the result. It may appear strange in the limit $y_2 - y_1 = 0$. It first looks as if the derived equation goes to infinity in this case, but in the original picture this situation makes the wedge into a rectangular brick and its resistance should be $\rho L/wy$. To expose an error or to resolve the apparent conflict we write $y_2 = y_1 + \delta$. Then

$$R = \frac{\rho L}{w\delta}\ln\left(\frac{y_1 + \delta}{y_1}\right) = \frac{\rho L}{w\delta}\ln\left(1 + \frac{\delta}{y_1}\right) = \frac{\rho L}{w\delta}\left(\frac{\delta}{y_1} - \frac{1}{2}\left(\frac{\delta}{y_1}\right)^2 + \frac{1}{3}\left(\frac{\delta}{y_1}\right)^3 - \ldots\right)$$

In the limit $\delta \to 0$ this becomes $R = \rho L\delta/w\delta y_1 = \rho L/wy_1$ as it should.

---

74. A more general definition of the temperature coefficient of resistivity is

$$\alpha = \frac{1}{\rho}\frac{d\rho}{dT}$$

where ρ is the resistivity at temperature T. (a) Assuming α is constant, show that

$$\rho = \rho_0 e^{\alpha(T - T_0)}$$

where ρ_0 is the resistivity at temperature T_0. (b) Using the series expansion $e^x \approx 1 + x$ for $x << 1$, show that the resistivity is given approximately by the expression $\rho = \rho_0[1 + \alpha(T - T_0)]$ for $\alpha(T - T_0) << 1$.

Solution

Conceptualize: The idea of a constant fractional rate of change of resistance with temperature is one model. The idea of a constant slope for the resistance-versus-temperature graph is another model. They are different in extreme cases but this problem will show that the two models can be consistent for small temperature changes.

Categorize: From the given equation for $d\rho/dT$ we are to evaluate the function $\rho(T)$. This means we are to solve a differential equation. Much of advanced physics proceeds in this pattern, as well as mathematical modeling in other fields.

Analyze:

(a) We are given $\alpha = \frac{1}{\rho}\frac{d\rho}{dT}$

Separating variables, $\int_{\rho_0}^{\rho} \frac{d\rho}{\rho} = \int_{T_0}^{T} \alpha \, dT$. On both sides we integrate from the physical situation at temperature T_0 to that at temperature T.

Integrating both sides, $\ln(\rho/\rho_0) = \alpha(T - T_0)$

Thus $\rho = \rho_0 e^{\alpha(T-T_0)}$ ■

(b) From the series expansion $e^x \approx 1 + x$, with x much less than 1,

$$\rho = \rho_0\left[1 + \alpha\left(T - T_0\right)\right]$$ ■

Finalize: The exponential function does not necessarily fit experimental data better than the linear function. The textbook's Figure 27.9 suggests that the linear-function model can be quite good.

28

Direct Current Circuits

EQUATIONS AND CONCEPTS

The **terminal voltage of a battery** will be less than the emf when the battery is providing current to an external circuit. This is due to the **internal resistance** of the battery. *The terminal voltage is equal to the emf when the current is zero (this is the open circuit voltage).*

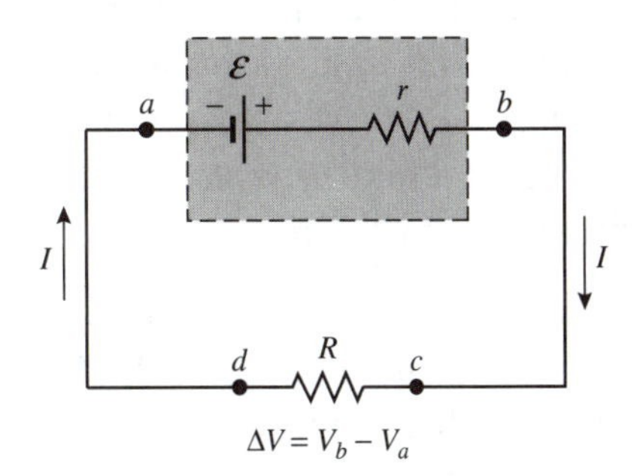

$$\Delta V = \mathcal{E} - Ir \qquad (28.1)$$

The **current** (I) **delivered by a battery** in a simple dc circuit depends on the value of the emf of the source ($\mathcal{E}$), the total load resistance in the circuit (R), and the internal resistance of the source (r).

$$I = \frac{\mathcal{E}}{R + r} \qquad (28.3)$$

The **total or equivalent resistance of a series combination of resistors** is equal to the sum of the individual resistances. *The resistors are connected so that there is one common circuit point per adjacent pair of resistors. For resistors combined in series, the current in each resistor equals the current delivered by the battery.*

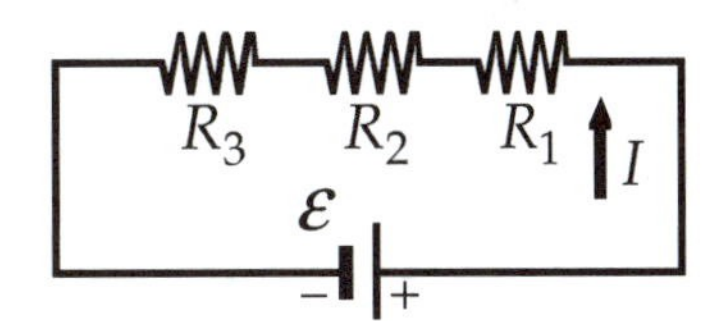

(resistors in series)

$$R_{eq} = R_1 + R_2 + R_3 + \cdots \qquad (28.6)$$

$$I = I_1 = I_2 = I_3 = \ldots$$

The **inverse of the equivalent resistance of several resistors connected in parallel** is equal to the sum of the inverses of the individual resistors. *The equivalent resistance is less than the smallest individual value of resistance in the group. Resistors in parallel are connected so that each resistor has two circuit points in common with all other resistors in the group. The potential difference is the same across each resistor in a parallel group and is equal to the potential difference across the battery.*

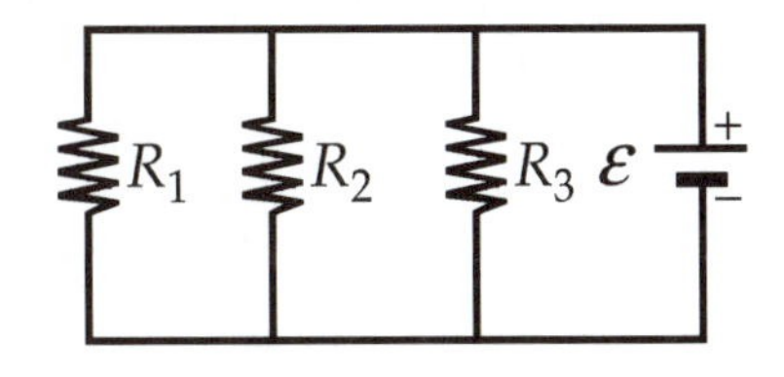

(resistors in parallel)

$$\frac{1}{R_{eq}} = \frac{1}{R_1} + \frac{1}{R_2} + \frac{1}{R_3} + \cdots \qquad (28.8)$$

$$\Delta V = \Delta V_1 = \Delta V_2 = \Delta V_3 = \ldots$$

Kirchhoff's rules can be used to analyze more complex multiloop circuits which cannot be reduced to a single loop by use of Ohm's law and the rules for combining resistors in series and parallel. *Review the procedure for applying Kirchhoff's rules as outlined in Suggestions, Skills, and Strategies.*

Junction rule. The sum of the currents entering any junction in a circuit must equal the sum of the currents leaving that junction.

$$\sum_{junction} I = 0 \qquad (28.9)$$

Loop rule. The algebraic sum of the potential differences across all circuit elements (e.g., resistors and sources of emf) around *any closed loop* in a circuit must be zero.

$$\sum_{\substack{closed\\loop}} \Delta V = 0 \qquad (28.10)$$

When a **capacitor is charged in series with a resistor** (see circuit inset in the figure) a quantity τ, **called the time constant of the circuit,** is used to describe the manner in which the charge on the capacitor varies with time. The charge on the capacitor increases from zero to 63.2% of its maximum value in a time interval equal to one time constant. *Also, during one time constant, the charging current decreases from its initial maximum value of $I_i = \mathcal{E}/R$ to 36.8% of I_i.*

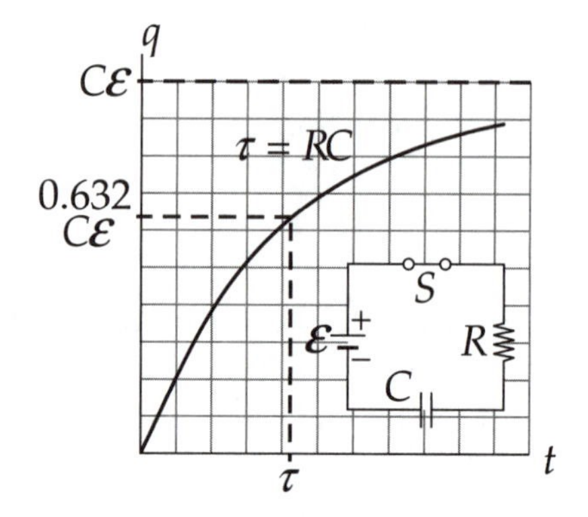

$$q(t) = C\mathcal{E}\left[1 - e^{-t/RC}\right] \qquad (28.14)$$

$$I(t) = \frac{\mathcal{E}}{R}e^{-t/RC} \qquad (28.15)$$

$$\tau = RC \qquad (28.16)$$

When a **capacitor is discharged through a resistor** (see circuit inset in figure), the charge and current decrease exponentially in time. In these equations, the initial current is $I_i = Q/RC$.

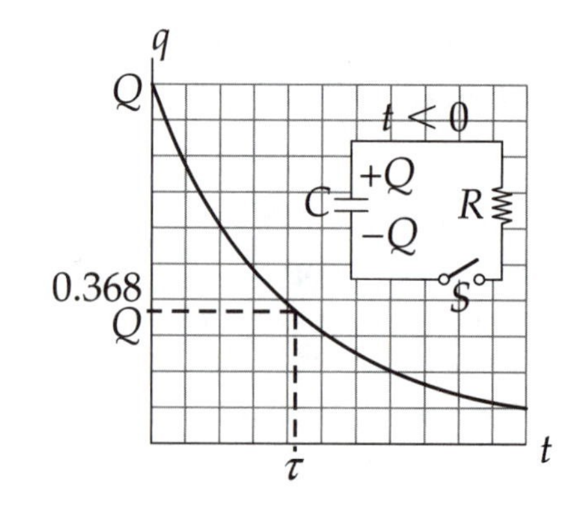

$$q(t) = Qe^{-t/RC} \qquad (28.18)$$

$$I(t) = -\frac{Q}{RC}e^{-t/RC} \qquad (28.19)$$

where $Q/RC = I_i$

SUGGESTIONS, SKILLS, AND STRATEGIES

PROBLEM-SOLVING STRATEGY FOR RESISTORS

- When two or more resistors are connected in series, the current is the same in each resistor; but the potential difference is not the same across each resistor (unless they have equal values of resistance). The resistors add directly to give the equivalent resistance of the series combination.

- When two or more resistors are connected in parallel, the potential difference is the same across each resistor. Since the current is inversely proportional to the resistance, the currents through them are not the same (unless they have equal values of resistance). The equivalent resistance of a parallel combination of resistors is found through reciprocal addition, and the equivalent resistance is always less than the smallest individual resistor.

- A multiloop circuit consisting of a series-parallel combination of resistors can often be reduced to a simple loop circuit containing only one equivalent resistor. To do so, repeatedly examine the circuit and replace any group of series or parallel resistors with a single resistor of equivalent value using Equations 28.6 and 28.8. Sketch a new circuit after each set of changes has been made. Continue this process until a single equivalent resistance is found for the entire circuit.

- If the current through, or the potential difference across, a resistor in the multiloop circuit is to be identified, start with the final equivalent circuit found in the last step above, and gradually work your way back through the circuits, using $\Delta V = IR$ to find the potential difference and current values for each resistor.

STRATEGY FOR USING KIRCHHOFF'S RULES

- First, draw the circuit diagram and assign labels and symbols to all the known and unknown quantities. You must assign directions to the currents in each part of the circuit. Do not be alarmed if you assume the incorrect direction of a current. The resulting value will be negative, but its magnitude will be correct. Although the assignment of current directions is arbitrary, you must stick with your assumption throughout a specific circuit as you apply Kirchhoff's rules.

- Apply the junction rule to all junctions in the circuit except one. *In general, the number of times the junction rule can be used is one fewer than the number of junction points in the circuit.*

- Now apply Kirchhoff's loop rule to as many loops in the circuit as are needed to solve for the unknowns. Remember, you must have as many equations (from applying the junction rule and the loop rule combined) as there are unknowns (values of I, R, and $\mathcal{E}$). *In order to apply the loop rule, you must correctly identify the increase or decrease in potential as you cross each circuit element in traversing the closed loop.* As you review the examples illustrated below, keep in mind the directions you have assumed for the currents, and watch out for signs!

Convenient rules which you may use to determine the increase or decrease in potential as you cross a resistor or source of emf in traversing a circuit loop are illustrated below.

- The potential decreases (changes by $-IR$) when a resistor is traversed in the direction of the current.

  $$V_b - V_a = -IR$$

- The potential increases (changes by $+IR$) when a resistor is traversed in the direction opposite the direction of the current.

  $$V_b - V_a = +IR$$

- The potential increases by $+\mathcal{E}$ when a source of emf is traversed in the direction of the emf (from – to +).

  $$V_b - V_a = \mathcal{E}$$

- The potential decreases by $-\mathcal{E}$ when a source of emf is traversed opposite to the direction of the emf (from + to –).

  $$V_b - V_a = -\mathcal{E}$$

- Finally, you must solve the equations simultaneously for the unknown quantities. Be careful in carrying out the algebraic steps and check your numerical answers for consistency. *Remember, a negative value for a current means that the actual direction of the current is opposite the direction assumed in applying Kirchhoff's rules to the circuit.*

As an illustration of the use of Kirchhoff's rules, consider a three-loop circuit which has the general form shown in the figure at the right. In this illustration, the actual circuit elements, R's and $\mathcal{E}$'s are not shown but assumed known. There are six possible different values of I in the circuit; therefore you will need six independent equations to solve for the six values of I. There are four junction points in the circuit (at points a, d, f, and h). The first rule applied at any three of these points will yield three equations. The circuit can be thought of as a group of three "blocks" as shown in the figure. Kirchhoff's second law, when applied to each of these loops (*abcda*, *ahfga*, and *defhd*), will yield three additional equations.

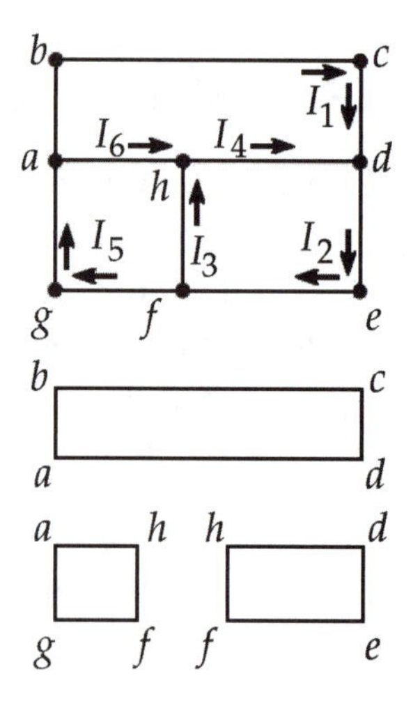

You can then solve the total of six equations simultaneously for the six values of I_1, I_2, I_3, I_4, I_5, and I_6. You can, of course, expect that the sum of the changes in potential difference around **any other closed loop** in the circuit will be zero (for example, *abcdefga* or *ahfedcba*); however, the equations found by applying Kirchhoff's second rule to these additional loops **will not be independent** of the six equations found previously.

REVIEW CHECKLIST

You should be able to:

- Calculate the terminal potential difference and internal resistance of a battery. (Section 28.1)
- Calculate the equivalent resistance of a group of resistors in parallel, series, or series-parallel combination. (Section 28.2)
- Use Ohm's law to calculate the current in a circuit and the potential difference between any two points in a single loop circuit or a circuit that can be reduced to an equivalent single-loop circuit. (Sections 28.1 and 28.2)
- Apply Kirchhoff's rules to solve multiloop circuits; that is, find the current and the potential difference between any two points. (Section 28.3)
- Calculate the charging (discharge) current $I(t)$ and the accumulated (residual) charge $q(t)$ during charging (and discharge) of a capacitor in an RC circuit. (Section 28.4)

ANSWER TO AN OBJECTIVE QUESTION

3. The terminals of a battery are connected across two resistors in series. The resistances of the resistors are not the same. Which of the following statements are correct? Choose all that are correct. (a) The resistor with the smaller resistance carries more current than the other resistor. (b) The resistor with the larger resistance carries less current than the other resistor. (c) The current in each resistor is the same. (d) The potential difference across each resistor is the same. (e) The potential difference is greatest across the resistor closest to the positive terminal.

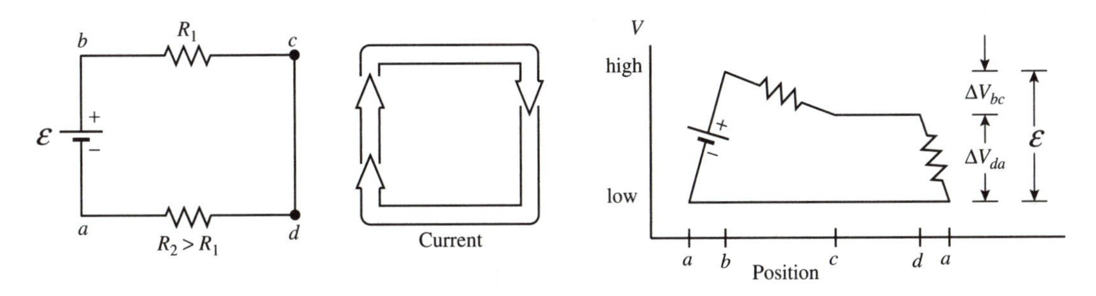

Figure OQ28.3

Answer You should learn the terminology and the standard symbols, so that from the two-sentence description of the circuit in the question, you can draw a circuit diagram, as in the first panel of our diagram here. And get into the habit of reasoning from the circuit diagram. The one correct answer is (c). There are no junctions in the circuit, so the current is the same everywhere, as in the second panel of the diagram, and statements (a) and (b) are false. If the second resistor has (say) twice the resistance of the first, the potential difference across it will be twice as large as across the first resistor. This relationship is described by $\Delta V = IR$, is shown in the third panel of the diagram, and makes statement (d) false. The *potential* is greatest at the positive terminal of the battery and at the end of the resistor connected to it; but the potential *difference* will be greater across the resistor of larger resistance, so statement (e) is false. The first panel of the diagram is the circuit schematic as it is usually drawn. It implies the information shown in the second and third panels of the diagram. Diagrams like these are not usually drawn, but you should learn to visualize them. The second panel shows the path of the current. The third panel is a graph of electric potential versus position in the circuit.

□ □ □ □

ANSWERS TO SELECTED CONCEPTUAL QUESTIONS

1. Is the direction of current in a battery always from the negative terminal to the positive terminal? Explain.

Answer No. If there is one battery in a circuit, the current inside it will be from its negative to its positive terminal. Whenever a battery is delivering electrically transmitted energy to a circuit, it will carry current in this direction. On the other hand, when another source of emf is charging the battery in question, it will have current pushed through it from its positive terminal to its negative terminal. If another source of emf has a certain special value in a particular circuit, our battery may carry no current.

□ □ □ □

3. Why is it possible for a bird to sit on a high-voltage wire without being electrocuted?

Answer The resistance of the short segment of wire between the bird's feet is so small that the potential difference $\Delta V = IR$ between the feet is negligible. In order for the bird to be electrocuted, a much larger potential difference would be required.

It could also be said like this: A small amount of current does go through the bird's legs and body, according to the rule of parallel resistors. However, since the resistance of the bird's feet is much higher than that of the wire between them, the amount of current that exists in the bird is not enough to harm it.

□ □ □ □

6. Referring to Figure CQ28.6, describe what happens to the lightbulb after the switch is closed. Assume the capacitor has a large capacitance and is initially uncharged. Also, assume the light illuminates when connected directly across the battery terminals.

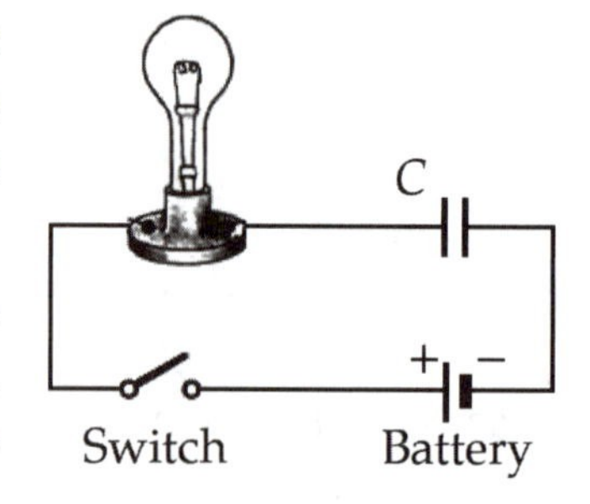

Figure CQ28.6

Answer The bulb will light up for a short time interval while the capacitor is being charged and there is a current in the circuit. As soon as the capacitor approaches full charge, the current in the circuit will drop nearly to zero, and the bulb will cease to glow.

□ □ □ □

8. (a) What advantage does 120-V operation offer over 240 V? (b) What disadvantages does it have?

Answer (a) Both 120-V and 240-V lines can deliver injurious or lethal shocks, but there is a somewhat better factor of safety with the lower voltage. To say it a different way, the insulation on a 120-V wire can be thinner. (b) On the other hand, a 240-V line carries less current to operate a device with the same power, so the conductor itself can be thinner. Finally, as we will see in Chapter 33, the last step-down transformer can also be somewhat smaller if it has to go down only to 240 volts from the high voltage of the main power line.

SOLUTIONS TO SELECTED END-OF-CHAPTER PROBLEMS

1. A battery has an emf of 15.0 V. The terminal voltage of the battery is 11.6 V when it is delivering 20.0 W of power to an external load resistor R. (a) What is the value of R? (b) What is the internal resistance of the battery?

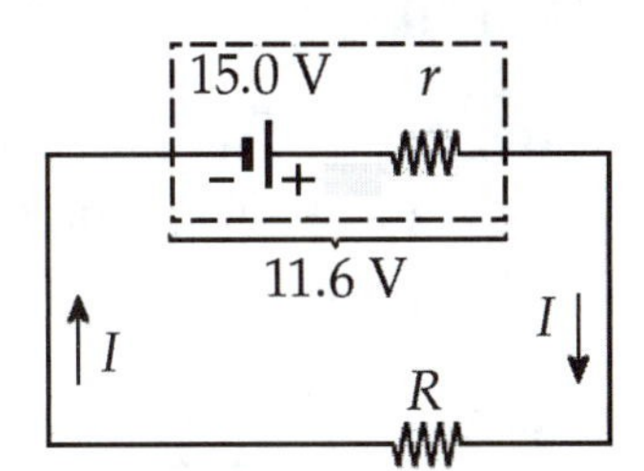

Solution

Conceptualize: The internal resistance of a battery usually is less than 1 Ω, with physically larger batteries having less resistance due to the larger anode and cathode areas. The voltage of this battery drops significantly (23%) when the load resistance is added, so a sizable amount of current must be drawn from the battery. If we assume that the internal resistance is about 1 Ω, then the current must be about 3 A to give the 3.4-V drop across the battery's internal resistance. If this is true, then the load resistance must be about $R \approx 12 \text{ V}/3 \text{ A} = 4\ \Omega$.

Categorize: We can find R precisely by using the power delivered to the load resistor when the voltage is 11.6 V. Then we can find the internal resistance of the battery by summing the electric potential differences around the circuit.

Analyze:

(a) Combining Joule's law, $P = I\Delta V$, and the definition of resistance, $\Delta V = IR$, gives

$$R = \frac{(\Delta V)^2}{P} = \frac{(11.6 \text{ V})^2}{20.0 \text{ W}} = 6.73\ \Omega \qquad \blacksquare$$

(b) The electromotive force of the battery must equal the voltage drops across the resistances: $\mathcal{E} = IR + Ir$, where $I = \Delta V/R$.

$$r = \frac{(\mathcal{E} - IR)}{I} = \frac{(\mathcal{E} - \Delta V)R}{\Delta V} = \frac{(15.0 \text{ V} - 11.6 \text{ V})(6.73\ \Omega)}{11.6 \text{ V}} = 1.97\ \Omega \qquad \blacksquare$$

Finalize: The resistance of the battery is larger than 1 Ω, but it is reasonable for an old battery or for a battery consisting of several small electric cells in series. The load resistance agrees reasonably well with our prediction, despite the fact that the battery's internal resistance was about twice as large as we assumed. Note that in our initial guess we did not consider the power delivered to the load resistance; however, this datum is required for an accurate solution.

---

9. Consider the circuit shown in Figure P28.9. Find (a) the current in the 20.0-Ω resistor and (b) the potential difference between points a and b.

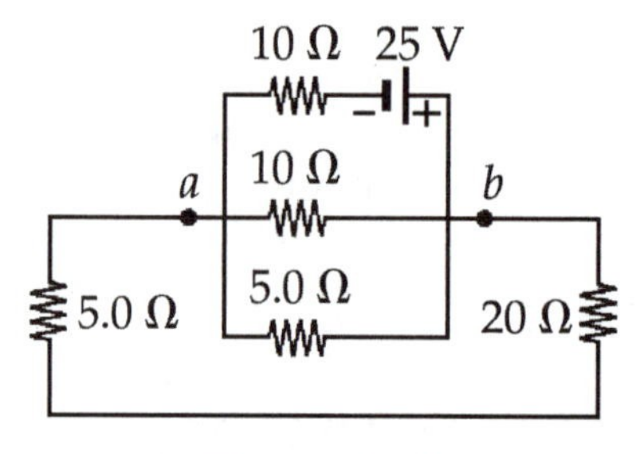

Figure P28.9

Solution

Conceptualize: The current will be from b to a through the 20-Ω and then the 5-Ω resistors, perhaps about half an amp. Then b will be at higher potential than a, perhaps by about 10 V.

Categorize: The problem might as well ask for the current in every resistor, because we must analyze the whole circuit to find the current in any one resistor. This circuit has only one power supply, so we can try ideas of equivalent resistances for series and parallel combinations.

Analyze: If we turn the given diagram on its side, we find that it is the same as Figure (a). The 20.0-Ω and 5.00-Ω resistors are in series, so the first reduction is as shown in (b).

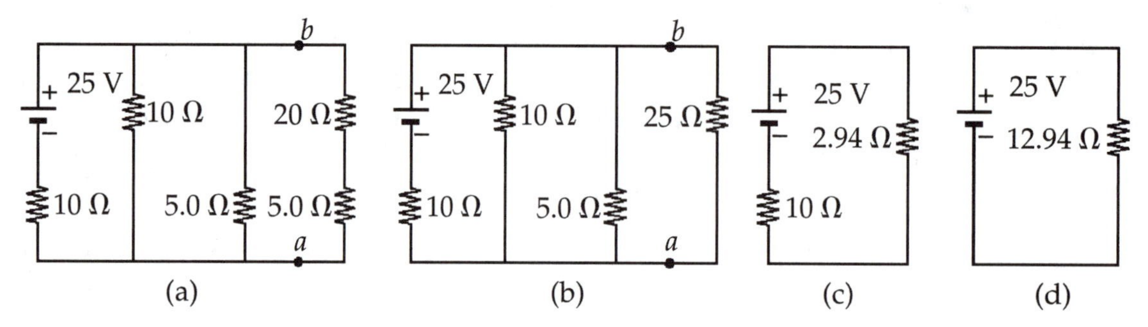

Next, since the 10.0-Ω, 5.00-Ω, and 25.0-Ω resistors are then in parallel, we can compute their equivalent resistance as

$$R_{eq} = \frac{1}{(1/10.0\ \Omega) + (1/5.00\ \Omega) + (1/25.0\ \Omega)} = 2.94\ \Omega$$

This is shown in Figure (c), which in turn reduces to the circuit shown in (d).

Next, we work backwards through the diagrams, applying $I = \Delta V/R$ and $\Delta V = IR$ alternately to each resistor, real or equivalent. The 12.94-Ω resistor is connected across 25.0 V, so the current through the voltage source in every diagram is

$$I = \frac{\Delta V}{R} = \frac{25.0\ \text{V}}{12.94\ \Omega} = 1.93\ \text{A}$$

In Figure (c), this 1.93 A goes through the 2.94-Ω equivalent resistor to give a voltage drop across this resistor of

$$\Delta V = IR = (1.93\ \text{A})(2.94\ \Omega) = 5.68\ \text{V}$$

From Figure (b), we see that this voltage drop is the same across ΔV_{ab}, the 10-Ω resistor, and the 5.00-Ω resistor.

(b) It happens that the answer to part (b) emerges first, as $\Delta V_{ab} = 5.68$ V. ■

(a) Since the current through the 20-Ω resistor is also the current through the 25.0-Ω line *ab*, this current is

$$I = \frac{\Delta V_{ab}}{R_{ab}} = \frac{5.68\text{ V}}{25.0\ \Omega} = 0.227\text{ A}$$ ■

Finalize: Each step in the solution is small, but you may be surprised at how many steps are required. We could not use $\Delta V = IR$ in the original circuit because there was not a single resistor for which we knew either ΔV or I. It really was necessary to draw all of the diagrams (a), (b), (c), and (d). Both answers are smaller than our guesses. Most of the battery current goes through the 5-Ω resistor in the center of the original diagram.

---

17. Calculate the power delivered to each resistor in the circuit shown in Figure P28.17.

2.00 Ω, 18.0 V, 1.00 Ω, 4.00 Ω, 3.00 Ω

Figure P28.17

Solution

Conceptualize: If the current in the battery is a few amperes, the resistor powers will add up to about 50 watts. The 4-Ω resistor will have twice the power of the 2-Ω, since these two carry the same current. Three times more power will be delivered to the 1-Ω resistor compared to the 3-Ω resistor, because the same potential difference is applied to both and the 1-Ω carries more current.

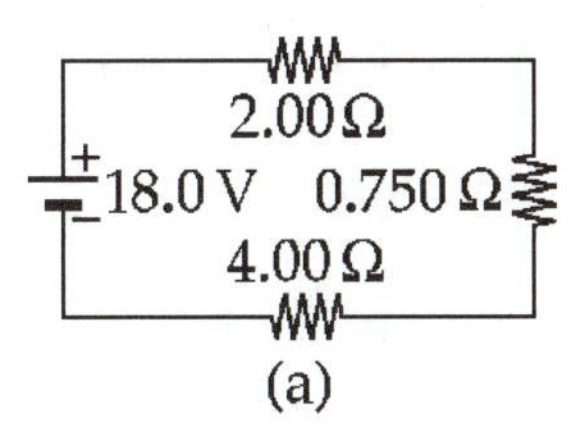

(a)

Categorize: The problem might just as well ask for the current in each resistor, as a step toward finding the power delivered to each. In turn, to find the current in each resistor, we . . .

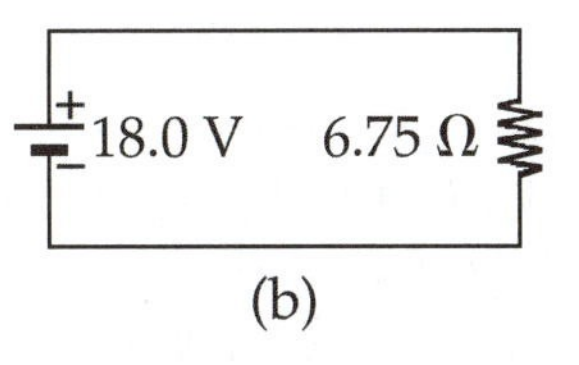

(b)

Analyze: . . . find the resistance seen by the battery. The given circuit reduces as shown in Figure (a), since

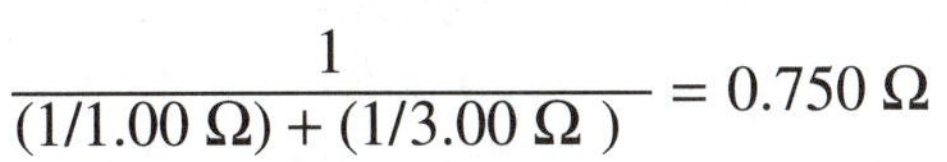

$$\frac{1}{(1/1.00\ \Omega) + (1/3.00\ \Omega)} = 0.750\ \Omega$$

In Figure (b), $I = 18.0\text{ V}/6.75\ \Omega = 2.67$ A

This is also the current in (a), so the 2.00-Ω and 4.00-Ω resistors convert powers

$$P_2 = I\Delta V = I^2R = (2.67\text{ A})^2\,(2.00\ \Omega) = 14.2\text{ W}$$ ■

and $$P_4 = I^2R = (2.67\text{ A})^2(4.00\ \Omega) = 28.4\text{ W}$$ ■

The voltage across the 0.750-Ω resistor in (a), and across both the 3.00-Ω and the 1.00-Ω resistors in Figure P28.17, is

$$\Delta V = IR = (2.67\ \text{A})(0.750\ \Omega) = 2.00\ \text{V}$$

Then for the 3.00-Ω resistor,

$$I = \frac{\Delta V}{R} = \frac{2.00\ \text{V}}{3.00\ \Omega}$$

and the power is

$$P_3 = I\Delta V = \left(\frac{2.00\ \text{V}}{3.00\ \Omega}\right)(2.00\ \text{V}) = 1.33\ \text{W}$$ ■

For the 1.00-Ω resistor,

$$I = \frac{2.00\ \text{V}}{1.00\ \Omega} \quad \text{and} \quad P_1 = \left(\frac{2.00\ \text{V}}{1.00\ \Omega}\right)(2.00\ \text{V}) = 4.00\ \text{W}$$ ■

Finalize: The total power of the resistors, 14.2 W + 28.4 W + 1.33 W + 4.00 W = 48 W agrees with our prediction and with the battery power 18 V(2.67 A) = 48.0 W. Every small step in the solution was necessary.

---

21. The circuit shown in Figure P28.21 is connected for 2.00 min. (a) Determine the current in each branch of the circuit. (b) Find the energy delivered by each battery. (c) Find the energy delivered to each resistor. (d) Identify the type of energy storage transformation that occurs in the operation of the circuit. (e) Find the total amount of energy transformed into internal energy in the resistors.

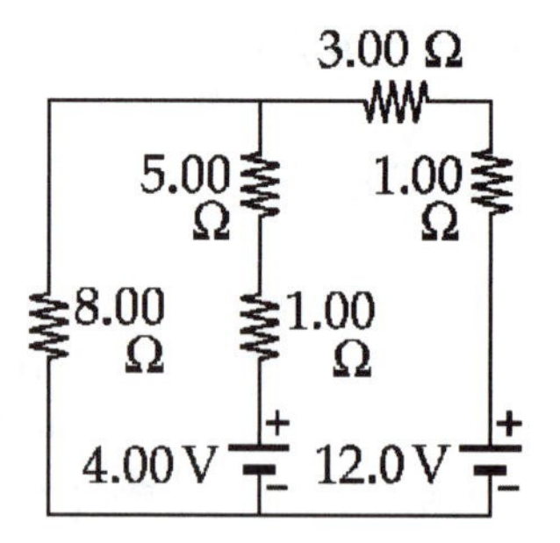

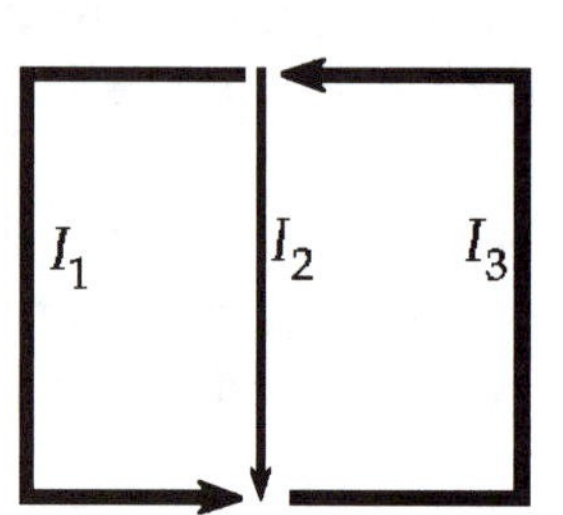

Figure P28.21

Solution

Conceptualize: We estimate a current of about 1 A in the 12-V battery and in the 8-Ω resistor, and a much smaller current in the 4-V battery. Then the total energy conversion might be about (25 J/s) (120 s) = 3 000 J.

Categorize: Many students might start their estimates, or even their solutions, with thinking about 12 V/4 Ω = 3 A and 4 V/6 Ω = 0.7 A, in the circuit diagram. This is quite unproductive—those voltages do not go with those resistances. You may wish to just carefully follow the steps of the Kirchhoff's-rules method. Or you may want to visualize a mountain landscape. The bottom line (wire) in the circuit diagram is at a uniform low potential which we naturally take as zero volts. The highest peak in the landscape is the positive pole of the 12-V battery. From this point, charge rolls downhill through the resistors and perhaps backward—from positive pole to negative pole—through the 4-V battery. After we find the three currents in part (a), it will in principle be easy to multiply voltage and current to find the power converted by each battery and resistor. Multiplying each of the seven numbers for power by 120 s will give the energy converted while the circuit is connected.

Analyze for part (a): We use the method of Kirchhoff's rules. First, we arbitrarily assign current directions and names, as shown in the second figure. The current rule then says that

$$I_3 - I_1 - I_2 = 0 \quad \text{or} \quad I_3 = I_1 + I_2 \qquad [1]$$

By the voltage rule, clockwise around the left-hand loop,

$$+I_1(8.00\ \Omega) - I_2(5.00\ \Omega) - I_2(1.00\ \Omega) - 4.00\ \text{V} = 0 \qquad [2]$$

Clockwise around the right-hand loop, combining the 1- and 5-Ω resistors and the 1- and 3-Ω resistors, the voltage rule reads

$$+4.00\ \text{V} + I_2(6.00\ \Omega) + I_3(4.00\ \Omega) - 12.0\ \text{V} = 0 \qquad [3]$$

We check that we have three independent equations in three unknowns. To solve, we substitute $(I_1 + I_2)$ for I_3, and reduce our three equations to the two equations

$$(8.00\ \Omega)I_1 - (6.00\ \Omega)I_2 - 4.00\ \text{V} = 0$$

and $\quad 4.00\ \text{V} + (6.00\ \Omega)I_2 + (4.00\ \Omega)(I_1 + I_2) - 12.0\ \text{V} = 0$

Solving the first of these equations for I_2 gives

$$I_2 = \frac{(8.00\ \Omega)I_1 - 4.00\ \text{V}}{6.00\ \Omega}$$

and rearranging the second of our pair of equations gives

$$I_1 + \frac{10.00\ \Omega}{4.00\ \Omega}I_2 = \frac{8.00\ \text{V}}{4.00\ \Omega}$$

We substitute for I_2 to get down to one equation in one unknown:

$$I_1 + 3.33I_1 - 1.67\ \text{V}/\Omega = 2.00\ \text{V}/\Omega$$

We solve for the current I_1, which is down the 8-Ω resistor:

$$I_1 = 0.846\ \text{V}/\Omega = 0.846\ \text{A} \qquad ■$$

If we were solving the three simultaneous equations by determinants or by calculator matrix inversion, we would have to do much more work to find I_2 and I_3. Our method of solving by substitution is more generally useful, and makes it easy to work out the remaining answers after the first, just as when the cat has kittens. We substitute again, putting the value for I_1 into equations we already solved for the other unknowns. Thus,

$$I_2 = \frac{(8.00\ \Omega)(0.846\ \text{A}) - 4.00\ \text{V}}{6.00\ \Omega} = 0.462\ \text{A} \quad \text{down in the middle branch} \qquad ■$$

and

$$I_3 = 0.846\ \text{A} + 0.462\ \text{A} = 1.31\ \text{A} \quad \text{up in the right-hand branch} \qquad ■$$

Finalize for part (a): Trust us: there is no method that is really simpler. When resistors are in series, like the 3 Ω and 1 Ω, their equivalent resistance will automatically appear in the loop equation. No resistors are in parallel with each other. At the first analysis step, you might naturally assume that the 4-V battery pushes current out of its positive terminal to go upward rather than downward in the middle branch. Get some really good practice by writing out an entirely separate solution proceeding from this assumption. Prove to yourself that it reaches the same physical answers as the solution here.

Analyze for parts (b) through (e):

(b) The power converted by a battery is $P = \Delta VI = U/\Delta t$, so the energy U converted is $U = (\Delta V)I\Delta t$. For the 4.00-V battery we have

$$U = (\Delta V)I\Delta t = (4.00\text{ V})(-0.462\text{ A})(120\text{ s}) = -222\text{ J} \quad \blacksquare$$

We have counted the current as negative because it is opposite in direction to the direction of increasing potential in this battery. For the 12.0-V battery,

$$U = (\Delta V)I\Delta t = (12.0\text{ V})(1.31\text{ A})(120\text{ s}) = 1.88\text{ kJ} \quad \blacksquare$$

(c) For a resistor (and only for a resistor) $\Delta V = IR$, so the energy converted $U = (\Delta V)I\Delta t$ becomes $U = I^2R\Delta t$. The energy delivered to the 8.00-Ω resistor is

$$U = I^2R\Delta t = (0.846\text{ A})^2(8.00\ \Omega)(120\text{ s}) = 687\text{ J} \quad \blacksquare$$

For the 5.00-Ω resistor, $U = (0.462\text{ A})^2(5.00\ \Omega)(120\text{ s}) = 128\text{ J}$ ■

For the 1.00-Ω resistor in the center branch,

$$U = (0.462\text{ A})^2(1.00\ \Omega)(120\text{ s}) = 25.6\text{ J} \quad \blacksquare$$

For the 3.00-Ω resistor, $U = (1.31\text{ A})^2(3.00\ \Omega)(120\text{ s}) = 616\text{ J}$ ■

For the 1.00-Ω resistor in the right-hand branch,

$$U = (1.31\text{ A})^2(1.00\ \Omega)(120\text{ s}) = 205\text{ J} \quad \blacksquare$$

(d) As the circuit carries current, the 12.0-V battery converts chemical energy into electrically transmitted energy. The 4.00-V battery is converting +222 J from electrically transmitted energy into chemical energy. All of the resistors are converting electrically transmitted energy into internal energy. The net transformation, in terms of energy storage, is from chemical energy into internal energy. ■

(e) The amount of energy transformed is altogether

$$687\text{ J} + 128\text{ J} + 25.6\text{ J} + 616\text{ J} + 205\text{ J} = 1.66\text{ kJ} \quad \blacksquare$$

Finalize for parts (b) through (e): Part (e) was about the internal energy created in just the resistors. We can count the same energy as the batteries convert it from chemical into electrically transmitted:

$$+1.88\text{ kJ} - 222\text{ J} = 1.66\text{ kJ}$$

Make sure that you do not think that the total energy converted is 1.66 kJ plus another 1.66 kJ. A hungry child does not have twice as much lunch money if he counts it at ten o'clock and again at eleven. Rather, the "net accumulation" of electrically transmitted energy is nonexistent, according to 1.66 kJ – 1.66 kJ = 0. Electrically transmitted energy is not stored energy and does not accumulate. Again, notice that $P = \Delta VI$ is true for both batteries and resistors, but $P = I^2R$ is true only for a resistor. A resistor can transform energy only from electrically transmitted into internal, but a battery with emf can transform energy either way between chemical and electrically transmitted. To be more complete, we add that a charging or discharging capacitor converts energy between electric potential energy and electrically transmitted energy. Later in the course we will see that an inductor converts energy between energy stored in its magnetic field and electrically transmitted energy.

---

22. For the circuit shown in Figure P28.22, calculate (a) the current in the 2.00-Ω resistor and (b) the potential difference between points *a* and *b*.

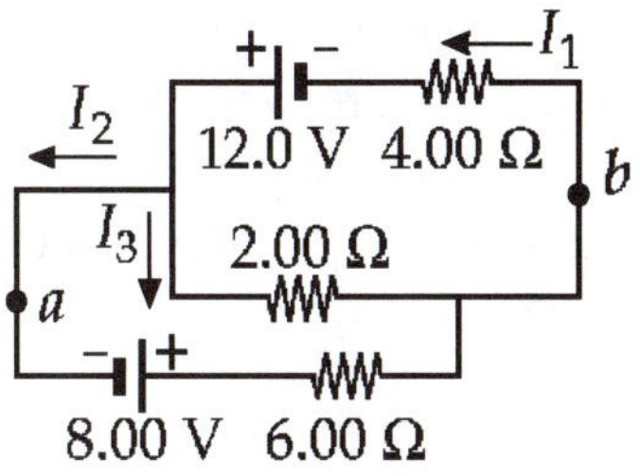

Figure P28.22

Solution

Conceptualize: It might be tempting to estimate I_3 = 12 V/6 Ω – 8 V/8 Ω = 1 A for the current. You need not worry about where this comes from—it is not a very good estimate. If the value of I_3 is positive, then *a* will be at higher voltage than *b* by $I_3(2\ \Omega)$, which may be about 2 V.

Categorize: The presence of two batteries suggests that we will use Kirchhoff's rules.

Analyze: Arbitrarily choose current directions as labeled in the figure.

(a) From the point rule for the junction below point *b*,

$$-I_1 + I_2 + I_3 = 0 \qquad [1]$$

Traversing the top loop counterclockwise gives the voltage loop equation

$$+12.0\text{ V} - (2.00\ \Omega)\,I_3 - (4.00\ \Omega)\,I_1 = 0 \qquad [2]$$

Traversing the bottom loop CCW,

$$+8.00\text{ V} - (6.00\ \Omega)\,I_2 + (2.00\ \Omega)I_3 = 0 \qquad [3]$$

From Equation [2], $$I_1 = \frac{12.0\text{ V} - (2.00\ \Omega)I_3}{4.00\ \Omega}$$

From Equation [3], $$I_2 = \frac{8.00\text{ V} + (2.00\ \Omega)I_3}{6.00\ \Omega}$$

Substituting both of these values into Equation [1], we find

$$-(3\text{ V} - 0.5\, I_3) + 1.33\text{ V} + 0.333\, I_3 + I_3 = 0$$

so $-1.67\text{ V} + 1.833\, I_3 - 0$

and the current in the 2.00-Ω resistor is $I_3 = 0.909$ A. ■

(b) Through the center wire, $V_a - (0.909\text{ A})(2.00\ \Omega) = V_b$

Therefore, $V_a - V_b = 1.82$ V, with $V_a > V_b$ ■

Finalize: When we are writing down a junction equation, a minus sign means out of the junction, contrasted to a plus sign for into the junction. We can call these accountant's signs. When we are writing down a loop equation, a minus sign means a drop in potential. The potential must drop when we go through a resistor in the direction of the current, just as the elevation drops while Huckleberry Finn floats down the Mississippi on a raft. After the equations are written down, the minus signs are governed just by the rules of algebra. The most common mistake is writing $-(3\text{ V} - 0.5\, I_3) = -3\text{ V} - 0.5\, I_3$, instead of $-(3\text{ V} - 0.5\, I_3) = -3\text{ V} + 0.5\, I_3$. You must keep your mind on what you are doing.

---

27. Taking $R = 1.00$ kΩ and $\mathcal{E} = 250$ V in Figure P28.27, determine the direction and magnitude of the current in the horizontal wire between a and e.

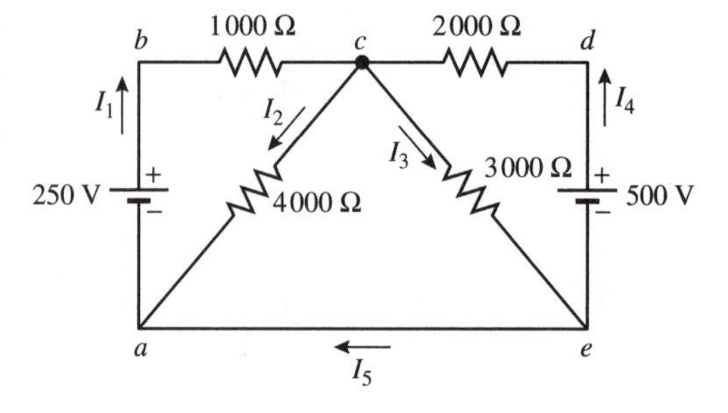

Figure P28.27

Solution

Conceptualize: Are you tempted to think that the wire ae carries no current, because no battery or resistor is in the branch that it forms? Or because it connects the low-potential terminals of two batteries? The negative terminal of one battery can be assigned the value zero volts, and then in this circuit the negative terminal of the other battery will also be at zero volts, but the zero-resistance wire will still carry a finite current. We can estimate its order of magnitude from the currents we might expect in the other branches, as a few hundred volts divided by a few thousand ohms, or on the order of a tenth of an amp.

Categorize: The 3 000-Ω and 4 000-Ω resistors are in parallel. But if we redrew an equivalent circuit diagram with a single equivalent resistance replacing them, the new circuit would not contain a branch with the current we are asked to find. For this reason, we choose to apply Kirchhoff's rules to the circuit as it stands. There are five branches, so there are five unknown currents. There are three junctions (one where four wires meet!) and we could think up many different trips around different large or small loops in the circuit. But we can minimize algebraic difficulties by carefully writing down a set of five equations before we think about solving them, and then doing just the algebra to solve for the one particular current required.

Analyze: We arbitrarily assign names I_1 through I_5 and possible directions to the currents in the different branches, as shown in our version of the diagram. We arbitrarily write junction equations just for junctions *a* and *c*. An equation for junction *e* would not be a mathematically independent equation.

Junction *a*: $-I_1 + I_2 + I_5 = 0$

Junction *c*: $+I_1 - I_2 - I_3 + I_4 = 0$

We need 5 – 2 = 3 loop equations. We choose arbitrarily to write them for clockwise trips around the loops *abca, acea*, and *cdec*. (These smallest loops are called meshes. If we wrote any other loop equation, such as for the perimeter *abcdea*, it would just be a linear combination of the three equations we have chosen.)

Loop *abca*: $+250\text{ V} - I_1\, 1.00\text{ k}\Omega - I_2\, 4.00\text{ k}\Omega = 0$

Loop *acea*: $+I_2\, 4.00\text{ k}\Omega - I_3\, 3.00\text{ k}\Omega = 0$

Loop *cdec*: $+I_4\, 2.00\text{ k}\Omega - 500\text{ V} + I_3\, 3.00\text{ k}\Omega = 0$

Our purpose is to solve for I_5. One sure-fire method is to eliminate other unknowns one by one. From the first equation, we substitute $I_1 = I_2 + I_5$ into each of the others, obtaining the four equations

$I_2 + I_5 - I_2 - I_3 + I_4 = 0$ or simplified $I_3 = I_5 + I_4$

$+250\text{ V} - I_2\, 1.00\text{ k}\Omega - I_5\, 1.00\text{ k}\Omega - I_2\, 4.00\text{ k}\Omega = 0$

$+I_2\, 4.00\text{ k}\Omega - I_3\, 3.00\text{ k}\Omega = 0$

and $+I_4\, 2.00\text{ k}\Omega - 500\text{ V} + I_3\, 3.00\text{ k}\Omega = 0$

Next we make the free choice to substitute $I_3 = I_5 + I_4$ into each of the others, to get the three equations

$+250\text{ V} - I_2\, 5.00\text{ k}\Omega - I_5\, 1.00\text{ k}\Omega = 0$

$+I_2\, 4.00\text{ k}\Omega - I_5\, 3.00\text{ k}\Omega - I_4\, 3.00\text{ k}\Omega = 0$

and $+I_4\, 2.00\text{ k}\Omega - 500\text{ V} + I_5\, 3.00\text{ k}\Omega + I_4\, 3.00\text{ k}\Omega = 0$

In just one more step, we can eliminate both the unknowns $I_2 = 0.050\,0\text{ A} - 0.200\, I_5$ and $I_4 = 0.100\text{ A} - 0.600\, I_5$ (from the first and third equations) by substituting these expressions into the second equation:

$4(0.0500\text{ A} - 0.200\, I_5) - 3\, I_5 - 3(0.100\text{ A} - 0.600\, I_5) = 0$

Clearing parentheses, $0.200\text{ A} - 0.800\, I_5 - 3.00\, I_5 - 0.300\text{ A} + 1.80\, I_5 = 0$

Gathering like terms, $-2.00\, I_5 = 0.100\text{ A}$

And at last solving, $I_5 = -0.0500\text{ A}$. The fact that it is negative means that it is opposite to the direction we assumed.

This current is +50.0 mA from *a* to *e*. ■

Finalize: We made a lot of arbitrary choices about which equations to write and what order to do the steps of algebra in. It is good practice to solve the problem again with a different set of choices—call the unknowns *u*, *w*, *x*, *y*, and *z*. You may need practice in working with sets of equations. At every stage, the number of equations must be equal to the number of remaining unknowns. Because we used the method of substitution, it would be easy to find all the other currents: equations like $I_4 = 0.100\text{ A} - 0.600\, I_5$ would give us the answers directly.

---

34. Consider a series *RC* circuit as in Figure P28.34 for which $R = 1.00\text{ M}\Omega$, $C = 5.00\ \mu\text{F}$, and $\mathcal{E} = 30.0\text{ V}$. Find (a) the time constant of the circuit and (b) the maximum charge on the capacitor after the switch is thrown closed. (c) Find the current in the resistor 10.0 s after the switch is thrown closed.

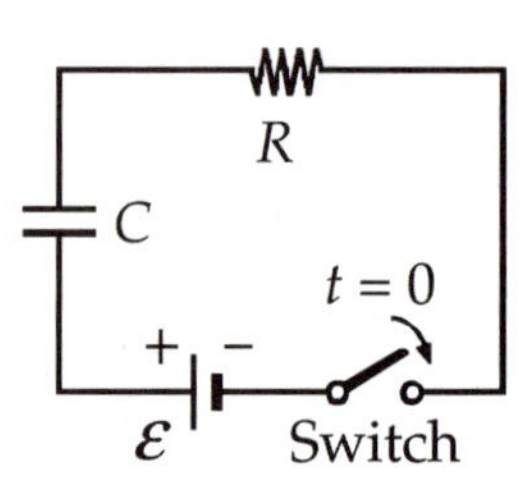

Figure P28.34

Solution

Conceptualize: The capacitor charges up after the switch is first thrown closed.

Categorize: The definition of the time constant, the definition of capacitance, and the equation for the time-dependent current in a charging-capacitor circuit will give the answers directly.

Analyze:

(a) The time constant is

$$\tau = RC = (1.00 \times 10^{6}\ \Omega)(5.00 \times 10^{-6}\text{ F}) = 5.00\ \Omega \cdot \text{F} = 5.00\text{ s}$$ ■

(b) After a long time interval, the capacitor is "charged to thirty volts," separating charges of

$$Q = C\Delta V = (5.00 \times 10^{-6}\text{ F})(30.0\text{ V}) = 150\ \mu\text{C}$$ ■

(c) $I = I_i e^{-t/\tau}$ where $I_i = \dfrac{\mathcal{E}}{R}$ and $\tau = RC = 5.00\text{ s}$

$$I = \frac{\mathcal{E}}{R} e^{-t/\tau} = \left(\frac{30.0\text{ V}}{1.00 \times 10^{6}\ \Omega}\right) e^{-10.0\text{ s}/5.00\text{ s}} \qquad I = 4.06 \times 10^{-6}\text{ A} = 4.06\ \mu\text{A}$$ ■

Finalize: The capacitor starts to charge instantly, at time zero, when the switch is closed. It takes an infinite time to finish charging. Still it has a well defined time constant of five seconds. We can think of this as a fairly long time interval, associated with the large resistance in the circuit. Learn to do the proof that RC has units of time:

$$\Omega \cdot \mathrm{F} = \frac{\mathrm{V}}{\mathrm{A}}\frac{\mathrm{C}}{\mathrm{V}} = \frac{\mathrm{C}}{\mathrm{C/s}} = \mathrm{s}$$

---

37. The circuit in Figure P28.37 has been connected for a long time. (a) What is the potential difference across the capacitor? (b) If the battery is disconnected from the circuit, over what time interval does the capacitor discharge to one tenth of its initial voltage?

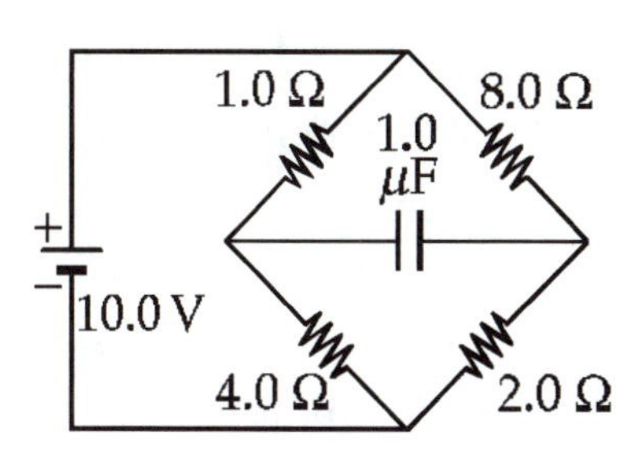

Figure P28.37

Solution

Conceptualize: The network of resistors contains voltage dividers so that the voltage across the capacitor will be a few volts, less than 10 V. The capacitor will discharge through an equivalent resistance on the order of a few ohms. With a microfarad capacitance, the time scale of its discharge is in microseconds.

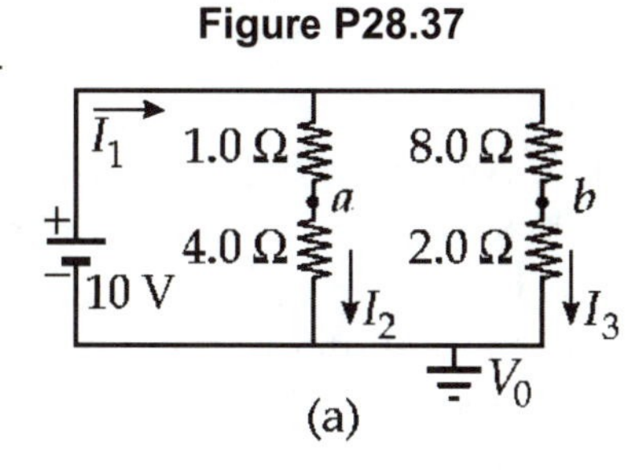

(a)

Categorize: We must use equivalent-resistance ideas in two quite different circuits to do the separate parts (a) and (b).

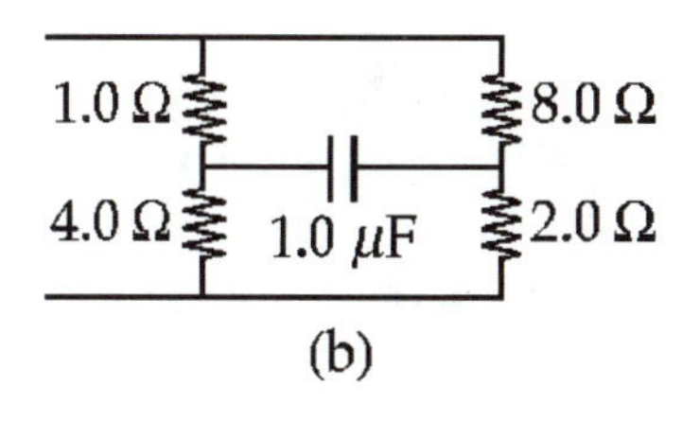

(b)

Analyze:

(a) After a long time interval, the capacitor branch will carry negligible current. The current is as shown in Figure (a). To find the voltage at point a, we first find the current, using the voltage rule:

$$10.0\ \mathrm{V} - (1.00\ \Omega)\, I_2 - (4.00\ \Omega)\, I_2 = 0$$

$$I_2 = 2.00\ \mathrm{A}$$

$$V_a - V_0 = (4.00\ \Omega)\, I_2 = 8.00\ \mathrm{V}$$

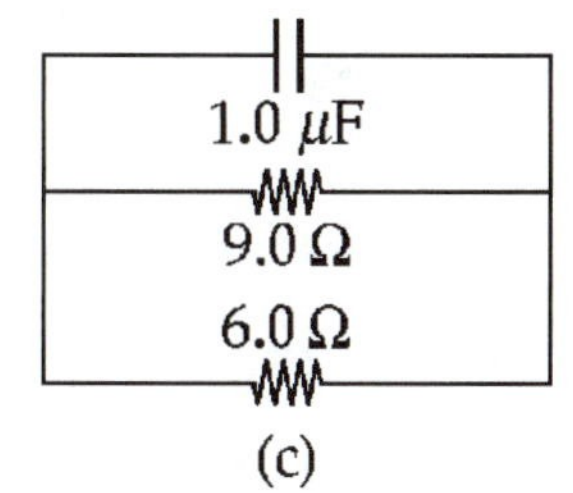

(c)

Similarly for the righthand branch,

$$10.0\ \mathrm{V} - (8.00\ \Omega) I_3 - (2.00\ \Omega)\, I_3 = 0$$

$$I_3 = 1.00\ \mathrm{A}$$

At point b, $V_b - V_0 = (2.00\ \Omega)\, I_3 = 2.00\ \mathrm{V}$

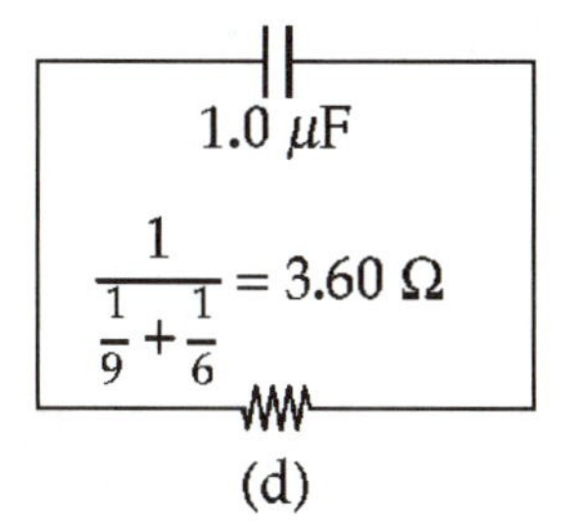

(d)

Thus, the voltage across the capacitor is

$$V_a - V_b = 8.00\ \mathrm{V} - 2.00\ \mathrm{V} = 6.00\ \mathrm{V} \qquad \blacksquare$$

(b) We suppose the battery is pulled out leaving an open circuit. We are left with Figure (b), which can be reduced to equivalent circuits (c) and (d).

From (d), we can see that the capacitor discharges through a 3.60-Ω equivalent resistance.

According to $q = Qe^{-t/RC}$

we calculate that $qC = QCe^{-t/RC}$

and $\Delta V = \Delta V_i e^{-t/RC}$

We proceed to solve for t: $V/V_i = e^{-t/RC}$ or $V_i/V = e^{+t/RC}$

Take natural logarithms of both sides: $\ln(V_i/V) = +t/RC$

so $t = RC\ln(V_i/V) = (3.60\ \Omega)(1.00\ \mu\text{F})\ln(V_i/0.100V_i)$

$t = 3.60\ \mu\text{s}\ln(10) = 3.60\ \mu\text{s}(2.30) = 8.29\ \mu\text{s}$ ■

Finalize: Our estimates were good. Every circuit diagram was necessary to draw.

---

42. An electric heater is rated at 1.50×10^3 W, a toaster at 750 W, and an electric grill at 1.00×10^3 W. The three appliances are connected to a common 120-V household circuit. (a) How much current does each draw? (b) If the circuit is protected with a 25.0-A circuit breaker, will the circuit breaker be tripped in this situation? Explain your answer.

Solution

Conceptualize: This is a very practical problem. An electric appliance is marked with the power that it converts, and just this information lets the householder figure out whether he or she can safely plug it in. Each appliance here will carry current of several amps. If all are turned on at once, they will likely pop the 25-A circuit breaker.

Categorize: We use just $P = I\Delta V$ to find the current in each appliance, and then use addition to find the total current.

Analyze:

(a) Heater: $$I = \frac{P}{\Delta V} = \left(\frac{1500\text{ W}}{120\text{ V}}\right)\left(\frac{1\text{ J/s}}{1\text{ W}}\right)\left(\frac{1\text{ V}}{1\text{ J/C}}\right)\left(\frac{1\text{ A}}{1\text{ C/s}}\right) = 12.5\text{ A}$$ ■

Toaster: $$I = \frac{750\text{ W}}{120\text{ V}} = 6.25\text{ A}$$ ■

Grill: $$I = \frac{1\,000\text{ W}}{120\text{ V}} = 8.33\text{ A}$$ ■

(b) Together in parallel they pass current 12.5 A + 6.25 A + 8.33 A = 27.1 A, so a 25.0-A circuit breaker will trip to switch all three off if the householder turns all three on at once. ■

Finalize: It is instructive to draw a circuit diagram for this case. The appliances are resistors in parallel, and that makes the addition of their currents meaningful. Your diagram can be a bit simpler than the text's Figure 28.19.

---

55. A rechargeable battery has an emf of 13.2 V and an internal resistance of 0.850 Ω. It is charged by a 14.7-V power supply for a time interval of 1.80 h. After charging, the battery returns to its original state as it delivers a constant current to a load resistor over 7.30 h. Find the efficiency of the battery as an energy storage device. (The efficiency here is defined as the energy delivered to the load during discharge divided by the energy delivered by the 14.7-V power supply during the charging process.)

Solution

Conceptualize: Gravitational energy can be stored in a counterweight by lifting it up; then very nearly 100% of the energy can be extracted by allowing the object to do work as it moves down. Nothing needs to get warm in that process. In the situation of this problem, the internal resistance of the battery will convert some electrically transmitted energy into internal energy both in the charging and in the discharging process. Thus the efficiency must be less than 100%. We might guess something more than 50%, if the internal resistance seems small.

Categorize. We will visualize the circuits used in the charging and discharging processes. We do not know the value of the load resistance, but knowing the time intervals for charging and discharging may somehow be just enough information. We use the same words "charging" and "discharging" as for a capacitor, but a battery is a very different device, which we model as having a constant emf, as containing an internal resistance, and as not storing charge but rather as boosting or lowering the electric potential energy of charge that goes through it, using its fund of chemical energy.

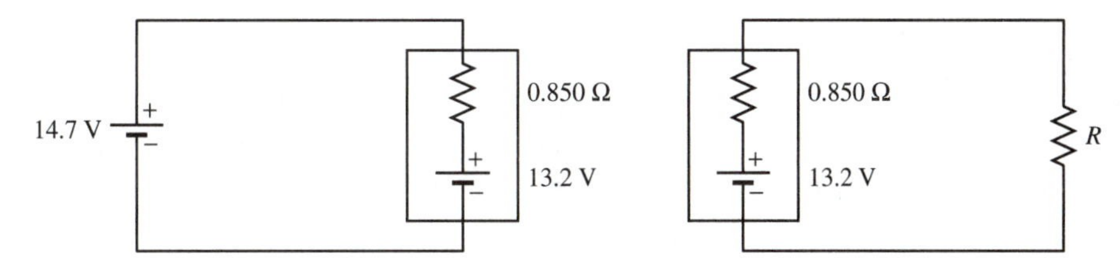

Figure P28.55

Analyze: The charging circuit is shown in the first frame of the diagram. Kirchhoff's loop rule gives $+14.7\text{ V} - 13.2\text{ V} - I\,(0.850\ \Omega) = 0$

so the charging current is $I = 1.5\text{ V}/0.850\ \Omega = 1.76\text{ A}.$

The charge passing through the battery as it charges is

$$q = I\Delta t = 1.76\text{ A}\,(1.80\text{ h}) = 3.18\text{ A}\cdot\text{h} = 11.4\text{ kC}$$

We can think of this charge as indexing a certain number of chemical reactions, producing a certain quantity of high-energy molecules in the battery. When the battery returns to its original state in discharging, we assume that the same number of reverse reactions uses up all of the high-energy chemical. In our model, the same charge passes through the battery in discharging, in the opposite direction. The discharge current is then

$$I = q/\Delta t = 3.18 \text{ A}\cdot\text{h}/7.30 \text{ h} = 0.435 \text{ A}$$

In the discharge circuit, the loop rule gives

$$13.2 \text{ V} - (0.435 \text{ A})(0.850 \ \Omega) - (0.435 \text{ A})R = 0$$

so the load resistance R is $12.8 \text{ V}/0.435 \text{ A} = 29.5 \ \Omega$.

Now we can get around to thinking about energy. The energy output of the 14.7-V power supply is $q\Delta V = (3.18 \text{ A}\cdot\text{h})(14.7 \text{ V}) = 46.7 \text{ W}\cdot\text{h} = 168 \text{ kJ}$

The energy delivered to the load during discharge is

$$q\Delta V = qIR = (3.18 \text{ A}\cdot\text{h})(0.435 \text{ A})(29.5 \ \Omega) = 40.8 \text{ W}\cdot\text{h} = 147 \text{ kJ}$$

The storage efficiency is $40.8 \text{ W}\cdot\text{h}/46.7 \text{ W}\cdot\text{h} = 0.873 = 87.3\%$ ■

Finalize: The efficiency is good; use of rechargeable batteries is compatible with efforts to save energy. Another way of stating our assumption about equal charges passing through the battery in charging and in discharging is that the same energy is converted by the emf. The limited efficiency shows up as 13.2 V being less than 14.7 V, and also the voltage across the load resistor, 12.8 V, being less than 13.2 V. In fact, the efficiency can be computed as $12.8 \text{ V}/14.7 \text{ V} = 87.3\%$.

---

60. Two resistors R_1 and R_2 are in parallel with each other. Together they carry total current I. (a) Determine the current in each resistor. (b) Prove that this division of the total current I between the two resistors results in less power delivered to the combination than any other division. It is a general principle that *current in a direct current circuit distributes itself so that the total power delivered to the circuit is a minimum.*

Solution

Conceptualize: Current does not all take the path of least resistance. The current will divide up so that the smaller resistance will carry somewhat more current. The result in part (b) makes it sound as if Nature is remarkably clever.

Categorize: Knowing that resistors in parallel have the same potential difference across them (according to Kirchhoff's loop rule), and that Kirchhoff's junction rule describes the division of the current at the junction between the resistors, we can do part (a). In (b), we will consider all possible current divisions, write down the total power converted by the two resistors, and use the calculus method of minimizing the power by setting its derivative equal to zero.

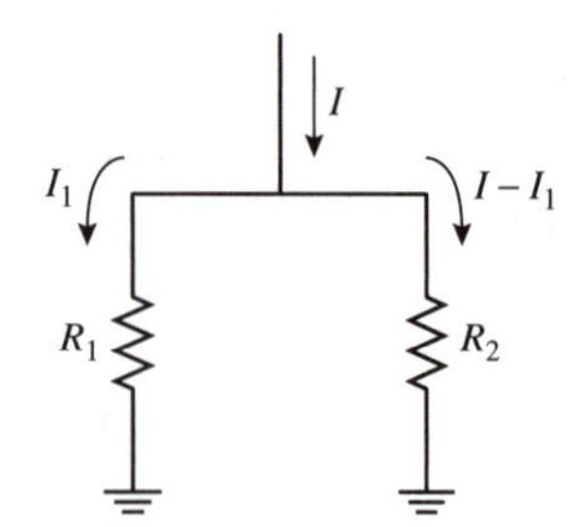

Figure P28.60

Analyze:

(a) In the diagram we could show the two resistors connected top end to top end and bottom end to bottom end with wires; we represent this connection instead by showing the bottom ends of both resistors connected to ground. The ground represents a conductor that is always at zero volts, and also can carry any current. Think of I, R_1, and R_2 as known quantities. We represent the current in R_1 as the unknown I_1. Then the current in the second resistor must by $I - I_1$. The total potential difference clockwise around the little loop containing both resistors must be zero:

$$-(I - I_1)R_2 + I_1R_1 = 0$$

We can already solve for I_1 in terms of the total current:

$$-IR_2 + I_1R_2 + I_1R_1 = 0 \qquad I_1 = IR_2/(R_1 + R_2) \qquad \blacksquare$$

Then the current in the second resistor is

$$I_2 = I - I_1 = I - IR_2/(R_1 + R_2) = I(R_1 + R_2 - R_2)/(R_1 + R_2)$$

$$I_2 = IR_1/(R_1 + R_2) \qquad \blacksquare$$

(b) Continue to think of I, R_1, and R_2 as known quantities and I_1 as an unknown. The power being converted by both resistors together is $P = I_1^2R_1 + (I - I_1)^2R_2$. Because the current is squared, the power would be extra large if all of the current went through either one of the resistors with zero current in the other. The minimum power condition must be with a more equitable division of current, and we find it by taking the derivative of P with respect to I_1 and setting the derivative equal to zero:

$$dP/dI_1 = 2\,I_1R_1 + 2(I - I_1)(0 - 1)R_2 = 0$$

Again we can solve directly for the real value of I_1 in

$$I_1R_1 - IR_2 + I_1R_2 = 0 \qquad \text{as} \qquad I_1 = IR_2/(R_1 + R_2) \qquad \blacksquare$$

So then again

$$I_2 = I - I_1 = IR_1/(R_1 + R_2) \qquad \blacksquare$$

This power-minimizing division of current is the same as the voltage-equalizing division of current that we found in part (a), so the proof is complete.

Finalize: Note that if R_2 is much greater than R_1, current I_2 will be much less than I_1, according to our answers.

Think of the two resistors as your coffeemaker and your toaster. Does Nature minimize your electric bill because she is thoughtful and kind? You can form your own opinion, but look back at Section 7.9 on energy diagrams and stability. There you saw that a bead sliding on a fixed curved wire tends to move to minimize its gravitational energy. The bead does not consciously have a purpose; rather, a component of the gravitational force pushes it down any slope. Surprisingly many physical laws can be stated in terms of minimizing some quantity. A grand example is Lagrange's principle of least action, studied in advanced mechanics. Chapter 35 mentions Fermat's principle of least time as describing the path of a ray of light.

63. The values of the components in a simple series RC circuit containing a switch (Fig. P28.34) are $C = 1.00\ \mu F$, $R = 2.00 \times 10^6\ \Omega$, and $\mathcal{E} = 10.0$ V. At the instant 10.0 s after the switch is thrown closed, calculate (a) the charge on the capacitor, (b) the current in the resistor, (c) the rate at which energy is being stored in the capacitor, and (d) the rate at which energy is being delivered by the battery.

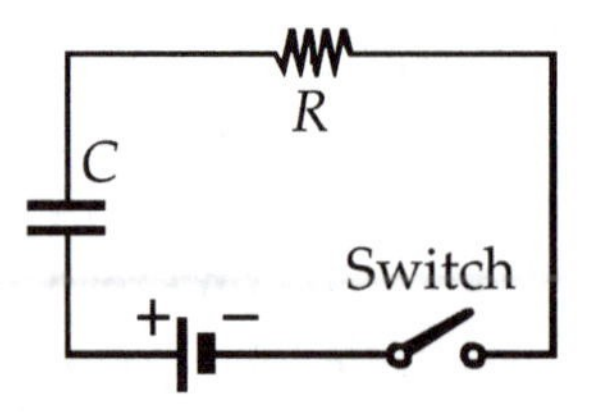

Figure P28.34

Solution

Conceptualize: The time constant of this circuit is $RC = 2\ M\Omega(1\ \mu F) = 2$ s. At 10 s we are mostly through with the charging process. The charge will be several microcoulombs, the current some microamps, and both powers some microwatts.

Categorize: We use the equation for the time-dependent charge on the capacitor. After that, general principles will tell us the current and rates of energy transfer.

Analyze:

(a) The charge on the capacitor at this instant is

$$q = C\mathcal{E}\left(1 - e^{-t/RC}\right) = \left(1.00 \times 10^{-6}\ \text{F}\right)(10.0\ \text{V})\left[1 - e^{-10.0\ \text{s}/\left(\left(2.00\times10^{6}\ \Omega\right)\left(1.00\times10^{-6}\ \text{F}\right)\right)}\right]$$

$$q = 9.93 \times 10^{-6}\ \text{C} = 9.93\ \mu C \qquad \blacksquare$$

(b) $$I = \frac{dq}{dt} = \frac{d}{dt}\left[C\mathcal{E}\left(1 - e^{-(t/RC)}\right)\right] = \left(\frac{\mathcal{E}}{R}\right)e^{-(t/RC)} = \left(\frac{10.0\ \text{V}}{2.00 \times 10^{6}\ \Omega}\right)e^{-(10.0/2.00)}$$

$$I = 3.37 \times 10^{-8}\ \text{A} \qquad \blacksquare$$

(c) Since the energy stored in the capacitor is $U = q^2/2C$, the rate of storing energy is

$$\frac{dU}{dt} = \frac{q}{C}\frac{dq}{dt} = \frac{qI}{C} = \left(\frac{9.93 \times 10^{-6}\ \text{C}}{1.00 \times 10^{-6}\ \text{F}}\right)\left(3.37 \times 10^{-8}\ \text{A}\right) = 3.34 \times 10^{-7}\ \text{W} \qquad \blacksquare$$

(d) The battery power output is

$$P_{\text{batt}} = I\mathcal{E} = (3.37 \times 10^{-8}\ \text{A})(10.0\ \text{V}) = 3.37 \times 10^{-7}\ \text{W} \qquad \blacksquare$$

Finalize: The charge is nearly up to its final value of 10 μC. The current is way down from its original maximum value of 10 V/2 MΩ = 5 μA. The battery power is always equal to the sum of the resistor power and the rate of energy addition to the capacitor. At the moment considered here the resistor power is only one percent of the other two terms.

---

66. Three identical 60.0-W, 120-V lightbulbs are connected across a 120-V power source as shown in Figure P28.66. Assuming the resistance of each lightbulb is constant (even though in reality the resistance might increase markedly with current), find (a) the total power supplied by the power source and (b) the potential difference across each lightbulb.

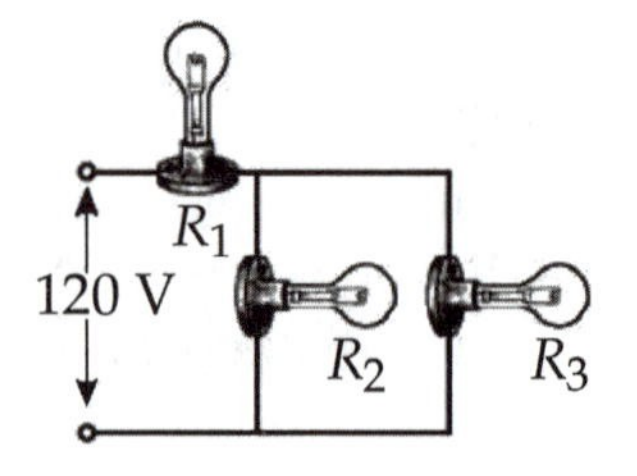

Figure P28.66

Solution

Conceptualize: The battery power is definitely less than 180 W. In fact, it will be less than 60 W, because the larger equivalent resistance seen by the power supply will draw less current than a single bulb.

Categorize: We find the resistance of one bulb, then the equivalent resistance across the power supply, then the current, and so the total power transferred.

Analyze: If the bulbs were connected individually across the power supply, or were all in parallel, the current in each would be

$$I = \frac{P}{\Delta V} = \frac{60.0 \text{ W}}{120 \text{ V}} = 0.500 \text{ A}$$

Then in any connection each bulb has resistance

$$R = \frac{\Delta V}{I} = \frac{120 \text{ V}}{0.500 \text{ A}} = 240 \ \Omega$$

In the given circuit, R_2 and R_3 have equivalent resistance

$$\frac{1}{(1/240 \ \Omega) + (1/240 \ \Omega)} = 120 \ \Omega$$

The three together have net resistance $240 \ \Omega + 120 \ \Omega = 360 \ \Omega$

The total current we calculate as $I = \dfrac{\Delta V}{R} = \dfrac{120 \text{ V}}{360 \ \Omega} = 0.333 \text{ A}$

(a) Thus the battery power delivered is $P = I\Delta V = (120 \text{ V})(0.333 \text{ A}) = 40.0 \text{ W}$ ■

(b) For bulb R_1, $\Delta V = IR$: $\Delta V = (0.333 \text{ A})(240 \ \Omega) = 80.0 \text{ V}$ ■

For bulb R_2 or R_3, the potential difference is $\Delta V = 120 \text{ V} - 80.0 \text{ V} = 40.0 \text{ V}$ ■

Finalize: Rating a bulb by power is useful for the purchaser to know how bright it will be and how costly to operate. The resistance is the most useful thing to know when we analyze a circuit containing the bulb.

---

71. In Figure P28.71, suppose the switch has been closed for a time interval sufficiently long for the capacitor to become fully charged. Find (a) the steady-state current in each resistor and (b) the charge Q on the capacitor. (c) The switch is now opened at $t = 0$. Write an equation for the current in R_2 as a function of time and (d) find the time interval required for the charge on the capacitor to fall to one-fifth its initial value.

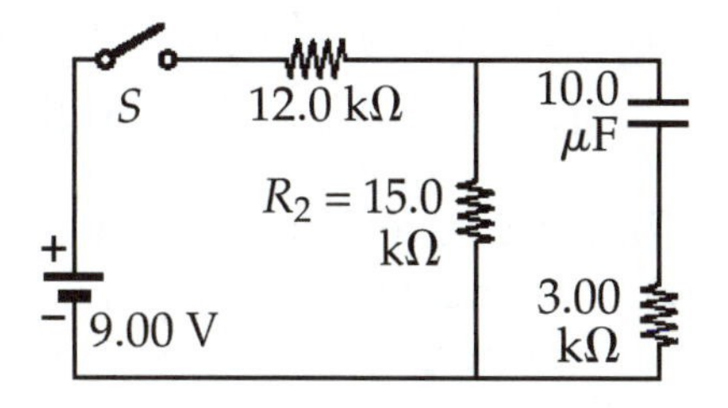

Figure P28.71

Solution

Conceptualize: The 12-kΩ and 15-kΩ resistors divide the 9 V across them, so the voltage across the capacitor will be about 5 V. Finding the equivalent resistance seen by the discharging capacitor will let us use the standard equations for its charge and the discharge current.

Categorize: We follow the current, seeing where it exists when the capacitor is fully charged and when it is discharging.

Analyze: We represent the resistors as $R_1 = 12.0\ \text{k}\Omega$, $R_2 = 15.0\ \text{k}\Omega$, and $R_3 = 3.00\ \text{k}\Omega$. We designate the respective currents they carry at some moment as I_1, I_2, and I_3.

(a) When the capacitor is fully charged, no current exists in it or in the 3.00-kΩ resistor:

$$I_3 = 0 \qquad \blacksquare$$

So the same current exists in resistors 1 and 2: $I_1 = I_2$, as given by the voltage rule,

$$+9.00\ \text{V} - (12.0\ \text{k}\Omega)\ I_1 - (15.0\ \text{k}\Omega)\ I_1 = 0$$

$$I_1 = 9.00\ \text{V}/27.0\ \text{k}\Omega = 0.333\ \text{mA} = I_2 \qquad \blacksquare$$

(b) For the right-hand loop, the voltage rule gives for the capacitor voltage:

$$(+15.0\ \text{k}\Omega)(0.333\ \text{mA}) - \Delta V_c - (3.00\ \text{k}\Omega)(0) = 0$$

Then $\Delta V_c = 5.00\ \text{V}$, and $Q = C\Delta V_c = (10.0\ \mu\text{F})(5.00\ \text{V}) = 50.0\ \mu\text{C}$ ■

(c) At $t = 0$, the current in R_1 drops to zero. The capacitor, charged to 5.00 V with top plate positive, will drive the same current $I_2 = I_3$ counterclockwise around the right-hand loop. At $t = 0$, its value is given by the equation $+5.00\ \text{V} - (15.0\ \text{k}\Omega)\ I_2 - (3.00\ \text{k}\Omega)\ I_2 = 0$.

So $I_2 = 5.00\ \text{V}/18.0\ \text{k}\Omega = 0.278\ \text{mA}$

Thereafter, it decays as it drains the capacitor's charge, with time constant

$$R_{eq}C = (18.0\ \text{k}\Omega)(10.0\ \mu\text{F}) = 180\ \text{ms}$$

So its equation is $I_2 = (0.278\ \text{mA})\, e^{-t/(180\ \text{ms})}$ or $I_2 = 0.278\ e^{-t/180}$, where I is in mA and t is in ms. ■

(d) The charge decays according to $q = Q_0 e^{-t/RC}$.

We solve for t step by step: $e^{+t/RC} = Q_0/q$

Taking natural logarithms, $t/RC = \ln(Q_0/q)$ and

$$t = RC\ln(Q_0/q) = 180\ \text{ms}\ \ln(Q_0/0.200\ Q_0) = 180\ \text{ms}\ \ln 5 = 180\ \text{ms}\ (1.61) = 290\ \text{ms} \qquad \blacksquare$$

Finalize: A circuit containing several power supplies, resistors, and capacitors can show somewhat complicated behavior. Here with just one power supply and just one capacitor we manage a complete analysis by faithfully following the idea of equivalent resistance.

29

Magnetic Fields

EQUATIONS AND CONCEPTS

The **magnetic force on a charged particle** moving in a magnetic field can be stated as a vector cross product. *The experimental observations of charged particles moving in a magnetic field are stated in Section 29.1 of the textbook and are summarized in Equation 29.1.*

$$\vec{\mathbf{F}}_B = q\vec{\mathbf{v}} \times \vec{\mathbf{B}} \quad (29.1)$$

The **magnitude of the magnetic force** depends on the angle between the velocity vector and the direction of the magnetic field, and will be maximum when the charge moves perpendicular to the direction of the magnetic field. In Equation 29.2, θ is the *smaller of two angles* between $\vec{\mathbf{v}}$ *and* $\vec{\mathbf{B}}$.

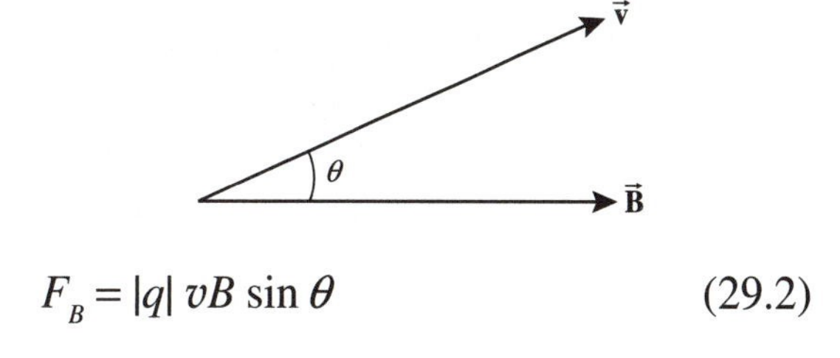

$$F_B = |q|\, vB \sin\theta \quad (29.2)$$

The **SI unit of the magnetic field** (magnetic intensity) is the tesla (T) or weber per square meter (Wb/m^2).

$$1\ \text{T} = 1\frac{\text{N}}{\text{C}\cdot\text{m/s}} = 1\frac{\text{N}}{\text{A}\cdot\text{m}} = 1\frac{\text{Wb}}{\text{m}^2}$$

The **direction of the magnetic force** on a charged particle moving with velocity $\vec{\mathbf{v}}$ can be determined by applying the right-hand rule (refer to figure): Hold your right hand open and point your fingers along the direction of $\vec{\mathbf{B}}$ while pointing your thumb in the direction of $\vec{\mathbf{v}}$. The force on a positive charge will be directed out of the palm of your hand. *If the charge is negative, the direction of the force will be reversed. A second version of the right-hand rule is illustrated in Suggestions, Skills, and Strategies.*

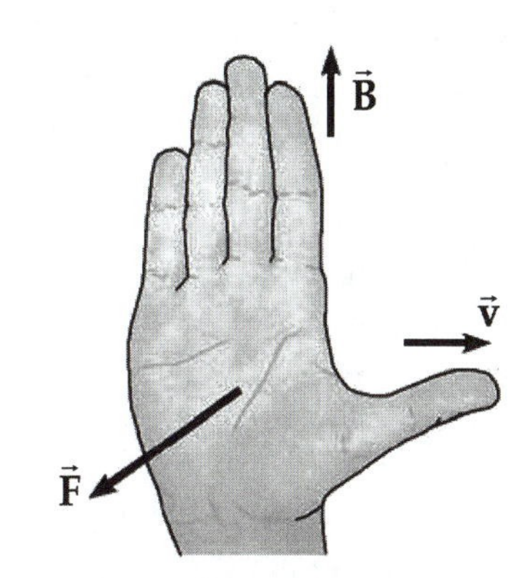

The **path of a charged particle entering a uniform magnetic** field with the *velocity vector initially perpendicular to the field* has the following characteristics:

The **path** of the particle will be circular and in a plane perpendicular to the direction of the field. The direction of rotation of the particle will be as determined by the right-hand rule.

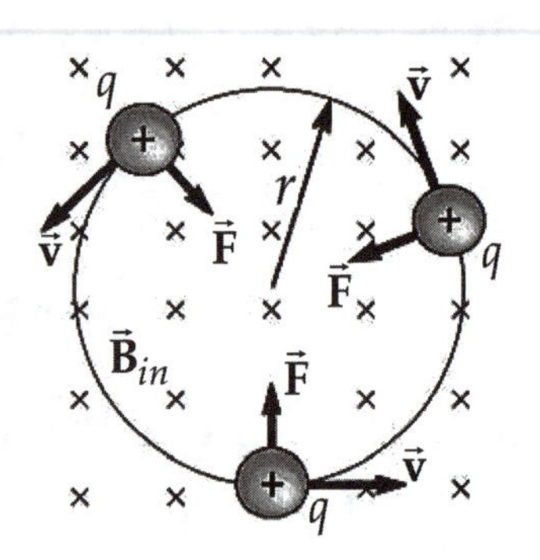

The **radius** of the circular path will be proportional to the linear momentum of the charged particle.

$$r = \frac{mv}{qB} \qquad (29.3)$$

The **angular speed** of the particle will be proportional to the ratio of charge to mass. *Note that the frequency, and therefore the period of rotation, will not depend on the radius of the path.*

$$\omega = \frac{qB}{m} \qquad (29.4)$$

$$T = \frac{2\pi m}{qB} \qquad (29.5)$$

If the **initial velocity is not perpendicular** to the field, the particle will move in a helical or spiral path. The axis of the spiral will be parallel to the magnetic field; and the "pitch" (distance between adjacent coils) will depend on the component of velocity parallel to the field.

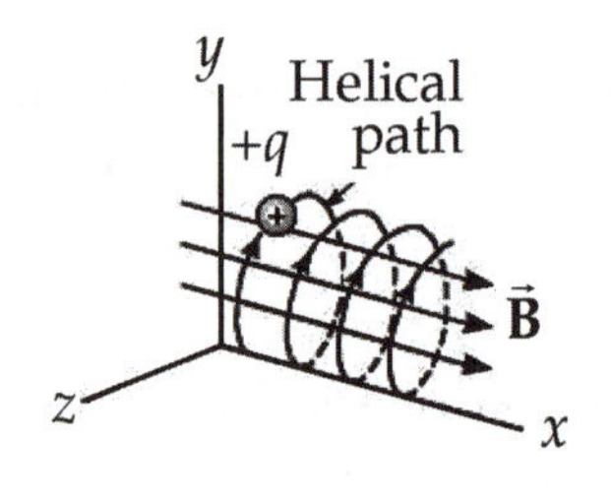

The **Lorentz force** is the total force experienced by a charged particle moving in a region where both an electric field and a magnetic field are present.

$$\vec{\mathbf{F}} = q\vec{\mathbf{E}} + q\vec{\mathbf{v}} \times \vec{\mathbf{B}} \qquad (29.6)$$

Applications of the motion of charged particles in a magnetic field include:

Velocity Selector—Consider a beam of *positively charged particles* directed into a region where uniform electric and magnetic fields are perpendicular to each other and perpendicular to the initial direction of the particle beam (see figure). The particles will experience an electric force directed toward the right side of the page and a magnetic force toward the left side of the page. Only those particles with a velocity $v = E/B$ will emerge through the slit.

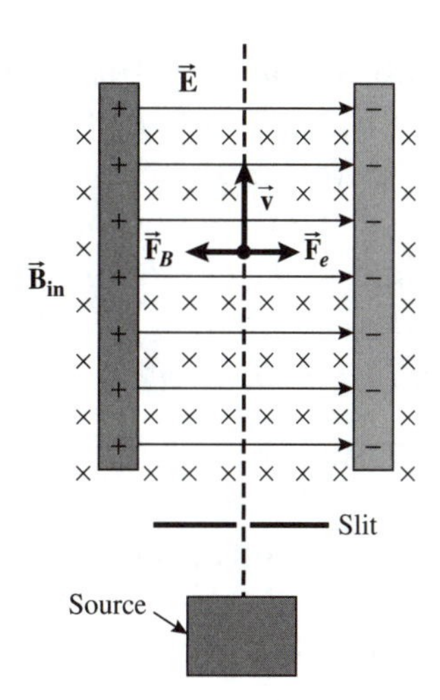

$$v = \frac{E}{B} \tag{29.7}$$

Mass Spectrometer—If an ion beam, after passing through a velocity selector, is directed perpendicularly into a *second uniform magnetic field* (B_0), the ratio of mass-to-charge of the isotopic species can be determined by measuring the radius of curvature of the beam in the second field.

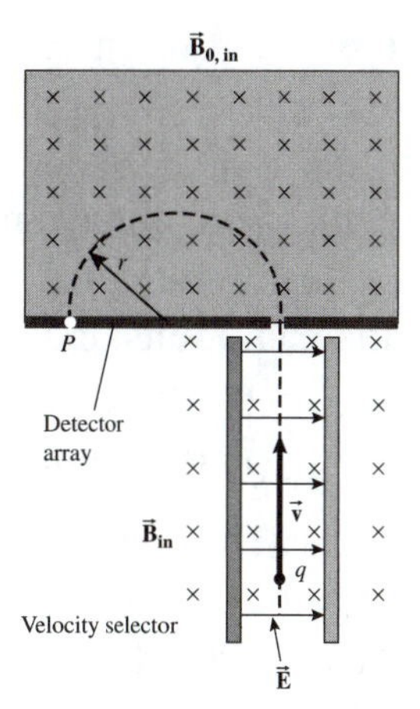

$$\frac{m}{q} = \frac{rB_0B}{E} \tag{29.8}$$

Cyclotron—The maximum kinetic energy acquired by an ion in a cyclotron depends on the radius of the "dees" and the intensity of the magnetic field. This relationship holds until the ion reaches relativistic energies (≈20 MeV). *This energy limit is true for ions of proton mass or greater, but is not true for electrons.*

$$K = \tfrac{1}{2}mv^2 = \frac{q^2B^2R^2}{2m} \tag{29.9}$$

The **magnetic force on a straight segment of current-carrying conductor** placed in a uniform magnetic field is given by Equation 29.10. The direction of $\vec{\mathbf{L}}$ is taken to be the direction of the current.

$$\vec{\mathbf{F}}_B = I\vec{\mathbf{L}} \times \vec{\mathbf{B}} \tag{29.10}$$

The **direction of the force** on the conductor is determined by the right-hand rule. In this case, your thumb must point in the direction of the current. *Use the right-hand rule to confirm that the direction of* $\vec{\mathbf{B}}$ *in the figure at right is into the plane of the page.*

The **magnitude of the force** on the conductor depends on the angle between the direction of the conductor and the direction of the field. *The magnetic force will be maximum when the conductor is directed perpendicular to the magnetic field.*

$$F_B = BIL \sin \theta$$

$$F_{B,\,\max} = BIL$$

The **magnetic force on a wire of arbitrary shape** is found by integrating Equation 29.12 over the length of the wire. The direction of $d\vec{\mathbf{s}}$ is that of an element of current. The magnetic force on a wire of any shape is equal to that of a straight wire connecting the end points and carrying the same current. *The net magnetic force on a closed current loop in a uniform field is zero.*

$$\vec{\mathbf{F}}_B = I\int_a^b d\vec{\mathbf{s}} \times \vec{\mathbf{B}} \qquad (29.12)$$

A **net torque will be exerted on a current loop** placed in an external magnetic field. In Equation 29.14, the area vector ($\vec{\mathbf{A}}$) is normal to the plane of the loop and has a magnitude equal to the area of the loop. *When the fingers of the right hand are curled around the loop in the direction of the current, the thumb points in the direction of* $\vec{\mathbf{A}}$.

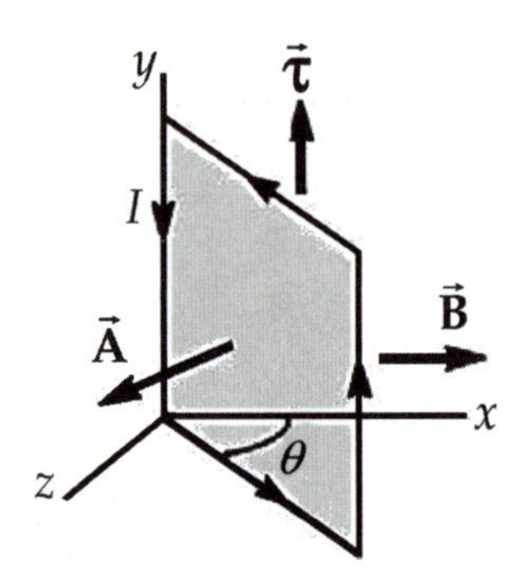

$$\vec{\tau} = I\vec{\mathbf{A}} \times \vec{\mathbf{B}} \qquad (29.14)$$

The **magnitude of the torque** depends on the angle between the direction of $\vec{\mathbf{B}}$ and the direction of $\vec{\mathbf{A}}$ and will be maximum when the magnetic field is parallel to the plane of the loop, that is, when $\vec{\mathbf{A}}$ is perpendicular to $\vec{\mathbf{B}}$.

$$\tau = IAB \sin\theta$$

$$\tau_{\max} = IAB \qquad (29.13)$$

The **direction of rotation of the loop** will be in the direction of decreasing values of θ (i.e., vector $\vec{\mathbf{A}}$ rotates toward the direction of the magnetic field). When $\vec{\mathbf{A}}$ becomes parallel to $\vec{\mathbf{B}}$, the torque on the loop will be zero. The loop shown in the figure will rotate counterclockwise as seen from above. *The direction of the vector torque is indicated by the thumb of the right hand when the fingers curl* $\vec{\mathbf{A}}$ *in to the direction of* $\vec{\mathbf{B}}$.

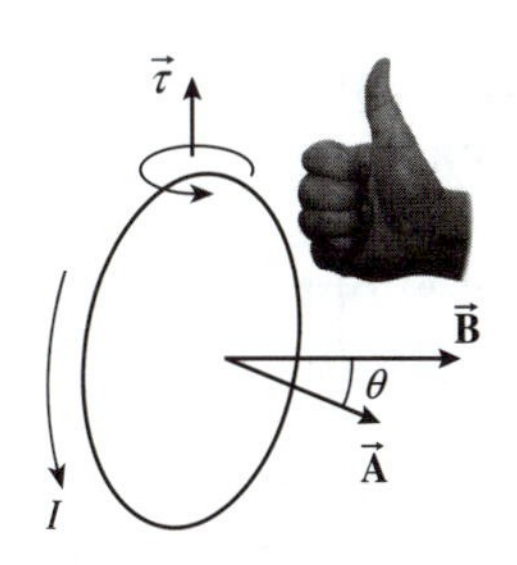

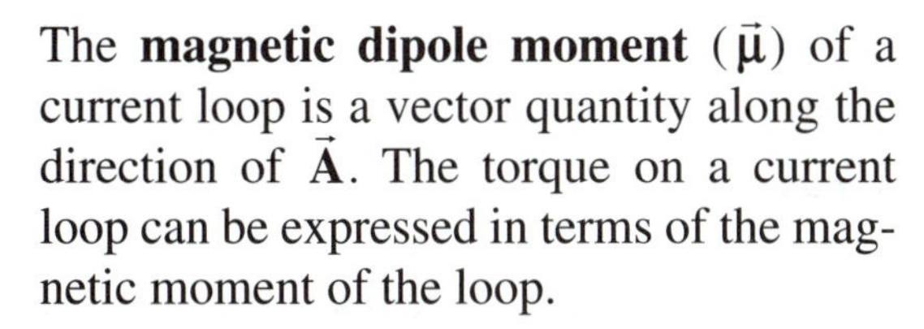

The **magnetic dipole moment** ($\vec{\mu}$) of a current loop is a vector quantity along the direction of $\vec{\mathbf{A}}$. The torque on a current loop can be expressed in terms of the magnetic moment of the loop.

$$\vec{\mu} \equiv I\vec{\mathbf{A}} \quad (29.15)$$

$$\vec{\tau} = \vec{\mu} \times \vec{\mathbf{B}} \quad (29.17)$$

The **potential energy of a magnetic dipole-magnetic field system** depends on the angle between the direction of the dipole moment and the direction of the magnetic field. *The system has minimum energy when* $\vec{\mu}$ *and* $\vec{\mathbf{B}}$ *are parallel.*

$$U = -\vec{\mu} \cdot \vec{\mathbf{B}} \quad (29.18)$$

The **Hall voltage** arises from the deflection of charge carriers in a current-carrying conductor when placed in a magnetic field. In Equation 29.21, d is the width of the conductor (the dimension that is perpendicular to the directions of both I and $\vec{\mathbf{B}}$), n is the charge-carrier density, and A is the cross-sectional area ($A = td$) of the conductor. *When the* **Hall Coefficient** *(1/nq) is known for a calibrated sample, the magnitude of the magnetic field can be determined by an accurate measurement of the Hall voltage* (ΔV_H)*. By measuring the voltage as + or –, the sign of the charge carriers can be determined.*

$$\Delta V_H = \frac{IBd}{nqA} \quad (29.21)$$

$$\Delta V_H = \frac{R_H IB}{t} \quad (29.22)$$

SUGGESTIONS, SKILLS, AND STRATEGIES

ILLUSTRATING THE DIRECTION OF MAGNETIC FIELDS

Graphical representation of vector quantities (e.g., velocity, magnetic field, magnetic force, and torque) is a useful technique.

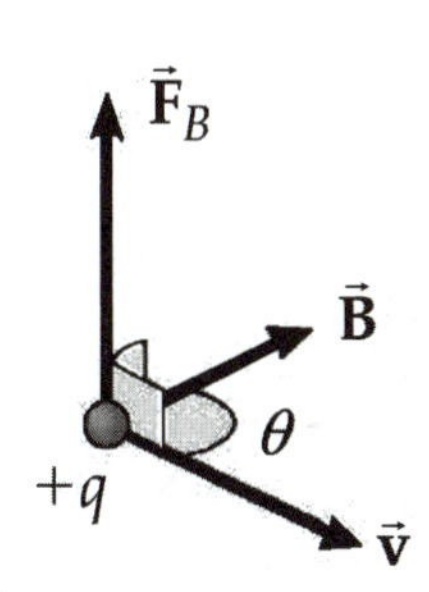

Vector lying in the plane of the page is shown by an arrow in the plane of the page.

Vector directed out of the page is shown by dots representing tips of arrows coming outward.

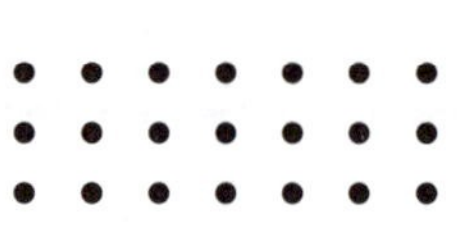

Vector directed into the page is shown by crosses representing the feathers of arrows going inward.

MAGNETIC FIELDS AND THE CROSS-PRODUCT

Equation 29.1, $\vec{\mathbf{F}} = q\vec{\mathbf{v}} \times \vec{\mathbf{B}}$, serves as the definition of the magnetic field vector $\vec{\mathbf{B}}$. If the vectors $\vec{\mathbf{v}}$ and $\vec{\mathbf{B}}$ are given in unit vector notation then $\vec{\mathbf{v}} \times \vec{\mathbf{B}}$ can be written as

$$\vec{\mathbf{v}} \times \vec{\mathbf{B}} = \hat{\mathbf{i}}\left(v_y B_z - v_z B_y\right) + \hat{\mathbf{j}}\left(v_z B_x - v_x B_z\right) + \hat{\mathbf{k}}\left(v_x B_y - v_y B_x\right)$$

The components of the magnetic force are:

$$F_x = q(v_y B_z - v_z B_y) \qquad F_y = q(v_z B_x - v_x B_z) \qquad F_z = q(v_x B_y - v_y B_x)$$

USING THE RIGHT-HAND RULE

Application of the right-hand rule as illustrated below applies to *positive charges;* results are reversed in the case of negative charges.

The right-hand rule is used to find the direction of the cross-product $\vec{\mathbf{F}} = q\vec{\mathbf{v}} \times \vec{\mathbf{B}}$. There are two versions of the right-hand rule; in order to avoid confusion, you should pick the version that suits you. In either version, note that the stated order of $\vec{\mathbf{v}}$ and $\vec{\mathbf{B}}$ is very important: $\vec{\mathbf{v}} \times \vec{\mathbf{B}} = -(\vec{\mathbf{B}} \times \vec{\mathbf{v}})$.

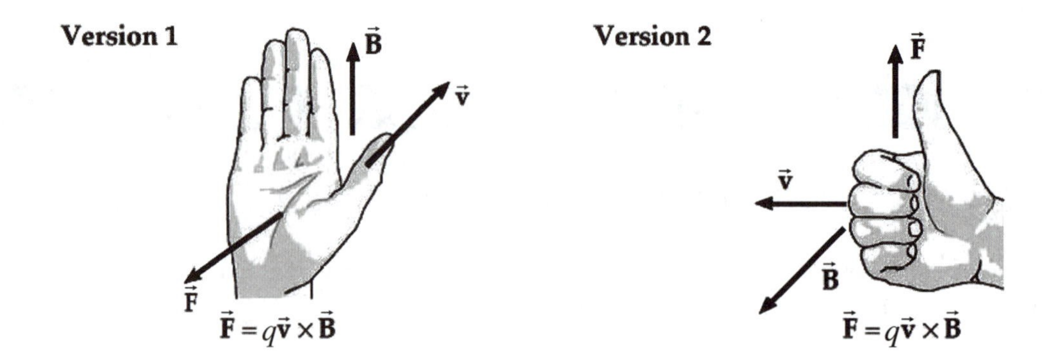

Version 1	**Version 2**
Hold your open right hand with your thumb pointing in the direction of $\vec{\mathbf{v}}$ (*the first named vector quantity*) and your fingers pointing in the direction of $\vec{\mathbf{B}}$ (*the second named vector quantity*). The force vector $\vec{\mathbf{F}}$ now is directed out of the palm of your hand. If your thumb and fingers are aligned, then the orientation of your hand is not determined; this is the case when the angle between $\vec{\mathbf{B}}$ and $\vec{\mathbf{v}}$ is zero degrees or 180 degrees and there is no force.	Orient your hand so that your fingers point in the direction of $\vec{\mathbf{v}}$ (*the first named vector quantity*) and then curl your fingers to point in the direction of $\vec{\mathbf{B}}$ (*the second named vector quantity*). Note that since your fingers cannot bend farther than 180º, you may have to flip your hand upside down to do this. Your thumb now points in the direction of $\vec{\mathbf{F}}$, where $\vec{\mathbf{F}}$ is perpendicular to both $\vec{\mathbf{v}}$ and $\vec{\mathbf{B}}$. If you do not need to curl your fingers because they already point along $\vec{\mathbf{B}}$, then the angle between $\vec{\mathbf{v}}$ and $\vec{\mathbf{B}}$ is zero and there is no force.

You may sometimes hear of a third version of the right-hand rule, involving the thumb, index finger, and middle finger. However, we recommend that you practice using one of the versions described above.

A different, but useful, right-hand rule applies in some specific situations. An example is when one of the quantities involves circulation (e.g., a current) and another a vector. The rule looks similar to version 2 of the cross product rule; curl your fingers around in the direction of the circulation, and your thumb will point in the direction of the vector. This rule applies, for example, when finding the direction of the area vector for a current loop.

REVIEW CHECKLIST

You should be able to:

- Use the defining equation for a magnetic field to determine the magnitude and direction of the magnetic force exerted on an electric charge moving in a region where there is a magnetic field. You should understand clearly the important differences between the forces exerted on electric charges by electric fields and those forces exerted on moving electric charges by magnetic fields. (Section 29.1)
- Calculate the period and radius of the circular orbit of a charged particle moving in a uniform magnetic field. (Section 29.2)
- Understand the essential features of the velocity selector and mass spectrometer, and make appropriate quantitative calculations regarding the operation of these instruments. (Section 29.3)
- Calculate the magnitude and direction of the magnetic force on a current-carrying conductor when placed in an external magnetic field. (Section 29.4)
- Determine the magnitude and direction of the torque exerted on a closed current loop in an external magnetic field. This includes correctly designating the direction of the area vector corresponding to a given current loop and incorporating the magnetic moment of the loop in the calculation of the torque on the loop. (Section 29.5)

ANSWER TO AN OBJECTIVE QUESTION

6. Answer each question yes or no. Assume the motions and currents mentioned are along the x axis and fields are in the y direction. (a) Does an electric field exert a force on a stationary charged object? (b) Does a magnetic field do so? (c) Does an electric field exert a force on a moving charged object? (d) Does a magnetic field do so? (e) Does an electric field exert a force on a straight current-carrying wire? (f) Does a magnetic field do so? (g) Does an electric field exert a force on a beam of moving electrons? (h) Does a magnetic field do so?

Answer (a) Yes, the definition of the electric field, represented by the equation $\vec{\mathbf{F}}_e = q\vec{\mathbf{E}}$, says that the electric field exerts a force on an object with charge.

(b) No, a stationary object has zero velocity and no magnetic force acts on it.

(c) Yes, a moving object feels an electric force with the same strength and direction that it would feel if it were stationary.

(d) Yes, the equation $\vec{\mathbf{F}}_B = q\vec{\mathbf{v}} \times \vec{\mathbf{B}}$, associated with the definition of the magnetic field, describes the magnetic force on a moving object with charge.

(e) No. An ordinary current-carrying wire has zero net charge. A metallic wire has equal numbers of (moving) conduction electrons and (stationary) ion cores in each cubic millimeter. So it feels zero force in an electric field.

(f) Yes. The moving electrons feel a magnetic force and the stationary positive charges feel no force, so the whole wire feels magnetic force $\vec{\mathbf{F}}_B = I\vec{\mathbf{L}} \times \vec{\mathbf{B}}$.

(g) and (h) Yes and yes. A beam of electrons, as in a cathode ray tube (such as the picture tube in a boxy television receiver or computer monitor), possesses net charge and consists of moving charges, so it feels electric and magnetic forces at the same time.

Notice that the only kind of charge is electric charge. There is no such thing as a magnetic charge. It is an object with electric charge that feels an electric force if it is in an electric field, and that feels a magnetic force if it is moving in a magnetic field. Any force is a vector measured in newtons and might have the effect of causing acceleration of the object on which it acts, or the effect of counterbalancing some other force. An electric field is an utterly different thing from a magnetic field, produced by a different source, measured in different units, and having different effects.

□ □ □ □

ANSWERS TO SELECTED CONCEPTUAL QUESTIONS

1. Two charged particles are projected in the same direction into a magnetic field perpendicular to their velocities. If the particles are deflected in opposite directions, what can you say about them?

Answer We know the magnetic field is the same for both particles. The velocity vectors are in the same direction, but one force vector is the negative of the other. From $\vec{\mathbf{F}}_B = q\vec{\mathbf{v}} \times \vec{\mathbf{B}}$, we conclude that the forces are in opposite directions because the charges have opposite signs.

□ □ □ □

3. Is it possible to orient a current loop in a uniform magnetic field such that the loop does not tend to rotate? Explain.

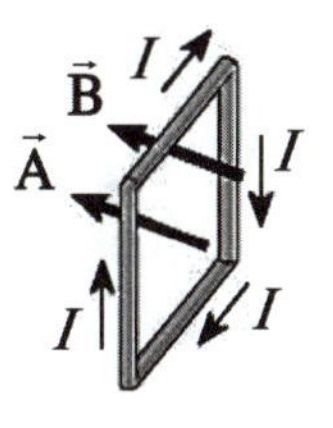

Answer Yes. If the magnetic field is perpendicular to the plane of the loop, the forces on opposite sides will be equal in magnitude, opposite in direction, and along the same lines of action, so they will produce no net torque about any axis.

5. How can a current loop be used to determine the presence of a magnetic field in a given region of space?

Answer The loop can be mounted free to rotate around an axis. The loop will rotate about this axis when placed in an external magnetic field at some arbitrary orientation. As the current through the loop is increased, the torque on it will increase.

SOLUTIONS TO SELECTED END-OF-CHAPTER PROBLEMS

2. Determine the initial direction of the deflection of charged particles as they enter the magnetic fields shown in Figure P29.2.

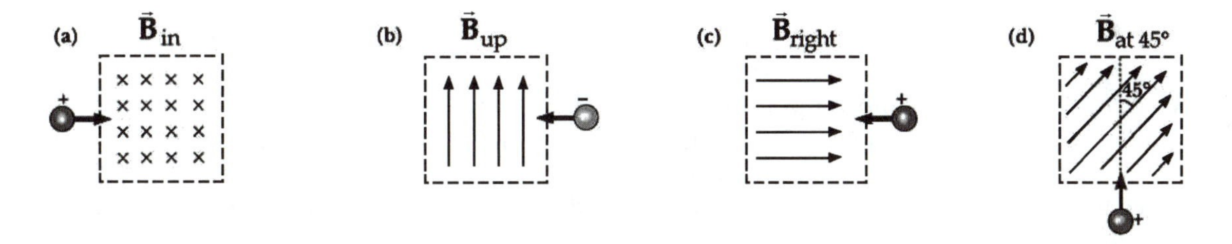

Figure P29.2

Solution

Conceptualize: The electric force exerted by an electric field on an electric charge is the same for any velocity of the charge. This problem is about how the magnetic force exerted by a magnetic field on an electric charge follows a quite different pattern.

Categorize: We must use the right-hand rule.

Analyze: Here we show both versions of the right-hand rule, open-handed and curled, for each cross product. Later on, we will draw just the curled-finger version. That version lets you work out any cross product, such as the torque on a magnetic moment, $\vec{\tau} = \vec{\mu} \times \vec{\mathbf{B}}$, without further memorization. If you are already familiar with the open-handed version of the rule, you can use the set of figures to practice the curled-finger version.

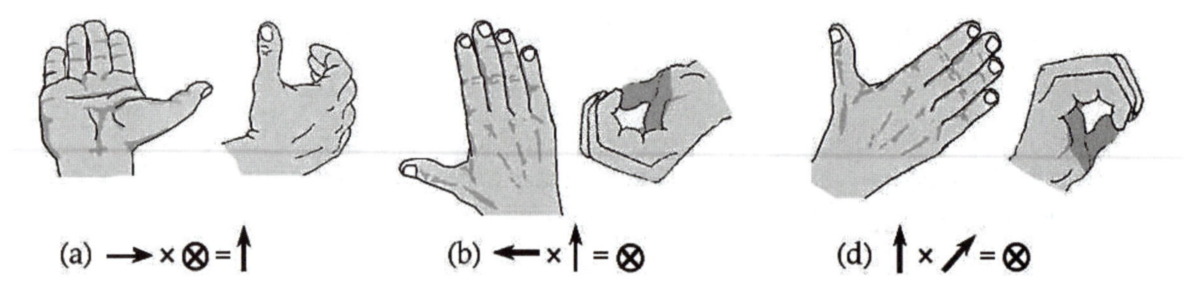

(a) By solution Figure (a), $\vec{\mathbf{v}} \times \vec{\mathbf{B}}$ is (right) × (away) = up. ■

(b) By solution Figure (b), $\vec{\mathbf{v}} \times \vec{\mathbf{B}}$ is (left) × (up) = away.

Since the charge is negative, $q\vec{\mathbf{v}} \times \vec{\mathbf{B}}$ is toward you. ■

(c) $\vec{\mathbf{v}} \times \vec{\mathbf{B}}$ is zero since the angle between $\vec{\mathbf{v}}$ and $\vec{\mathbf{B}}$ is 180° and $\sin 180° = 0$.

There is no deflection. ■

(d) $\vec{\mathbf{v}} \times \vec{\mathbf{B}}$ is (up) × (up and right), or away from you. ■

Finalize: The cross product has a direction, and it also has a size. In parts (a) and (b) the factor $\sin 90° = 1$ makes magnitude of the magnetic force equal to $|qvB|$. In part (c) it is zero. In part (d), where $\sin 45° = 0.707$, the magnitude of the force is $0.707\,|qvB|$.

---

6. A proton moving at 4.00×10^6 m/s through a magnetic field of magnitude 1.70 T experiences a magnetic force of magnitude 8.20×10^{-13} N. What is the angle between the proton's velocity and the field?

Solution

Conceptualize: The largest possible size for the force is $qvB = (1.6 \times 10^{-19})(4 \times 10^6)\,(1.7)$ N. The factor by which the given force is less than this…

Categorize: … will be the sine of the angle between velocity and field, so from it we can find the angle.

Analyze: The magnitude of the force on a moving charge in a magnetic field is $F_B = qvB \sin\theta$, so

$$\theta = \sin^{-1}\left[\frac{F_B}{qvB}\right]$$

$$\theta = \sin^{-1}\frac{8.20 \times 10^{-13}\text{ N}}{(1.60 \times 10^{-19}\text{ C})(4.00 \times 10^6\text{ m/s})(1.70\text{ T})} = 48.9° \text{ or } 131°$$ ■

Finalize: Both possible answers are equally realistic. To visualize the situation, you may suppose the field is horizontally to the right, the velocity is to the right at 48.9° above the horizontal or alternatively to the left at 48.9° above the horizontal, and the force exerted on the proton is horizontally away from you.

---

8. A proton moves with a velocity of $\vec{\mathbf{v}} = (2\hat{\mathbf{i}} - 4\hat{\mathbf{j}} + \hat{\mathbf{k}})$ m/s in a region in which the magnetic field is $\vec{\mathbf{B}} = (\hat{\mathbf{i}} + 2\hat{\mathbf{j}} - 3\hat{\mathbf{k}})$ T. What is the magnitude of the magnetic force this particle experiences?

Solution

Conceptualize: The velocity is very low for an experimental proton, and the field is very large. This is a mathematical exercise about computing a force of some attonewtons.

Categorize: The problem asks only for the magnitude, not the direction of the force that the field exerts. But the practical way to find the size of the force is to work it out in unit-vector notation and then to use the three-dimensional Pythagorean theorem to find its magnitude.

Analyze: The force on a charged particle is proportional to the vector product of the velocity and the magnetic field:

$$\vec{\mathbf{F}}_B = q\vec{\mathbf{v}} \times \vec{\mathbf{B}} = (1.60 \times 10^{-19}\text{ C})\left[(2\hat{\mathbf{i}} - 4\hat{\mathbf{j}} + \hat{\mathbf{k}})(\text{m/s}) \times (\hat{\mathbf{i}} + 2\hat{\mathbf{j}} - 3\hat{\mathbf{k}})\text{T}\right]$$

Since 1 C · m · T/s = 1 N, we can write this in determinant form as:

$$\vec{\mathbf{F}}_B = (1.60 \times 10^{-19}\text{ N})\begin{vmatrix} \hat{\mathbf{i}} & \hat{\mathbf{j}} & \hat{\mathbf{k}} \\ 2 & -4 & 1 \\ 1 & 2 & -3 \end{vmatrix}$$

Expanding the determinant as described in Equation 11.8, we have

$$\vec{\mathbf{F}}_{B,x} = (1.60 \times 10^{-19}\text{ N})\left[(-4)(-3) - (1)(2)\right]\hat{\mathbf{i}}$$

$$\vec{\mathbf{F}}_{B,y} = (1.60 \times 10^{-19}\text{ N})\left[(1)(1) - (2)(-3)\right]\hat{\mathbf{j}}$$

$$\vec{\mathbf{F}}_{B,z} = (1.60 \times 10^{-19}\text{ N})\left[(2)(2) - (1)(-4)\right]\hat{\mathbf{k}}$$

Again in unit-vector notation,

$$\vec{\mathbf{F}}_B = (1.60 \times 10^{-19}\text{ N})(10\hat{\mathbf{i}} + 7\hat{\mathbf{j}} + 8\hat{\mathbf{k}}) = \left(16.0\hat{\mathbf{i}} + 11.2\hat{\mathbf{j}} + 12.8\hat{\mathbf{k}}\right) \times 10^{-19}\text{ N}$$

$$\left|\vec{\mathbf{F}}_B\right| = \left(\sqrt{16.0^2 + 11.2^2 + 12.8^2}\right) \times 10^{-19}\text{ N} = 23.4 \times 10^{-19}\text{ N}$$ ■

Finalize: Unit-vector notation makes it as convenient as possible to work out and write down the answer. It is possible to visualize the answer in space and verify that the right-hand rule applies. Suppose you are facing a vertical television screen, which we take as the *xy* plane. The proton velocity is to the right, steeply down, and a little toward you. The field is to the right, steeply up, and more sharply away from you. Then the force is to the right, upward, and toward you. We could state its direction by giving the angle between the force and the *z* axis, and the angle, say, between its projection on the *xy* plane and the *x* axis. The unit-vector expression contains this information as well.

9. A proton moves perpendicular to a uniform magnetic field $\vec{\mathbf{B}}$ at a speed of 1.00×10^7 m/s and experiences an acceleration of 2.00×10^{13} m/s^2 in the positive x direction when its velocity is in the positive z direction. Determine the magnitude and direction of the field.

Solution

Conceptualize: The magnetic field exerts a magnetic force on the proton, and the force causes the acceleration. Ten million meters per second is a perfectly reasonable speed for a proton in a vacuum. If the computed field magnitude is anything between a microtesla and a tesla, then we can infer that 10^{13} m/s^2 is a reasonable acceleration.

Categorize: We will use the particle under acceleration model to find the net force on the proton, which is the magnetic force. Then the particle-in-a-magnetic field model will tell us the field.

Analyze: By Newton's second law,

$\Sigma F = ma = (1.67 \times 10^{-27}\text{ kg})(2.00 \times 10^{13}\text{ m/s}^2) = 3.34 \times 10^{-14}$ N in the $+x$ direction.
The magnetic force is $F = 3.34 \times 10^{-14}\text{ N} = qvB \sin 90°$

Rearranging,

$$B = \frac{F}{qv} = \frac{3.34 \times 10^{-14}\text{ N}}{\left(1.60 \times 10^{-19}\text{ C}\right)\left(1.00 \times 10^{7}\text{ m/s}\right)} = 2.09 \times 10^{-2}\text{ T} \qquad \blacksquare$$

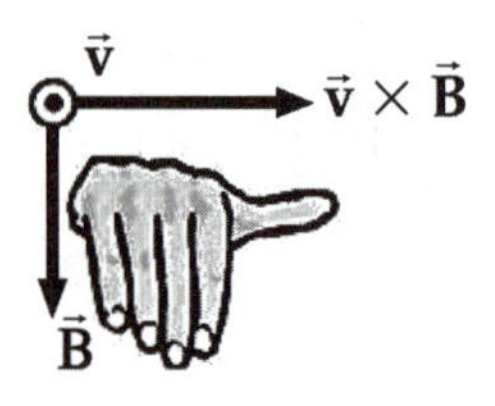

By the right-hand rule, $\vec{\mathbf{B}}$ must be in the $-y$ direction. $\blacksquare$

This yields a force on the proton in the $+x$ direction when $\vec{\mathbf{v}}$ points in the $+z$ direction.

Finalize: Twenty milliteslas is an easy field to produce with an electromagnet. In the student laboratory you may do an experiment like this, but with electrons rather than protons. The experimenter chooses to fire the particle at a ninety degree angle to the field. Then nature makes the direction of the force at ninety degrees to both the velocity and the field. Note that one tesla is one newton-second per coulomb-meter.

---

13. A proton (charge $+e$, mass m_p), a deuteron (charge $+e$, mass $2m_p$), and an alpha particle (charge $+2e$, mass $4m_p$) are accelerated from rest through a common potential difference ΔV. Each of the particles enters a uniform magnetic field $\vec{\mathbf{B}}$, with its velocity in a direction perpendicular to $\vec{\mathbf{B}}$. The proton moves in a circular path of radius r_p. In terms of r_p, determine (a) the radius r_d of the circular orbit for the deuteron and (b) the radius r_α for the alpha particle.

Solution

Conceptualize: In general, particles with greater speed, more mass, and less charge will have larger radii as they move in a circular path due to a constant magnetic field. Since the effects of mass and charge have opposite influences on the path radius, it is not obvious which particle will have the largest radius, but the low-mass proton should move in the smallest-radius, sharpest curve. Since the masses and charges of the three particles are alike within a factor of four, we should expect that the radii also fall within a similar range.

Categorize: The radius of each particle's path can be found by applying Newton's second law, where the force causing the centripetal acceleration is the magnetic force $\vec{\mathbf{F}}_B = q\vec{\mathbf{v}} \times \vec{\mathbf{B}}$. The speeds of the particles can be found from the kinetic energy resulting from the change in electric potential given.

Analyze: An electric field changes the speed of each particle according to $(K + U)_i = (K + U)_f$. Therefore, noting that the particles start from rest, we can write

$$q\Delta V = \tfrac{1}{2}mv^2$$

After they are fired, the particles have the magnetic field change their direction as described by $\Sigma\vec{\mathbf{F}} = m\vec{\mathbf{a}}$:

$$qvB\sin 90° = \frac{mv^2}{r} \quad \text{thus} \quad r = \frac{mv}{qB} = \frac{m}{qB}\sqrt{\frac{2q\Delta V}{m}} = \frac{1}{B}\sqrt{\frac{2m\Delta V}{q}}$$

For the protons, $r_p = \dfrac{1}{B}\sqrt{\dfrac{2m_p\Delta V}{e}}$

(a) For the deuterons, $r_d = \dfrac{1}{B}\sqrt{\dfrac{2(2m_p)\Delta V}{e}} = \sqrt{2}r_p$ ■

(b) For the alpha particles, $r_\alpha = \dfrac{1}{B}\sqrt{\dfrac{2(4m_p)\Delta V}{2e}} = \sqrt{2}r_p$ ■

Finalize: Somewhat surprisingly, the radii the deuterons' and alpha particles' trajectories are the same and are only 41% greater than for the protons.

---

19. A cosmic-ray proton in interstellar space has an energy of 10.0 MeV and executes a circular orbit having a radius equal to that of Mercury's orbit around the Sun (5.80×10^{10} m). What is the magnetic field in that region of space?

Solution

Conceptualize: A very big orbit implies that the magnetic field in this interstellar space is a tiny fraction of a tesla.

Categorize: Think of the proton as having accelerated through a potential difference $\Delta V = 10^7$ V. We use the energy version of the isolated system model, applied to the particle and the electric field that made it speed up from rest. Then we use the particle in a magnetic field model and the particle in uniform circular motion model.

Analyze: By conservation of energy for the proton-electric-field system in the process that set the proton moving, its kinetic energy is

$$E = \tfrac{1}{2}mv^2 = e\Delta V \qquad \text{so its speed is} \qquad v = \sqrt{\frac{2e\Delta V}{m}}$$

Now Newton's second law for its circular motion in the magnetic field is

$$\Sigma F = ma \qquad \text{becoming} \qquad \frac{mv^2}{R} = evB\sin 90^\circ$$

So $B = \dfrac{mv}{eR} = \dfrac{m}{eR}\sqrt{\dfrac{2e\Delta V}{m}} = \dfrac{1}{R}\sqrt{\dfrac{2m\Delta V}{e}}$

And $B = \dfrac{1}{5.80 \times 10^{10}\text{ m}}\sqrt{\dfrac{2\left(1.67\times10^{-27}\text{ kg}\right)\left(10^7\text{ V}\right)}{1.60\times10^{-19}\text{C}}} = 7.88\times10^{-12}\text{ T}$ ■

Finalize: This is a reasonable answer. Some matter far away created this field that exists in a space empty of matter, and exists unchanged when the proton enters, to exert a force on the proton.

---

24. A cyclotron designed to accelerate protons has a magnetic field of magnitude 0.450 T over a region of radius 1.20 m. What are (a) the cyclotron frequency and (b) the maximum speed acquired by the protons?

Solution

Conceptualize: Cyclotrons are useful, for example, to produce radioactive nuclides for medical treatments. We expect the angular frequency to be some millions of radians per second. The maximum speed may be a considerable fraction of the speed of light.

Categorize: We apply directly the textbook account of a cyclotron producing circular motion. This is a substitution problem.

Analyze:

(a) The name "cyclotron frequency" refers to the angular frequency or angular speed

$$\omega = \frac{qB}{m}$$

For protons, $\omega = \dfrac{\left(1.60\times10^{-19}\text{ C}\right)\left(0.450\text{ T}\right)}{1.67\times10^{-27}\text{ kg}} = 4.31\times10^{7}\text{ rad/s}$ ■

(b) The path radius is $R = \dfrac{mv}{Bq}$

Just before the protons escape, their speed is

$$v = \frac{BqR}{m} = \frac{(0.450\text{ T})(1.60\times10^{-19}\text{ C})(1.20\text{ m})}{1.67\times10^{-27}\text{ kg}} = 5.17\times10^{7}\text{ m/s}$$ ■

Finalize: What people call the cyclotron frequency is an angular speed or angular frequency. The frequency in hertz would be $\omega/2\pi$, so it is some megahertz, a radio frequency rather than an audio frequency. The speed is just ωr. It is not a large fraction of the speed of light so there is not a large relativistic correction, and we have not used the expression for relativistic momentum from chapter 39.

---

29. The picture tube in an old black-and-white television uses magnetic deflection coils rather than electric deflection plates. Suppose an electron beam is accelerated through a 50.0-kV potential difference and then through a region of uniform magnetic field 1.00 cm wide. The screen is located 10.0 cm from the center of the coils and is 50.0 cm wide. When the field is turned off, the electron beam hits the center of the screen. What field magnitude is necessary to deflect the beam to the side of the screen? Ignore relativistic corrections.

Solution

Conceptualize: Electrons are the smallest-mass particles of any that have measured mass. They are the easiest particles to deflect. A field of some milliteslas should do the trick.

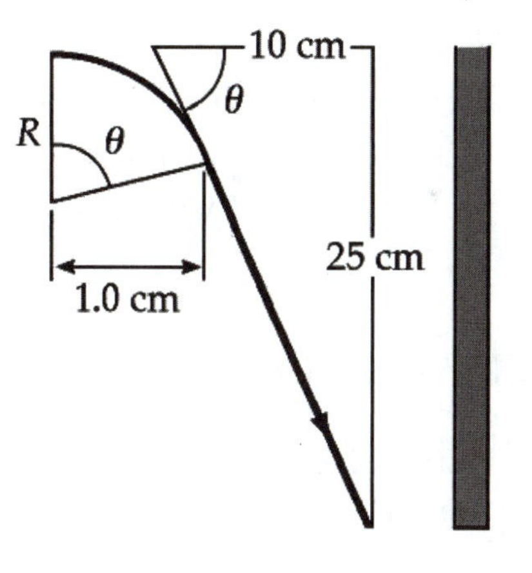

Categorize: We must do some geometry to quantify the angle of deflection and relate it to the radius of curvature of the electrons' path, an arc of a circle, when they are in the field.

Analyze: The beam is deflected by the angle

$$\theta = \tan^{-1}\left(\frac{25.0\text{ cm}}{10.0\text{ cm}}\right) = 68.2^\circ$$

The two angles θ shown are equal because their sides are perpendicular, right side to right side and left side to left side. The radius of curvature of the electrons' path in the field is

$$R = \frac{1.00\text{ cm}}{\sin 68.2^\circ} = 1.077\text{ cm}$$

Now $\frac{1}{2}mv^2 = |q|\Delta V$

so $v = \sqrt{\dfrac{2|q|\Delta V}{m}} = \sqrt{\dfrac{2(1.60\times10^{-19}\text{C})(50\ 000\text{ V})}{9.11\times10^{-31}\text{kg}}} = 1.33\times10^{8}\text{m/s}$

$\Sigma\vec{\mathbf{F}} = m\vec{\mathbf{a}}$ becomes $\dfrac{mv^2}{R} = |q|vB\sin 90°$

$$B = \frac{mv}{|q|R} = \frac{(9.11 \times 10^{-31}\ \text{kg})(1.33 \times 10^{8}\ \text{m/s})}{(1.60 \times 10^{-19}\ \text{C})(1.077 \times 10^{-2}\ \text{m})} = 70.1\ \text{mT}$$ ■

Finalize: The remarkably useful theorem about equal angles with perpendicular sides is in Appendix B.3 of the textbook, with a diagram that can help you to identify the right and left sides of the two angles marked θ in the diagram here. In the problem answer, several milliteslas is what we estimated. Do not endanger yourself, but if you look inside the case of an old television set you can see the magnetic coils, with many turns of fine copper wire, outside the throat of the picture tube. The speed is well over one-tenth of the speed of light, so using equations from Chapter 39 would give a noticeably more accurate answer, according to the theory of relativity.

---

35. A wire having a mass per unit length of 0.500 g/cm carries a 2.00-A current horizontally to the south. What are (a) the direction and (b) the magnitude of the minimum magnetic field needed to lift this wire vertically upward?

Solution

Conceptualize: (a) Since $I = 2.00$ A south, $\vec{\mathbf{B}}$ must be to the east to make $\vec{\mathbf{F}}$ upward according to the right-hand rule for currents in a magnetic field. ■

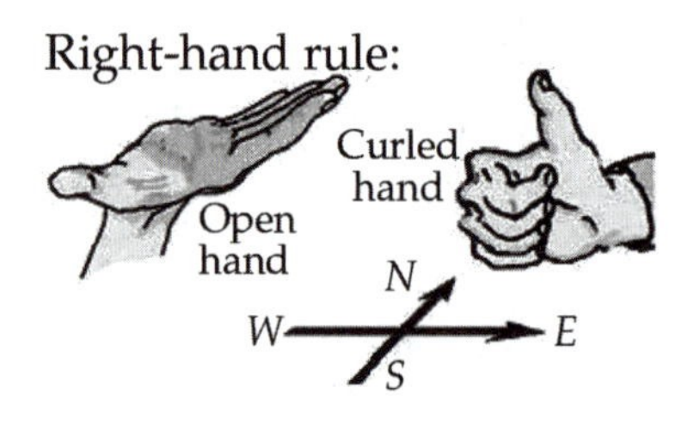

As before, in viewing the diagrams get used to using the curled-hand version of the right-hand rule for later applications.

The magnitude of $\vec{\mathbf{B}}$ should be significantly greater than the Earth's magnetic field (~50 μT), since we do not typically see electric currents making wires levitate.

Categorize: The force on a current-carrying wire in a magnetic field is $\vec{\mathbf{F}}_B = I\vec{\mathbf{L}} \times \vec{\mathbf{B}}$, from which we can find $\vec{\mathbf{B}}$.

Analyze: (b) With I to the south and $\vec{\mathbf{B}}$ to the east, the force on the wire is $F_B = ILB \sin 90°$, which must counterbalance the weight of the wire, mg.

So, $$B = \frac{F_B}{IL} = \frac{mg}{IL} = \frac{g}{I}\left(\frac{m}{L}\right) = \left(\frac{9.80\ \text{m/s}^2}{2.00\ \text{A}}\right)\left(0.500\frac{\text{g}}{\text{cm}}\right)\left(\frac{10^2\ \text{cm/m}}{10^3\ \text{g/kg}}\right) = 0.245\ \text{T}$$ ■

Finalize: The required magnetic field is about 5 000 times stronger than the Earth's magnetic field. Thus it was reasonable to ignore the Earth's magnetic field in this problem. In other situations the Earth's field can have a significant effect. If the field were not straight to the east horizontally, it could lift the wire but it would have to be stronger, so we have successfully found the minimum field. In our description of the magnetic force on a wire carrying electric current, we attribute the vector direction to the length of the wire and not to the current. Current cannot be a vector because at a junction in an electric circuit currents add as scalars and not as vectors.

---

41. A strong magnet is placed under a horizontal conducting ring of radius r that carries current I as shown in Figure P29.41. If the magnetic field $\vec{\mathbf{B}}$ makes an angle θ with the vertical at the ring's location, what are (a) the magnitude and (b) the direction of the resultant magnetic force on the ring?

Figure P29.41

Solution

Conceptualize: You might have learned that a uniform electric field exerts zero force on an electric dipole, but a nonuniform electric field does exert a force on an electric dipole. The current chapter points out that a uniform magnetic field exerts a torque but zero force on a magnetic dipole. This problem illustrates that a nonuniform magnetic field exerts a net force on a magnetic dipole. The force considered here is important in some simple applications.

Categorize: We evaluate in symbolic terms the magnetic force on one bit of the current loop, and then the force on the whole loop.

Analyze: (a) and (b) The magnetic force on each bit of ring is inward and upward, at an angle θ above the radial line, according to

$$\left|d\vec{\mathbf{F}}\right| = I\left|d\vec{\mathbf{s}} \times \vec{\mathbf{B}}\right| = I\,ds\,B$$

The radially inward components tend to squeeze the ring, but cancel out as forces. The upward components $I\,ds\,B\sin\theta$ all add to

$$\vec{\mathbf{F}} = I(2\pi r)\,B\,\sin\theta \text{ up} \qquad \blacksquare$$

Finalize: The magnetic moment of the ring is down. The current loop as a magnetic dipole has a downward-facing north pole, which is repelled by the upward-pointing (and upward-weakening) field coming out of the north pole of the external magnet shown in the textbook picture. This problem is a model for the force that one magnet exerts on another magnet.

---

47. A rectangular coil consists of $N = 100$ closely wrapped turns and has dimensions $a = 0.400$ m and $b = 0.300$ m. The coil is hinged along the y axis, and its plane makes an angle $\theta = 30.0°$ with the x axis (Fig. P29.47). (a) What is the magnitude of the torque exerted on the coil by a uniform magnetic field $B = 0.800$ T directed in the positive x direction when the current is $I = 1.20$ A in the direction shown? (b) What is the expected direction of rotation of the coil?

Solution

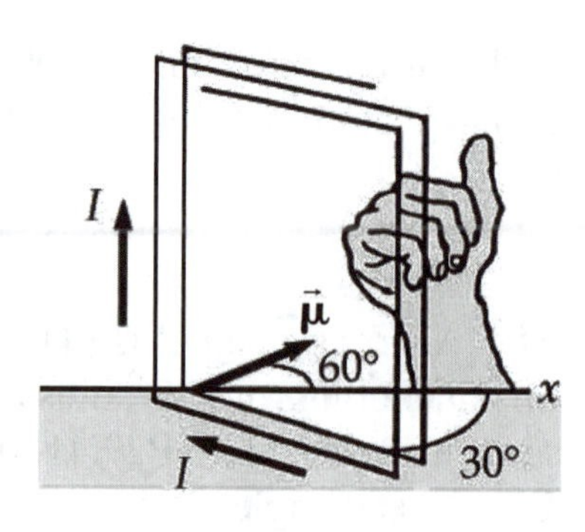

Conceptualize: A magnetic field twists a compass needle, exerting a torque tending to align the needle with the field. Here we have a model for that phenomenon.

Categorize: We use the result proved in the chapter for the torque on a loop of current in a magnetic field.

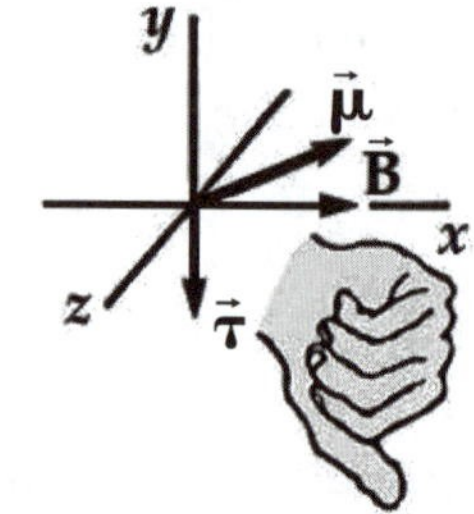

Figure P29.47

Analyze: The magnetic moment of the coil is $\mu = NIA$, perpendicular to its plane and making a $\phi = 60°$ angle with the x axis as shown to the right. The torque on the dipole is then

$$\vec{\tau} = \vec{\mu} \times \vec{B} = NBAI\sin\phi \text{ down}$$

(a) having a magnitude $\tau = NBAI\sin\phi$

$$\tau = (100)(0.800\text{ T})(0.400 \times 0.300\text{ m}^2)(1.20\text{ A})(\sin 60.0°) = 9.98\text{ N}\cdot\text{m}$$ ■

Note that ϕ is the angle between the magnetic moment and the $\vec{B}$ field.

(b) We model the coil as a rigid body under a net torque; the coil will rotate so as to align the magnetic moment with the $\vec{B}$ field. Looking down along the y axis, we will see the coil rotate in the clockwise direction. ■

Finalize: This is a big coil in a strong field, so it feels a rather large torque. This is a model for the functioning of an electric motor.

---

51. In an experiment designed to measure the Earth's magnetic field using the Hall effect, a copper bar 0.500 cm thick is positioned along an east–west direction. Assume $n = 8.46 \times 10^{28}$ electrons/m^3 and the plane of the bar is rotated to be perpendicular to the direction of $\vec{B}$. If a current of 8.00 A in the conductor results in a Hall voltage of 5.10×10^{-12} V, what is the magnitude of the Earth's magnetic field at this location?

Solution

Conceptualize: The Earth's magnetic field is about 50 μT (see Table 29.1), so we should expect a result of that order of magnitude.

Categorize: The magnetic field can be found from the Hall effect voltage:

$$\Delta V_H = \frac{IB}{nqt} \quad \text{or} \quad B = \frac{nqt\Delta V_H}{I}$$

Analyze: From the Hall voltage,

$$B = \frac{(8.46 \times 10^{28}\text{ e}^-/\text{m}^3)(1.60 \times 10^{-19}\text{ C/e}^-)(0.005\,00\text{ m})(5.10 \times 10^{-12}\text{ V})}{8.00\text{ A}}$$

$$B = 4.31 \times 10^{-5}\text{ T} = 43.1\ \mu\text{T}$$ ■

Finalize: The calculated magnetic field is slightly less than we guessed and is reasonable considering that the Earth's local magnetic field varies in magnitude as well as in direction.

---

55. A particle with positive charge $q = 3.20 \times 10^{-19}$ C moves with a velocity $\vec{\mathbf{v}} = (2\hat{\mathbf{i}} + 3\hat{\mathbf{j}} - \hat{\mathbf{k}})$ m/s through a region where both a uniform magnetic field and a uniform electric field exist. (a) Calculate the total force on the moving particle (in unit-vector notation), taking $\vec{\mathbf{B}} = (2\hat{\mathbf{i}} + 4\hat{\mathbf{j}} + \hat{\mathbf{k}})$ T and $\vec{\mathbf{E}} = (4\hat{\mathbf{i}} - \hat{\mathbf{j}} - 2\hat{\mathbf{k}})$ V/m. (b) What angle does the force vector make with the positive x axis?

Solution

Conceptualize: The particle could be a doubly ionized calcium atom. The speed is slow for a vacuum tube. The electric field is very weak and the magnetic field strong, but we have a good exercise in computing a force of some attonewtons.

Categorize: Compare this problem with Problem 8. Here we just add in an electric contribution to the total force. In this problem we are explicitly asked for a direction angle, while Problem 8 asked just for the magnitude. But both problems are done by finding the force in unit-vector notation.

Analyze: The total force is the Lorentz force,

(a) $$\vec{\mathbf{F}} = q\vec{\mathbf{E}} + q\vec{\mathbf{v}} \times \vec{\mathbf{B}} = q\left(\vec{\mathbf{E}} + \vec{\mathbf{v}} \times \vec{\mathbf{B}}\right)$$

$$\vec{\mathbf{F}} = q\left[\left(4\hat{\mathbf{i}} - \hat{\mathbf{j}} - 2\hat{\mathbf{k}}\right)\text{V/m} + \left(2\hat{\mathbf{i}} + 3\hat{\mathbf{j}} - \hat{\mathbf{k}}\right)\text{m/s} \times \left(2\hat{\mathbf{i}} + 4\hat{\mathbf{j}} + \hat{\mathbf{k}}\right)\text{T}\right]$$

$$\vec{\mathbf{F}} = q\left[\left(4\hat{\mathbf{i}} - \hat{\mathbf{j}} - 2\hat{\mathbf{k}}\right)\text{V/m} + \left(7\hat{\mathbf{i}} - 4\hat{\mathbf{j}} + 2\hat{\mathbf{k}}\right)\text{T} \cdot \text{m/s}\right]$$

$$\vec{\mathbf{F}} = q\left[\left(11\hat{\mathbf{i}} - 5\hat{\mathbf{j}}\right)\text{V/m}\right] = q\left[\left(11\hat{\mathbf{i}} - 5\hat{\mathbf{j}}\right)\text{N/C}\right]$$

$$\vec{\mathbf{F}} = \left(3.20 \times 10^{-19}\ \text{C}\right)\left[\left(11\hat{\mathbf{i}} - 5\hat{\mathbf{j}}\right)\ \text{N/C}\right] = \left(3.52\hat{\mathbf{i}} - 1.60\hat{\mathbf{j}}\right) \times 10^{-18}\ \text{N} \qquad \blacksquare$$

(b) $$\vec{\mathbf{F}} \cdot \hat{\mathbf{i}} = F \cos\theta = F_x: \qquad \theta = \cos^{-1}\left(F_x / F\right) = \cos^{-1}(3.52/3.87) = 24.4° \qquad \blacksquare$$

Finalize: The problem is solved by straightforward evaluation, but contains somewhat sophisticated ideas. To prove that a tesla-meter per second is the same as a volt per meter, think about the unit equations N = C · m · T/s from $F = qvB\sin\theta$, J = N · m from $W = F\Delta r \cos\theta$, and J = V · C from $U = qV$. To evaluate the cross product you can use a determinant, or you can use the cross-product multiplication table in the textbook's equations 11.7:

$$\left(2\hat{\mathbf{i}} + 3\hat{\mathbf{j}} - \hat{\mathbf{k}}\right) \times \left(2\hat{\mathbf{i}} + 4\hat{\mathbf{j}} + \hat{\mathbf{k}}\right) =$$

$$= 2\hat{\mathbf{i}} \times 2\hat{\mathbf{i}} + 2\hat{\mathbf{i}} \times 4\hat{\mathbf{j}} + 2\hat{\mathbf{i}} \times 1\hat{\mathbf{k}} + 3\hat{\mathbf{j}} \times 2\hat{\mathbf{i}} + 3\hat{\mathbf{j}} \times 4\hat{\mathbf{j}} + 3\hat{\mathbf{j}} \times 1\hat{\mathbf{k}} - 1\hat{\mathbf{k}} \times 2\hat{\mathbf{i}} - 1\hat{\mathbf{k}} \times 4\hat{\mathbf{j}} - 1\hat{\mathbf{k}} \times 1\hat{\mathbf{k}} =$$

$$= 0 \quad + 8\hat{\mathbf{k}} \quad - 2\hat{\mathbf{j}} \quad - 6\hat{\mathbf{k}} \quad + 0 \quad + 3\hat{\mathbf{i}} \quad - 2\hat{\mathbf{j}} \quad + 4\hat{\mathbf{i}} \quad + 0 \quad =$$

$$= 7\hat{\mathbf{i}} - 4\hat{\mathbf{j}} + 2\hat{\mathbf{k}}$$

---

56. Heart-lung machines and artificial kidney machines employ blood pumps. The blood is confined to an electrically insulating tube, cylindrical in practice but represented here for simplicity as a rectangle of width w and height h. Figure P29.56 shows a rectangular section of blood within the tube. Two electrodes fit into the top and the bottom of the tube. The potential difference between them establishes an electric current through the blood, with current density J over a section of length L shown in Figure P29.56. A perpendicular magnetic field exists in the same region. (a) Explain why this arrangement produces on the liquid a force that is directed along the length of the pipe. (b) Show that the section of liquid in the magnetic field experiences a pressure increase JLB. (c) After the blood leaves the pump, is it charged? (d) Is it carrying current? (e) Is it magnetized? (The same electromagnetic pump can be used for any fluid that conducts electricity, such as liquid sodium in a nuclear reactor.)

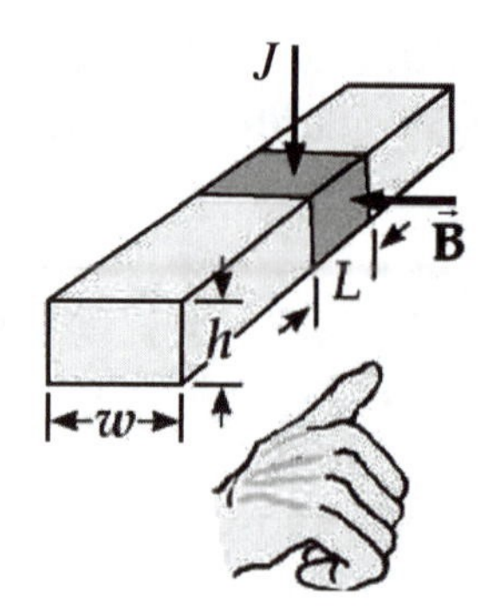

Figure P29.56

Solution

Conceptualize: Electricity is economically important because (for one thing) it is so easy to convert electrically transmitted energy into mechanical energy with high efficiency, using a magnetic force. This conversion happens in a spinning electric motor, and also in this electromagnetic pump. A mechanical pump could mangle blood cells. The simplicity of design makes this pump dependable. The blood is easily kept uncontaminated; the tube is simple to clean or cheap to replace.

Categorize: We think about the magnetic force acting on the electric current between the electrodes. We need not think about the motion of the blood.

Analyze:

(a) We define vector $\vec{\mathbf{h}}$ as downward, in the direction of the current. By the right-hand rule, the electric current carried by the material experiences a force $I\vec{\mathbf{h}} \times \vec{\mathbf{B}}$ along the pipe, into the page. ■

(b) The blood, containing ions and electrons, flows along the pipe transporting no net charge. But inside the section of length L, electrons drift upward to constitute downward electric current $J(\text{area}) = JLw$. The current feels magnetic force $I\vec{\mathbf{h}} \times \vec{\mathbf{B}} = JLwhB \sin 90°$. This force along the pipe axis can make the fluid move, exerting pressure

$$\frac{F}{\text{area}} = \frac{JLwhB}{hw} = JLB$$ ■

The hand in the figure shows that the fluid moves away from you, into the page.

(c) Charge moves within the fluid inside the length L, but charge does not accumulate. The fluid is not charged after it leaves the pump. (d) It is not current-carrying and (e) it is not magnetized. ■

Finalize: An electric force and a magnetic force can act on the same electric charge. Here an ion in the blood feels a vertical electric force making it participate in the vertical electric current, and then also a horizontal magnetic force. Make sure you know that a magnetic charge or a magnetic current does not exist.

---

77. Consider an electron orbiting a proton and maintained in a fixed circular path of radius $R = 5.29 \times 10^{-11}$ m by the Coulomb force. Treat the orbiting particle as a current loop. Calculate the resulting torque when the electron-proton system is placed in a magnetic field of 0.400 T directed perpendicular to the magnetic moment of the loop.

Solution

Conceptualize: We model the proton as standing still, because its mass is so much larger than that of the electron. Since the mass of the electron is very small ($\sim 10^{-30}$ kg), we might expect that the torque on the orbiting electron will be very small as well, perhaps $\sim 10^{-30}$ N · m.

Categorize: The torque on a current loop that is perpendicular to a magnetic field can be found from $|\vec{\tau}| = IAB \sin\theta$. The magnetic field is given, $\theta = 90°$, the area of the loop can be found from the radius of the circular path, and the current can be found from the centripetal acceleration that results from the Coulomb force that attracts the electron to the proton.

Analyze: The area of the loop is

$$A = \pi r^2 = \pi(5.29 \times 10^{-11}\text{ m})^2 = 8.79 \times 10^{-21}\text{ m}^2$$

If v is the speed of the electron, then the period of its circular motion will be

$$T = \frac{2\pi r}{v}$$

and the effective current due to the orbiting electron is

$$I = \frac{\Delta Q}{\Delta t} = \frac{e}{T}$$

Applying Newton's second law with the Coulomb force causing the centripetal acceleration

gives $$\sum F = \frac{k_e q^2}{R^2} = \frac{mv^2}{R}$$

so that $$v = q\sqrt{\frac{k_e}{mR}} = \frac{2\pi R}{T} \qquad \text{and} \qquad T = 2\pi\sqrt{\frac{mR^3}{q^2 k_e}}$$

$$T = 2\pi\sqrt{\frac{(9.11 \times 10^{-31}\text{ kg})(5.29 \times 10^{-11}\text{ m})^3}{(1.60 \times 10^{-19}\text{ C})^2(8.99 \times 10^9\text{ N} \cdot \text{m}^2/\text{C}^2)}} = 1.52 \times 10^{-16}\text{ s}$$

The torque is $|\vec{\tau}| = \left(\frac{q}{T}\right)AB$

$$|\vec{\tau}| = \left(\frac{1.60 \times 10^{-19}\ \text{C}}{1.52 \times 10^{-16}\ \text{s}}\right)\left[\pi(5.29 \times 10^{-11}\text{m})^2\right](0.400\ \text{T})$$

$$= (9.25 \times 10^{-24}\ \text{A} \cdot \text{m}^2)(0.400\ \text{N/A} \cdot \text{m}) = 3.70 \times 10^{-24}\ \text{N} \cdot \text{m}$$ ■

Finalize: The torque is certainly small, but a million times larger than we guessed. This torque will cause the atom to precess with a frequency proportional to the applied magnetic field. This problem is about Niels Bohr's model for the ground state of a hydrogen atom. The magnetic moment we computed is called a Bohr magneton. A similar process on the nuclear, rather than the atomic, level leads to nuclear magnetic resonance (NMR), which is used for magnetic resonance imaging (MRI) scans employed for medical diagnostic testing, described in Section 44.8.

30

Sources of the Magnetic Field

EQUATIONS AND CONCEPTS

The **Biot-Savart law** is used to calculate the magnetic field at a point in space which is a distance r from a current I. *The vector $d\vec{\mathbf{B}}$ is perpendicular to both $d\vec{\mathbf{s}}$ (which points in the direction of the current) and to the unit vector $\hat{\mathbf{r}}$ (which points from the current element toward the point where the field is to be determined).* You should confirm that at point P shown in the figure, the direction of $d\vec{\mathbf{B}}$ due to element $d\vec{\mathbf{s}}$ is directed *out* of the page; but the direction of $d\vec{\mathbf{B}}$ at point P' is *into* the page.

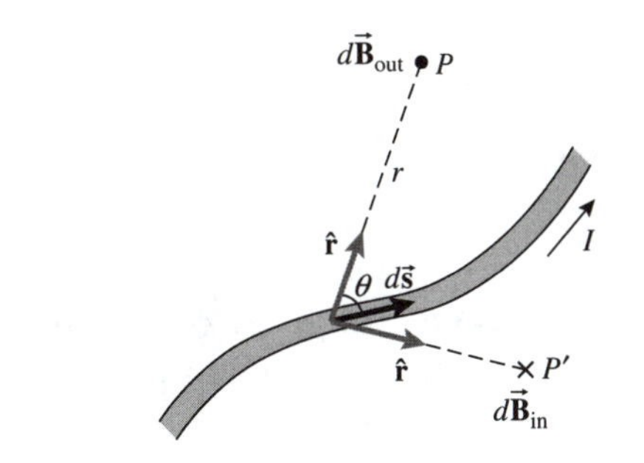

$$d\vec{\mathbf{B}} = \frac{\mu_0}{4\pi}\frac{I\, d\vec{\mathbf{s}} \times \hat{\mathbf{r}}}{r^2} \tag{30.1}$$

$$\mu_0 = 4\pi \times 10^{-7}\ \mathrm{T \cdot m/A} \tag{30.2}$$

The **total magnetic field** at a point in the vicinity of a current of finite length is found by integrating the Biot-Savart law equation over the entire length of the current path. *Remember, the integrand is a vector quantity.*

$$\vec{\mathbf{B}} = \frac{\mu_0 I}{4\pi}\int \frac{d\vec{\mathbf{s}} \times \hat{\mathbf{r}}}{r^2} \tag{30.3}$$

The **magnitudes of the magnetic fields** calculated by applying Equation 30.3 to several current geometries are shown below.

B at a perpendicular **distance a from a straight wire of finite length**, where θ_1 and θ_2 are as shown in the figure to the right:

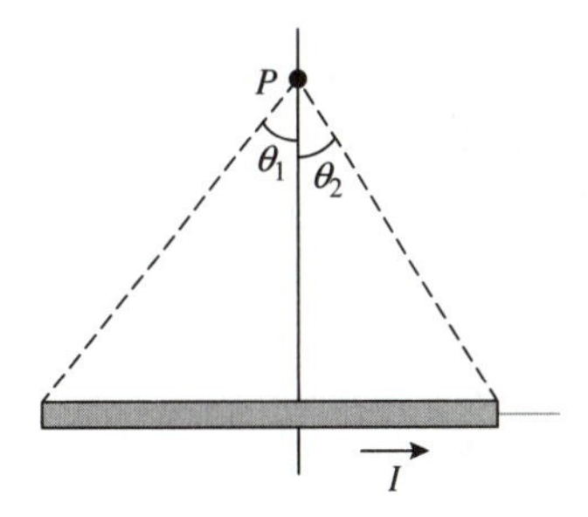

$$B = \frac{\mu_0 I}{4\pi a}(\sin\theta_1 - \sin\theta_2) \tag{30.4}$$

B at a **distance a from a long straight conductor**, carrying a current I. *In this case the length of the conductor is assumed to be infinite compared to the distance a:*

$$B = \frac{\mu_0 I}{2\pi a} \tag{30.5}$$

B at the **center of an arc of radius a** which subtends an angle θ (in radians) at the center of the arc:

$$B = \frac{\mu_0 I}{4\pi a}\theta \tag{30.6}$$

B_x **at any point x along the axis of a circular loop** of radius a:

$$B_x = \frac{\mu_0 I a^2}{2\left(a^2 + x^2\right)^{3/2}} \tag{30.7}$$

B **at the center of a current loop** of radius a. This corresponds to $x = 0$ in Equation 30.7 above:

$$B = \frac{\mu_0 I}{2a} \tag{30.8}$$

B **on the axis of a current loop at a distance much greater than the radius:**

$$B \approx \frac{\mu_0 I a^2}{2x^3} \qquad (x >> a) \tag{30.9}$$

The **magnetic force per unit length** between very long parallel conductors depends on the distance a between the conductors and the magnitudes of the two currents. Parallel conductors carrying currents in the same direction attract each other; and parallel conductors carrying currents in opposite directions repel each other. *The forces on the two conductors will be equal in magnitude regardless of the relative magnitude of the two currents.*

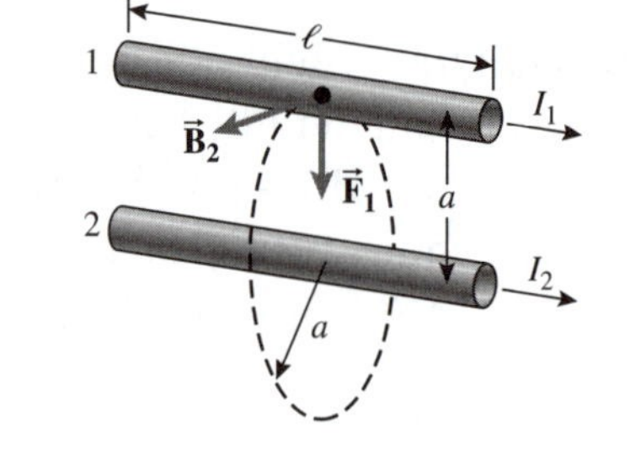

$$\frac{F_B}{\ell} = \frac{\mu_0 I_1 I_2}{2\pi a} \tag{30.12}$$

Ampère's law describes the creation of magnetic fields by continuous current pathways. *The integral of the tangential component of the magnetic field around any closed path is equal to μ_0 times the total current passing through a surface bounded by the closed path.*

$$\oint \vec{\mathbf{B}} \cdot d\vec{\mathbf{s}} = \mu_0 I \tag{30.13}$$

Current geometries with a high degree of symmetry are good candidates for Ampère's law. Results for two current geometries are shown on the following page.

Magnetic field inside a toroid having N total turns and at a distance r from the center of the toroid.

$$B = \frac{\mu_0 NI}{2\pi r} \quad (30.16)$$

Magnetic field at a point within a solenoid of N total turns and length ℓ. In Equation 30.17, n equals the number of turns per unit length of the solenoid $(n = N/\ell)$.

$$B = \mu_0 \frac{N}{\ell} I = \mu_0 nI \quad (30.17)$$

The **magnetic flux** through a surface is the integral of the normal component of the field over the surface. *The magnetic flux through any closed surface is equal to zero.*

$$\Phi_B \equiv \int \vec{\mathbf{B}} \cdot d\vec{\mathbf{A}} \quad (30.18)$$

The **direction of the magnetic field due to current in a long wire** is determined by using the right-hand rule. Hold the conductor in your right hand with your thumb pointing in the direction of the conventional current. Your fingers will wrap around the wire in the direction of the magnetic field lines. *The magnetic field is tangent to the circular field lines at every point in the region around the conductor.*

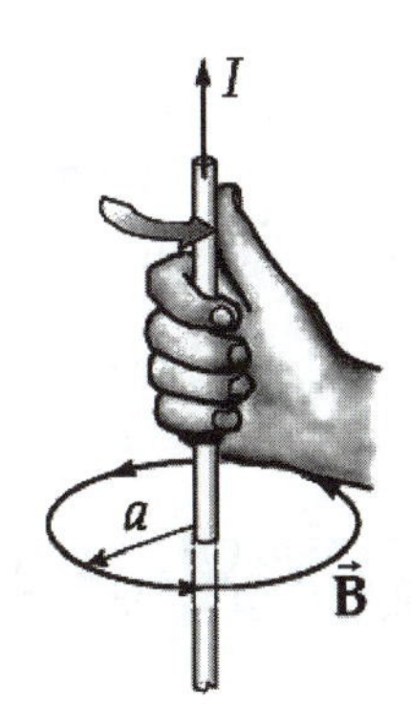

The **direction of the magnetic field at the center of a current loop** is perpendicular to the plane of the loop and directed in the sense given by the right-hand rule.

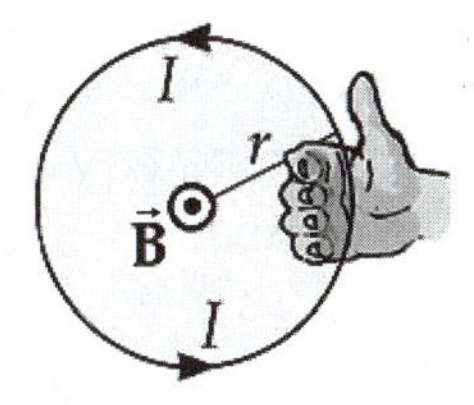

Magnetic field directed out of the page.

Within a solenoid, the magnetic field is parallel to the axis of the solenoid and pointing in a sense determined by applying the right-hand rule to one of the coils.

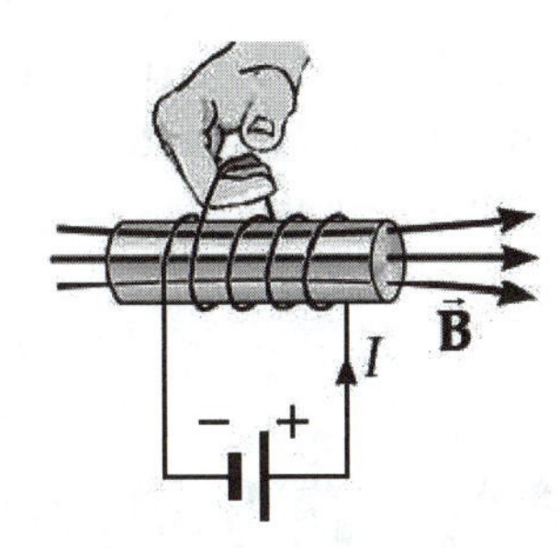

The **magnitude of the orbital magnetic moment** (μ) of an electron is proportional to its orbital angular momentum (L). The orbital angular momentum is quantized and equals integer multiples (including zero) of $\hbar$ where $\hbar = h/2\pi$.

$$\mu = \left(\frac{e}{2m_e}\right)L \quad (30.22)$$

The direction of $\vec{\boldsymbol{\mu}}$ is opposite the direction of $\vec{\mathbf{L}}$.

The **Bohr magneton** is numerically equal to the spin magnetic moment of an electron.

$$\mu_B = \frac{e\hbar}{2m_e} = 9.27\times10^{-24}\ \text{J/T} \tag{30.25}$$

REVIEW CHECKLIST

You should be able to:

- Use the Biot-Savart law to calculate the magnetic field ($\vec{\mathbf{B}}$) at a specified point in the vicinity of a current element; and by integration find the total magnetic field due to a number of important geometric configurations. Use of the Biot-Savart law must include a clear understanding of the direction of the magnetic field contribution relative to the direction of the current element which produces it and the direction of the vector which locates the point at which the field is to be calculated. (Section 30.1)
- Calculate the force between two parallel current-carrying conductors. (Section 30.2)
- Use Ampère's law to calculate the magnetic field due to steady current configurations which have a sufficiently high degree of symmetry such as a long straight conductor, a long solenoid, and a toroidal coil. (Sections 30.3 and 30.4)
- Calculate the magnetic flux through a surface area placed in either a uniform or non-uniform magnetic field. (Section 30.5)

ANSWER TO AN OBJECTIVE QUESTION

1. What creates a magnetic field? Choose every correct answer. (a) A stationary object with electric charge, (b) a moving object with electric charge, (c) a stationary conductor carrying electric current, (d) a difference in electric potential, (e) a charged capacitor disconnected from a battery and at rest. *Note:* In Chapter 34, we will see that a changing electric field also creates a magnetic field.

Answer The correct choices are (b) and (c). By contrast, if the question asked what creates an electric field, definitely correct answers would be (a), (b), and (e); and a difference in potential (d) can only exist in a space that also contains an electric field. Notice that a moving charged object creates both an electric field and a magnetic field. Both a stationary and a moving current-carrying conductor create magnetic fields. But a typical conductor has zero net charge, and creates no electric field.

□ □ □ □

ANSWERS TO SELECTED CONCEPTUAL QUESTIONS

1. Explain why two parallel wires carrying currents in opposite directions repel each other.

Answer The diagram will help you understand this result. The magnetic field $\vec{\mathbf{B}}_2$ created by wire 2 at the position of wire 1 is directed out of the paper. Hence, the magnetic force on wire 1, given by $I_1\vec{\mathbf{L}}_1 \times \vec{\mathbf{B}}_2$, must be directed to the left since $\vec{\mathbf{L}}_1 \times \vec{\mathbf{B}}_2$ is down × out of the paper = to the left. Likewise, you can show that the magnetic force on wire 2 due to the field of wire 1 is directed towards the right.

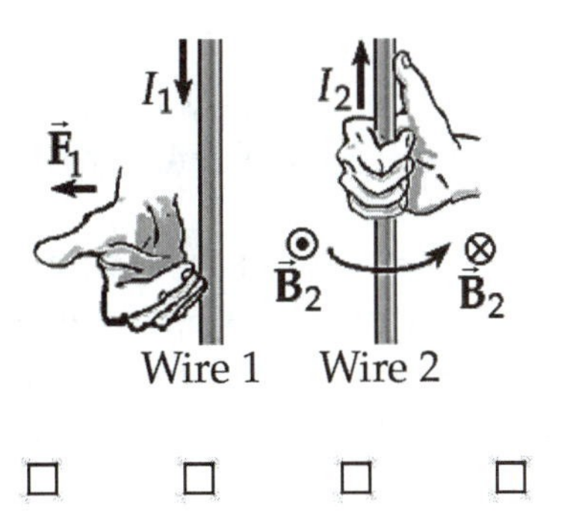

□ □ □ □

4. A hollow copper tube carries a current along its length. Why is $B = 0$ inside the tube? Is B nonzero outside the tube?

Answer Let us apply Ampère's law to the closed path labeled 1 in this figure. Since there is no current through this path, and because of the symmetry of the configuration, we see that the magnetic field inside the tube must be zero. On the other hand, the net current through the path labeled 2 is I, the current carried by the conductor. Therefore, the field outside the tube is nonzero.

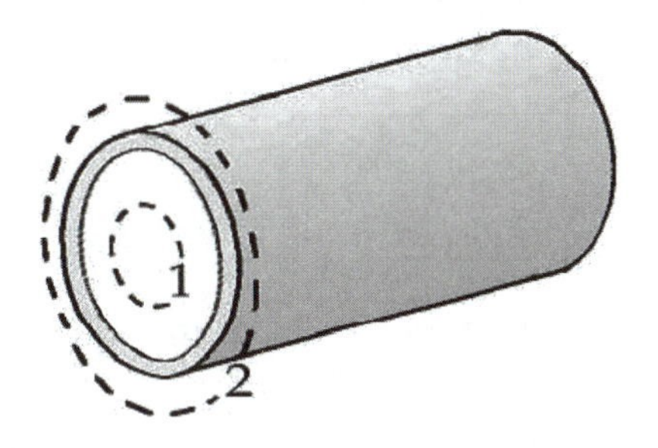

□ □ □ □

SOLUTIONS TO SELECTED END-OF-CHAPTER PROBLEMS

5. (a) A conducting loop in the shape of a square of edge length $\ell = 0.400$ m carries a current $I = 10.0$ A as shown in Figure P30.5. Calculate the magnitude and direction of the magnetic field at the center of the square. (b) **What If?** If this conductor is reshaped to form a circular loop and carries the same current, what is the value of the magnetic field at the center?

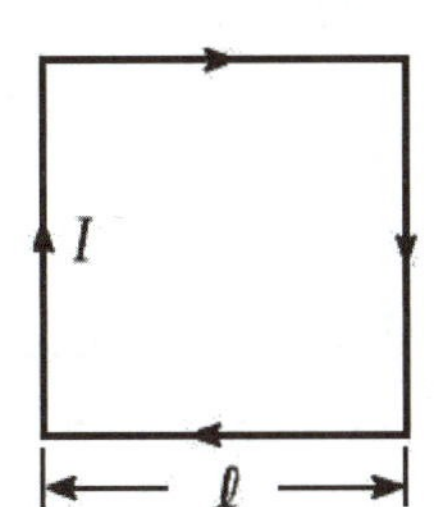

Figure P30.5

Solution

Conceptualize: As shown in the second drawing of the square, the magnetic field at the center is directed into the page from the clockwise current. If we consider the sides of the square to be sections of four infinite wires, then we could expect the magnetic field at the center of the square to be a little less than four times the strength of the field at a point $\ell/2$ away from an infinite wire with current I.

$$B < 4\frac{\mu_0 I}{2\pi a} = 4\frac{\left(4\pi \times 10^{-7}\ \text{T}\cdot\text{m/A}\right)(10.0\ \text{A})}{2\pi(0.200\ \text{m})} = 40.0\ \mu\text{T}$$

Forming the wire into a circle should not greatly change the magnetic field at the center since the average distance of the wire from the center will not be much different.

Categorize: Each side of the square is simply a section of a thin, straight conductor, so the solution derived from the Biot-Savart law in Example 30.1 can be applied to part (a) of this problem. For part (b), the Biot-Savart law can also be used to derive the equation for the magnetic field at the center of a circular current loop as shown in Example 30.3.

Analyze:

(a) We use Equation 30.4 for the field created by each side of the square. Each side contributes a field away from you at the center, so together they produce this magnetic field:

$$B = \frac{4\mu_0 I}{4\pi a}\left(\sin\frac{\pi}{4} - \sin\left(-\frac{\pi}{4}\right)\right) = \frac{4\left(4\pi \times 10^{-7}\ \text{T}\cdot\text{m/A}\right)(10.0\ \text{A})}{4\pi(0.200\ \text{m})}\left(\frac{\sqrt{2}}{2} + \frac{\sqrt{2}}{2}\right)$$

so at the center of the square,

$$\vec{\mathbf{B}} = 2.00\sqrt{2} \times 10^{-5}\ \text{T} = 28.3\ \mu\text{T} \quad \text{perpendicularly into the page}$$ ■

(b) As in the first part of the problem, the direction of the magnetic field will be into the page. The new radius is found from the length of wire: $4\ell = 2\pi R$, so $R = 2\ell/\pi = 0.255$ m. Equation 30.8 gives the magnetic field at the center of a circular current loop:

$$B = \frac{\mu_0 I}{2R} = \frac{(4\pi \times 10^{-7}\ \text{T}\cdot\text{m/A})(10.0\ \text{A})}{2(0.255\ \text{m})} = 2.47 \times 10^{-5}\ \text{T} = 24.7\ \mu\text{T}$$ ■

Caution! If you use your calculator, it may not understand the keystrokes [4] [×] [π] [EXP] [+/–] [7]. To get the right answer, you may need to use [4] [EXP] [+/–] [7] [×] [π].

Finalize: The magnetic field in part (a) is less than 40 μT as we predicted. Also, the magnetic fields from the square and circular loops are similar in magnitude, with the field from the circular loop being about 15% less than from the square loop.

Quick Tip: A simple way to use your right hand to find the magnetic field due to a current loop is to curl the fingers of your right hand in the direction of the current. Your extended thumb will then point in the direction of the magnetic field within the loop or solenoid. This is also the direction of the magnetic moment of the current loop.

6. An infinitely long wire carrying a current I is bent at a right angle as shown in Figure P30.6. Determine the magnetic field at point P, located a distance x from the corner of the wire.

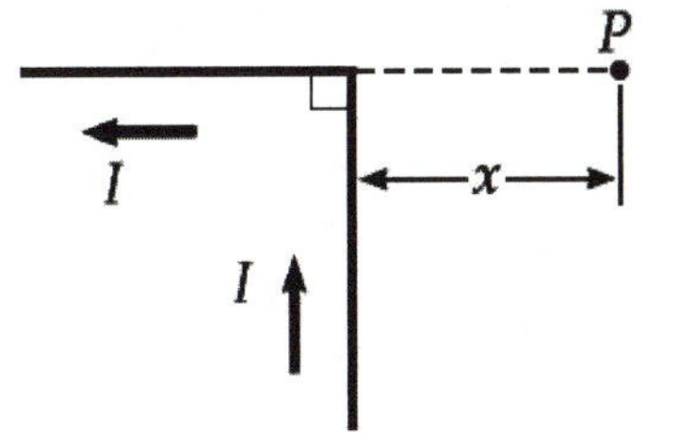

Figure P30.6

Solution

Conceptualize: We will add as vectors the field created at P by the current in the vertical section of wire, and the field created at P by the horizontal section. The answer should be directly proportional to μ_0 and directly proportional to I, should decrease as x increases, and should contain some proportionality constant we cannot guess in advance.

Categorize: For the vertical section, we can use half of the field created by an infinitely long straight wire. For the horizontal section this $\mu_0 I/2\pi a$ result does not apply, because the point P is not perpendicularly off to the side of the horizontal section. Instead, we will reason from the Biot-Savart law.

Analyze: The vertical section of wire constitutes one half of an infinitely long straight wire at distance x from P, so it creates a field equal to

$$B = \tfrac{1}{2}\left(\frac{\mu_0 I}{2\pi x}\right)$$

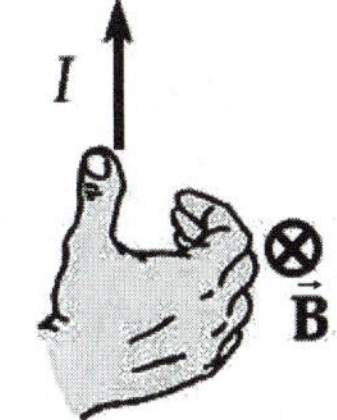

Hold your right hand with extended thumb in the direction of the current; the field is away from you, into the paper.

For each bit of the horizontal section of wire $d\vec{\mathbf{s}}$ is to the left and $\hat{\mathbf{r}}$ is to the right, so $d\vec{\mathbf{s}} \times \hat{\mathbf{r}} = 0$. The horizontal current produces zero field at P. Thus,

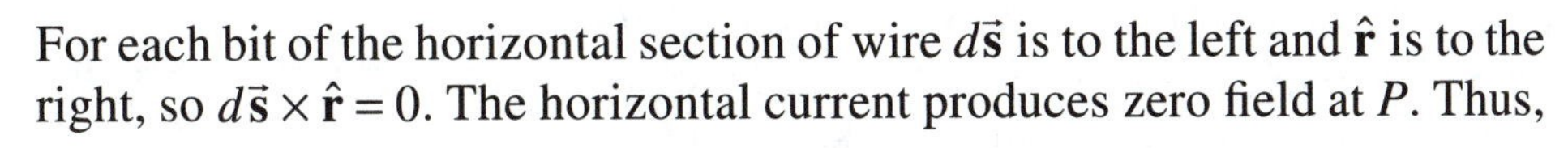

$$\vec{\mathbf{B}} = \frac{\mu_0 I}{4\pi x} \quad \text{into the paper}$$ ■

Finalize: The dependence on x turns out to be simply inverse proportionality. The proportionality constant is $1/4\pi$.

An electric current creating a magnetic field is a phenomenon different from any we have studied before, and the field direction is at least as important as the magnitude. (Make sure your light saber is pointing away from you when you switch it on!) The magnetic field is generally in the last direction you could guess. Get plenty of practice with the right-hand rules. In this problem reasoning about vector directions contributes to knowing the magnitude of the field.

---

12. One long wire carries current 30.0 A to the left along the x axis. A second long wire carries current 50.0 A to the right along the line ($y = 0.280$ m, $z = 0$). (a) Where in the plane of the two wires is the total magnetic field equal to zero? (b) A particle with a charge of $-2.00\ \mu\text{C}$ is moving with a velocity of $150\hat{\mathbf{i}}$ Mm/s along the line ($y = 0.100$ m, $z = 0$). Calculate the vector magnetic force acting on the particle. (c) **What If?** A uniform electric field is applied to allow this particle to pass through this region undeflected. Calculate the required vector electric field.

Solution

Conceptualize: In an experiment or demonstration, you may see current-carrying wires causing deflection of moving charged particles in a vacuum tube. We estimate that the magnetic fields of the two currents here may add to zero at a location some tens of centimeters below the 30-A wire. In part (b) the force should be a small fraction of a newton. Figuring out its direction will be as important as calculating its magnitude. In part (c) the strength of the electric field may be some kilonewtons per coulomb.

Categorize: In part (a), we must find a location, farther away from the 50-A current and closer to the 30-A current, where the fields which the currents create separately are equal in magnitude and opposite in direction. In part (b) we will compute the magnitude and identify the direction of the magnetic field each current creates at the location between them; then add the fields as vectors; and then find the magnetic force on the moving charge by working out a cross product. Part (c) reviews the definition of the electric field by asking us for an extra electric field that will exert a counterbalancing force on the same moving charge.

Analyze:

(a) Above the pair of wires, the field out of the page of the 50 A current will be stronger than the $(-\hat{\mathbf{k}})$ field of the 30 A current, so they cannot add to zero. Between the wires, both produce fields into the page. The magnetic fields can only add to zero below the wires, at coordinate $y = -|y|$. Here we represent the total field as

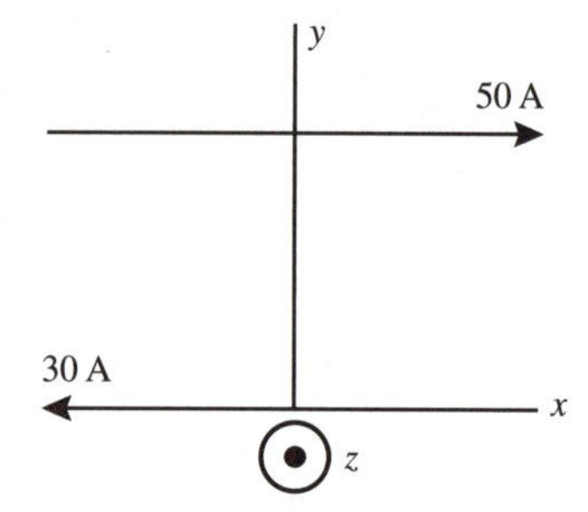

$$\vec{\mathbf{B}} = \frac{\mu_0 I}{2\pi r} + \frac{\mu_0 I}{2\pi r}$$ which requires y to satisfy

$$0 = \frac{\mu_0}{2\pi}\left[\frac{50\text{ A}}{(|y| + 0.28\text{ m})}\left(-\hat{\mathbf{k}}\right) + \frac{30\text{ A}}{|y|}\left(\hat{\mathbf{k}}\right)\right] \quad \text{or}$$

$$50|y| = 30(|y| + 0.28\text{ m})$$

$$50(-y) = 30(0.28\text{ m} - y)$$

$$-20y = 30(0.28\text{ m})$$

So the net field is zero all along the line at $y = -0.420$ m ■

(b) At $y = 0.1$ m the total field is $\vec{\mathbf{B}} = \dfrac{\mu_0 I}{2\pi r} + \dfrac{\mu_0 I}{2\pi r}$

$$\vec{\mathbf{B}} = \frac{4\pi \times 10^{-7}\ \mathrm{T\cdot m/A}}{2\pi}\left(\frac{50\ \mathrm{A}}{(0.28 - 0.10)\ \mathrm{m}}\left(-\hat{\mathbf{k}}\right) + \frac{30\ \mathrm{A}}{0.10\ \mathrm{m}}\left(-\hat{\mathbf{k}}\right)\right)$$

$$= 1.16 \times 10^{-4}\ \mathrm{T}\left(-\hat{\mathbf{k}}\right) = 116\ \mu\mathrm{T}\ \text{into the plane of the paper}$$

The force on the particle is then

$$\vec{\mathbf{F}} = q\vec{\mathbf{v}} \times \vec{\mathbf{B}} = \left(-2 \times 10^{-6}\mathrm{C}\right)\left(150 \times 10^{6}\ \mathrm{m/s}\right)\left(\hat{\mathbf{i}}\right) \times \left(1.16 \times 10^{-4}\ \mathrm{N\cdot s/C\cdot m}\right)\left(-\hat{\mathbf{k}}\right)$$

$$= 3.47 \times 10^{-2}\mathrm{N}\left(-\hat{\mathbf{j}}\right) \qquad \blacksquare$$

$$= 34.7\ \mathrm{mN}\ \text{downward toward the bottom of the page in the diagram}$$

(c) We require the electric force to be the same size but in the opposite direction to the magnetic force, according to

$$\vec{\mathbf{F}}_e = 3.47 \times 10^{-2}\mathrm{N}\left(+\hat{\mathbf{j}}\right) = q\vec{\mathbf{E}} = \left(-2 \times 10^{-6}\mathrm{C}\right)\vec{\mathbf{E}}$$

So $\vec{\mathbf{E}} = -1.73 \times 10^{4}\,\hat{\mathbf{j}}\ \mathrm{N/C} = 17.3\ \mathrm{kN/C}$ toward the bottom of the page ■

Finalize: In part (a), we might think of the zero-net-field location as remarkably far away, compared to where +50 μC and −30 μC charged particles would create zero net electric field. That extra distance happens because the magnetic field of a straight wire decreases as $1/r$ and not in the inverse-square-law pattern of $1/r^2$.

We could think of part (b) as requiring us to use a right hand three times, to find the magnetic fields of both currents and also the direction of the magnetic force. The field is never toward the wire or away from it, and never in the direction of the current or the opposite direction. The magnetic force exerted on the extra moving charge, in contrast, is toward the 30-A current and away from the 50-A current. The negative charge moving to the right constitutes an element of current to the left, which is attracted by the 30-A current in that same direction and repelled by the oppositely directed 50-A current. Notice that the forces that the two wires exert on each other have nothing to do with this problem at all.

Use part (c) to remind yourself of how different electric and magnetic forces are. The electric force does not depend on the velocity of the charge, but the magnetic force does. The electric force on a negative charge is directly opposite to the electric field direction, while the magnetic force is perpendicular to the magnetic field. The kilonewton-per-coulomb electric field and the microtesla magnetic field here exert forces of equal size.

---

23. In Figure P30.23, the current in the long, straight wire is $I_1 = 5.00$ A and the wire lies in the plane of the rectangular loop, which carries the current $I_2 = 10.0$ A. The dimensions in the figure are $c = 0.100$ m, $a = 0.150$ m, and $\ell = 0.450$ m. Find the magnitude and direction of the net force exerted on the loop by the magnetic field created by the wire.

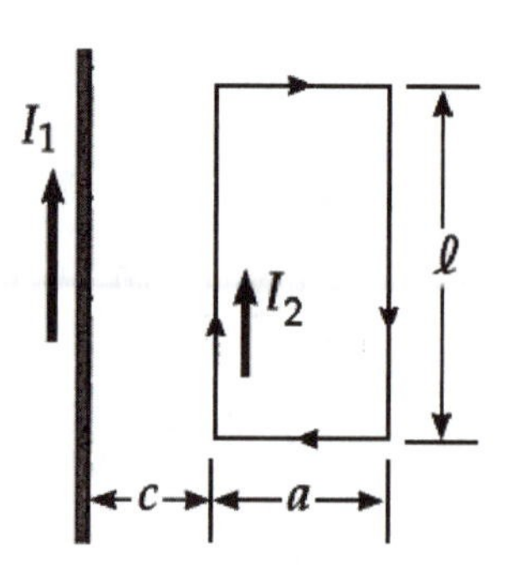

Figure P30.23

Solution

Conceptualize: There are forces in opposite directions on the loop, but we must remember that the magnetic field is stronger near the wire than it is farther away. By symmetry the forces exerted on sides 2 and 4 (the horizontal segments of length a) are equal and opposite, and therefore add to zero. The magnetic field in the plane of the loop is directed into the page to the right of I_1. By the right-hand rule, $\vec{\mathbf{F}} = I\vec{\mathbf{L}} \times \vec{\mathbf{B}}$ is directed toward the **left** for side 1 of the loop and a smaller force is directed toward the **right** for side 3. Therefore, we should expect the net force to be to the left, possibly in the μN range for the currents and distances given.

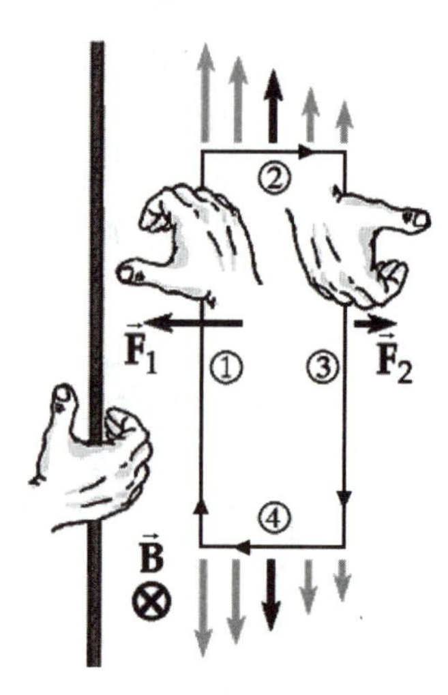

Categorize: The magnetic force between two parallel wires can be found from Equation 30.11, which can be applied to sides 1 and 3 of the loop to find the net force resulting from these opposing force vectors.

Analyze:

$$\vec{\mathbf{F}} = \vec{\mathbf{F}}_1 + \vec{\mathbf{F}}_2 = \frac{\mu_0 I_1 I_2 \ell}{2\pi}\left(\frac{1}{c+a} - \frac{1}{c}\right)\hat{\mathbf{i}} = \frac{\mu_0 I_1 I_2 \ell}{2\pi}\left(\frac{-a}{c(c+a)}\right)\hat{\mathbf{i}}$$

$$\vec{\mathbf{F}} = \frac{\left(4\pi \times 10^{-7}\ \text{N/A}^2\right)(5.00\ \text{A})(10.0\ \text{A})(0.450\ \text{m})}{2\pi}\left(\frac{-0.150\ \text{m}}{(0.100\ \text{m})(0.250\ \text{m})}\right)\hat{\mathbf{i}}$$

$$\vec{\mathbf{F}} = \left(-2.70 \times 10^{-5}\hat{\mathbf{i}}\right)\text{N} \quad \text{or} \quad \vec{\mathbf{F}} = 2.70 \times 10^{-5}\ \text{N} \ \text{ toward the left}$$ ■

Finalize: The net force is to the left and in the μN range as we expected. The symbolic representation of the net force on the loop shows that the net force would be zero if either current disappeared, if either dimension of the loop became very small ($a \to 0$ or $\ell \to 0$), or if the magnetic field were uniform ($c \to \infty$).

---

32. Four long, parallel conductors carry equal currents of $I = 5.00$ A. Figure P30.32 is an end view of the conductors. The current direction is into the page at points A and B and out of the page at C and D. Calculate (a) the magnitude and (b) the direction of the magnetic field at point P, located at the center of the square of edge length $\ell = 0.200$ m.

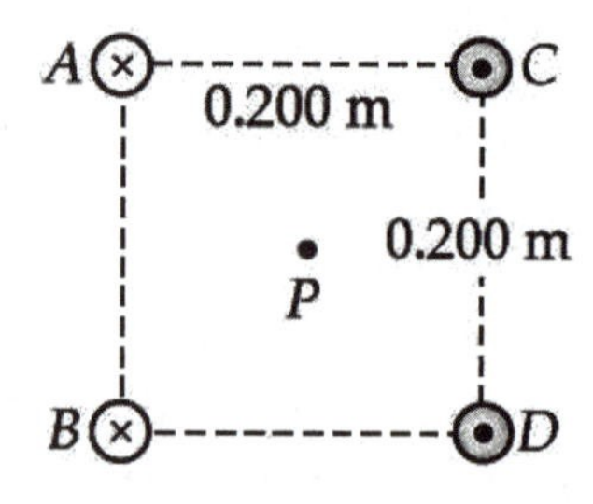

Figure P30.32

Solution

Conceptualize: To get a field in the millitesla range we would typically need a coil with many turns. Here we expect a microtesla field. The currents point out of the page and into the page, so the field will not. Rather, the field will lie in the plane of the page. If A and B were positive charges and C and D were negative, their net electric field would be to the right. But the total magnetic field will not be right or left.

Categorize: We find the field created at P by each wire, and do a vector addition of the four fields.

Analyze: Each wire is distant from P by (0.200 m) cos 45.0° = 0.141 m. Each wire produces a field at P of the same magnitude:

$$B = \frac{\mu_0 I}{2\pi a} = \frac{\left(2.00\times10^{-7}\ \mathrm{T\cdot m/A}\right)(5.00\ \mathrm{A})}{0.141\ \mathrm{m}} = 7.07\ \mu\mathrm{T}$$

Carrying currents away from you, the left-hand wires produce fields at P of 7.07 μT, in the following directions, determined by the right-hand rule:

A: to the bottom of the diagram and left, at 225°

B: to the bottom of the diagram and right, at 315°.

Carrying currents toward you, the wires to the right also produce fields at P of 7.07 μT, in the following directions:

C: downward and to the right, at 315°

D: downward and to the left, at 225°.

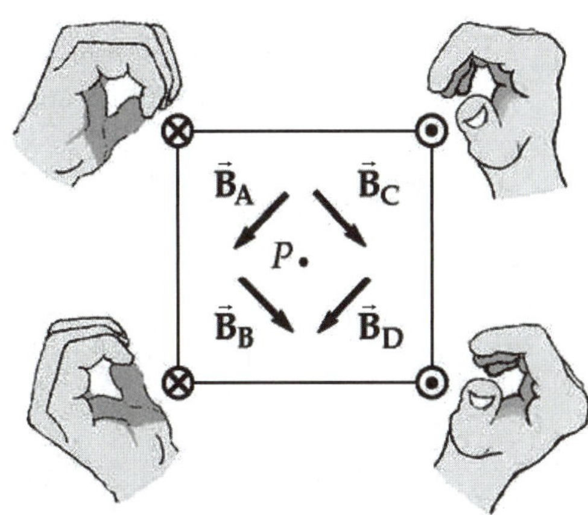

The diagram shows the four contributions to the total field at P. Vector addition shows that the total field is (a) 4(7.07 μT) sin 45.0° = 20.0 μT (b) toward the bottom of the page. ■

Finalize: The calculation would be significantly more complicated to find the field at other points away from the center of the array of wires. If you need to review vector addition by the component method from Chapter 3, do so. After that, you need to practice using your right hand to identify directions of magnetic fields. Describing the field as "clockwise" for A and B and "counterclockwise" for C and D is not definite enough. The field has a single direction at any particular point.

---

34. A packed bundle of 100 long, straight, insulated wires forms a cylinder of radius R = 0.500 cm. If each wire carries 2.00 A, what are (a) the magnitude and (b) the direction of the magnetic force per unit length acting on a wire located 0.200 cm from the center of the bundle? (c) **What If?** Would a wire on the outer edge of the bundle experience a force greater or smaller than the value calculated in parts (a) and (b)? Give a qualitative argument for your answer.

Solution

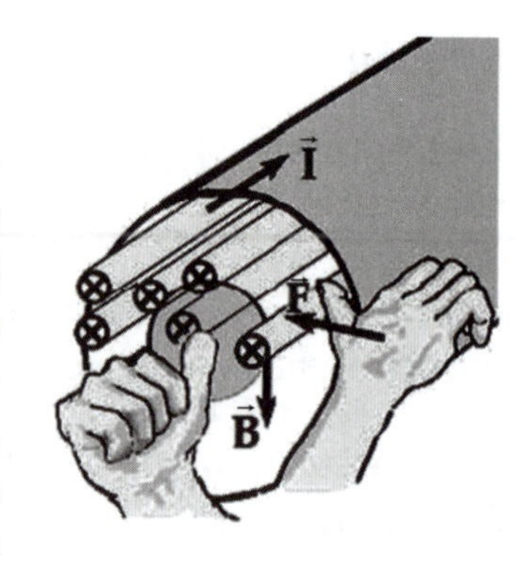

Conceptualize: The force **on** one wire comes from its interaction with the magnetic field created **by** the other ninety-nine wires. According to Ampère's law, at a distance r from the center, only the wires enclosed inside a radius r contribute to this net magnetic field; the other wires outside the radius produce magnetic field vectors in opposite directions that cancel out at r. Therefore, the magnetic field (and also the force on a particular wire at radius r) will be greater for larger radii within the bundle, and will decrease for distances beyond the radius of the bundle, as shown in the graph. Applying $\vec{\mathbf{F}} = I\vec{\mathbf{L}} \times \vec{\mathbf{B}}$, the magnetic force on a single wire will be directed toward the center of the bundle, so that all the wires tend to attract each other. This is already the answer to part (b). ■

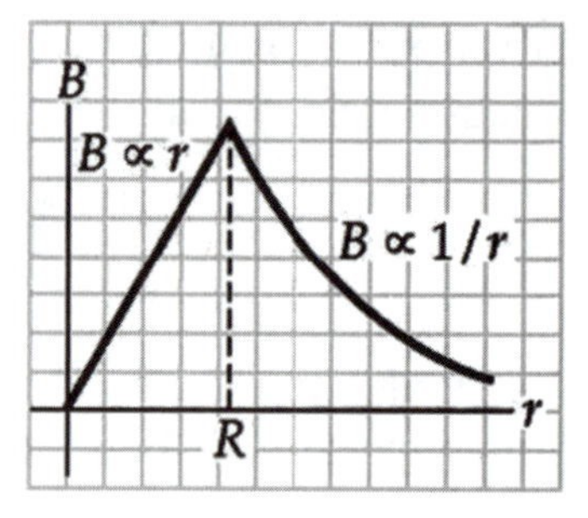

Categorize: Using Ampère's law, we can find the magnetic field at any radius, so that the magnetic force $\vec{\mathbf{F}} = I\vec{\mathbf{L}} \times \vec{\mathbf{B}}$ on a single wire can then be calculated.

Analyze:

(a) Ampère's law is used to derive Equation 30.15, which we can use to find the magnetic field at $r = 0.200$ cm from the center of the cable:

$$B = \frac{\mu_0 I r}{2\pi R^2} = \frac{\left(4\pi \times 10^{-7}\ \text{T}\cdot\text{m/A}\right)(99)(2.00\ \text{A})\left(0.200 \times 10^{-2}\ \text{m}\right)}{2\pi\left(0.500 \times 10^{-2}\ \text{m}\right)^2} = 3.17 \times 10^{-3}\ \text{T}$$

This field points tangent to a circle of radius 0.200 cm and exerts a force $\vec{\mathbf{F}} = I\vec{\mathbf{L}} \times \vec{\mathbf{B}}$ toward the center of the bundle, on the single hundredth wire. The force per length is

$$\frac{F}{\ell} = IB\sin\theta = (2.00\ \text{A})(3.17 \times 10^{-3}\ \text{T})(\sin 90°) = 6.34\ \text{mN/m}$$ ■

(c) As is shown above in the graph taken from the text, the magnetic field increases linearly as a function of r until it reaches a maximum at the outer surface of the cable. Therefore, the force on a single wire at the outer radius $r = 0.5$ cm would be greater than at $r = 0.2$ cm by a factor of 5/2. ■

Finalize: We did not estimate the expected magnitude of the force, but 200 amperes is a lot of current. It would be interesting to see if the magnetic force that pulls together the individual wires in the bundle is enough to hold them against their own weight: If we assume that the insulation accounts for about half the volume of the bundle, then a single copper wire in this bundle would have a cross sectional area of about

$$(1/2)(0.01)\pi(0.500\ \text{cm})^2 = 4 \times 10^{-7}\ \text{m}^2$$

with a weight per unit length of

$$\rho g A = (8\,920\ \text{kg/m}^3)(9.8\ \text{N/kg})(4 \times 10^{-7}\ \text{m}^2) = 0.03\ \text{N/m}$$

Therefore, the outer wires experience an inward magnetic force that is about half the magnitude of their own weight. If placed on a table, this bundle of wires would form a loosely held mound without the outer sheathing to hold them together.

---

36. A long, cylindrical conductor of radius R carries a current I as shown in Figure P30.36. The current density J, however, is not uniform over the cross section of the conductor but is a function of the radius according to $J = br$, where b is a constant. Find an expression for the magnetic field magnitude B (a) at a distance $r_1 < R$ and (b) at a distance $r_2 > R$, measured from the center of the conductor.

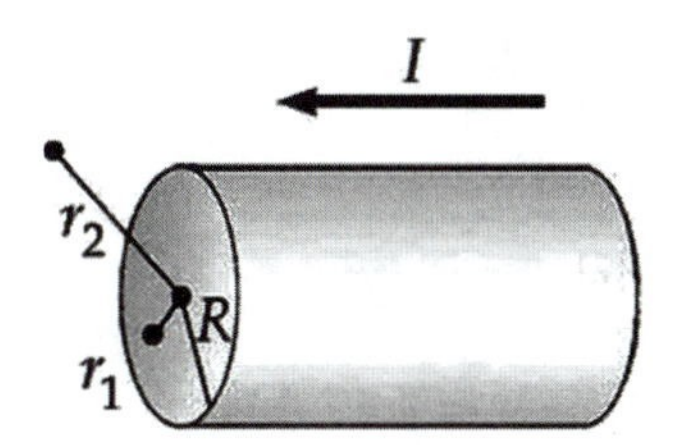

Figure P30.36

Solution

Conceptualize: The current density increases toward the outside of the bar. Then the magnetic field should be zero along the axis and should increase with r, proportional to some power greater than the first power. Outside the bar, the whole current creates magnetic field, inversely proportional to r.

Categorize: We will use Ampère's law to find the field, separately inside and outside the bar. We must do an integral to find the current inside radius r_1 within the bar in terms of b and r_1. We must do another integral to find the whole current in terms of b and R.

Analyze: Take a circle of radius r_1 or r_2 to apply $\oint \vec{\mathbf{B}} \cdot d\vec{\mathbf{s}} = \mu_0 I$, where for nonuniform current density $I = \int J dA$. In this case $\vec{\mathbf{B}}$ is parallel to $d\vec{\mathbf{s}}$ and the direction of J is straight through the area element dA, so Ampère's law gives

$$\oint B ds = \mu_0 \int J dA$$

(a) For $r_1 < R$, $2\pi r_1 B = \mu_0 \int_0^{r_1} br(2\pi r dr) = \mu_0 2\pi b \left[\frac{r_1^3}{3} - 0\right]$ and $B = \frac{1}{3}\left(\mu_0 b r_1^2\right)$ (inside) ■

(b) For $r_2 > R$, $2\pi r_2 B = \mu_0 \int_0^R br(2\pi r dr)$ and $B = \dfrac{\mu_0 b R^3}{3r_2}$ (outside) ■

Finalize: Viewed from the left end, the direction of the field, inside and out, is along a tangent to the counterclockwise circular field line passing through any particular point. The field is proportional to r^{-1} outside the bar, as predicted. It turns out to be proportional to r^2 inside the bar. The "inside" and "outside" answers must agree on the same value for the field at the surface of the bar. And they do, both giving $\mu_0 bR^2/3$.

---

39. A long solenoid that has 1 000 turns uniformly distributed over a length of 0.400 m produces a magnetic field of magnitude 1.00×10^{-4} T at its center. What current is required in the windings for that to occur?

Solution

Conceptualize: We estimate on the order of one amp. The field is many microteslas. It is larger than fields typically produced by a single wire, mostly (we think) because of the many turns and not because the current is very large.

Categorize: We use just the equation derived in the chapter text for the field of a solenoid.

Analyze: The magnetic field at the center of a solenoid is $B = \mu_0 \frac{N}{\ell} I$.

So $$I = \frac{B\ell}{\mu_0 N} = \frac{(1.00 \times 10^{-4}\ \text{T}\cdot\text{A})(0.400\ \text{m})}{(4\pi \times 10^{-7}\ \text{T}\cdot\text{m})(1\ 000)} = 31.8\ \text{mA}$$ ■

Finalize: The answer is thirty times smaller than our guess. This solenoid looks like a roll for wrapping paper. It could have any cross-sectional area, so long as its radius is small compared to its 40-cm length. Winding the wire onto the form would require a machine or else be quite tedious. Once the coil is made, the problem is perfectly realistic. The field could be displayed by iron filings or by an oscillating compass needle, as described in Problem 72 of Chapter 29.

---

45. A cube of edge length $\ell = 2.50$ cm is positioned as shown in Figure P30.45. A uniform magnetic field given by $\vec{\mathbf{B}} = (5\hat{\mathbf{i}} + 4\hat{\mathbf{j}} + 3\hat{\mathbf{k}})$ T exists throughout the region. (a) Calculate the magnetic flux through the shaded face. (b) What is the total flux through the six faces?

Solution

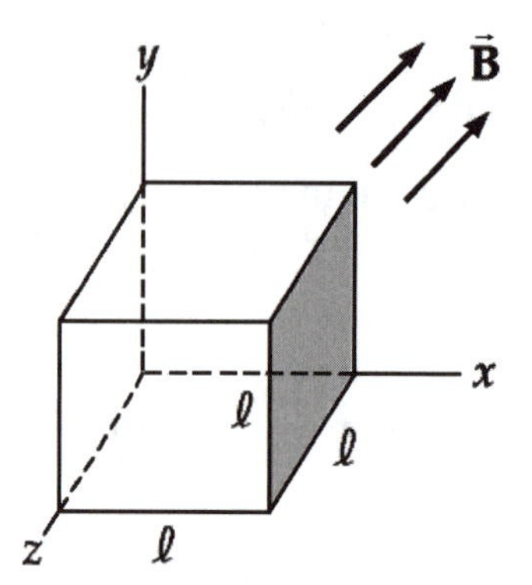

Figure P30.45

Conceptualize: For part (a) we estimate some tesla-centimeter-squared's, which is some 10^{-4} T · m². For part (b) we estimate zero: the field is uniform so every bit of flux counted as negative where it goes in will also be counted as positive where it comes out.

Categorize: We use the definition of magnetic flux.

Analyze: The flux is defined as $\Phi_B = \vec{\mathbf{B}}\cdot\vec{\mathbf{A}}$

(a) $$\Phi_B = B_x A_x + B_y A_y + B_z A_z$$

The shaded square's area is in the yz plane, so it counts as an x component of area. Here $A_y = A_z = 0$

so $\Phi_B = (5.00\ \text{T})(0.025\ 0\ \text{m})^2 = 3.12 \times 10^{-3}\ \text{T}\cdot\text{m}^2$ ■

(b) For a closed surface, $\oint \vec{\mathbf{B}}\cdot d\vec{\mathbf{A}} = 0$ so $\Phi_B = 0$ ■

Finalize: We could have obtained the answer to (b) by adding up three positive and three negative fluxes for the six faces of the cube. But the answer must be zero not just because the field is uniform, but because of the nature of magnetic fields as opposed to electric fields.

---

49. The magnetic moment of the Earth is approximately 8.00×10^{22} A · m^2. Imagine that the planetary magnetic field were caused by the complete magnetization of a huge iron deposit with density 7 900 kg/m^3 and approximately 8.50×10^{28} iron atoms/m^3. (a) How many unpaired electrons, each with a magnetic moment of 9.27×10^{-24} A · m^2, would participate? (b) At two unpaired electrons per iron atom, how many kilograms of iron would be present in the deposit?

Solution

Conceptualize: We know that much of the Earth is not iron, so if the situation described provides an accurate model, then the iron deposit must certainly be less than the mass of the Earth ($M_{\text{Earth}} = 5.97 \times 10^{24}$ kg). One mole of iron has a mass of 55.8 g and contributes $2(6.02 \times 10^{23})$ unpaired electrons, so we should expect the total number of unpaired electrons to be less than 10^{50}.

Categorize: The Bohr magneton μ_B is the measured value for the magnetic moment of a single unpaired electron. Therefore, we can find the number of unpaired electrons by dividing the magnetic moment of the Earth by μ_B. We can then use the density of iron to find the mass of the iron atoms that each contribute two electrons.

Analyze:

(a) The Bohr magneton is

$$\mu_B = \left(9.27 \times 10^{-24}\ \frac{\text{J}}{\text{T}}\right)\left(\frac{\text{N}\cdot\text{m}}{1\ \text{J}}\right)\left(\frac{1\ \text{T}}{\text{N}\cdot\text{s/C}\cdot\text{m}}\right)\left(\frac{1\ \text{A}}{\text{C/s}}\right) = 9.27 \times 10^{-24}\ \text{A}\cdot\text{m}^2$$

The number of unpaired electrons is

$$N = \frac{8.00 \times 10^{22}\ \text{A}\cdot\text{m}^2}{9.27 \times 10^{-24}\ \text{A}\cdot\text{m}^2} = 8.63 \times 10^{45}\ \text{e}^- \quad \blacksquare$$

(b) Each iron atom has two unpaired electrons, so the number of iron atoms required is

$$\tfrac{1}{2}N = \tfrac{1}{2}(8.63 \times 10^{45}) = 4.31 \times 10^{45}\ \text{iron atoms}$$

Thus, $$M_{\text{Fe}} = \frac{(4.31 \times 10^{45}\ \text{atoms})(7\,900\ \text{kg/m}^3)}{8.50 \times 10^{28}\ \text{atoms/m}^3} = 4.01 \times 10^{20}\ \text{kg} \quad \blacksquare$$

Finalize: The calculated answers seem reasonable based on the limits we expected. From the data in this problem, the iron deposit required to produce the magnetic moment would only be about 1/15 000 the mass of the Earth and would form a sphere 500 km in diameter. Although this is certainly a large amount of iron, it is much smaller than the inner core of the Earth, which is estimated to have a diameter of about 3 000 km.

---

53. A very long, thin strip of metal of width w carries a current I along its length as shown in Figure P30.53. The current is distributed uniformly across the width of the strip. Find the magnetic field at point P in the diagram. Point P is in the plane of the strip at distance b away from its edge.

Solution

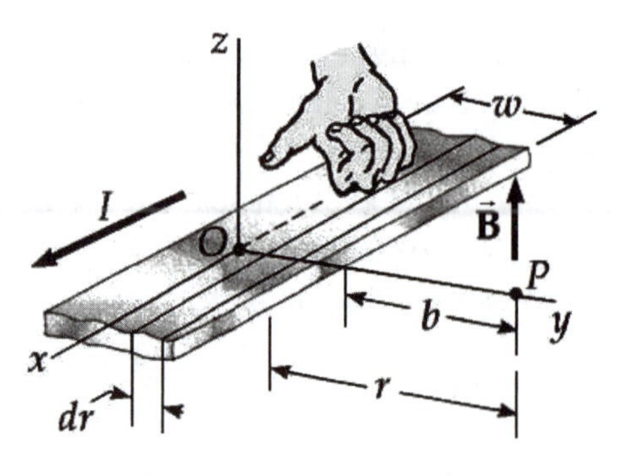

Figure P30.53 (modified)

Conceptualize: The field direction will not be parallel or antiparallel to the current. It will not be toward the strip or away from it. It must be in the $+z$ or $-z$ direction. Get out your right hand. Hold it with the extended thumb pointing toward you, in the direction of the current in the diagram. Then the field at a point directly to the right is straight up. ■

The field magnitude should be proportional to μ_0 and proportional to I. It should decrease as b increases.

Categorize: We think of the strip as made of infinitely many filaments, each acting as a long straight wire to create field at P. We identify the field dB due to one filament and then integrate to find the field of them all.

Analyze: Consider a long filament in the strip, which has width dr and is a distance r from point P. The magnetic field at a distance r from a long conductor is

$$B = \frac{\mu_0 I}{2\pi r}$$

Thus, the field due to the filament is $d\vec{\mathbf{B}} = \frac{\mu_0 dI}{2\pi r}\hat{\mathbf{k}}$ where $dI = I\left(\frac{dr}{w}\right)$

so
$$\vec{\mathbf{B}} = \int_b^{b+w} \frac{\mu_0}{2\pi r}\left(I\frac{dr}{w}\right)\hat{\mathbf{k}} = \frac{\mu_0 I}{2\pi w}\int_b^{b+w}\frac{dr}{r}\hat{\mathbf{k}} = \frac{\mu_0 I}{2\pi w}\ln\left(1+\frac{w}{b}\right)\hat{\mathbf{k}}$$ ■

Finalize: The field has the predicted proportionalities to μ_0 and I. For b large compared to w, the fact that the current is a strip rather than just a wire should be unimportant, so the field should be inversely proportional to b. And it is, because the condition $b >> w$ makes $\ln(1 + w/b)$ take the form $w/b - (w/b)^2/2 + (w/b)^3/3 - \cdots \approx w/b$, so the field becomes $\frac{\mu_0 I}{2\pi w}\frac{w}{b}\hat{\mathbf{k}} = \frac{\mu_0 I}{2\pi b}\hat{\mathbf{k}}$ as it should be for a thin wire.

Note: Compare Problem 55 and its solution with Problem 56 and its solution. They contain the same steps of physical analysis and reasoning, but in Problem 56 the numerical "plug and chug" steps are not included.

55. A nonconducting ring of radius 10.0 cm is uniformly charged with a total positive charge 10.0 μC. The ring rotates at a constant angular speed 20.0 rad/s about an axis through its center, perpendicular to the plane of the ring. What is the magnitude of the magnetic field on the axis of the ring 5.00 cm from its center?

Solution

Conceptualize: The "static" charge on the ring constitutes an electric current as the ring rotates. The current will only be in the microamp range, so the field may be on the order of 10^{-12} T.

Categorize: We will find the current and then use the equation for the magnetic field on the axis of a current loop.

Analyze: The time interval required for the 10.0-μC charge to go past a point is the period of rotation,

$$T = \frac{\theta}{\omega} = \frac{2\pi \text{ rad}}{20 \text{ rad/s}} = 0.314 \text{ s}$$

The current is

$$I = \frac{q}{T} = \frac{10.0 \times 10^{-6} \text{ C}}{0.314 \text{ s}} = 3.18 \times 10^{-5} \text{ A}$$

The current has the shape of a flat compact circular coil with one turn, so the magnetic field it creates is

$$B = \frac{\mu_0 I R^2}{2\left(x^2 + R^2\right)^{3/2}}$$

Substituting numerical values for these variables,

$$B = \frac{\left(4\pi \times 10^{-7} \frac{\text{T}\cdot\text{m}}{\text{A}}\right)\left(3.18 \times 10^{-5} \text{ A}\right)(0.100 \text{ m})^2}{2\left((0.050\,0 \text{ m})^2 + (0.100 \text{ m})^2\right)^{3/2}}$$

$$B = \frac{4.00 \times 10^{-13} \text{ T}\cdot\text{m}^3}{2\left(1.40 \times 10^{-3} \text{ m}^3\right)} = 1.43 \times 10^{-10} \text{ T}$$ ■

On the side of the ring from which it is seen as turning counterclockwise, the magnetic field is along the axis away from its center. On the other side of the ring, the magnetic field is toward the center of the ring, in the same direction in space.

Finalize: We have found the magnetic field at 5 cm from the center. It would be stronger at the center itself. The ring also creates an electric field, which is zero at the center and points away from the center, in opposite directions on opposite sides of the axis.

---

56. A nonconducting ring of radius R is uniformly charged with a total positive charge q. The ring rotates at a constant angular speed ω about an axis through its center, perpendicular to the plane of the ring. What is the magnitude of the magnetic field on the axis of the ring a distance $R/2$ from its center?

Solution

Conceptualize: The "static" charge on the ring constitutes an electric current as the ring rotates. The current will be proportional to q and to ω. Then the field will be proportional to both of these quantities and to μ_0. It will decrease as R increases.

Categorize: We will find the current and then use the equation for the magnetic field on the axis of a current loop.

Analyze: The time interval required for the charge q to go past a point is the period of rotation,

$$T = \frac{\theta}{\omega} = \frac{2\pi}{\omega}$$

The current is

$$I = \frac{q}{T} = \frac{q}{2\pi/\omega} = \frac{q\omega}{2\pi}$$

The current has the shape of a flat compact circular coil with one turn, so the magnetic field it creates is

$$B = \frac{\mu_0 IR^2}{2\left(x^2 + R^2\right)^{3/2}}$$

In this case, the distance x is equal to $R/2$, so

$$B = \frac{\mu_0 q\omega R^2}{4\pi\left(R^2/4 + R^2\right)^{3/2}} = \frac{\mu_0 q\omega R^2}{4\pi\left(5R^2/4\right)^{3/2}} = \frac{2\mu_0\, q\omega}{5\sqrt{5}\pi R}$$ ■

On the side of the ring from which it is seen as turning counterclockwise, the magnetic field is along the axis away from its center. On the other side of the ring, the magnetic field is toward the center of the ring, in the same direction in space.

Finalize: The result has the predicted proportionalities. Its dependence on R turns out to be simply an inverse proportionality. We have found the magnetic field at $R/2$ from the center. It would be stronger at the center itself. The ring also creates an electric field, which is zero at the center and points away from the center, in opposite directions on opposite sides of the axis.

---

72. A wire is formed into the shape of a square of edge length L (Fig. P30.72). Show that when the current in the loop is I, the magnetic field at point P, a distance x from the center of the square along its axis is

$$B = \frac{\mu_0 IL^2}{2\pi\left(x^2 + L^2/4\right)\sqrt{x^2 + L^2/2}}$$

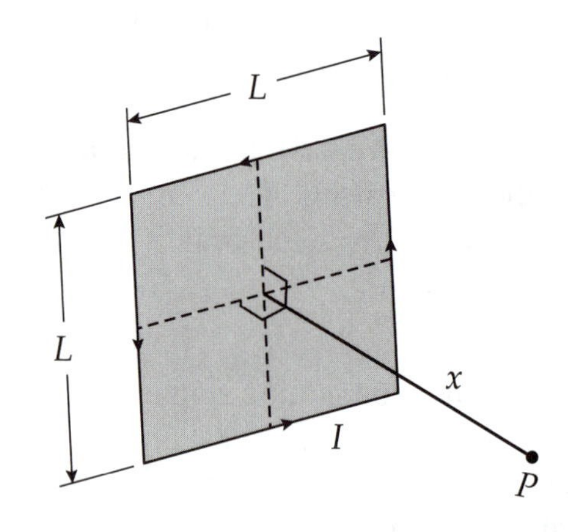

Figure P30.72

Solution

Conceptualize: The proportionalities to μ_0 and I are reasonable. The field decreases as x increases (reasonable), displaying proportionality to x^{-3} for large x (interesting).

Categorize: We find the field due to one side. Then we do a vector addition of the fields of the four sides.

Analyze: Consider the top side of the square. The distance from its center to point P is

$$a = \sqrt{x^2 + (L/2)^2}$$

Equation 30.4 describes the field it creates at point P. The distance from one of the corners of the square to P is

$$\sqrt{x^2 + (L/2)^2 + (L/2)^2} = \sqrt{x^2 + L^2/2}$$

Thus $\sin\theta_1 = \dfrac{L}{2\sqrt{x^2 + L^2/2}}$ and $\sin\theta_2 = -\sin\theta_1 = -\dfrac{L}{2\sqrt{x^2 + L^2/2}}$

Then, $B_{\text{top}} = \dfrac{\mu_0 I}{4\pi a}(\sin\theta_1 - \sin\theta_2)$

$$B_{\text{top}} = \frac{\mu_0 I}{4\pi\sqrt{x^2 + L^2/4}}\left(\frac{L}{2\sqrt{x^2 + L^2/2}} + \frac{L}{2\sqrt{x^2 + L^2/2}}\right)$$

or $$B_{\text{top}} = \frac{\mu_0 IL}{4\pi\sqrt{x^2 + L^2/4}\sqrt{x^2 + L^2/2}}$$

The component of this field along the direction of x is

$$B_{\text{top}}\cos\phi = B_{\text{top}}\frac{L/2}{a}$$

Each of the four sides of the square produces this same size field along the direction of x, with the other components adding to zero. We assemble our results and find that the net magnetic field points away from the center of the square, with magnitude

$$B = 4B_{\text{top}}\frac{L}{2a} = \frac{\mu_0 IL^2}{2\pi\left(x^2 + L^2/4\right)\sqrt{x^2 + L^2/2}}$$ ■

Finalize: We can investigate two limiting forms. Far from the square loop, L^2 is negligible compared with x^2 so we have $B \approx \mu_0 IL^2/2\pi x^3$. Here L^2 is the area of the loop, IL^2 is its magnetic moment μ, and the field is $\mu_0\mu/2\pi x^3$, in agreement with Equation 30.10.

At the center of the square, $x = 0$ and the field is $B = \dfrac{\mu_0 IL^2}{2\pi L^3/4\sqrt{2}} = \dfrac{\sqrt{2}}{\pi}\dfrac{\mu_0 I}{L/2}$. This agrees with the result of Problem 5(a).

31

Faraday's Law

EQUATIONS AND CONCEPTS

The **total magnetic flux** through a loop located in a uniform magnetic field depends on the angle between the direction of the magnetic field and the normal to the surface of the loop. *The maximum flux through the loop occurs when the magnetic field is perpendicular to the plane of the loop.* The unit of magnetic flux is the weber, Wb.

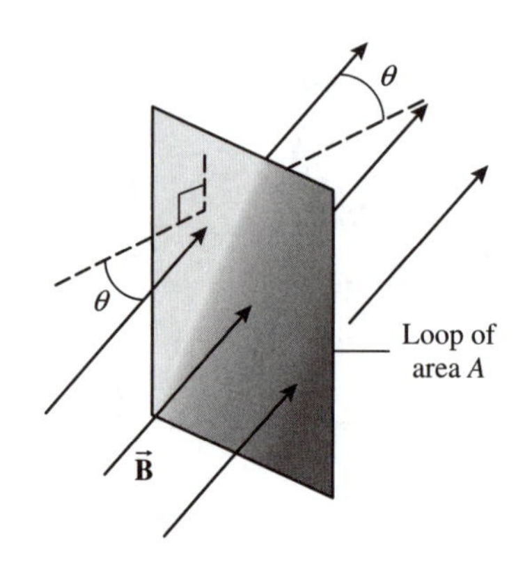

$$\Phi_B = BA\cos\theta$$
$$\Phi_{B\max} = BA$$

For a **non-uniform magnetic field**, the magnetic flux through an area is the integral of the normal component of the magnetic field over the area.

$$\Phi_B = \int \vec{\mathbf{B}} \cdot d\vec{\mathbf{A}}$$

Faraday's law of induction states that the emf induced in a circuit is proportional to the rate of change of magnetic flux through the circuit. *The minus sign is included to indicate the polarity of the induced emf, which can be found by use of Lenz's law.*

$$\mathcal{E} = -\frac{d\Phi_B}{dt} \quad \text{(single loop)} \quad (31.1)$$

$$\mathcal{E} = -N\frac{d\Phi_B}{dt} \quad (N \text{ loops}) \quad (31.2)$$

An **induced emf** will be generated in a circuit when any one or more of the following change with time:

- magnitude of $\vec{\mathbf{B}}$
- area enclosed by the circuit
- angle θ between $\vec{\mathbf{B}}$ and the normal

$$\mathcal{E} = -\frac{d}{dt}(BA\cos\theta) \quad (31.3)$$

Lenz's law states that, in a closed circiuit, the polarity of the induced emf will produce a current whose associated magnetic field opposes the change in the flux through the circuit. In the coil on the left, the magnetic flux through the loop increases as the N-pole of the magnet approaches. The induced current creates a magnetic field directed toward the left with flux as shown in the right-hand coil. *That is, the induced current tends to maintain the original flux through the circuit.*

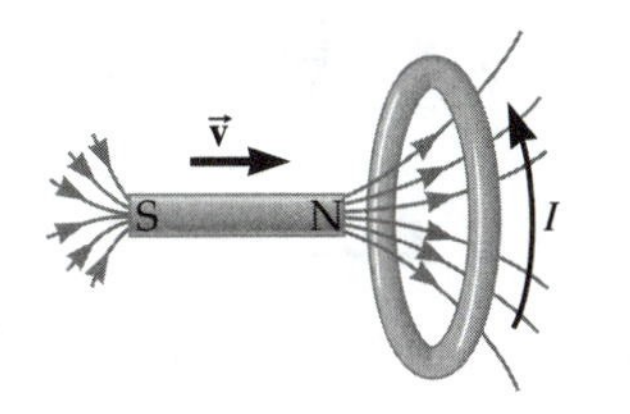

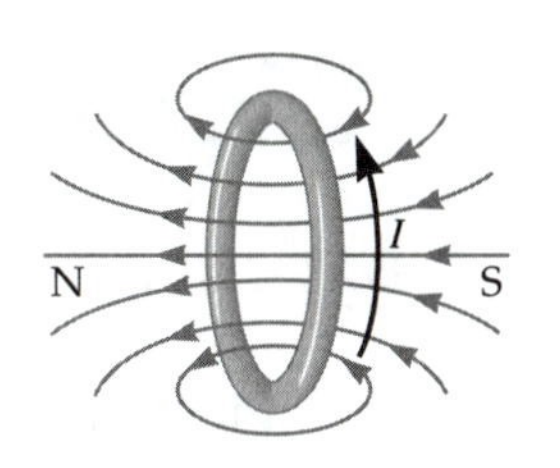

A **motional emf** is induced in a conductor moving through a magnetic field. The induced emf is given by Equation 31.5 when a conductor of length ℓ moves with speed v perpendicular to a constant magnetic field of magnitude B. *Recall the significance of the negative sign as required by Lenz's law.*

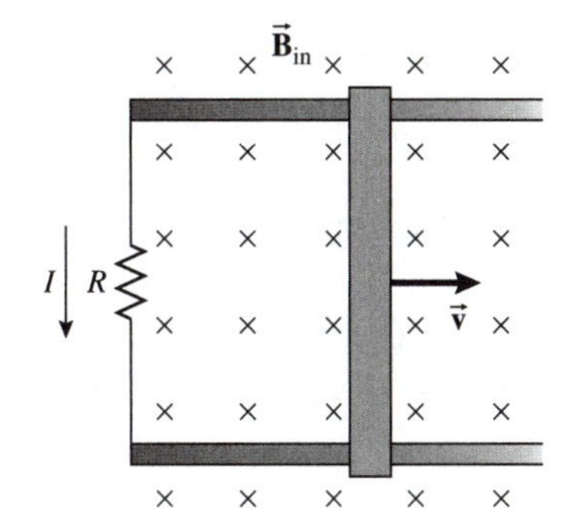

$$\mathcal{E} = -B\ell v \quad (31.5)$$

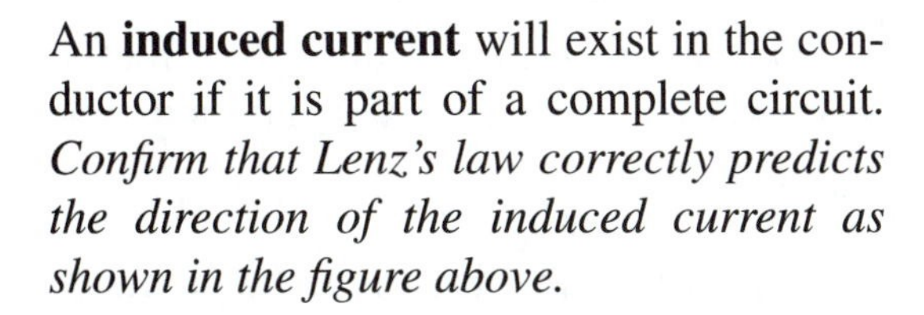

An **induced current** will exist in the conductor if it is part of a complete circuit. *Confirm that Lenz's law correctly predicts the direction of the induced current as shown in the figure above.*

$$I = \frac{|\mathcal{E}|}{R} = \frac{B\ell v}{R} \quad (31.6)$$

Faraday's law in a more general form can be written as the integral of the electric field around a closed path. The induced electric field is a nonconservative field that is generated by a changing magnetic field.

$$\oint \vec{\mathbf{E}} \cdot d\vec{\mathbf{s}} = -\frac{d\Phi_B}{dt} \quad (31.9)$$

$$\oint \vec{\mathbf{E}} \cdot d\vec{\mathbf{s}} = -\frac{d}{dt}\int \vec{\mathbf{B}} \cdot d\vec{\mathbf{A}}$$

A **loop rotating in a magnetic field** produces a sinusoidally varying emf. *For a given loop, the maximum value of the induced emf is proportional to the angular velocity of the loop.*

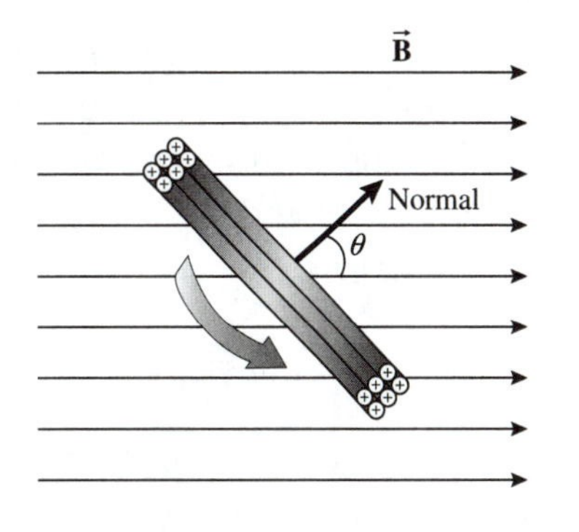

$$\mathcal{E} = NAB\omega \sin \omega t \quad (31.10)$$

$$\mathcal{E}_{max} = NAB\omega \quad (31.11)$$

SUGGESTIONS, SKILLS, AND STRATEGIES

CALCULATING AN INDUCED EMF

It is important to distinguish clearly between the instantaneous value of emf induced in a circuit and the average value of the emf induced in the circuit over a finite time interval. To calculate the average induced emf, it is often useful to write Equation 31.2 as

$$\mathcal{E}_{\text{avg}} = -N\left(\frac{d\Phi_B}{dt}\right)_{\text{avg}} = -N\frac{\Delta\Phi_B}{\Delta t} \quad \text{or} \quad \mathcal{E}_{\text{avg}} = -N\left(\frac{\Phi_{B,f} - \Phi_{B,i}}{\Delta t}\right)$$

where the subscripts i and f refer to the magnetic flux through the circuit at the beginning and end of the time interval Δt. For a circuit or coil in a plane, $\Phi_B = NBA\cos\theta$, where θ is the angle between the vector normal to plane of the circuit (conducting coil) and the direction of the magnetic field.

Equation 31.3 can be used to calculate the *instantaneous value of an induced emf.* For a multi-turn coil, the induced emf is

$$\mathcal{E} = -N\frac{d}{dt}(BA\cos\theta)$$

where in a particular case B, A, θ, or any combination of those parameters can be time dependent while the others remain constant. The expression resulting from the differentiation is then evaluated using the known or calculated values of B, A, and θ and their time derivatives.

Use the right-hand rule when applying Lenz's law to determine the direction of an induced current. *Remember, the induced current will create a magnetic field along a direction which opposes any change in the flux through the circuit.*

AN EXAMPLE OF LENZ'S LAW

Consider two single-turn, concentric coils *lying in the plane of the paper* as shown in the figure. The outer coil is part of a circuit containing a resistor (R), a battery ($\mathcal{E}$), and switch (S). The inner coil is not part of the circuit containing the battery. When the switch is moved from **"open"** to **"closed"** the direction of the induced current in the inner coil can be predicted by Lenz's law.

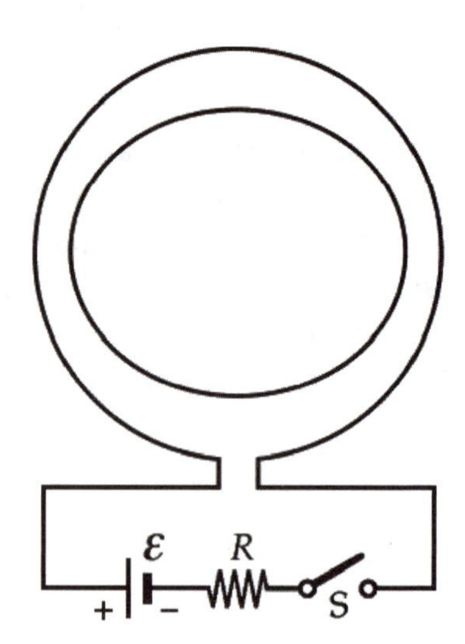

Consider the following steps:

(1) When the switch is in the **"open"** position as shown, there will be no current in either coil.

(2) When the switch is moved to the **"closed"** position, there will be a clockwise current in the outer coil.

(3) Magnetic field lines due to current in the outer coil will be directed into the page through the area enclosed by the coil. *Use the right-hand rule (first application) to confirm this.* Magnetic flux into the page will penetrate the entire area enclosed by the outer coil, *including the area of the inner coil.*

(4) By Faraday's law, the increasing flux produces an induced emf (and current) in the inner coil.

(5) Lenz's law requires that the induced current have a direction which will tend to maintain the initial flux condition (which in this case was zero). By using the right-hand rule (second application), you should be able to determine that the direction of the induced current in the inner coil must be counterclockwise, contributing to a flux out of the page. Try this!

As a second example you should follow steps similar to those above to predict the direction of the induced current in the inner coil when the switch is moved from **"closed"** to **"open."**

REVIEW CHECKLIST

You should be able to:

- Calculate the emf (or current) induced in a circuit when the magnetic flux through the circuit is changing in time. The variation in flux might be due to a change in (a) the area of the circuit, (b) the magnitude of the magnetic field, (c) the direction of the magnetic field, or (d) the orientation or location of the circuit in the magnetic field. (Section 31.1)

- Calculate the emf induced between the ends of a conducting bar as it moves through a region where there is a constant magnetic field (motional emf). (Section 31.2)

- Apply Lenz's law to determine the direction of an induced emf or current. You should also understand that Lenz's law is a consequence of the law of conservation of energy. (Section 31.3)

- Calculate the electric field at various points in a charge-free region when the time variation of the magnetic field over the region is specified. (Section 31.4)

- Calculate the maximum and instantaneous values of the sinusoidal emf generated in a conducting loop rotating in a constant magnetic field. (Section 31.5)

ANSWER TO AN OBJECTIVE QUESTION

6. The bar in Figure OQ31.6 moves on rails to the right with a velocity $\vec{\mathbf{v}}$, and the uniform, constant magnetic field is directed out of the page. Which of the following statements are correct? More than one statement may be correct. (a) The induced current in the loop is zero. (b) The induced current in the loop is clockwise. (c) The induced current in the loop

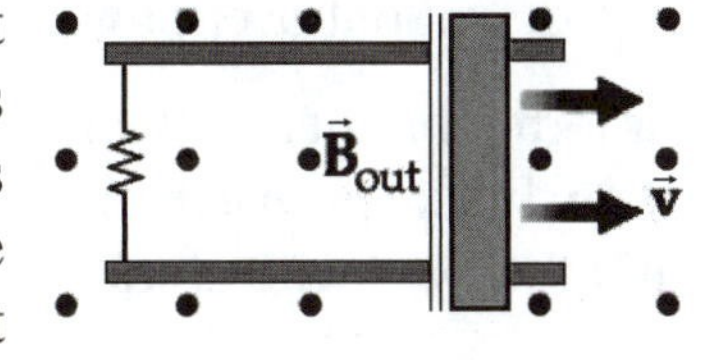

Figure OQ31.6

is counterclockwise. (d) An external force is required to keep the bar moving at constant speed. (e) No force is required to keep the bar moving at constant speed.

Answer The correct statements are (b) and (d). The externally-produced magnetic field is out of the paper. As the area A enclosed by the loop increases, the external flux increases according to $\Phi_B = BA\cos\theta = BA$.

As predicted by Lenz's law, due to the increase in flux through the loop, free electrons will produce a current to create a magnetic flux inside the loop to oppose the change in flux.

In this case, the magnetic field due to the current must point into the paper in order to oppose the increasing magnetic flux of the external field coming out of the paper. By the right-hand rule (with your thumb pointing in the direction of the current along each wire), the current must be clockwise, so statement (b) is correct.

Next, the clockwise current is downward in the picture in the bar. The external magnetic field perpendicularly out of the page exerts on this current an $i\,\vec{\mathbf{L}} \times \vec{\mathbf{B}}$ magnetic force in the direction down cross toward you, which is left in the diagram. A magnetic braking force acts on the bar. The bar would coast to a stop if some outside agent did not exert on it a counterbalancing force to the right to keep the bar's speed constant and to provide energy to maintain the current. Thus answer (d) is also correct.

□ □ □ □

ANSWER TO A CONCEPTUAL QUESTION

5. In a hydroelectric dam, how is energy produced that is then transferred out by electrical transmission? That is, how is the energy of motion of the water converted to energy that is transmitted by AC electricity?

Answer As the water falls, it gains kinetic energy. It is then forced to pass through a turbine (a water wheel), transferring some of its energy to the rotor of a large AC electric generator.

The rotor of the generator is supplied with a small amount of DC current, which powers electromagnets in the rotor. Because the rotor is spinning, the electromagnets then create a magnetic flux that changes with time, according to the equation $\Phi_B = BA\cos\omega t$.

Coils of wire that are placed near the rotor then experience an induced emf described by the equation $\mathcal{E} = -N\,d\Phi_B/dt$.

Finally, a small amount of this electricity is used to supply the rotor with its DC current; the rest is sent out over power lines to supply customers with electricity.

In terms of energy, the hydroelectric generating station is a nonisolated system in steady state. It takes in mechanical energy and puts out a precisely equal quantity of energy, nearly all of it by electrical transmission.

□ □ □ □

SOLUTIONS TO SELECTED END-OF-CHAPTER PROBLEMS

6. A strong electromagnet produces a uniform magnetic field of 1.60 T over a cross-sectional area of 0.200 m^2. A coil having 200 turns and a total resistance of 20.0 Ω is placed around the electromagnet. The current in the electromagnet is then smoothly reduced until it reaches zero in 20.0 ms. What is the current induced in the coil?

Solution

Conceptualize: A strong magnetic field turned off in a short time interval (20.0 ms) will produce a large emf, maybe on the order of 1 kV. With only 20.0 Ω of resistance in the coil, the induced current produced by this emf will probably be larger than 10 A but less than 1 000 A.

Categorize: According to Faraday's law, if the magnetic field is reduced uniformly, then a constant emf will be produced. The definition of resistance can be applied to find the induced current from the emf.

Analyze: The induced voltage is

$$\mathcal{E} = -N\frac{d(\vec{\mathbf{B}}\cdot\vec{\mathbf{A}})}{dt} = -N\left(\frac{0 - B_i A\cos\theta}{\Delta t}\right)$$

$$\mathcal{E} = \frac{(+200)(1.60\text{ T})(0.200\text{ m}^2)(\cos 0°)}{20.0\times10^{-3}\text{ s}}\left(\frac{\text{N}\cdot\text{s/C}\cdot\text{m}}{1\text{ T}}\right)\left(\frac{1\text{ V}\cdot\text{C}}{\text{N}\cdot\text{m}}\right) = 3\,200\text{ V}$$

(**Note:** The unit conversions come from $\vec{\mathbf{F}} = q\vec{\mathbf{v}}\times\vec{\mathbf{B}}$ and $U = qV$.)

$$I = \frac{\mathcal{E}}{R} = \frac{3\,200\text{ V}}{20.0\ \Omega} = 160\text{ A}$$ ■

Finalize: This is a large current, as we expected. The positive sign means that the current in the coil is in the same direction as the current in the electromagnet.

---

9. An aluminum ring of radius $r_1 = 5.00$ cm and resistance 3.00×10^{-4} Ω is placed around one end of a long air-core solenoid with 1 000 turns per meter and radius $r_2 = 3.00$ cm as shown in Figure P31.9. Assume the axial component of the field produced by the solenoid is one-half as strong over the area of the end of the solenoid as at the center of the solenoid. Also assume the solenoid produces negligible field outside its cross-sectional area. The

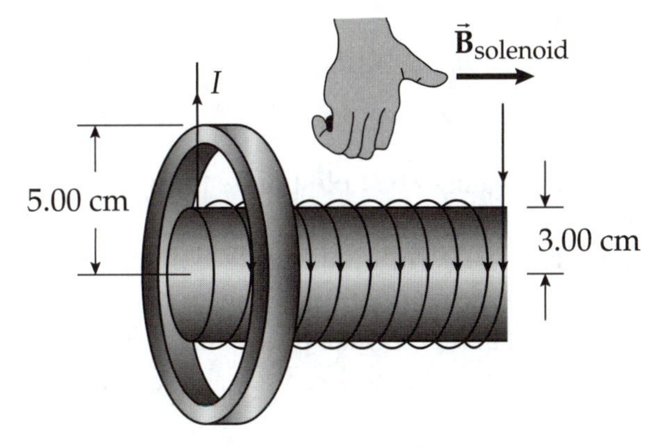

Figure P31.9 (modified)

current in the solenoid is increasing at a rate of 270 A/s. (a) What is the induced current in the ring? At the center of the ring, what are (b) the magnitude and (c) the direction of the magnetic field produced by the induced current in the ring?

Solution

Conceptualize: Many turns, a rapid change in the solenoid current, and low resistance in the ring suggest that the current will be many amps. But the ring's field it creates will likely be much less than the solenoid field. If the solenoid current were 50 A at one instant, the solenoid field would be $\mu_0 nI = 63$ mT to the right at the center of the solenoid.

Categorize: We use Faraday's law. In $BA\cos\theta$ the factor that is changing is the magnetic field. We can find its rate of change from the rate of change of the solenoid current. Then the definition of resistance will tell us the current in the ring from the induced emf. At last, the ring current will let us compute its constant field.

Analyze: The symbol for the radius of the ring is r_1, and we use R to represent its resistance. The emf induced in the ring is

$$\mathcal{E} = -\frac{d}{dt}(BA\cos\theta) = -\frac{d}{dt}(0.500\mu_0 nIA\cos 0°) = -0.500\mu_0 nA\frac{dI}{dt}$$

Note that A must be interpreted as the area $A = \pi r_2^2$ of the solenoid, where the field is strong:

$$\mathcal{E} = -0.500\,(4\pi\times 10^{-7}\text{ T}\cdot\text{m/A})(1\,000\text{ turns/m})[\pi(0.030\,0\text{ m})^2]\,(270\text{ A/s})$$

$$\mathcal{E} = \left(-4.80\times 10^{-4}\frac{\text{T}\cdot\text{m}^2}{\text{s}}\right)\left(\frac{1\text{ N}\cdot\text{s}}{\text{C}\cdot\text{m}\cdot\text{T}}\right)\left(\frac{1\text{ V}\cdot\text{C}}{\text{N}\cdot\text{m}}\right) = -4.80\times 10^{-4}\text{ V}$$

(a) The negative sign means that the current in the ring is counterclockwise, opposite to the current in the solenoid. Its magnitude is

$$I_{\text{ring}} = \frac{|\mathcal{E}|}{R} = \frac{0.000\,480}{0.000\,300} = 1.60\text{ A} \quad ■$$

(b) $$B_{\text{ring}} = \frac{\mu_0 I_{\text{ring}}}{2r_1} = \frac{4\pi\times 10^{-7}\text{T}\cdot\text{m}\,(1.60\text{ A})}{2\quad \text{A}(0.050\,0\text{ m})} = 2.01\times 10^{-5}\text{ T} \quad ■$$

(c) The solenoid's field points to the right, and is increasing, so B_{ring} points to the left. ■

Finalize: The ring current is not really very large. Here 270 A/s is not a large rate of change in current, compared to what a switch can cause. The ring field is indeed small compared to an estimate of the solenoid field. Note well that the solenoid current must keep increasing steadily to keep the ring current constant.

14. A long solenoid has $n = 400$ turns per meter and carries a current given by $I = 30.0\,(1 - e^{-1.60t})$ where I is in amperes and t is in seconds. Inside the solenoid and coaxial with it is a coil that has a radius of 6.00 cm and consists of a total of $N = 250$ turns of fine wire (Fig. P31.14). What emf is induced in the coil by the changing current?

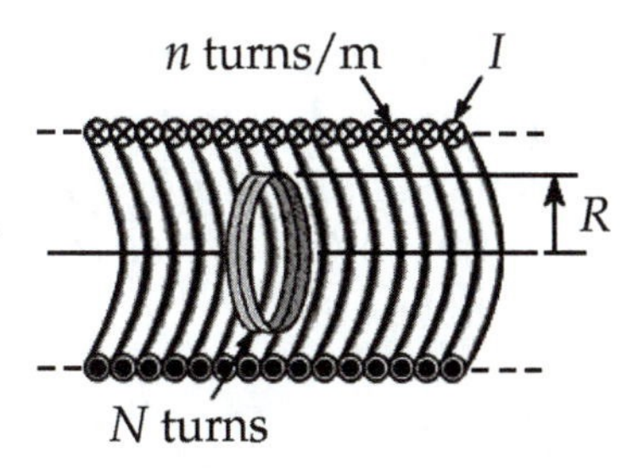

Figure P31.14

Solution

Conceptualize: The solenoid and the flat coil are not part of the same circuit. The transfer of energy between them is mediated by the magnetic field created by the solenoid. The solenoid current starts from zero, increases most rapidly at first, and then asymptotically approaches 30 A as its final value. We think of its rate of change to estimate that the induced voltage will start from its largest value, on the order of one volt, at time zero, and then decay exponentially toward zero.

Categorize: The solenoid current tells us the solenoid field. Faraday's law tells us the induced emf.

Analyze: The solenoid creates a magnetic field

$$B = \mu_0 n I = (4\pi \times 10^{-7}\ \mathrm{N/A^2})(400\ \mathrm{turns/m})(30.0\ \mathrm{A})(1 - e^{-1.60\,t})$$

$$B = (1.51 \times 10^{-2}\ \mathrm{N/m \cdot A})(1 - e^{-1.60\,t})$$

The magnetic flux through one turn of the flat coil is $\Phi_B = \int B\, dA \cos\theta$, but since $dA \cos\theta$ refers to the area perpendicular to the flux, and the magnetic field is uniform over the area A of the flat coil, this integral simplifies to

$$\Phi_B = B \int dA = B(\pi R^2) = (1.51 \times 10^{-2}\ \mathrm{N/m \cdot A})(1 - e^{-1.60\,t})[\pi(0.060\,0\ \mathrm{m})^2]$$

$$\Phi_B = (1.71 \times 10^{-4}\ \mathrm{N \cdot m/A})(1 - e^{-1.60\,t})$$

The emf generated in the N-turn coil is $\mathcal{E} = -N\, d\Phi_B/dt$. Because t has the standard unit of seconds, the factor 1.60 must have the unit $\mathrm{s^{-1}}$.

$$\mathcal{E} = -(250)\left(1.71 \times 10^{-4}\,\frac{\mathrm{N \cdot m}}{\mathrm{A}}\right)\frac{d\left(1 - e^{-1.60\,t}\right)}{dt} = -\left(0.042\,6\,\frac{\mathrm{N \cdot m}}{\mathrm{A}}\right)(1.60\ \mathrm{s^{-1}})e^{-1.60t}$$

$\mathcal{E} = -(6.82 \times 10^{-2}\ \mathrm{N \cdot m/C})e^{-1.60\,t} = -(6.82 \times 10^{-2})\, e^{-1.60\,t}$ where $\mathcal{E}$ is in volts and t is in seconds ■

Finalize: The induced emf shows the predicted exponential decay. The maximum voltage is smaller than we expected. The current in the solenoid is never changing very fast. The minus sign indicates that the emf will produce counterclockwise current in the smaller coil, opposite to the direction of the increasing current in the solenoid.

15. A coil formed by wrapping 50 turns of wire in the shape of a square is positioned in a magnetic field so that the normal to the plane of the coil makes an angle of 30.0° with the direction of the field. When the magnetic field is increased uniformly from 200 μT to 600 μT in 0.400 s, an emf of magnitude 80.0 mV is induced in the coil. What is the total length of the wire in the coil?

Solution

Conceptualize: If we assume that this square coil is some convenient size between 1 cm and 1 m across, then the total length of wire would be between 2 m and 200 m.

Categorize: The changing magnetic field will produce an emf in the coil according to Faraday's law of induction. The constant area of the coil can be found from the change in flux required to produce the emf.

Analyze: Faraday's law, $\mathcal{E} = -N\dfrac{d\Phi_B}{dt}$

becomes here $\mathcal{E} = -N\dfrac{d}{dt}\left(BA\cos\theta\right) = -NA\cos\theta\dfrac{dB}{dt}$

The magnitude of the emf is $|\mathcal{E}| = NA\cos\theta\left(\dfrac{\Delta B}{\Delta t}\right)$

The area is $A = \dfrac{|\mathcal{E}|}{N\cos\theta\left(\dfrac{\Delta B}{\Delta t}\right)}$

$$A = \frac{80.0 \times 10^{-3}\ \text{V}}{50\left(\cos 30.0^\circ\right)\left(\dfrac{600 \times 10^{-6}\ \text{T} - 200 \times 10^{-6}\ \text{T}}{0.400\ \text{s}}\right)} = 1.85\ \text{m}^2$$

Each side of the coil has length $d = \sqrt{A}$, so the total length of the wire is

$$L = N(4d) = 4N\sqrt{A} = (4)(50)\sqrt{1.85\ \text{m}^2} = 272\ \text{m}$$ ■

Finalize: The total length of wire is slightly longer than we predicted. With $d = 1.36$ m, a person could easily step through this large coil! As a bit of foreshadowing to a future chapter on AC circuits, an even bigger coil with more turns could be hidden in the ground below high-power transmission lines so that a significant amount of power could be "stolen" from the electric utility. There is a story of one man who did this and was arrested when investigators finally found the reason for a large power loss in the transmission lines!

---

23. Figure P31.23 shows a top view of a bar that can slide on two frictionless rails. The resistor is $R = 6.00\ \Omega$, and a 2.50-T magnetic field is directed perpendicularly downward, into the paper. Let $\ell = 1.20$ m. (a) Calculate the applied force required to move the bar to the right at a constant speed of 2.00 m/s. (b) At what rate is energy delivered to the resistor?

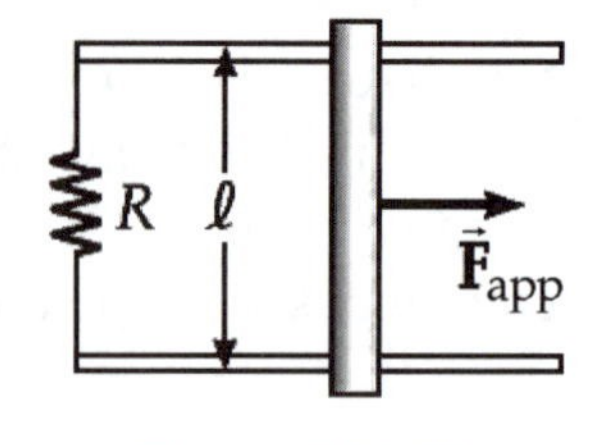

Figure P31.23

Solution

Conceptualize: The motion of the bar induces an emf in the rectangle. The emf produces a current in the circuit, including the bar. Then the magnetic field exerts a force on the bar, and an outside agent must counterbalance this force to keep the bar moving steadily.

Categorize: We use the rigid body in equilibrium model, together with the definition of resistance and Faraday's law.

Analyze:

(a) At constant speed, the net force on the moving bar equals zero, or

$$\left|\vec{\mathbf{F}}_{\mathbf{app}}\right| = I\left|\vec{\mathbf{L}} \times \vec{\mathbf{B}}\right|$$

where the current in the bar is $I = \mathcal{E}/R$

and the motional emf is $\mathcal{E} = B\ell v$

Therefore, $F_{\text{app}} = \left(\dfrac{B\ell v}{R}\right)\ell B = \dfrac{B^2\ell^2 v}{R}$

$$F_{\text{app}} = \frac{(2.50\text{ T})^2(1.20\text{ m})^2(2.00\text{ m/s})}{6.00\ \Omega} = 3.00\text{ N to the right} \quad \blacksquare$$

(b) The outside agent pulling the bar must provide input power

$$P = F_{\text{app}}v = (3.00\text{ N})(2.00\text{ m/s}) = 6.00\text{ W} \quad \blacksquare$$

Finalize: In terms of energy, the circuit is a nonisolated system in steady state. The energy, six joules every second, delivered to the circuit by work done by the applied force is delivered to the resistor by electrical transmission. The energy can leave the resistor as energy transferred by heat into the surrounding air. When we studied an electric circuit consisting of a resistor connected across a battery, we did not consider the details of the chemical reaction in the battery and the battery's loss of chemical energy. The circuit in this problem is equally simple electrically. The outside agent pulling the bar through the magnetic field takes the place of the battery. This is the first circuit in which we can account completely for the energy transformations. The power could also be computed as $\mathcal{E}I = I^2/R$.

---

27. The *homopolar generator,* also called the *Faraday disk,* is a low-voltage, high-current electric generator. It consists of a rotating conducting disk with one stationary brush (a sliding electrical contact) at its axle and another at a point on its circumference as shown in Figure P31.27. A uniform magnetic field is applied perpendicular to the plane of the disk. Assume the field is 0.900 T, the angular speed is 3.20×10^3 rev/min, and the radius of the disk is 0.400 m. Find the emf generated between the brushes. When superconducting coils are used to produce a large magnetic field, a homopolar generator can have a power output of several megawatts. Such a generator is useful, for example, in purifying metals by electrolysis. If a voltage is applied to the output terminals of the generator, it runs in reverse as a *homopolar motor* capable of providing great torque, useful in ship propulsion.

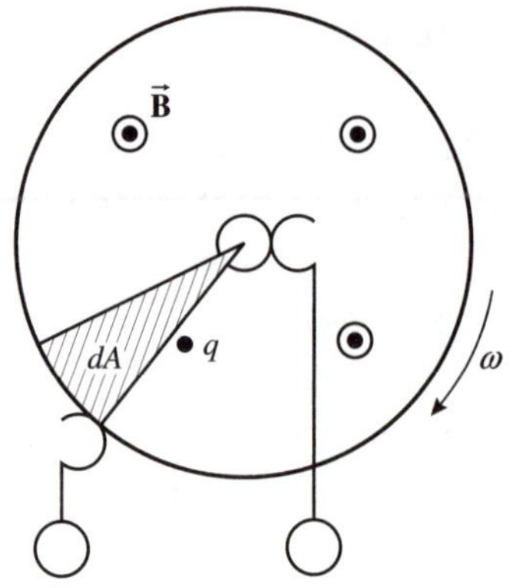

Figure P31.27

Solution

Conceptualize: Faraday did this work before the formulation of the idea of energy converted from one form to another, but the generator is clearly described as taking in mechanical work from the agent turning the disk, and converting (most of) it into energy put out by electric transmission.

Categorize: We are not asked to evaluate an amount of energy, but only the emf described by Faraday's law. For this motional emf, we can apply the law by thinking of magnetic flux changing because area changes.

Analyze: In time dt, the disk turns by angle $d\theta = \omega dt$

The outer brush slides over distance $rd\theta$

The radial line to the outer brush sweeps over area $dA = \frac{1}{2} rrd\theta = \frac{1}{2} r^2 \omega dt$

The emf generated is $\varepsilon = -N \frac{d}{dt} \vec{\mathbf{B}} \cdot \vec{\mathbf{A}}$

giving $\varepsilon = -(1) B \cos 0° \frac{dA}{dt} = -B\left(\frac{1}{2} r^2 \omega\right)$

(We could think of this result as following from the example in the chapter text about the rotating bar.)

The magnitude of the emf is

$$|\varepsilon| = B\left(\frac{1}{2} r^2 \omega\right) = (0.9\ \text{N} \cdot \text{s/C} \cdot \text{m})\left[\frac{1}{2}(0.4\ \text{m})^2 (3\,200\ \text{rev/min})\right]\left(\frac{2\pi\ \text{rad/rev}}{60\ \text{s/min}}\right)$$

$$|\varepsilon| = 24.1\ \text{V}$$ ■

A free positive charge q, represented in our version of the diagram, turning with the disk, feels a magnetic force $q\vec{\mathbf{v}} \times \vec{\mathbf{B}}$ radially outward. Thus the outer contact is positive. ■

Finalize. As the statement of the problem suggests, this is a relatively low-voltage generator. The resistance across the radius of a copper disk can be very low, so depending on the rest of the circuit and on the agent turning the crank, the generator can produce high current. Observe that the homopolar generator has no commutator and produces a voltage constant in time: it generates DC with no ripple.

The following story should be true even if it isn't: As a special honor, the Prime Minister of Britain toured Faraday's laboratory and politely observed the varied demonstrations without understanding much. Jowls quivering, he then asked, "Your experiments, Professor Faraday, are most interesting. But can you tell me, of what *practical value* is this electricity?" The humble scientist replied, "I suppose, Sir, that someday you may come to put a tax upon it."

---

30. A rectangular coil with resistance R has N turns, each of length ℓ and width w as shown in Figure P31.30. The coil moves into a uniform magnetic field $\vec{\mathbf{B}}$ with constant velocity $\vec{\mathbf{v}}$. What are the magnitude and direction of the total magnetic force on the coil (a) as it enters the magnetic field, (b) as it moves within the field, and (c) as it leaves the field?

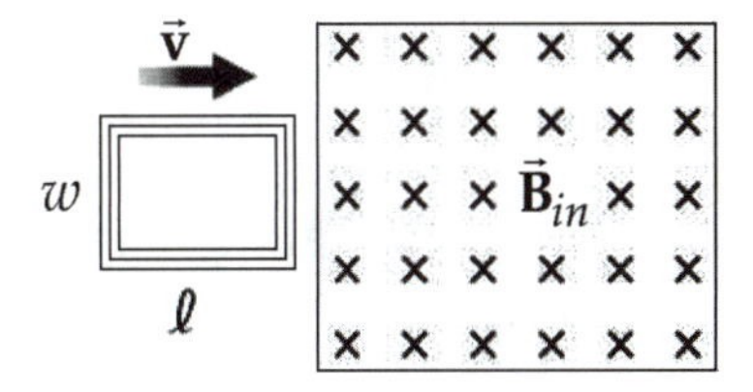

Figure P31.30

Solution

Conceptualize: While it is entering and leaving the field region, the magnetic flux through the coil will be changing, so an emf is induced in it. The emf causes current in the coil. The magnetic field exerts a force on the current. It is reasonable for the force to increase for larger values of N, w, v, and B, and to be inversely proportional to R. We predict the force will be proportional to the square of the magnetic field, because the field causes the emf and acts again to cause the force.

Categorize: We use Faraday's law, the definition of resistance, and the law describing the magnetic force on an electric current. We will need two right-hand rules, for the emf and for the magnetic force.

Analyze:

(a) Call x the distance that the leading edge has penetrated into the strong field region. The flux (Bwx away from you) through the coil increases in time so a voltage of magnitude

$$|\mathcal{E}| = N\frac{d}{dt}Bwx = NBw\frac{dx}{dt} = NBwv$$

is induced in the coil, tending to produce counterclockwise current so that its own field will be toward you.

The current in the coil is $|I| = \dfrac{|\mathcal{E}|}{R} = \dfrac{NBwv}{R}$

The current upward in the leading edge experiences a force

$$\vec{\mathbf{F}} = N\,I\,\vec{\mathbf{L}} \times \vec{\mathbf{B}} = N\left(\frac{NBwv}{R}\right)w\hat{\mathbf{j}} \times B\left(-\hat{\mathbf{k}}\right) = \left(\frac{N^2B^2w^2v}{R}\right)\left(-\hat{\mathbf{i}}\right)$$ ■

(b) The flux through the coil is constant when it is wholly within the high-field region. The induced emf, induced current, and magnetic force are zero. ■

(c) As the coil leaves the field, the away-from-you flux it encloses decreases. To oppose this change, the coil carries clockwise current to make some away-from-you field of its own. Again,

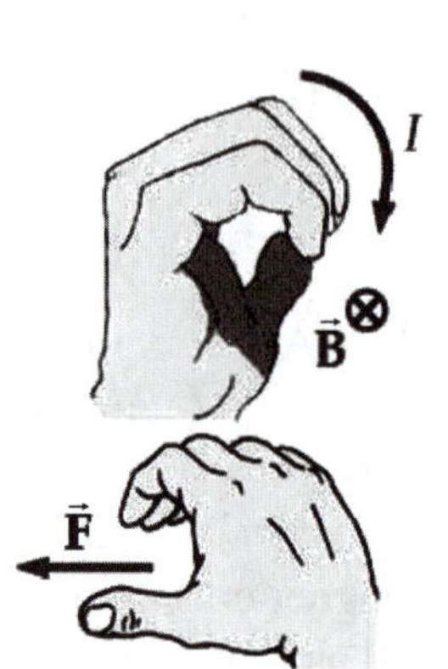

$$|I| = \frac{NBwv}{R}$$

Now the trailing edge carries upward current to experience a force of

$$\vec{\mathbf{F}} = \frac{N^2B^2w^2v}{R} \text{ to the left}$$ ■

Finalize: The current in the coil is in opposite directions when the coil enters and leaves the field. The magnetic force acts on opposite sides of the coil to exert a backward or retarding force in the same direction, to the left, in both cases. We have here a model for eddy-current damping. The proportionalities of the force to the square of N and the square of w, as well as to the square of B, are interesting. The force is proportional only to the first power of v and the minus-first power of R, as predicted. We can say it is proportional to the zeroth power of ℓ.

---

34. A magnetic field directed into the page changes with time according to $B = 0.030\,0t^2 + 1.40$, where B is in teslas and t is in seconds. The field has a circular cross section of radius $R = 2.50$ cm (see Fig. P31.33). When $t = 3.00$ s and $r_2 = 0.020\,0$ m, what are (a) the magnitude and (b) the direction of the electric field at point P_2?

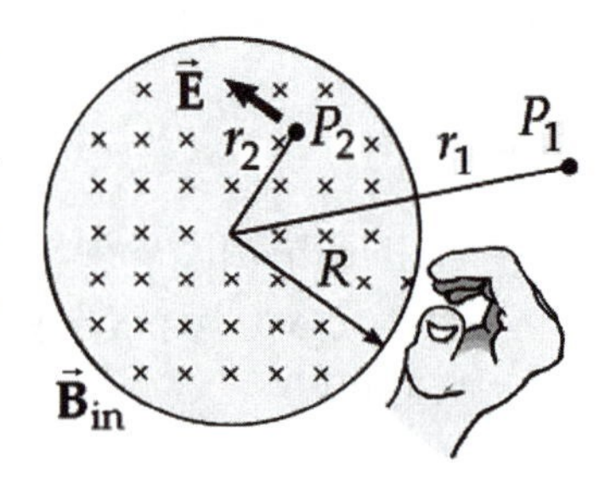

Figure P31.33

Solution

Conceptualize: The electric field is not created by an electric charge, but by a changing magnetic field. There need not be a test charge or a conducing loop at point P_2, but if there were it would respond to the electric field created . . .

Categorize: . . . according to the general form of Faraday's law,

$$\oint \vec{\mathbf{E}} \cdot d\vec{\mathbf{s}} = -\frac{d\Phi_B}{dt}$$

Analyze: Consider a circular integration path of radius r_2. Due to symmetry, Faraday's law becomes

$$E(2\pi r_2) = -\frac{d}{dt}(BA) = -A\left(\frac{dB}{dt}\right)$$

$$|E| = \frac{A}{2\pi r_2}\frac{d}{dt}(0.030\,0t^2 + 1.40) = \frac{\pi r_2^{\,2}}{2\pi r_2}(0.060\,0t) = \frac{r_2}{2}(0.060\,0t)$$

At $t = 3.00$ s, the field is

(a) $E = \frac{1}{2}(0.020\,0\text{ m})(0.060\,0\text{ T/sec})(3.00\text{ sec}) = 1.80 \times 10^{-3}\text{ N/C}$ ■

(b) If there were a circle of wire of radius r_2, it would enclose increasing magnetic flux due to a magnetic field away from you. It would carry counterclockwise current to make its own magnetic field toward you, to oppose the change. Even without the wire and current, the counterclockwise electric field that would cause the current is lurking. At point P_2, it is upward and to the left, perpendicular to r_2. ■

Finalize: This is a very weak electric field. We do not have large area, multiple turns in a coil, or sudden change working in our advantage to create it. The large 1.4-T field is constant in time, so it does not contribute to making the electric field large.

---

36. A 100-turn square coil of side 20.0 cm rotates about a vertical axis at 1.50×10^3 rev/min, as indicated in Figure P31.36. The horizontal component of the Earth's magnetic field at the coil's location is 2.00×10^{-5} T. (a) Calculate the maximum emf induced in the coil by this field. (b) What is the orientation of the coil with respect to the magnetic field when the maximum emf occurs?

Solution

Conceptualize: The magnetic field is constant and so is the area, but the flux of field through the area of the coil changes because of the $\cos\theta$ factor. This device is an AC electric generator. The maximum emf may be on the order of millivolts.

Categorize: The chapter text uses Faraday's law to analyze a coil rotating steadily in a magnetic field. We use its result.

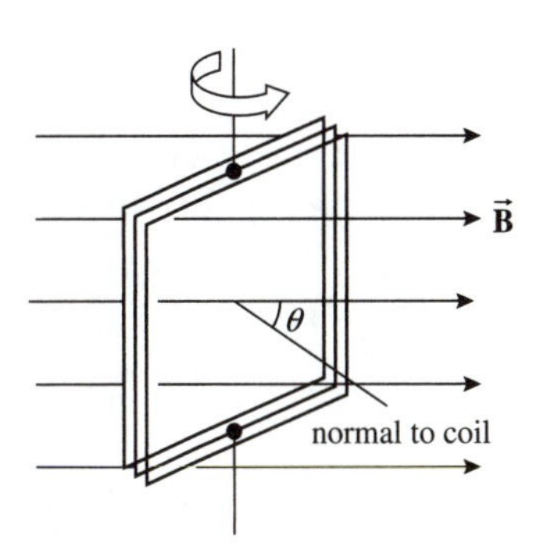

Figure P31.36

Analyze:

(a) By Equation 31.10, $\mathcal{E} = NBA\omega\sin\theta$

The maximum value of $\sin\theta$ is 1, so the maximum value of the voltage is

$$\mathcal{E}_{\max} = NBA\omega$$

With these data, $\mathcal{E}_{max} = (100)(2.00 \times 10^{-5}\ \text{T})(0.200\ \text{m})^2(1\ 500\ \text{rev}/60\ \text{s})(2\pi\ \text{rad}/1\ \text{rev})$

$$\mathcal{E}_{max} = 12.6\ \text{mV}$$ ■

(b) $|\mathcal{E}|$ is equal to $\mathcal{E}_{max}$ when $|\sin\theta|$ is equal to 1 or $\theta = \pm\,\pi/2$.

Therefore, at maximum emf, the normal to the coil is perpendicular to the field, and the plane of the coil is parallel to the field. The flux is changing most rapidly when the flux through the coil is zero. ■

Finalize: This spinning coil can indicate the strength of the Earth's magnetic field. By contrast, a tabletop electromagnet could produce a larger field and an alternating emf of greater amplitude in the spinning coil. Look back through the problems you have done to see how they together show that $\mathcal{E} = -N\dfrac{d}{dt}(BA\cos\theta)$ describes situations where B changes, where A changes, and where θ changes.

---

57. The plane of a square loop of wire with edge length $a = 0.200$ m is oriented vertically and along an east-west axis. The Earth's magnetic field at this point is of magnitude $B = 35.0\ \mu$T and directed northward at 35.0° below the horizontal. The total resistance of the loop and the wires connecting it to a sensitive ammeter is 0.500 Ω. If the loop is suddenly collapsed by horizontal forces as shown in Figure P31.57, what total charge enters one terminal of the ammeter?

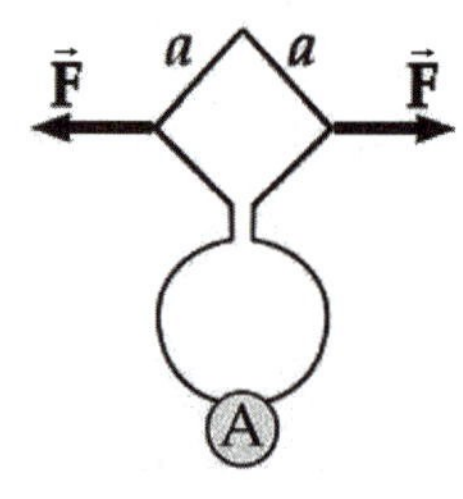

Figure P31.57

Solution

Conceptualize: For the situation described, the maximum current is probably less than 1 mA. So if the loop is closed in 0.1 s, then the total charge would be less than

$$Q = I\Delta t \sim (1\ \text{mA})(0.1\ \text{s}) = 100\ \mu\text{C}$$

Categorize: We do not know how quickly the loop is collapsed, but we can find the total charge by integrating the change in magnetic flux due to the change in area of the loop (from a^2 to 0).

Analyze: The normal to the loop is horizontally north, at 35.0° to the magnetic field. We assume that 0.500 Ω is the total resistance around the circuit, including the ammeter.

$$Q = \int I dt = \int \frac{\mathcal{E}dt}{R} = \frac{1}{R}\int -\left(\frac{d\Phi_B}{dt}\right) dt = -\frac{1}{R}\int d\Phi_B = -\frac{1}{R}\int d(BA\cos\theta)$$

$$= -\frac{B\cos\theta}{R}\int_{A_1=a^2}^{A_2=0} dA$$

$$Q = -\left[\frac{B\cos\theta}{R}A\right]_{A_1=a^2}^{A_2=0} = \frac{B\cos\theta\, a^2}{R} = \frac{(35.0\times10^{-6}\ \mathrm{T})(\cos 35.0°)(0.200\ \mathrm{m})^2}{0.500\ \Omega}$$

$$= 2.29\times10^{-6}\ \mathrm{C}$$ ■

Finalize: The total charge is less than the maximum charge we predicted, so the answer seems reasonable. It is interesting that this charge can be calculated without knowing either the current or the time to collapse the loop.

---

63. A rectangular coil of 60 turns, dimensions 0.100 m by 0.200 m and total resistance 10.0 Ω, rotates with angular speed 30.0 rad/s about the y axis in a region where a 1.00-T magnetic field is directed along the x axis. The time $t = 0$ is chosen to be at an instant when the plane of the coil is perpendicular to the direction of $\vec{\mathbf{B}}$. Calculate (a) the maximum induced emf in the coil, (b) the maximum rate of change of magnetic flux through the coil, (c) the induced emf at $t = 0.050\,0$ s, and (d) the torque exerted by the magnetic field on the coil at the instant when the emf is a maximum.

Solution

Conceptualize: We have an AC electric generator. The maximum emf may be on the order of a hundred volts.

Categorize: Faraday's law will tell us the emf as a function of time. From it we can determine answers (a), (b), and (c). The definition of resistance will tell us then the maximum current in the coil and $\vec{\boldsymbol{\tau}} = \vec{\boldsymbol{\mu}} \times \vec{\mathbf{B}}$ will tell us the torque that must be counterbalanced by an outside agent to keep the generator turning.

Analyze: Let θ represent the angle between the perpendicular to the coil and the magnetic field. Then $\theta = 0$ at $t = 0$ and $\theta = \omega t$ at all later times.

(a)
$$\mathcal{E} = -N\frac{d}{dt}(BA\cos\theta) = -NBA\frac{d}{dt}(\cos\omega t) = +NBA\omega\sin\omega t$$

The maximum value of $\sin\theta$ is 1, so the maximum voltage is

$$\mathcal{E}_{\max} = NBA\omega = (60)(1.00\ \mathrm{T})(0.020\,0\ \mathrm{m}^2)(30.0\ \mathrm{rad/s}) = 36.0\ \mathrm{V}$$ ■

(b)
$$\frac{d\Phi_B}{dt} = \frac{d}{dt}(BA\cos\theta) = -BA\omega\sin\omega t$$

The minimum value of $\sin\theta$ is -1, so the maximum of $d\Phi_B/dt$ is

$$\left(\frac{d\Phi_B}{dt}\right)_{\max} = +BA\omega = (1.00\ \mathrm{T})(0.020\,0\ \mathrm{m}^2)(30.0\ \mathrm{rad/s}) = 0.600\ \mathrm{T\cdot m^2/s}$$ ■

(c) $$\mathcal{E} = NBA\omega \sin \omega t = (36.0 \text{ V})\sin\left[(30.0 \text{ rad/s})(0.050\,0 \text{ s})\right]$$

$$\mathcal{E} = (36.0 \text{ V})\sin(1.50 \text{ rad}) = (36.0 \text{ V})(\sin 85.9°) = 35.9 \text{ V}$$ ■

(d) The emf is maximum when $\theta = 90°$, and $\vec{\boldsymbol{\tau}} = \vec{\boldsymbol{\mu}} \times \vec{\mathbf{B}}$, so

$$\tau_{\max} = \mu B \sin 90° = NIAB = N\mathcal{E}_{\max}\frac{AB}{R}$$

and $$\tau_{\max} = (60)(36.0 \text{ V})\frac{(0.020\,0 \text{ m}^2)(1.00 \text{ T})}{10.0\ \Omega} = 4.32 \text{ N} \cdot \text{m}$$ ■

Finalize: From the torque it is just one more step to find the maximum mechanical power input to the generator, $P_{\max} = \tau\omega = 4.32 \text{ N} \cdot \text{m } 30 \text{ rad/s} = 130 \text{ W}$. This can also be counted as the electric power output, $\mathcal{E}I = \mathcal{E}^2/R = (36 \text{ V})^2/10\ \Omega = 130 \text{ W}$. It is an important part of your education to turn the crank on a generator, feeling how hard it is to light a light bulb and how lower electric resistance makes it still harder.

---

67. The magnetic flux through a metal ring varies with time t according to $\Phi_B = at^3 - bt^2$, where Φ_B is in webers, $a = 6.00 \text{ s}^{-3}$, $b = 18.0 \text{ s}^{-2}$, and t is in seconds. The resistance of the ring is $3.00\ \Omega$. For the interval from $t = 0$ to $t = 2.00$ s, determine the maximum current induced in the ring.

Solution

Conceptualize: The given equation says that the flux is zero at time zero and goes through negative values immediately thereafter. But what really counts is its rate of change . . .

Categorize: . . . which we will find by doing the derivative in Faraday's law. To identify the extreme value for emf we will differentiate again and use the calculus idea that the derivative of a continuous function is zero when the function is a maximum. Knowing the time when the emf is a maximum, we will substitute to evaluate that maximum and at last use the definition of resistance to find the current.

Analyze: Substituting the given values, $\Phi_B = 6.00\,t^3 - 18.0\,t^2$.

Therefore, the emf induced is $$\mathcal{E} = -\frac{d\Phi_B}{dt} = -18.0t^2 + 36.0t$$

The maximum $\mathcal{E}$ occurs when

$$\frac{d\mathcal{E}}{dt} = -36.0t + 36.0 = 0,$$ which gives $t = 1.00$ s.

The maximum current at $t = 1.00$ s is

$$I_{\max} = \frac{\mathcal{E}}{R} = \frac{(-18.0 + 36.0)\text{V}}{3.00\ \Omega} = 6.00 \text{ A}$$ ■

Finalize: In many problems we can think of the answer as a number. But Faraday's law wants a function, magnetic flux as it depends on time, as input data, and returns a function, emf as it depends on time, as the answer. The function $\mathcal{E}(t)$ is much richer than a single number, but numbers can be pulled from it.

---

69. A long, straight wire carries a current given by $I = I_{max} \sin(\omega t + \phi)$. The wire lies in the plane of a rectangular coil of N turns of wire, as shown in Figure P31.69. The quantities I_{max}, ω, and ϕ are all constants. Assume $I_{max} = 50.0$ A, $\omega = 200\pi \text{ s}^{-1}$, $N = 100$, $h = w = 5.00$ cm, and $L = 20.0$ cm. Determine the emf induced in the coil by the magnetic field created by the current in the straight wire.

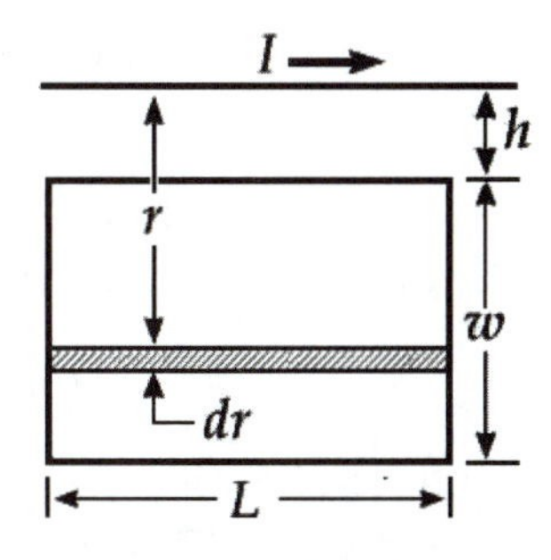

Figure P31.69 (Modified)

Solution

Conceptualize: If the current were constant in time, it would create a magnetic field at the coil but induce no voltage in the coil. The time-varying current acts across empty space to create changing flux through the rectangle, and so to induce an emf in it, perhaps with an amplitude of a few volts.

Categorize: The magnetic field varies in time and also in space between the top and the bottom of the rectangle. We will identify the flux through a ribbon of width dr running horizontally across the rectangle, and do an integral to find the total flux. Then a time derivative in Faraday's law will tell us the emf.

Analyze: The coil is the boundary of a rectangular area. The magnetic field produced by the current in the straight wire is perpendicular to the plane of the area at all points. The magnitude of the field is

$$B = \frac{\mu_0 I}{2\pi r}$$

Thus the flux through the rectangle is

$$\Phi_B = \frac{\mu_0 L}{2\pi} I \int_h^{h+w} \frac{dr}{r} = \frac{\mu_0 L}{2\pi} I_{max} \ln\left(\frac{h+w}{h}\right) \sin(\omega t + \phi)$$

Finally, the induced emf is

$$\mathcal{E} = -N\frac{d\Phi_B}{dt} = -\frac{\mu_0 N L}{2\pi} I_{max}\, \omega \ln\left(\frac{h+w}{h}\right) \cos(\omega t + \phi)$$

$$\mathcal{E} = -\left(4\pi \times 10^{-7} \frac{\text{T} \cdot \text{m}}{\text{A}}\right) \frac{(100)(0.2 \text{ m})}{2\pi} (50 \text{ A})\left(200\pi \text{ s}^{-1}\right) \ln\left(\frac{10 \text{ cm}}{5 \text{ cm}}\right) \cos(\omega t + \phi)$$

$\mathcal{E} = -(87.1 \text{ mV})\cos(200\pi t \text{ rad/s} + \phi)$
$= -87.1 \cos(200\pi t + \phi)$ where $\mathcal{E}$ is in millivolts and t is in seconds ■

Finalize: With these numbers the sinusoidally varying emf has an amplitude only on the order of a tenth of a volt. Practical transformers have many turns in the primary as well as the secondary.

Related Comment: The factor $\sin(\omega t + \phi)$ in the expression for the current in the straight wire does not change appreciably when ωt changes by 0.1 rad or less. Thus, the current does not change appreciably during a time interval

$$\Delta t < \frac{0.100}{200\pi \text{ s}^{-1}} = 1.59 \times 10^{-4} \text{ s}$$

We define a critical length,

$$c\Delta t = (3.00 \times 10^8 \text{ m/s})(1.59 \times 10^{-4} \text{ s}) = 4.77 \times 10^4 \text{ m}$$

equal to the distance to which field changes could be propagated during an interval of 1.59×10^{-4} s. This length is so much larger than any dimension of the loop or its distance from the wire that, although we consider the straight wire to be infinitely long, we can also safely ignore the field propagation effects in the vicinity of the loop. Moreover, the phase angle can be considered to be constant along the wire in the vicinity of the loop. If the angular frequency ω were much larger, say $200\pi \times 10^5 \text{ s}^{-1}$, the corresponding critical length would be only 48 cm. In this situation, propagation effects would be important and the above expression of $\mathcal{E}$ would require modification. As a rough rule, we can consider field propagation effects for circuits of laboratory size to be negligible for frequencies, $f = \omega/2\pi$, that are less than about 10^6 Hz.

72. A bar of mass m and resistance R slides without friction in a horizontal plane, moving on parallel rails as shown in Figure P31.72. The rails are separated by a distance d. A battery that maintains a constant emf $\mathcal{E}$ is connected between the rails, and a constant magnetic field $\vec{\mathbf{B}}$ is directed perpendicularly out of the page. Assuming the bar starts from rest at time $t = 0$, show that at time t it moves with a speed

$$v = \frac{\mathcal{E}}{Bd}\left(1 - e^{-B^2 d^2 t/mR}\right)$$

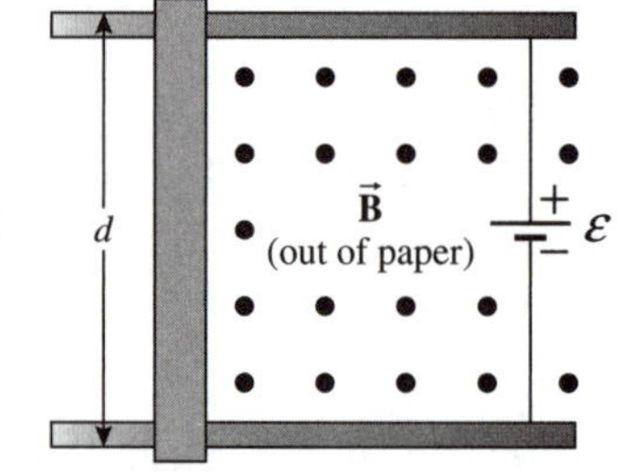

Figure P31.72

Solution

Conceptualize: The battery establishes a counterclockwise current in the circuit, a current that in the bar is downward toward the bottom of the page. Then the external magnetic field exerts a force on the bar in the direction down × toward you = left. The bar accelerates toward the left. Now the magnetic flux through the rectangle changes, so an extra generated or induced emf appears, to oppose the battery. Its presence will make the current somewhat smaller, to make the magnetic force smaller, to reduce the acceleration of the bar. But the leftward velocity of the bar keeps increasing gradually, until the induced emf approaches balancing the battery voltage. The current and the magnetic force on the bar approach zero. The bar approaches a terminal speed.

That's a lot of causally linked simultaneous physical processes, and …

Categorize: … we must use the laws describing all of them. Mathematics can follow the whole process of v increasing, first rapidly and then approaching a limit. We use Kirchhoff's loop rule, the definition describing how a magnetic field exerts a force, Newton's second law for the bar as a particle under a net force as it approaches equilibrium, Faraday's law, and the definition of acceleration.

Analyze: Mass m, distance d, battery voltage $\mathcal{E}$, and magnetic field B are all constants. Think of them as known. Time t is a variable, upon which depend the unknown functions I, $\mathcal{E}_{\text{induced}}$, magnetic force F, speed v, and acceleration dv/dt.

From Kirchhoff's loop rule, the current is $I = \dfrac{\mathcal{E} + \mathcal{E}_{\text{induced}}}{R}$

where the induced emf is $\mathcal{E}_{\text{induced}} = -\dfrac{d}{dt}(BA) = -Bvd$

(Because this equation contains a minus sign, we were careful to show a plus sign in the equation for current—don't count a minus sign twice.)

The only force on the bar is the magnetic force, causing its speed to change according to

Newton's second law $F = m\dfrac{dv}{dt} = IBd$

We rearrange and substitute to eliminate unknowns:

$$\frac{dv}{dt} = \frac{IBd}{m} = \frac{Bd}{mR}\left(\mathcal{E} + \mathcal{E}_{\text{induced}}\right)$$

$$\frac{dv}{dt} = \frac{Bd}{mR}(\mathcal{E} - Bvd)$$

We have used all of the physical principles to express the unknown function v in terms of the known quantities. Now we proceed with mathematics instead of physics. The equation involves dv/dt as well as v. It is a differential equation. To solve it, we define a new unknown function $u = \mathcal{E} - Bvd$. Then $\dfrac{du}{dt} = -Bd\dfrac{dv}{dt}$ and the differential equation becomes $-\dfrac{1}{Bd}\dfrac{du}{dt} = \dfrac{Bd}{mR}u$

Now we "separate variables," getting the unknown u on one side in

$$\int_{u_0}^{u} \frac{du}{u} = -\int_{0}^{t} \frac{(Bd)^2}{mR} dt$$

We have chosen to integrate from the release point $t = 0$ to any particular later time $t = t$. You know how to integrate to get

$$\ln\frac{u}{u_0} = -\frac{(Bd)^2}{mR}t \qquad \text{or} \qquad \frac{u}{u_0} = e^{-B^2 d^2 t/mR}$$

Now we put the answer in terms or v instead of u.

Since $v = 0$ when $t = 0$, we have $u_0 = \mathcal{E}$

and our definition was $u = \mathcal{E} - Bvd$

Substitution gives $\mathcal{E} - Bvd = \mathcal{E} e^{-B^2 d^2 t/mR}$

So we can solve for the speed: $v = \dfrac{\mathcal{E}}{Bd}\left(1 - e^{-B^2 d^2 t/mR}\right)$ ■

Finalize: The equation we have proved says that at $t = 0$, the speed is $v = (\mathcal{E}/Bd)(1 - e^0) = (\mathcal{E}/Bd)(1 - 1) = 0$. This is a true statement, a "boundary condition" that we included in the solution. As t increases without limit, the speed keeps increasing, approaching asymptotically a definite limit, $v_{max} = (\mathcal{E}/Bd)(1 - e^{-\infty}) = (\mathcal{E}/Bd)(1 - 0) = \mathcal{E}/Bd$. Check that this equation is dimensionally correct:

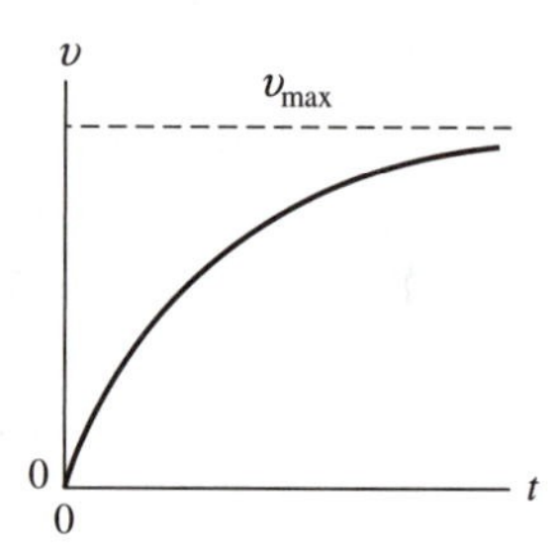

$$V/T \cdot m = (N \cdot m/C)/(N \cdot s/C \cdot m)m = m/s$$

In between, the speed shows a pattern of rapid increase followed by slower and slower increase, as sketched in the graph.

You have seen this shape twice before: For a falling object feeling a force of air resistance proportional to its speed in Section 6.4, and for the charge on a capacitor charging up in an *RC* circuit with a battery, in Section 28.4. You will see this shape again, for the current in a coil after it is connected to a battery to form an *RL* circuit, in Section 32.2.

You can profitably study this example repeatedly. First work to understand all of the physical processes described in the Conceptualize step. Then identify how each is expressed in an equation. After that, find a pattern in the mathematics of eliminating unknown functions by substitution and at last solving the differential equation. In a math class you may solve differential equations by doing indefinite integrals and dealing with "constants of integration." We prefer to show definite integrals with physically identifiable limits.

If Lenz's law were not true, if the equation to be solved were $\dfrac{dv}{dt} = \dfrac{Bd}{mR}(\mathcal{E} + Bvd)$ with a plus sign instead of a minus sign, then the speed would increase exponentially, like a population of bacteria with unlimited resources. This pattern of speed increase could not sustain itself physically and would violate conservation of energy for the circuit-field system.

32

Inductance

EQUATIONS AND CONCEPTS

A **self-induced emf** is present in a circuit element (coil, solenoid, toroid, coaxial cable, etc.) when the current in the circuit changes in time. *The inductance (L) of a given device depends on its physical characteristics, and is a measure of the opposition of the device to a change in the current. An **inductor** is a circuit element which has a large inductance.*

$$\mathcal{E}_L = -L\frac{dI}{dt} \tag{32.1}$$

The SI **unit of inductance** is the henry (H). A rate of change of current of 1 ampere per second in an inductor of 1 henry will produce a self-induced emf of 1 volt.

$$1\,\text{H} = 1\frac{\text{V}\cdot\text{s}}{\text{A}} = 1\,\Omega\cdot\text{s}$$

The **inductance of a circuit element** can be expressed as the ratio of magnetic flux to current (Equation 32.2) or the ratio of induced emf to rate of change of current (Equation 32.3).

$$L = \frac{N\Phi_B}{I} \tag{32.2}$$

$$L = -\frac{\mathcal{E}_L}{dI/dt} \tag{32.3}$$

The **inductance of a uniformly wound solenoid** is proportional to the square of the number of turns per unit of length, n. In Equations 32.4 and 32.5, A is the cross-sectional area of a solenoid of length ℓ and volume V.

$$L = \mu_0\frac{N^2}{\ell}A \tag{32.4}$$

$$L = \mu_0 n^2 A\ell = \mu_0 n^2 V \tag{32.5}$$

A **series *RL* circuit** is shown in the figure at right. The circuit elements are a battery, resistor, inductor, and switch. *The inductor opposes any change in the current in the circuit.*

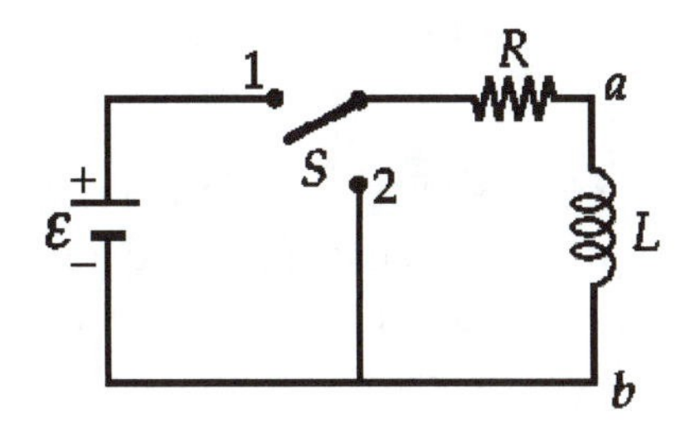

Graphs illustrating the manner in which current increases and decays in this circuit are shown on the following page.

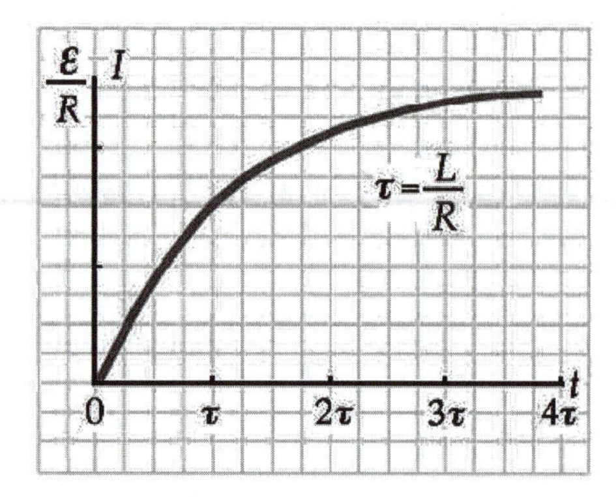

Increase of current (from zero) in the *RL* circuit when the switch is moved from the "Open" position to position 1.

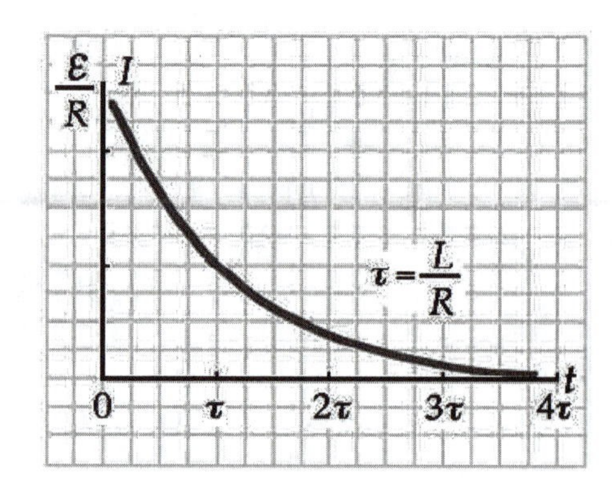

Exponential decay of current (from maximum) in the *RL* circuit when the switch is moved from position 1 to position 2.

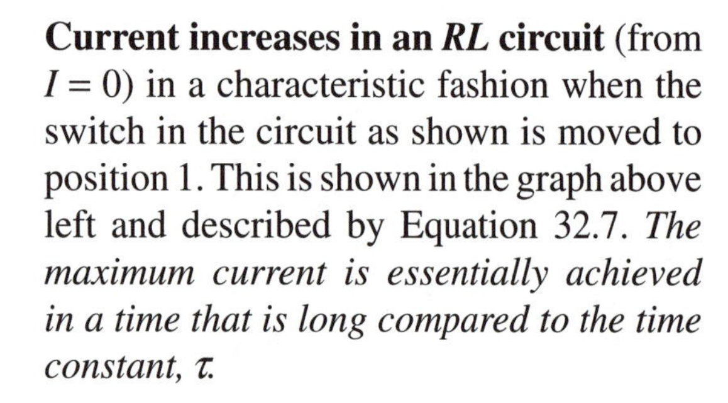

Current increases in an *RL* circuit (from $I = 0$) in a characteristic fashion when the switch in the circuit as shown is moved to position 1. This is shown in the graph above left and described by Equation 32.7. *The maximum current is essentially achieved in a time that is long compared to the time constant, τ.*

$$I = \frac{\mathcal{E}}{R}\left(1 - e^{-t/\tau}\right) \quad (32.7)$$

The **time constant of the circuit** is the time required for the current to reach 63.2% of its final value.

$$\tau = L/R \quad (32.8)$$

Current will decay exponentially (from $I_i = \mathcal{E}/R$) when the switch in the circuit is moved from position 1 to position 2. The curve of the decay is shown in the right-hand graph above and is described by Equation 32.10.

$$I = \frac{\mathcal{E}}{R}e^{-t/\tau} = I_i e^{-t/\tau} \quad (32.10)$$

where $I_i = \dfrac{\mathcal{E}}{R}$

The **stored energy** (U) in the magnetic field of an inductor is proportional to the square of the current in the inductor.

$$U = \tfrac{1}{2}LI^2 \quad (32.12)$$

The **energy density** (u_B) is the energy per unit volume stored in a magnetic field. *This expression is valid for any region of space in which a magnetic field exists.*

$$u_B = \frac{U}{V} = \frac{B^2}{2\mu_0} \quad (32.14)$$

Mutual inductance (M) is characteristic of a system of two nearby circuits (e.g., coils). The value of the mutual inductance in a particular case depends on the geometry of each circuit and on their orientation with respect to each other. *Mutual inductance is a shared property of a pair of circuits.* In the equations shown, M_{12} is the mutual inductance of coil 2 with respect to coil 1; Φ_{12} is the magnetic flux through coil 2 due to the current in coil 1. Corresponding notations apply for M_{21}. The unit of mutual inductance is the henry (H).

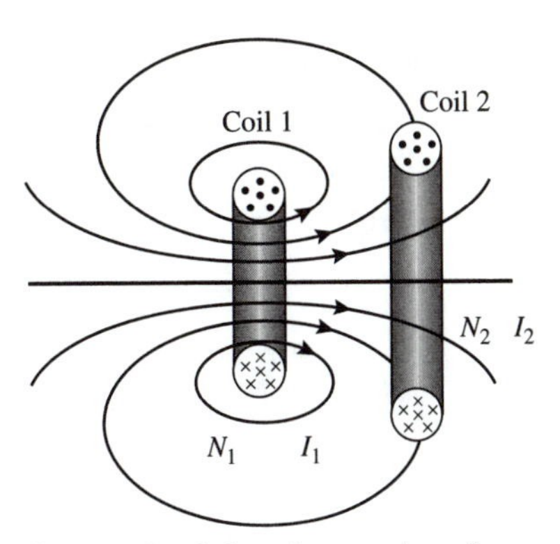

Cross-sectional view of two nearby coils.

$$M_{12} = \frac{N_2\Phi_{12}}{I_1} \quad (32.15)$$

$$M_{21} = \frac{N_1\Phi_{21}}{I_2}$$

$$M_{12} = M_{21}$$

The **emf induced in one coil** is always proportional to the rate at which the current in the other coil is changing.

$$\mathcal{E}_2 = -M_{12}\frac{dI_1}{dt} \quad (32.16)$$

$$\mathcal{E}_1 = -M_{21}\frac{dI_2}{dt} \quad (32.17)$$

A **series *LC* circuit** is shown in the figure at right. The charged capacitor can be connected by switch S to the inductor and the circuit is assumed to have zero resistance. When the switch S is closed, the circuit exhibits oscillation with transfer of energy between the electric field of the capacitor and the magnetic field of the inductor.

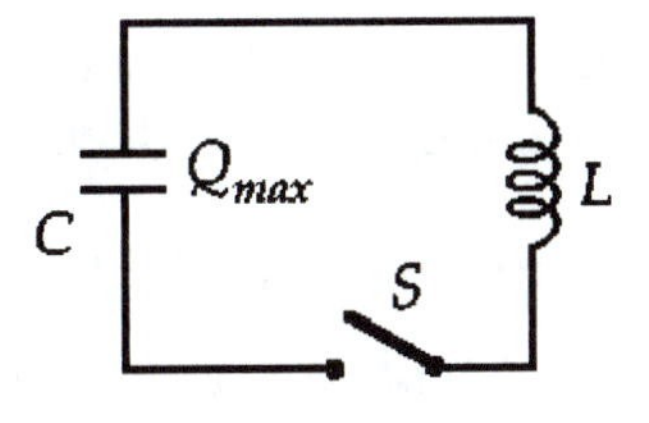

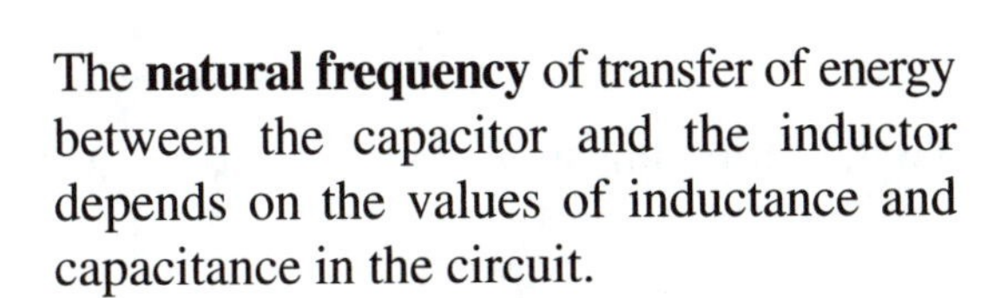
The **natural frequency** of transfer of energy between the capacitor and the inductor depends on the values of inductance and capacitance in the circuit.

$$\omega = \frac{1}{\sqrt{LC}} \quad (32.22)$$

The **charge on the capacitor and the current in the inductor** vary sinusoidally in time and are 90° out of phase. When the charge on the capacitor is maximum, the current is instantaneously zero; and when the charge on the capacitor is zero, the current is maximum.

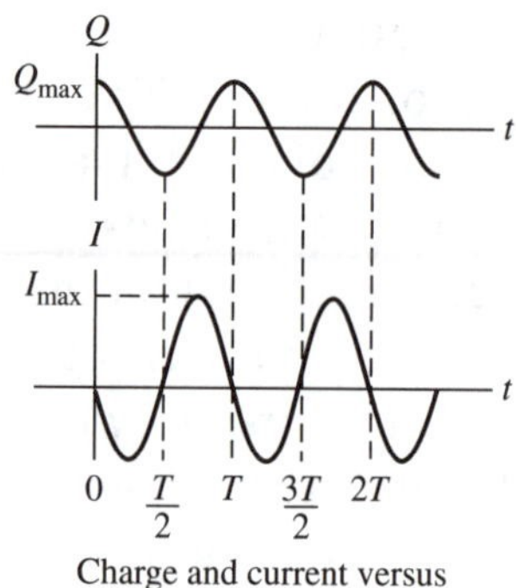

Charge and current versus time in an LC circuit.

Q varies between $+Q_{max}$ and $-Q_{max}$

$$Q = Q_{max} \cos \omega t \qquad (32.24)$$

I varies between $+I_{max}$ and $-I_{max}$

$$I = -I_{max} \sin \omega t \qquad (32.25)$$

Equation 32.24 and Equation 32.25 are valid for $Q = Q_{max}$ when $t = 0$.

The **total energy in an** $\boldsymbol{LC}$ **circuit** is shared between the electric field of the capacitor and the magnetic field of the inductor. If the system has zero resistance and electromagnetic radiation is ignored, the total energy of the system remains constant.

$$U = U_C + U_L \qquad (32.26)$$

$$U = \frac{Q^2_{\ max}}{2C} \cos^2 \omega t + \tfrac{1}{2} L I^2_{max} \sin^2 \omega t$$

SUGGESTIONS, SKILLS, AND STRATEGIES

Equation 32.2, $L = N\Phi_B/I$, and Equation 32.12, $U = \frac{1}{2}LI^2$, provide two different approaches for the calculation of the inductance L for a particular device.

In order to use Equation 32.2 to calculate L, take the following steps:

- Assume a current I to exist in the conductor for which you wish to calculate L (coil, solenoid, coaxial cable, or other device).
- Calculate the magnetic flux through the appropriate cross section using $\Phi_B = \int \vec{\mathbf{B}} \cdot d\vec{\mathbf{A}}$. Remember that in many cases, $\vec{\mathbf{B}}$ will not be uniform over the area.
- Calculate L directly from the defining Equation 32.2.

In order to use Equation 32.12 to calculate L, take the following steps:

- Assume a current I in the conductor.
- Find an expression for B for the magnetic field produced by I.
- Use Equation 32.14, $u_B = B^2/2\mu_0$, and integrate this value of u_B over the appropriate volume to find the total energy stored in the magnetic field of the inductor $U_L = \int u_B dV$.
- Substitute this value of U_L into Equation 32.12 and solve for L.

REVIEW CHECKLIST

You should be able to:

- Calculate the inductance of a device of suitable geometry. (Section 32.1)
- Calculate the magnitude and direction of the self-induced emf in a circuit containing one or more inductive elements when the current changes with time. (Section 32.1)
- Determine the current, rate of change of current, time constant, and self-induced emf in an *LR* circuit. (Section 32.2)
- Calculate the total magnetic energy stored in a magnetic field. You should be able to perform this calculation if (1) you are given the values of the inductance of the device with which the field is associated and the current in the circuit, or (2) given the value of the magnetic field throughout the region of space in which the magnetic field exists. In the latter case, you must integrate the expression for the energy density u_B over an appropriate volume. (Section 32.3)
- Determine the mutual inductance of two coils and calculate the emf induced by mutual inductance in one coil due to a time-varying current in a nearby coil. (Section 32.4)
- Calculate the natural frequency of oscillation of an *LC* circuit. Calculate the value of total energy and determine charge and current as a function of time. (Section 32.5)

ANSWER TO AN OBJECTIVE QUESTION

6. If the current in an inductor is doubled, by what factor is the stored energy multiplied? (a) 4 (b) 2 (c) 1 (d) 1/2 (e) 1/4

Answer (a). The energy stored in an inductor carrying a current I is given by $U = \frac{1}{2}LI^2$. Therefore, doubling the current will quadruple the energy stored in the inductor.

□ □ □ □

ANSWERS TO SELECTED CONCEPTUAL QUESTIONS

1. The current in a circuit containing a coil, a resistor, and a battery has reached a constant value. (a) Does the coil have an inductance? (b) Does the coil affect the value of the current?

Answer (a) The coil has an inductance regardless of the nature of the current in the circuit. Inductance depends only on the coil geometry and its construction. (b) Since the current is constant, the self-induced emf in the coil is zero, and the coil does not affect the steady-state current. (We assume the resistance of the coil is negligible.)

□ □ □ □

8. After the switch is closed in the LC circuit shown in Figure CQ32.8, the charge on the capacitor is sometimes zero, but at such instants the current in the circuit is not zero. How is this behavior possible?

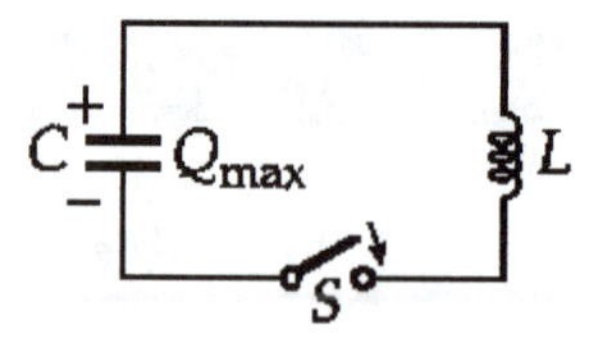

Figure CQ32.8

Answer When the capacitor is fully discharged, the current in the circuit is a maximum. The inductance of the coil is making the charge continue to flow, while the capacitor charge passes through the value zero on its way between positive and negative values. At such an instant, the magnetic field of the coil contains all the energy that was originally stored in the charged capacitor.

SOLUTIONS TO SELECTED END-OF-CHAPTER PROBLEMS

3. A 2.00-H inductor carries a steady current of 0.500 A. When the switch in the circuit is opened, the current is effectively zero after 10.0 ms. What is the average induced emf in the inductor during this time interval?

Solution

Conceptualize: This is a physically large coil with a rapidly changing current. We estimate several volts of self-induced emf.

Categorize: We apply the definition of self-inductance.

Analyze: The self-induced emf at any instant is

$$\varepsilon_L = -L\frac{dI}{dt}$$

Its average value is

$$\varepsilon_{L,\text{ave}} = -L\left(\frac{I_f - I_i}{t}\right) = (-2.00\text{ H})\left(\frac{0 - 0.500\text{ A}}{1.00\times 10^{-2}\text{ s}}\right)\left(\frac{\text{V}\cdot\text{s/A}}{1\text{ H}}\right) = +100\text{ V} \quad \blacksquare$$

Finalize: For the milliseconds while the current is changing, a large voltage is induced in the coil, tending to keep the current at its original value. In mechanics, the inertia of an object describes its tendency to maintain its original velocity. But inertia is not a force; self-induction produces a real electromotive force just like that of a battery.

---

7. A 10.0-mH inductor carries a current $I = I_{\max}\sin\omega t$, with $I_{\max} = 5.00$ A and $f = \omega/2\pi =$ 60.0 Hz. What is the self-induced emf as a function of time?

Solution

Conceptualize: The AC current keeps changing, so new positive and negative pulses of AC voltage are induced in the coil. The amplitude may be a few volts.

Categorize: We use the definition of self-inductance.

Analyze: $\mathcal{E}_L = -L\dfrac{dI}{dt} = -L\dfrac{d}{dt}\left(I_{\max}\sin\omega t\right)$

$$\mathcal{E}_L = -L\omega I_{\max}\cos\omega t = -(0.010\ 0\ \text{H})\left(120\pi\ \text{s}^{-1}\right)(5.00\text{A})\cos(120\pi\ t)$$

$$\mathcal{E}_L = -(18.8)\cos(377\,t) \qquad \text{where } \mathcal{E} \text{ is in volts and } t \text{ is in seconds} \qquad \blacksquare$$

Finalize: The answer is not a number but an infinity of numbers making up a function, which is the voltage as it depends on time.

---

10. An inductor in the form of a solenoid contains 420 turns and is 16.0 cm in length. A uniform rate of decrease of current through the inductor of 0.421 A/s induces an emf of 175 μV. What is the radius of the solenoid?

Solution

Conceptualize: This is a laboratory tabletop situation. The rate of change in current can be thought of as 0.421 milliamps per millisecond. We expect a radius on the order of a centimeter.

Categorize: We represent the inductance of the solenoid symbolically in terms of its dimensions. Then the definition of inductance will relate the induced voltage to how fast the current is changing. We will be able to solve for the radius as the single unknown.

Analyze: The inductance is $L = \dfrac{\mu_0 N^2 A}{\ell}$ with $A = \pi r^2$. The induced emf as a function of time is $\mathcal{E}_L = -L\dfrac{dI}{dt}$. By substitution we have

$$\mathcal{E}_L = -L\frac{dI}{dt} = -\frac{\mu_0 N^2 \pi r^2}{\ell}\frac{dI}{dt} \quad \text{and} \quad r = \left(\frac{-\mathcal{E}_L \ell}{\mu_0 N^2 \pi\, dI/dt}\right)^{1/2}$$

$$\text{Then } r = \left(\frac{-(175\times10^{-6}\ \text{V})(0.16\ \text{m})}{(4\pi\times10^{-7}\ \text{N/A}^2)(420)^2\pi(-0.421\ \text{A/s})}\right)^{1/2} = 9.77\ \text{mm} \qquad \blacksquare$$

Finalize: We did not write down the inductance of the coil, but it is about 416 μH. We were right about the order of magnitude of the radius. The value of the rate of change of current is negative, to represent the current as decreasing.

---

14. A 12.0-V battery is connected into a series circuit containing a 10.0-Ω resistor and a 2.00-H inductor. In what time interval will the current reach (a) 50.0% and (b) 90.0% of its final value?

Solution

Conceptualize: The time constant for this circuit is $\tau = L/R = 0.2$ s. This means that in 0.2 s, the current will reach $1 - 1/e = 63\%$ of its final value, as shown in the graph. We can see from this graph that the time interval to reach 50% of I_{max} should be slightly less than the time constant, perhaps about 0.15 s, and the time interval to reach $0.9I_{max}$ should be about $2.5\tau = 0.5$ s.

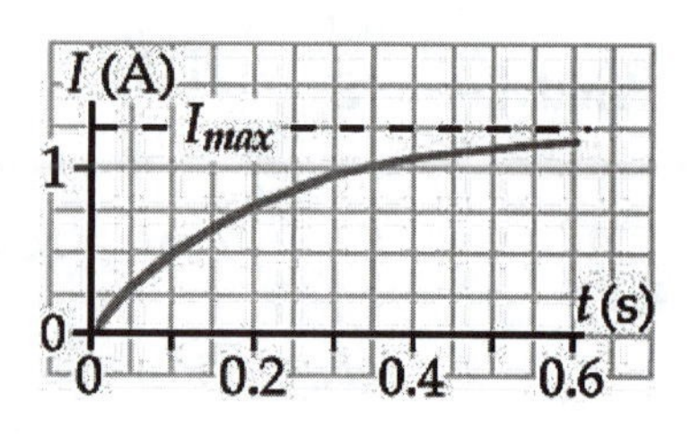

Categorize: The precise time intervals can be found from the equation that describes the rising current in the graph shown and gives the current as a function of time for a known emf, resistance, and time constant.

Analyze: At time t after connecting the circuit, $I(t) = \dfrac{\mathcal{E}\left(1 - e^{-t/\tau}\right)}{R}$ where, after a long time, the current is $I_{max} = \dfrac{\mathcal{E}\left(1 - e^{-\infty}\right)}{R} = \dfrac{\mathcal{E}}{R}$

(a) At 50% of this maximum value, $I(t) = 0.500\, I_{max} = I_{max}(1 - e^{-t/\tau})$

This then yields $0.500 = 1 - e^{-t/\tau}$ or $e^{-t/\tau} = 0.500$ or $e^{+t/\tau} = 2.00$

The natural logarithm is the inverse of the exponential function, so we proceed to

$t/\tau = \ln 2$ and $t = \tau \ln 2 = (L/R) \ln 2 = (2\text{ H}/10\ \Omega)\, 0.693 = 0.139$ s ■

(b) Similarly, to reach 90% of I_{max}, we have

$0.900 = 1 - e^{-t/\tau}$ or $e^{-t/\tau} = 0.100$ or $e^{+t/\tau} = 10.0$

so $t = \tau \ln 10 = 0.200\text{ s}\,(2.30) = 0.461$ s ■

Finalize: The calculated time intervals agree reasonably well with our predictions. We must be careful to avoid confusing the equation for the rising current with the similar equation for the falling current in a circuit without a battery. Checking our answers against predictions is a good way to prevent such mistakes.

---

23. For the *RL* circuit shown in Figure P32.16, let the inductance be 3.00 H, the resistance 8.00 Ω, and the battery emf 36.0 V. (a) Calculate $\Delta V_R/\mathcal{E}_L$, that is, the ratio of the potential difference across the resistor to the emf across the inductor when the current is 2.00 A. (b) Calculate the emf across the inductor when the current is 4.50 A.

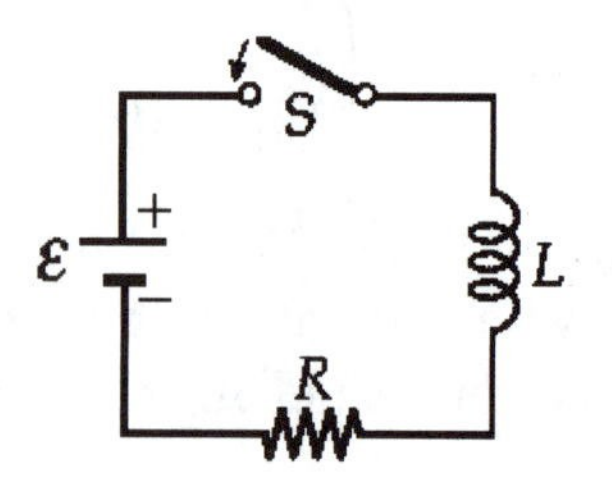

Figure P32.16

Solution

Conceptualize: The voltage across the resistor is proportional to the current, $\Delta V_R = IR$, while the voltage across the inductor is proportional to the **rate of change** in the current, $\mathcal{E}_L = -LdI/dt$. When the switch is first closed, the voltage across the inductor will be large as it opposes the sudden change in current. As the current approaches its steady state value, the voltage across the resistor increases, and the inductor's emf decreases.

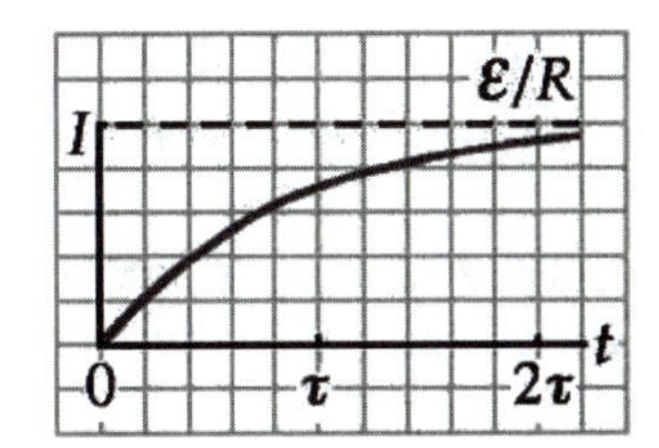

The maximum current will be $\mathcal{E}/R = 4.50$ A, so when $I = 2.00$ A, the resistor and inductor will share similar voltages at this mid-range current. When $I = 4.50$ A, the entire circuit voltage will be across the resistor, and the voltage across the inductor will be zero.

Categorize: We can use the definition of resistance to calculate the voltage across the resistor for each current. We will find the voltage across the inductor by using Kirchhoff's loop rule.

Analyze:

(a) When $I = 2.00$ A, the voltage across the resistor is

$$\Delta V_R = IR = (2.00\text{ A})(8.00\ \Omega) = 16.0\text{ V}$$

Kirchhoff's loop rule tells us that the sum of the changes in potential around the loop must be zero:

$$\mathcal{E} - \Delta V_R - \mathcal{E}_L = 36.0\text{ V} - 16.0\text{ V} - \mathcal{E}_L = 0$$

so $\mathcal{E}_L = 20.0$ V and $\dfrac{\Delta V_R}{\mathcal{E}_L} = \dfrac{16.0\text{ V}}{20.0\text{ V}} = 0.800$ ■

(b) Similarly, for $I = 4.50$ A, $\Delta V_R = IR = (4.50\text{ A})(8.00\ \Omega) = 36.0$ V

$$\mathcal{E} - \Delta V_R - \mathcal{E}_L = 36.0\text{ V} - 36.0\text{V} - \mathcal{E}_L = 0$$

So $\mathcal{E}_L = 0$ ■

Finalize: We see that when $I = 2.00$ A, $\Delta V_R < \mathcal{E}_L$, but they are similar in magnitude, as expected. Also as predicted, the voltage across the inductor goes to zero when the current reaches its maximum value. A worthwhile exercise would be to consider the ratio of these voltages for several different times if the battery is taken out of the circuit to "turn off" the current; for that exercise the answers would be very simple.

---

27. A 140-mH inductor and a 4.90-Ω resistor are connected with a switch to a 6.00-V battery as shown in Figure P32.27. (a) After the switch is first thrown to a (connecting the battery), what time interval elapses before the current reaches 220 mA? (b) What is the current in the inductor 10.0 s after the switch is closed? (c) Now the switch is quickly thrown from a to b. What time interval elapses before the current in the inductor falls to 160 mA?

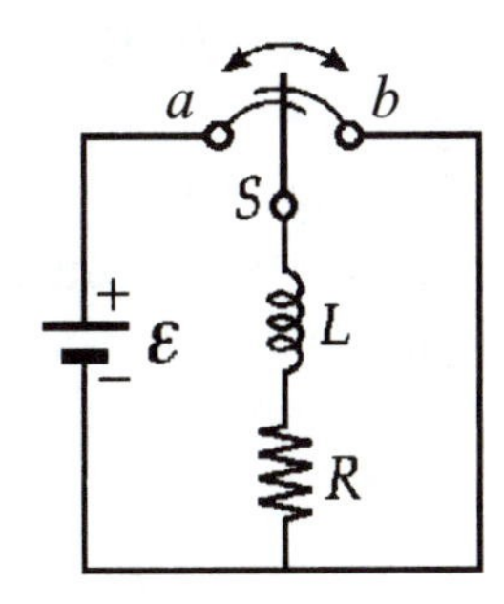

Figure P32.27

Solution

Conceptualize: The resistor limits the *amount* of current. The inductor limits the *rate of increase* of current after it is switched on, and the *rate of decrease* after it is switched off. Noting that the time constant is $L/R = 0.14\text{ H}/4.9\ \Omega = 28.6$ ms, we estimate some milliseconds as the answers to (a) and (c). The time interval is so long in part (b) that the current will be more than one amp, nearly at its final value of 6 V/4.9 Ω = 1.22 A.

Categorize: We use the equations derived in the chapter for the increasing current in an *LR* circuit when the circuit is switched on and for the exponentially decaying current when it is switched off.

Analyze:

(a) The equation for current buildup is obtained by combining Equations 32.7 and 32.8:

$$I = \frac{\mathcal{E}\left(1 - e^{-Rt/L}\right)}{R}$$

We proceed step by step to solve for t in terms of the other quantities, all of which are given:

$IR/\mathcal{E} = 1 - e^{-Rt/L}$ so $e^{-Rt/L} = 1 - IR/\mathcal{E}$ and $-Rt/L = \ln(1 - IR/\mathcal{E})$

$t = -(L/R)\ln(1 - IR/\mathcal{E})$

$t = -(0.140\text{ H}/4.90\ \Omega)\ln[1 - (0.220\text{ A})(4.90\ \Omega)/6.00\text{ V}] = -(0.028\,6\text{ s})\ln[0.820]$

$t = -(0.028\,6\text{ s})(-0.198) = 5.66\text{ ms}$ ■

(b) We now make the general equation refer to a different instant. The current after ten seconds is

$$I = \frac{6.00\text{ V}}{4.90\ \Omega}\left(1 - e^{(-35.0\text{ s}^{-1})(10.0\text{ s})}\right) = 1.22\text{ A}(1 - e^{-350}) = 1.22\text{ A}$$ ■

(c) The equation for current decrease after the battery is removed is $I = \frac{\mathcal{E}}{R}e^{-Rt/L}$

We solve for t: $\frac{IR}{\mathcal{E}} = e^{-Rt/L}$ or $\frac{\mathcal{E}}{IR} = e^{+Rt/L}$

Then $\ln(\mathcal{E}/IR) = Rt/L$ and $t = (L/R)\ln(\mathcal{E}/IR)$

Substituting,

$t = (0.140\text{ H}/4.90\ \Omega)\ln[6.00\text{ V}/(0.160\text{ A}\cdot 4.90\ \Omega)]$

$t = (0.0286\text{ s})\ln 7.65 = 58.1\text{ ms}$ ■

Finalize: Part (a) is about an instant early in the current buildup. Part (b) is very late in the current buildup, reminding us that we can think of the current as essentially constant after many time constants have passed. Part (c) is about an instant fairly late in the current decay process.

---

29. An air-core solenoid with 68 turns is 8.00 cm long and has a diameter of 1.20 cm. When the solenoid carries a current of 0.770 A, how much energy is stored in its magnetic field?

Solution

Conceptualize: This is a fairly small solenoid carrying a fairly small current, so the stored energy should be a very small fraction of a joule.

Categorize: We use the derived result for the self-inductance of a solenoid with length much larger than its radius, and then the equation for energy stored in an inductor.

Analyze: For a solenoid of length ℓ, the inductance is $L = \dfrac{\mu_0 N^2 A}{\ell}$

Thus, since $U_B = \frac{1}{2} LI^2 = \dfrac{\mu_0 N^2 A I^2}{2\ell}$ the stored energy is

$$U_B = \frac{(4\pi \times 10^{-7}\ \text{N/A}^2)(68)^2 \pi (6.00 \times 10^{-3}\ \text{m})^2 (0.770\ \text{A})^2}{2\,(0.080\,0)\ \text{m}} = 2.44 \times 10^{-6}\ \text{J} \quad \blacksquare$$

Finalize: We did not have to write it down, but the solenoid's inductance is indeed small, 8.21 μH. One direct way to make it larger is to add more turns, since the inductance is proportional to the square of the number of turns. Does it make sense to you that the magnetic field in a solenoid does not depend on its radius, while the inductance is proportional to the radius squared? A solenoid of larger cross-sectional area passes more flux of its own field.

---

31. On a clear day at a certain location, a 100-V/m vertical electric field exists near the Earth's surface. At the same place, the Earth's magnetic field has a magnitude of 0.500×10^{-4} T. Compute the energy densities of (a) the electric field and (b) the magnetic field.

Solution

Conceptualize: These are typical values for fields we routinely encounter. The fields contain energy, but no energy input is required just to maintain the fields. From its small size in the standard SI unit, we might guess that the magnetic field has lower energy density.

Categorize: We substitute into standard equations for energy per volume in electric and magnetic fields.

Analyze:

(a) The energy density in the electric field is

$$u_E = \frac{\epsilon_0 E^2}{2} = \frac{\left(8.85 \times 10^{-12}\ \text{C}^2/\text{N}\cdot\text{m}^2\right)(100\ \text{N/C})^2(1\ \text{J/N}\cdot\text{m})}{2} = 44.2\ \text{nJ/m}^3 \quad \blacksquare$$

(b) The energy density in the magnetic field is

$$u_B = \frac{B^2}{2\mu_0} = \frac{\left(5.00 \times 10^{-5}\ \text{T}\right)^2}{2\left(4\pi \times 10^{-7}\ \text{T}\cdot\text{m/A}\right)} = 995 \times 10^{-6}\ \text{T}\cdot\text{A/m}$$

We display explicitly how the units work out:

$$u_B = (995 \times 10^{-6}\ \text{T}\cdot\text{A/m})(1\ \text{N}\cdot\text{s/T}\cdot\text{C}\cdot\text{m})(1\ \text{J/N}\cdot\text{m}) = 995\ \mu\text{J/m}^3 \quad \blacksquare$$

Finalize: Contrary to our guess, the magnetic energy density is 22 500 times greater than that in the electric field. Unknown in detail, electro-thermal-mechanical currents in dense matter deep within the Earth create its magnetic field. The electric field in this problem is incidental to weather, notably distant storms.

---

37. Two solenoids A and B, spaced close to each other and sharing the same cylindrical axis, have 400 and 700 turns, respectively. A current of 3.50 A in solenoid A produces an average flux of 300 μWb through each turn of A and a flux of 90.0 μWb through each turn of B. (a) Calculate the mutual inductance of the two solenoids. (b) What is the inductance of A? (c) What emf is induced in B when the current in A changes at the rate of 0.500 A/s?

Solution

Conceptualize: The problem does not tell us the length or area of either solenoid. The magnetic flux produced by a particular current must be enough to determine the inductances.

Categorize: Equation 32.15 expresses the definition of mutual inductance, which we will use in (a). The chain of logic in the text between Equations 32.1 and 32.2 is so short and sure that we could take Equation 32.2 as the definition of self-inductance; anyway, we use it in this problem. Then part (c) will be a direct application of the result of (b).

Analyze:

(a) The mutual inductance of the coils is

$$M_{12} = \frac{N_2\Phi_{12}}{I_1} = \frac{(700)(90 \times 10^{-6}\text{ Wb})}{3.50\text{ A}} = 18.0\text{ mH} \qquad \blacksquare$$

(b) The inductance of coil A is

$$L = \frac{N\Phi_B}{I} = \frac{(400)(300 \times 10^{-6}\text{ Wb})}{3.50\text{ A}} = 34.3\text{ mH} \qquad \blacksquare$$

(c) The emf induced in the other coil is

$$\mathcal{E}_2 = -M_{12}\frac{dI_1}{dt} = -\left(18.0 \times 10^{-3}\text{ H}\right)(0.500\text{ A/s}) = -9.00\text{ mV} \qquad \blacksquare$$

Finalize: These coils might be the size of soup cans. It is good to have a reminder of the remarkable fact that an electric circuit can influence another circuit nearby with its magnetic field, and it can influence itself in the same way.

---

42. A 1.05-μH inductor is connected in series with a variable capacitor in the tuning section of a short wave radio set. What capacitance tunes the circuit to the signal from a transmitter broadcasting at 6.30 MHz?

Solution

Conceptualize: It is difficult to predict a value for the capacitance without doing the calculations, but we might expect a typical value in the μF or pF range.

Categorize: We want the resonance frequency of the circuit to match the broadcasting frequency, and for a simple *LC* circuit, the resonance frequency only depends on the magnitudes of the inductance and capacitance.

Analyze: The resonance frequency is $f_0 = \dfrac{1}{2\pi\sqrt{LC}}$

Thus, $$C = \frac{1}{(2\pi f_0)^2 L} = \frac{1}{\left[2\pi(6.30 \times 10^6 \text{ Hz})\right]^2 (1.05 \times 10^{-6} \text{ H})} = 608 \text{ pF}$$ ■

Finalize: This is indeed a typical capacitance, so our calculation appears reasonable. You probably would not hear any familiar music on this broadcast frequency. The frequency range for FM radio broadcasting is 88.0 MHz to 108.0 MHz, and AM radio is 535 kHz to 1 605 kHz. The 6.30 MHz frequency falls in the Maritime Mobile SSB Radiotelephone range, so you might hear a ship captain instead of Top 40 tunes! This and other information about the radio frequency spectrum can be found on the National Telecommunications and Information Administration (NTIA) website, which at the time of this writing is at http://www.ntia.doc.gov/osmhome/allochrt.html.

---

48. An *LC* circuit like that in Figure CQ32.8 consists of a 3.30-H inductor and an 840-pF capacitor that initially carries a 105-μC charge. The switch is open for $t < 0$ and is then thrown closed at $t = 0$. Compute the following quantities at $t = 2.00$ ms: (a) the energy stored in the capacitor, (b) the energy stored in the inductor, and (c) the total energy in the circuit.

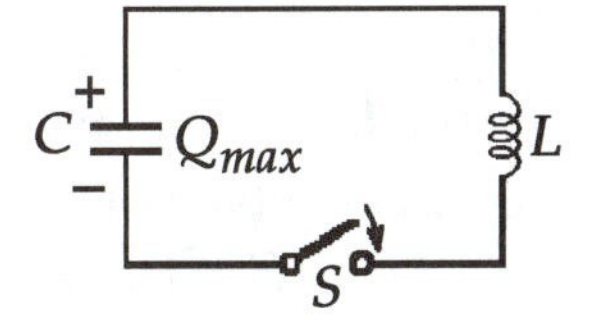

Figure CQ32.8

Solution

Conceptualize: The circuit is an oscillator with frequency

$$\frac{1}{2\pi\sqrt{LC}} = \frac{1}{2\pi\sqrt{3.3 \times 840 \times 10^{-12}}} = 3.02 \text{ kHz} \text{ and so with period } \frac{1}{3\,023/\text{s}} = 331\ \mu\text{s}.$$

Several cycles of oscillation have time to occur during the 2 ms, but we need not count them, because . . .

Categorize: . . . the derived equation for the oscillating charge on the capacitor will tell us all we need about the 2-ms instant. Part (c) is especially easy to answer because energy is the same at all instants, including time zero.

Analyze: At $t = 0$ the capacitor charge is at its maximum value, so $\phi = 0$ in

$$Q = Q_{\text{max}} \cos(\omega t + \phi) = Q_{\text{max}} \cos\left(\frac{t}{\sqrt{LC}}\right)$$

Substituting the given information, the charge at 2 ms is

$$Q = (105 \times 10^{-6}\ \text{C})\cos\left(\frac{2.00 \times 10^{-3}\ \text{s}}{\sqrt{(3.30\ \text{H})(840 \times 10^{-12}\ \text{F})}}\right)$$

$$Q = (105 \times 10^{-6}\ \text{C})(\cos 38.0\ \text{rad}) = 1.01 \times 10^{-4}\ \text{C}$$

(a) Then the energy in the capacitor is

$$U_C = \frac{Q^2}{2C} = \frac{\left(1.01 \times 10^{-4}\ \text{C}\right)^2}{2\left(840 \times 10^{-12}\ \text{F}\right)} = 6.03\ \text{J}$$ ■

(c) The constant total energy is that originally of the capacitor:

$$U = \frac{Q^2_{\text{max}}}{2C} = \frac{\left(1.05 \times 10^{-4}\ \text{C}\right)^2}{2\left(840 \times 10^{-12}\ \text{F}\right)} = 6.56\ \text{J}$$ ■

(b) So the inductor's energy is the remaining

$$U_L = 6.56\ \text{J} - 6.03\ \text{J} = 0.529\ \text{J}$$ ■

Finalize: We could also find answer (b) from

$$\tfrac{1}{2}LI^2 = \tfrac{1}{2}L\left(\frac{d}{dt}Q_{\text{max}}\cos\omega t\right)^2 = \tfrac{1}{2}LQ^2_{\text{max}}\,\omega^2\sin^2\omega t$$

Six joules of energy of oscillation for an electric circuit is perfectly possible, but it might be more convenient with a larger capacitor. This circuit's capacitor was originally charged to $\Delta V = Q/C = 105 \times 10^{-6}\ \text{C}/840 \times 10^{-12}\ \text{F} = 125\ 000\ \text{V}$, so there might be some sparking.

---

53. Consider an LC circuit in which $L = 500$ mH and $C = 0.100\ \mu$F. (a) What is the resonance frequency ω_0? (b) If a resistance of 1.00 kΩ is introduced into this circuit, what is the frequency of the damped oscillations? (c) By what percentage does the frequency of the damped oscillations differ from the resonance frequency?

Solution

Conceptualize: The coil-and-capacitor circuit will oscillate without a power supply if it is connected with the capacitor originally charged. The values of the inductance and capacitance determine the angular frequency, which could be some thousands of radians per second in this case. The introduction of a lot of resistance would strongly damp the electrical vibrations or prevent them altogether. But it may be that one kilohm is a small enough resistance that that it will depress the frequency just a bit.

Categorize: The problem asks for "frequency," which would ordinarily mean frequency f in hertz. It gives the symbol ω that stands for angular frequency $\omega = 2\pi f$ in radians per second. So we will interpret the problem as asking about angular frequency throughout. We use the chapter's identifications of the oscillation angular frequencies of coil-capacitor circuits without resistance and with resistance.

Analyze:

(a) The angular frequency of undamped oscillations is

$$\omega_0 = \frac{1}{\sqrt{LC}} = \frac{1}{\sqrt{(0.50\text{ H})\left(1.00\times10^{-7}\text{ F}\right)}} = 4.47\times10^3\text{ rad/s} \qquad \blacksquare$$

(b) With the resistance present, the angular frequency is

$$\omega_d = \sqrt{\frac{1}{LC} - \left(\frac{R}{2L}\right)^2} = \sqrt{\frac{1}{(0.500\text{ H})\left(1.00\times10^{-7}\text{ F}\right)} - \left(\frac{1.00\times10^3\ \Omega}{2(0.500\text{ H})}\right)^2}$$

$$\omega_d = 4.36\times10^3\text{ rad/s} \qquad \blacksquare$$

(c) $$\frac{\Delta\omega}{\omega_0} = \frac{4.36-4.47}{4.47} = -0.025\,3 = -2.53\% \qquad \blacksquare$$

Thus, the damped frequency is 2.53% lower than the undamped frequency.

Finalize: The situation described is a good tabletop experiment. A frequency meter would read in part (a) $f = \omega/2\pi = 712$ Hz, or the period read from an oscilloscope is $1/f = 1.40$ ms. You can walk into a hobby electronics shop and buy a 1-kΩ resistor and a 0.1-μF capacitor. If the clerk does not know about an inductor with self-inductance in henrys, ask for a "choke coil."

---

66. At $t = 0$, the open switch in Figure P32.66 is thrown closed. We wish to find a symbolic expression for the current in the inductor for time $t > 0$. Let this current be called I and choose it to be downward in the inductor in Figure P32.66. Identify I_1 as the current to the right through R_1 and I_2 as the current downward through R_2. (a) Use Kirchhoff's junction rule to find a relation among the three currents. (b) Use Kirchhoff's loop rule around the left loop to find another relationship. (c) Use Kirchhoff's loop rule around the outer loop to find a third relationship. (d) Eliminate I_1 and I_2 among the three equations to find an equation involving only the current I. (e) Compare the equation in part (d) with Equation 32.6 in the text. Use this comparison to rewrite Equation 32.7 in the text for the situation in this problem and show that

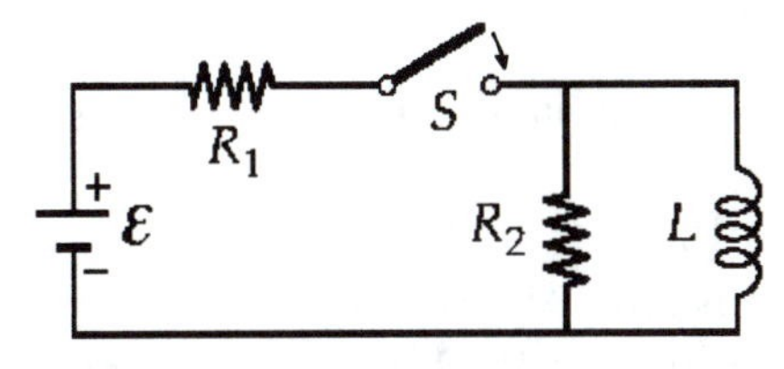

Figure P32.66

$$I(t) = \frac{\mathcal{E}}{R_1}\left[1 - e^{-(R'/L)t}\right]$$

where $R' = R_1R_2/(R_1 + R_2)$.

Solution

Conceptualize: The current in the inductor will start from zero at $t = 0$ and increase to asymptotically approach a final value. The junctions make this circuit more complicated than the LR circuit considered in the chapter text. To find the time-dependent current we must . . .

Categorize: . . . use first principles, in the form of Kirchhoff's rules, and then compare with the solution of the simpler circuit in the chapter text. The equation we are to prove is reasonable in that it starts from $I = 0$ at $t = 0$. Long after the switch is closed, the current will be constant, the voltage across the inductor will be zero, and no current will go through R_2. Then the current will approach the limiting value $\mathcal{E}/R_1$, and the equation we are to prove gets this right.

Analyze:

(a) With I the downward current through the inductor and I_2 the downward current through R_2, Kirchhoff's junction rule says that $I_1 = I + I_2$ is the current in R_1.

(b) Left-hand loop: $\mathcal{E} - (I + I_2)R_1 - I_2R_2 = 0$

(c) Outside loop: $\mathcal{E} - (I + I_2)R_1 - L\dfrac{dI}{dt} = 0$

(d) Eliminate I_2, obtaining $I\underbrace{\dfrac{R_1R_2}{R_1 + R_2}}_{R'} + L\dfrac{dI}{dt} = \underbrace{\dfrac{R_2}{R_1 + R_2}\mathcal{E}}_{\mathcal{E}'}$

(e) Thus, the equation has the form $\mathcal{E}' - IR' - L\dfrac{dI}{dt} = 0$

This is of the same form as Equation 32.6, so the reasoning following that equation in the text shows that the solution is the same form as Equation 32.7,

$$I = \frac{\mathcal{E}'}{R'}\left[1 - e^{-R't/L}\right] \quad \text{with} \quad \frac{\mathcal{E}'}{R'} = \frac{\mathcal{E}R_2/(R_1 + R_2)}{R_1R_2/(R_1 + R_2)} = \frac{\mathcal{E}}{R_1}$$

and we have $I(t) = \dfrac{\mathcal{E}}{R_1}\left[1 - e^{-(R't/L)}\right]$ ■

Finalize: The current grows in a pattern like that in a circuit with an inductor and a single resistor. The time constant is L/R' with R' being equal to the equivalent resistance of R_1 and R_2 in parallel. We would not guess this from the original circuit, but the Kirchhoff equations demonstrate it.

---

76. In Figure P32.76, the battery has emf $\mathcal{E} = 18.0$ V and the other circuit elements have values $L = 0.400$ H, $R_1 = 2.00$ kΩ, and $R_2 = 6.00$ kΩ. The switch is closed for $t < 0$ and steady-state conditions are established. The switch is then opened at $t = 0$. (a) Find the emf across L immediately after $t = 0$. (b) Which end of the coil, a or b, is at the higher voltage? (c) Make graphs of the currents in R_1 and in R_2 as a function

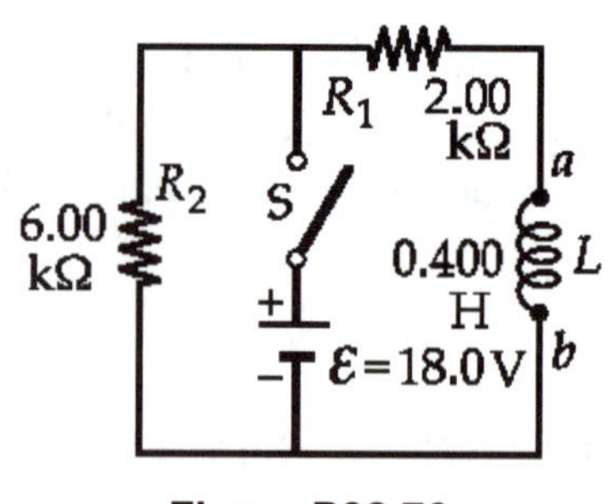

Figure P32.76

of time, treating the steady-state directions as positive. Show values before and after $t = 0$. (d) At what moment after $t = 0$ does the current in R_2 have the value 2.00 mA?

Solution

Conceptualize: Before time $t = 0$, the current downward in R_2 is equal to 18.0 V/6.00 kΩ = 3.00 mA, and the current is clockwise in R_1 and the coil, with the value 18.0 V/2.00 kΩ = 9.00 mA. The inductor forbids any instantaneous change in the current in it. At and immediately after $t = 0$, the current in the inductor is still 9.00 mA down. This must be the instantaneous current in R_2, now upward, and also in R_1, to the right. As time goes on the current will decay toward zero in the outer loop as a series circuit with equivalent resistance $R_1 + R_2$.

Categorize: Kirchhoff's loop rule will tell us the voltage across the inductor at time zero. The equation for current decay in an *LR* circuit will let us sketch the current graphs and find when the current has dropped to 2 mA.

Analyze:

(a) Just after $t = 0$, the current in the outer loop is 9.00 mA clockwise but decreasing. Note that $\mathcal{E}_0 = V_b - V_a$. Kirchhoff's loop rule for the outer loop gives (clockwise):

$$V_b - (9.00\text{ mA})(6.00\text{ k}\Omega) - (9.00\text{ mA})(2.00\text{ k}\Omega) = V_a$$

$$\mathcal{E}_0 = V_b - V_a = 72.0\text{ V}$$ ■

(b) Because this answer is positive, point b is at the higher potential. ■

(c) The currents in R_1 and R_2 are shown below. After $t = 0$, the current in R_1 decreases from an initial value of 9.00 mA according to $I = I_0 e^{-Rt/L}$. Taking the original current direction as positive in each resistor, the current decreases from +9.00 mA (to the right) to zero in R_1. In R_2 the current jumps from +3.00 mA (downward) to –9.00 mA (upward) and then decreases in magnitude to zero. The time constant of each decay is 0.4 H/8 000 Ω = 50 μs. Thus we draw each current dropping to $1/e$ = 36.8% of its original value = 3.3 mA at the 50 μs instant. ■

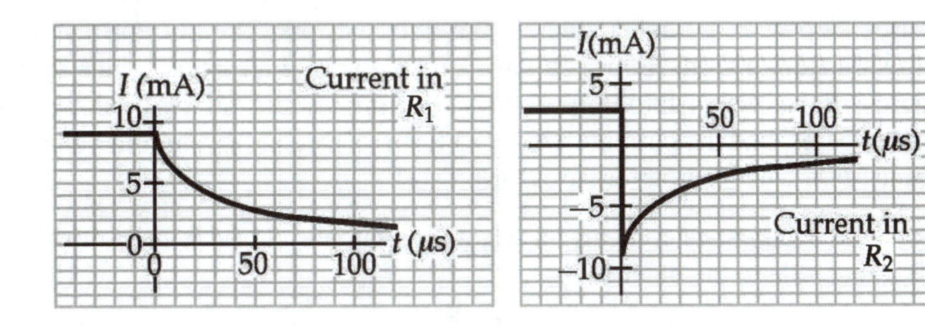

(d) Resistance R_1 establishes the original value of the current in the outer loop, but the series combination of R_1 and R_2 establishes the time constant. The current decreases according to $I = (\mathcal{E}/R_1)e^{-Rt/L}$ where $R = R_1 + R_2$ represents the equivalent series resistance in the discharge loop.

Solving for t, we have $R_1 I/\mathcal{E} = e^{-Rt/L}$ so $\mathcal{E}/R_1 I = e^{+Rt/L}$

and $\ln(\mathcal{E}/R_1 I) = Rt/L$ Then at last $t = (L/R)\ln(\mathcal{E}/R_1 I)$

Substituting,

$$t = (0.400\ \text{H}/8\,000\ \Omega)\ \ln[18.0\ \text{V}/(2\,000\ \Omega)(0.002\,00\ \text{A})]$$

$$t = 50.0\ \mu\text{s}\ \ln 4.50 = 75.2\ \mu\text{s}$$

■

Finalize: 2 mA is a bit less than 37% of 9 mA, so it is reasonable that the time interval is a bit more than one time constant. Note again that the current in a coil cannot change instantaneously, but the voltage across a coil can change instantly. The current in the wires on both sides of a capacitor can change instantaneously, but the voltage across a capacitor cannot undergo an instant jump. Both current and voltage can change instantly for a resistor.

---

77. To prevent damage from arcing in an electric motor, a discharge resistor is sometimes placed in parallel with the armature. If the motor is suddenly unplugged while running, this resistor limits the voltage that appears across the armature coils. Consider a 12.0-V DC motor with an armature that has a resistance of 7.50 Ω and an inductance of 450 mH. Assume the magnitude of the self-induced emf in the armature coils is 10.0 V when the motor is running at normal speed. (The equivalent circuit for the armature is shown in Fig. P32.77.) Calculate the maximum resistance R that limits the voltage across the armature to 80.0 V when the motor is unplugged.

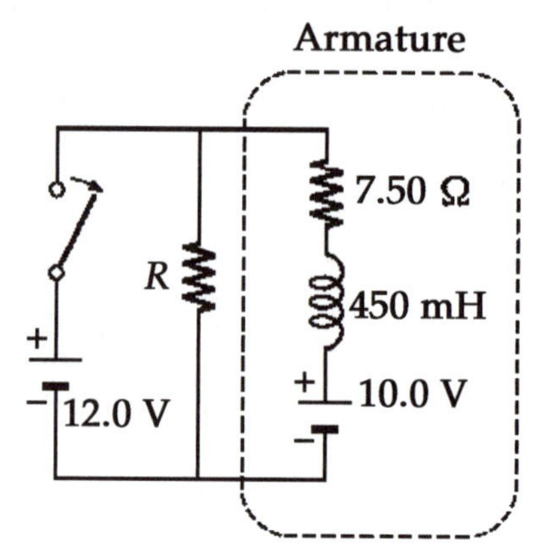

Figure P32.77

Solution

Conceptualize: We should expect R to be significantly greater than the resistance of the armature coil, for otherwise a large portion of the source current would be diverted through R and much of the total power would be wasted on producing internal energy in this discharge resistor.

Categorize: When the motor is unplugged, the 10-V back emf will still exist for an instant because the motor's inertia will keep it spinning. Now the circuit is reduced to a simple series loop with an emf, inductor, and two resistors. The current that was going through the armature coil must now go through the discharge resistor, which will create a voltage across R that we wish to limit to 80 V. As time passes, the motor slows down, the back emf decreases, and the current is reduced from its original maximum value. The current becomes zero when the motor stops turning.

Analyze: The steady-state coil current when the switch is closed is found from applying Kirchhoff's loop rule to the outer loop:

$$+12.0\ \text{V} - I(7.50\ \Omega) - 10.0\ \text{V} = 0$$

so $I = \dfrac{2.00\ \text{V}}{7.50\ \text{k}\Omega} = 0.267\text{A}$

We then require that $\Delta V_R = 80.0\ \text{V} = (0.267\ \text{A})\,R$

so $R = \dfrac{\Delta V_R}{I} = \dfrac{80.0\ \text{V}}{0.267\ \text{A}} = 300\ \Omega$

■

Finalize: As we expected, this discharge resistance is considerably greater than the coil's resistance. Note that while the motor is running, the discharge resistor transforms $P = (12\text{ V})^2/300\ \Omega = 0.48$ joule in every second, from electrically transmitted energy into internal energy. The power wasted is 0.48 W. The source delivers energy at the rate of $P = I\Delta V = [0.267\text{ A} + (12\text{ V}/300\ \Omega)](12\text{ V}) = 3.68\text{ W}$, so the discharge resistor wastes about 13% of the total power. The mechanical output power is $10\text{ V}(0.267\text{ A}) = 2.67\text{ W}$. This motor could lift a 26.7-N weight at the rate of 0.1 m/s.

33

Alternating-Current Circuits

EQUATIONS AND CONCEPTS

A **series alternating current circuit** with a sinusoidal source of emf is shown in the figure to the right. The rectangle [ ] in the circuit represents the circuit element(s) which, in a particular case, may be a resistor (R), a capacitor (C), an inductor (L), or some combination of the above. The AC source frequency determines the current frequency.

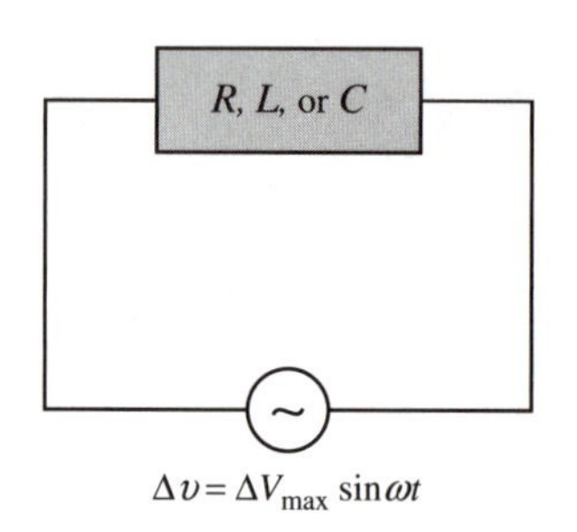

Δv = instantaneous voltage
ΔV_{max} = maximum voltage

When a **resistor** (R) **is the circuit element**, the current and voltage across the resistor are *in phase*, and I_{max} is the maximum current.

$$\Delta v_R = I_{max} R \sin \omega t \qquad (33.3)$$

$$i_R = I_{max} \sin \omega t \qquad (33.1)$$

When an **inductor** (L) **is the circuit element**, the *current lags the voltage across the inductor* by 90°.

$$\Delta v_L = \Delta V_{max} \sin \omega t \qquad (33.6)$$

$$i_L = \frac{\Delta V_{max}}{\omega L} \sin\left(\omega t - \frac{\pi}{2}\right) \qquad (33.8)$$

When a **capacitor** (C) **is the circuit element**, *the current leads the voltage across the capacitor* by 90°.

$$\Delta v_C = \Delta V_{max} \sin \omega t$$

$$i_C = \omega C \Delta V_{max} \sin\left(\omega t + \frac{\pi}{2}\right) \qquad (33.16)$$

Inductive reactance (X_L), capacitive reactance (X_C), and resistance (R) act to limit current in an RLC circuit. *The inductive reactance increases with increasing frequency, while the capacitive reactance decreases with increasing frequency.* Resistance is frequency independent.

$$X_L \equiv \omega L \qquad (33.10)$$

$$X_C \equiv \frac{1}{\omega C} \qquad (33.18)$$

The **maximum value of the current** (or current amplitude) through each element is proportional to the amplitude of the AC voltage across the element. *In the case of an inductor and a capacitor, the maximum value of the current depends also on the angular frequency of the source of emf.*

Resistor: $$I_{max} = \frac{\Delta V_{max}}{R} \quad (33.2)$$

Inductor: $$I_{max} = \frac{\Delta V_{max}}{X_L} \quad (33.11)$$

Capacitor: $$I_{max} = \frac{\Delta V_{max}}{X_C} \quad (33.19)$$

The **series *RLC* circuit**, as shown at right, includes a resistor, inductor, capacitor, and a sinusoidally varying voltage source. Instantaneous current (i) has the same amplitude and phase at every point in the series *RLC* circuit. *The phase angle (ϕ) is the degree by which the current lags the applied voltage.*

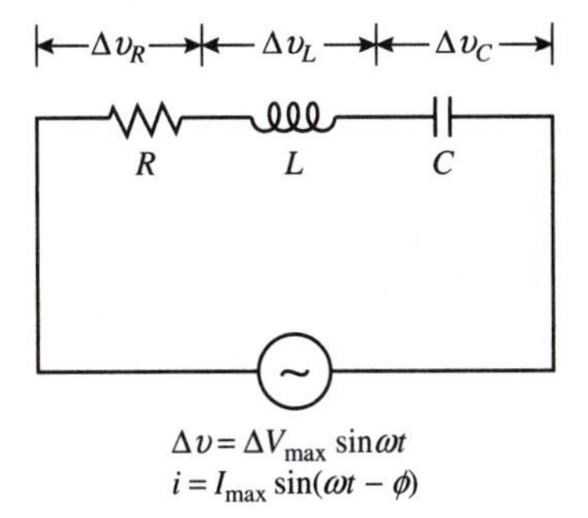

The **phase relationships** among the instantaneous voltages across R, L, and C can be shown *relative to the common current phase.* Compare Equations 33.21, 33.22, and 33.23 with Equations 33.1, 33.8, and 33.16.

$$\Delta v_R = I_{max} R \sin(\omega t) \quad (33.21)$$

$$\Delta v_L = I_{max} X_L \sin\left(\omega t + \frac{\pi}{2}\right) \quad (33.22)$$

$$\Delta v_C = I_{max} X_C \sin\left(\omega t - \frac{\pi}{2}\right) \quad (33.23)$$

The **maximum voltage** across each circuit element can be written in the form of an Ohm's law expression.

$\Delta V_{max} = I_{max} R$ (resistor)

$\Delta V_{max} = I_{max} X_L$ (inductor)

$\Delta V_{max} = I_{max} X_C$ (capacitor)

Impedance (Z) is a circuit parameter determined by the values of R, X_L, and X_C; and therefore impedance is frequency dependent. The SI unit of impedance is the ohm (Ω).

$$Z \equiv \sqrt{R^2 + \left(X_L - X_C\right)^2} \quad (33.25)$$

The **maximum current** in an *RLC* series circuit depends on the values of ΔV_{max} and Z.

$$I_{max} = \frac{\Delta V_{max}}{Z} \quad (33.26)$$

The **phase angle** (ϕ) can be determined from the *impedance triangle*. It is a measure of the phase difference between the *applied voltage* and the current in the circuit.

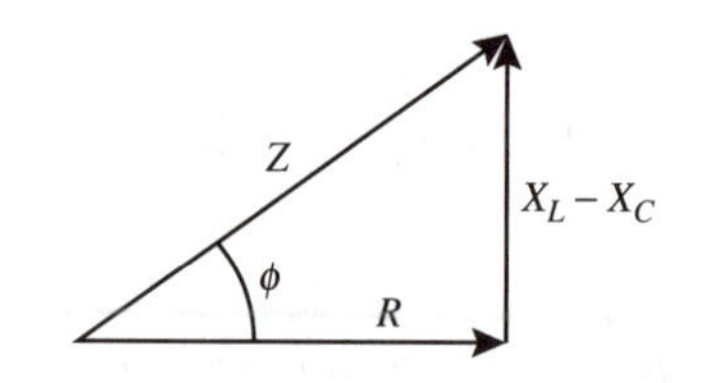

$$\phi = \tan^{-1}\left(\frac{X_L - X_C}{R}\right) \qquad (33.27)$$

The **average power** delivered by a generator (source of emf) to an *RLC* series circuit is directly proportional to cos ϕ, the **power factor** of the circuit. *There is zero power loss in ideal inductors and capacitors; the average power delivered by the source is converted to internal energy in the resistor.*

$$P_{avg} = I_{rms}\Delta V_{rms}\cos\phi \qquad (33.31)$$

$\cos\phi$ = power factor

$$P_{avg} = I_{rms}^2 R \qquad (33.32)$$

Root-mean-square (rms) values of current and voltage are those values displayed by AC voltmeters and ammeters.

$$I_{rms} = \frac{I_{max}}{\sqrt{2}} = 0.707 I_{max} \qquad (33.4)$$

$$\Delta V_{rms} = \frac{\Delta V_{max}}{\sqrt{2}} = 0.707\,\Delta V_{max} \qquad (33.5)$$

The **resonance frequency** (ω_0) is that frequency for which $X_L = X_C$. *At this frequency* ($Z = R$), *the current has its maximum value and the phase angle equals zero.*

$$\omega_0 = \frac{1}{\sqrt{LC}} \qquad (33.35)$$

A **transformer** consists of a primary coil of N_1 turns and a secondary coil of N_2 turns wound on a common core.

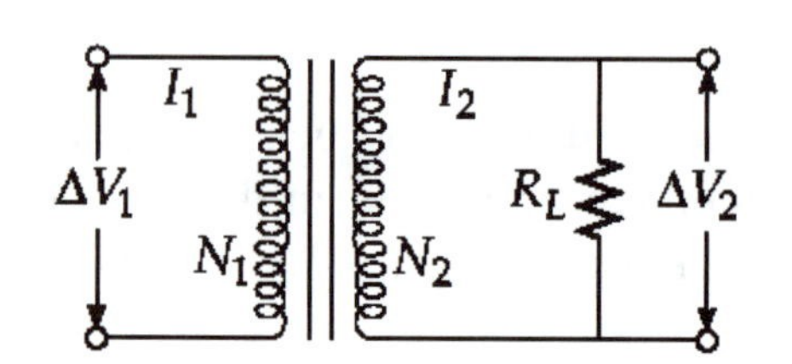

v_1 and v_2 are instantaneous values

V_1 and V_2 are rms values

I_1 = rms current in the primary

I_2 = rms current in the secondary

R_L = load resistor

In the ideal transformer, the ratio of voltages is equal to the ratio of turns, and the ratio of currents is equal to the inverse of the ratio of turns.

$$\Delta v_2 = \frac{N_2}{N_1}\Delta v_1 \qquad (33.41)$$

$$I_1\Delta V_1 = I_2\Delta V_2 \qquad (33.42)$$

SUGGESTIONS, SKILLS, AND STRATEGIES

A **phasor diagram** which describes the AC circuit of Figure (a) below is shown in Figure (b). Each phasor has a length which is proportional to the magnitude of the voltage or current which it represents and rotates counterclockwise about the common origin with a frequency which equals the frequency (ω) of the alternating source.

The direction of the phasor which represents the current, shown as a lightly shaded arrow in Figure (b), in the circuit is used as the *reference direction* to establish the correct phase differences among the phasors, which represent the voltage drops across the resistor, inductor, and capacitor. The *instantaneous values* Δv_R, Δv_L, Δv_C, and i are given by the *projection onto the vertical axis of the corresponding phasor.*

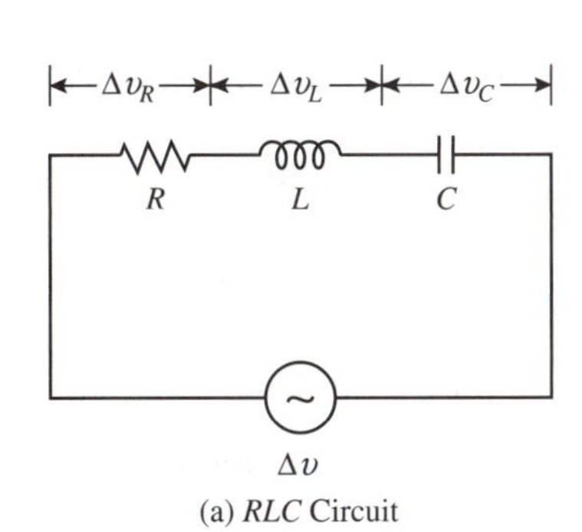

(a) *RLC* Circuit

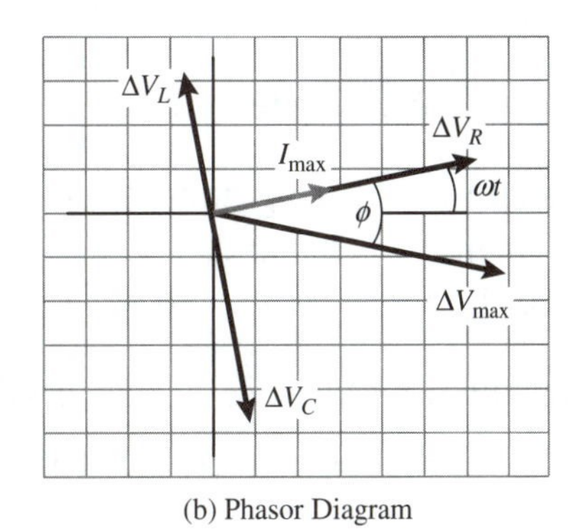

(b) Phasor Diagram

Consider the phasor diagram in Figure (b), assume counterclockwise rotation, and note that:

(1) ΔV_L **leads the current by 90°** (ΔV_L reaches its maximum 90° before ΔV_R).

(2) ΔV_C **lags the current by 90°** (ΔV_C reaches maximum 90° later than ΔV_R).

(3) ΔV_R *is greater than* ΔV_L: The **maximum voltage across the resistor is greater** than the maximum voltage across the inductor as seen by the lengths of the respective vectors.

(4) v_R *is less than* v_L: The **instantaneous voltage across the resistor is less** than the instantaneous voltage across inductor as seen by the projections of the vectors onto the vertical axis.

Remember, the instantaneous values are the projections of the maximum values onto the vertical axis.

Also notice that as time increases and the phasors rotate counterclockwise, maintaining their constant relative phase, the voltage amplitudes (ΔV_R, ΔV_L, and ΔV_C) will remain constant in magnitude; but the instantaneous values (Δv_R, Δv_L, and Δv_C) will vary sinusoidally with time. For the case shown in Figure (b), the phase angle ϕ is negative (this is because

$X_C > X_L$ and therefore $\Delta V_C > \Delta V_L$); hence, the current in the circuit leads the applied voltage in phase.

The **maximum voltage** across an *RLC* circuit (ΔV_{max}) is found by adding ΔV_R, ΔV_L, and ΔV_C as vector quantities. Therefore: **The maximum applied voltage does not equal the sum of the individual maximum voltages**:

$$\Delta V_{max} \neq \Delta V_{R,max} + \Delta V_{L,max} + \Delta V_{C,max}$$

The **instantaneous voltage** (Δv) across each circuit component is the projection of the corresponding phasor onto the vertical axis. Therefore: **The instantaneous applied voltage does equal the sum of the instantaneous voltages across the individual components:**

$$\Delta v = \Delta v_R + \Delta v_L + \Delta v_C$$

The following procedures are recommended when solving alternating current problems:

- The first step in analyzing alternating current circuits is to calculate as many of the unknown quantities such as X_L and X_C as possible. (Note that when calculating X_C, the capacitance should be expressed in farads, rather than, say, microfarads).
- Apply the equation $\Delta V = IZ$ to that portion of the circuit of interest. That is, if you want to know the voltage drop across the combination of an inductor and a resistor, the equation reduces to $\Delta V = I\sqrt{R^2 + X_L^{\,2}}$.

REVIEW CHECKLIST

You should be able to:

- Calculate instantaneous values of current and voltage for an AC source applied to a resistor, inductor, or capacitor. (Sections 33.1, 33.2, 33.3, and 33.4)
- Calculate, for an *RLC* circuit (when values of resistance, inductance, capacitance, and the characteristics of the generator are known): (i) the instantaneous and rms voltage across each circuit component, (ii) the instantaneous and rms current in the circuit, (iii) the phase angle by which the current leads or lags the voltage, (iv) the power expended in the circuit, and (v) the resonance frequency of the circuit. (Sections 33.5, 33.6, and 33.7)
- Understand the description and analysis of AC circuits based on phasor diagrams. (Sections 33.2, 33.3, 33.4, and 33.5)
- Make calculations involving voltage and current ratios; input and output power; load resistance; and efficiency of a transformer. (Section 33.8)
- Sketch circuit diagrams for high-pass and low-pass filter circuits; calculate the ratio of output to input voltage for given circuit parameters. (Section 33.9)

ANSWER TO AN OBJECTIVE QUESTION

8. **(i)** When a particular inductor is connected to a source of sinusoidally varying emf with constant amplitude and a frequency of 60.0 Hz, the rms current is 3.00 A. What is the rms current if the source frequency is doubled? (a) 12.0 A (b) 6.00 A (c) 4.24 A (d) 3.00 A (e) 1.50 A **(ii)** Repeat part (i) assuming the load is a capacitor instead of an inductor. **(iii)** Repeat part (i) assuming the load is a resistor instead of an inductor.

Answer **(i)** The frequency of the oscillating current in an AC circuit is set by the power supply. Doubling the frequency will double the impedance $X_L = 2\pi fL$ of an inductive load, to cut the current in half. From 3.00 A the current is reduced to 1.50 A, answer (e). Observe that the value of the original frequency is unnecessary information.

(ii) Doubling the frequency will cut in half the impedance $X_C = 1/(2\pi fC)$ of a capacitor as the load. From 3.00 A the current will double to 6.00 A, answer (b).

(iii) The impedance R of a purely resistive load remains constant at all frequencies, so the current is unchanged, answer (d).

□ □ □ □

ANSWERS TO SELECTED CONCEPTUAL QUESTIONS

4. (a) Does the phase angle in an *RLC* series circuit depend on frequency? (b) What is the phase angle for the circuit when the inductive reactance equals the capacitive reactance?

Answer (a) Yes. Since the phase angle is a function of the reactance, which depends on frequency, it must be frequency dependent. (b) The phase angle is zero when the inductive reactance is equal to the capacitive reactance.

□ □ □ □

8. Will a transformer operate if a battery is used for the input voltage across the primary? Explain.

Answer No. A voltage can only be induced in the secondary coil if the magnetic flux through the core changes in time.

□ □ □ □

SOLUTIONS TO SELECTED PROBLEMS

5. The current in the circuit shown in Figure P33.5 equals 60.0% of the peak current at $t = 7.00$ ms. What is the lowest source frequency that gives this current?

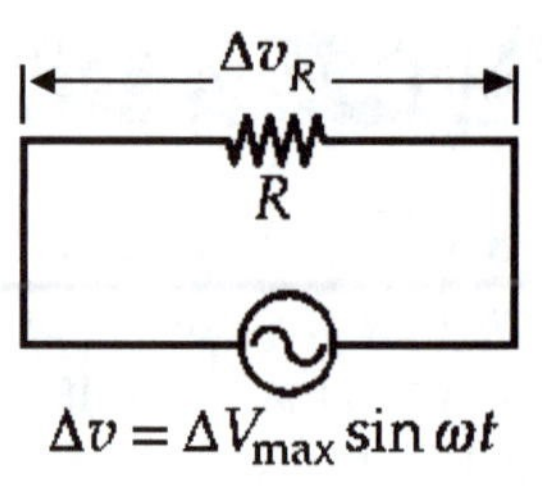

Figure P33.5

Solution

Conceptualize: The figure includes an equation specifying a pattern of sinusoidal variation of voltage in this circuit. The voltage and so the current start from zero at time zero, and the problem is about identifying another particular point in the time variation.

Categorize: In this purely resistive circuit the current is in phase with the voltage and reaches 60% of its maximum value when the voltage is $0.6\Delta V_{max}$. We use the equation given in the figure.

Analyze: The current as a function of time is $i = \dfrac{\Delta v}{R} = \left(\dfrac{\Delta V_{max}}{R}\right)\sin \omega t$

Given the value of t, we want to identify a point with

$$0.600\frac{\Delta V_{max}}{R} = \frac{\Delta V_{max}}{R}\sin(\omega t)$$

or $\omega t = \sin^{-1} 0.600$

To find the lowest frequency we choose the smallest angle satisfying this relation:

$$\omega\,(7.00 \text{ ms}) = 36.9° = 0.644 \text{ rad}$$

Thus, $\omega = 91.9 \text{ rad/s}$ and $f = \dfrac{\omega}{2\pi} = 14.6 \text{ cycle/s}$ ■

Finalize: Infinitely many higher frequencies exist that would satisfy the condition. Their current oscillations would go through one or more maxima before getting to 60% of the maximum value at 7 ms.

---

10. In a purely inductive AC circuit, as shown in Figure P33.10, $\Delta V_{max} = 100$ V. (a) The maximum current is 7.50 A at 50.0 Hz. Calculate the inductance L. (b) **What If?** At what angular frequency ω is the maximum current 2.50 A?

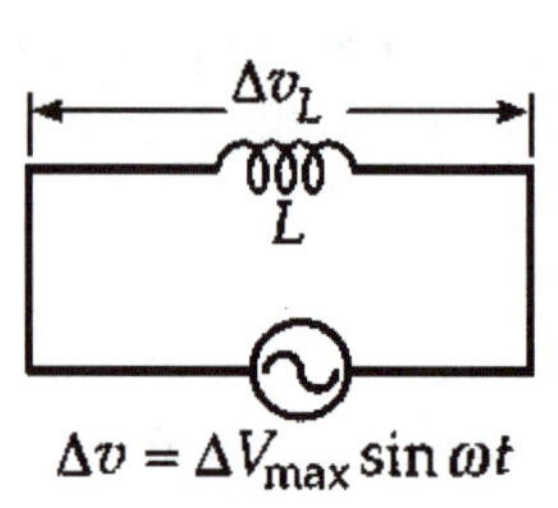

Figure P33.10

Solution

Conceptualize: The current is limited to 7.5 A not by a resistance but by the impedance of the inductor, which will be on the order of ten ohms. To have this much impedance at a frequency of only 50 Hz, the coil's inductance will be several millihenrys. It has the same inductance at a higher frequency, but more impedance, to bring the current down to only 2.5 A.

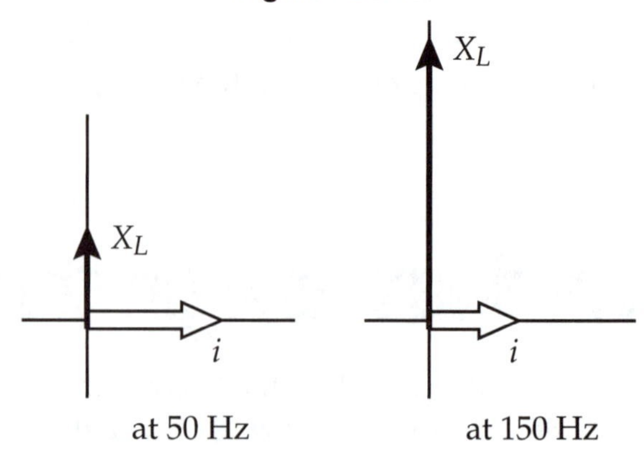

Categorize: We use just the definition of inductive reactance and its relation to inductance.

Analyze: We find the reactance from $I_{max} = \Delta V_{max}/X_L$

(a) $X_L = \dfrac{\Delta V_{max}}{I_{max}} = \dfrac{100\text{ V}}{7.50\text{ A}} = 13.3\ \Omega = \omega L$ Then the inductance is

$$L = \frac{X_L}{\omega} = \frac{13.3\ \Omega}{2\pi\left(50.0\text{ s}^{-1}\right)} = 42.4\text{ mH} \qquad \blacksquare$$

(b) At the new frequency we have a new larger reactance,

$$X_L = \frac{\Delta V_{max}}{I_{max}} = \frac{100\text{ V}}{2.50\text{ A}} = 40.0\ \Omega = \omega L$$

but the inductance is the same, so the new angular frequency is

$$\omega = \frac{40.0\ \Omega}{42.4\text{ mH}} = 942\text{ rad/s} \qquad \blacksquare$$

Finalize: A frequency 3 times higher makes the inductive reactance 3 times larger.

---

11. For the circuit shown in Figure P33.10, $\Delta V_{max} = 80.0$ V, $\omega = 65.0\pi$ rad/s, and $L = 70.0$ mH. Calculate the current in the inductor at $t = 15.5$ ms.

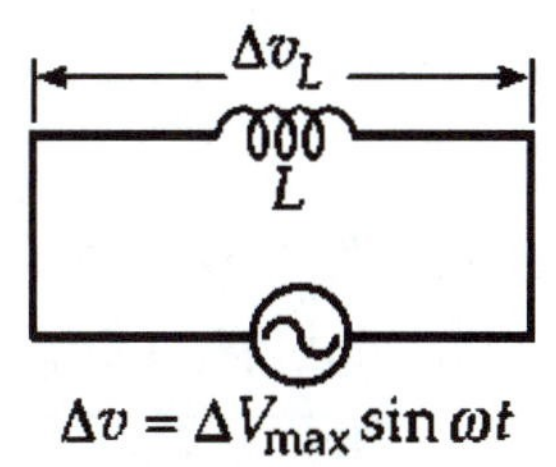

Figure P33.10

Solution

Conceptualize: We estimate a few amps, but we will not know whether it is positive or negative until we know how far into a cycle 15.5 ms takes us.

Categorize: We compute the coil's reactance at this frequency and use the definition of inductive reactance to find the current amplitude. The equation in the figure tells us about the time variation of the voltage, but the current is 90° out of phase with the voltage in this circuit.

Analyze: The inductive reactance is

$$X_L = \omega L = (65.0\ \pi\text{s}^{-1})(70.0 \times 10^{-3}\text{ V} \cdot \text{s/A}) = 14.3\ \Omega$$

The amplitude of the current is $I_{max} = \dfrac{\Delta V_{max}}{X_L} = \dfrac{80.0\text{ V}}{14.3\ \Omega} = 5.60\text{A}$

Textbook Equation 33.7 lets us evaluate the current:

$$i = -I_{max} \cos \omega t = -(5.60\text{ A})\cos\left[(65.0\ \pi\text{ s}^{-1})(0.015\ 5\text{ s})\right]$$

$$i = -(5.60\text{ A}) \cos (3.17\text{ rad}) = +5.60\text{ A} \qquad \blacksquare$$

Finalize: The frequency is 32.5 Hz, and the period 30.8 ms. So at 15.5 ms we are about halfway through with the first cycle. The voltage is close to zero but the inductor current is close to its maximum value. Reminder: when the units divide out to 1, the angle is in radians, not degrees.

---

21. What maximum current is delivered by an AC source with $\Delta V_{max} = 48.0$ V and $f = 90.0$ Hz when connected across a 3.70-μF capacitor?

Solution

Conceptualize: We estimate on the order of one amp. The capacitor's reactance at this particular frequency limits the value of current.

Categorize: This problem is in standard cause-and-effect order. The power supply sets the voltage amplitude and the frequency. The capacitor is in itself characterized by its capacitance. We begin by computing its reactance at this frequency and then use the definition of reactance to find the current.

Analyze: We combine the steps in the equation

$$I_{max} = \frac{\Delta V_{max}}{X_C} = \Delta V_{max}\,\omega C = \Delta V_{max}(2\pi f C)$$

$$I_{max} = (48.0\text{ V})(2\pi)(90.0\text{ Hz})(3.70 \times 10^{-6}\text{ F}) = 0.100\text{ A} = 100\text{ mA} \quad \blacksquare$$

Finalize: It is worthwhile to follow the units. V · Hz · F is V · (1/s)(C/V) = C/s = A. You could think of what we have done as computing the current from $I_{max} = \frac{\Delta V_{max}}{Z}$ with $Z = \sqrt{R^2 + (X_L - X_C)^2}$ but with $R = 0$ and $X_L = 0$ for this circuit modeled as without resistance or inductance.

---

25. An inductor ($L = 400$ mH), a capacitor ($C = 4.43$ μF), and a resistor ($R = 500$ Ω) are connected in series. A 50.0-Hz AC source produces a peak current of 250 mA in the circuit. (a) Calculate the required peak voltage ΔV_{max}. (b) Determine the phase angle by which the current leads or lags the applied voltage.

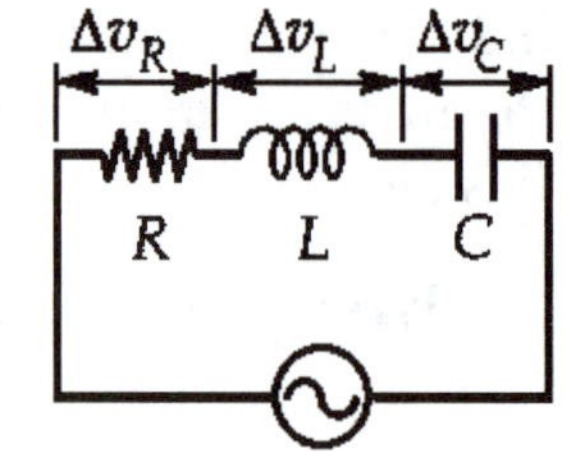

Solution

Conceptualize: Depending on the frequency, the total impedance will be a little or a lot more than 500 Ω. Then the voltage amplitude will be a little or a lot more than (0.25 A) (500 Ω) = 125 V. About the phase angle, all we know in advance is that it must be strictly between +90° and −90°.

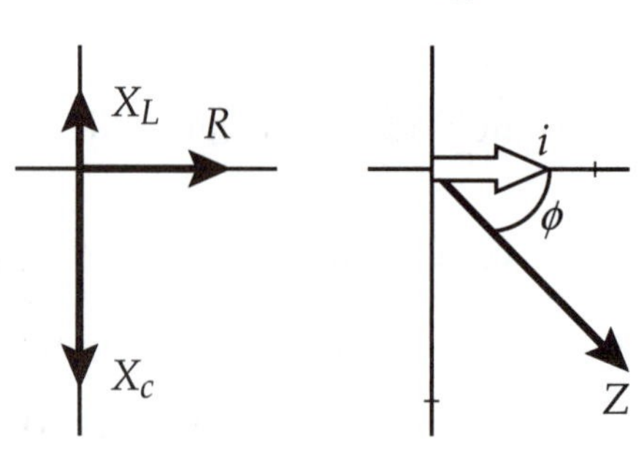

Categorize: We find the impedance of the inductor and of the capacitor. We choose to assign to the impedance components themselves the phase-space directions that belong to the voltages across the various circuit elements, as our diagrams show. We combine the impedances by vector addition to find the total impedance, and that is the proportionality constant between current and voltage. The direction of the total impedance in phase space is the angle between current and voltage.

Analyze: We first find the impedances of the inductor and the capacitor:

$$X_L = \omega L = 2\pi(50.0\text{ Hz})(400\times10^{-3}\text{ H}) = 126\ \Omega$$

and
$$X_C = \frac{1}{\omega C} = \frac{1}{2\pi(50.0\text{ Hz})\left(4.43\times10^{-6}\text{ F}\right)} = 719\ \Omega$$

Then, we substitute these values into the equation for a series *LRC* circuit:

$$\Delta V_{\max} = I_{\max}Z = I_{\max}\sqrt{R^2+\left(X_L - X_C\right)^2}$$

Thus,
$$Z = \sqrt{(500\ \Omega)^2 + (126\ \Omega - 719\ \Omega)^2} = 776\ \Omega$$

and

(a) $\Delta V_{\max} = I_{\max}Z = (0.250\text{ A})(776\ \Omega) = 194\text{ V}$ ■

(b) $\tan\phi = \dfrac{X_L - X_C}{R}$ so $\phi = \tan^{-1}\left(\dfrac{126-719}{500}\right) = -49.9°$ ■

The current **leads** the voltage by 49.9°.

Finalize: The relatively large value of the capacitive reactance means that this is a capacitor-dominated circuit. We are below the resonance frequency. The voltage lags the current. Study the phasor addition diagram showing 126 Ω straight up, 500 Ω to the right, and 719 Ω straight down. Their resultant is 776 Ω at –49.9°. If you learn the pattern of resistance to the right, X_L up and X_C down, you do not need to memorize the equations $Z = \sqrt{R^2+\left(X_L - X_C\right)^2}$ and $\tan\phi = \left(X_L - X_C\right)/R$. The equations just describe the phasor addition.

---

29. An *RLC* circuit consists of a 150-Ω resistor, a 21.0-μF capacitor, and a 460-mH inductor, connected in series with a 120-V, 60.0-Hz power supply. (a) What is the phase angle between the current and the applied voltage? (b) Which reaches its maximum earlier, the current or the voltage?

Solution

Conceptualize: As soon as we find the reactances of the coil and the capacitor at this frequency we will be able to answer question (b). If X_L is the larger, the voltage ↗ will lead the current →.

If X_C is the larger, the voltage will lag ↘ and the current → will reach its maximum earlier. If the two values are fairly close together, the magnitude of the phase angle will be fairly small. If one reactance is much larger than the other, the phase angle will be close to 90° in size.

Categorize: The problem does not ask for the value of the current, but we must still do most of the standard steps toward finding it. We must find the reactances and think about how they add with the resistance to a total impedance.

Analyze: The reactance of the inductor is

$$X_L = \omega L = 2\pi f L = 2\pi\,(60.0\ \text{s}^{-1})(0.460\ \text{H}) = 173\ \Omega$$

The reactance of the capacitor is

$$X_C = \frac{1}{\omega C} = \frac{1}{2\pi f C} = \frac{1}{2\pi\left(60.0\ \text{s}^{-1}\right)\left(21.0\times10^{-6}\ \text{F}\right)} = 126\ \Omega$$

(a) $\tan\phi = \dfrac{X_L - X_C}{R} = \dfrac{173\ \Omega - 126\ \Omega}{150\ \Omega} = 0.314$

so $\phi = 0.304\ \text{rad} = 17.4°$ ■

(b) Since $X_L > X_C$, ϕ is positive, so Δv leads the current. This means that the power-supply or total voltage goes through each maximum, zero-crossing, and minimum earlier in time than the current does. ■

Finalize: In phase space, imagine that the current is always along the x axis. The power-supply voltage is in the direction of Z. If it is above the x axis, ϕ is positive, the circuit is inductor-dominated, and the voltage leads the current. If Z is below the x axis, ϕ is negative, the circuit is capacitor-dominated, and the voltage lags behind the current in time.

---

36. An AC voltage of the form $\Delta v = 100 \sin 1\,000t$, where Δv is in volts and t is in seconds, is applied to a series RLC circuit. Assume the resistance is 400 Ω, the capacitance is 5.00 μF, and the inductance is 0.500 H. Find the average power delivered to the circuit.

Solution

Conceptualize: Comparing $\Delta v = 100 \sin (1\,000\ t)$ with $\Delta v = \Delta V_{max} \sin \omega t$, we see that $\Delta V_{max} = 100$ V and $\omega = 1\,000\ \text{s}^{-1}$. Only the resistor takes electrically transmitted energy out of the circuit, but the capacitor and inductor will impede the charge flow (that is, the current), and therefore reduce the voltage across the resistor. Because of this impedance, the average power delivered to the resistor must be less than the power the source would deliver if it were connected directly across the resistor:

$$P_{av} = \frac{(\Delta V_{max})^2}{2R} = \frac{(100\text{ V})^2}{2(400\ \Omega)} = 12.5\text{ W}$$

Categorize: The average power delivered to the resistor can be found from $P_{av} = I^2_{rms}R$, where $I_{rms} = \Delta V_{rms}/Z$.

Analyze: The rms voltage of the power supply is

$$\Delta V_{rms} = \frac{100\text{ V}}{\sqrt{2}} = 70.7\text{ V}$$

In order to calculate the impedance, we first need the capacitive and inductive reactances:

$$X_C = \frac{1}{\omega C} = \frac{1}{(1\,000\text{ s}^{-1})(5.00\times10^{-6}\text{ F})} = 200\ \Omega$$

$$X_L = \omega L = (1\,000\text{ s}^{-1})(0.500\text{ H}) = 500\ \Omega$$

Next, $Z = \sqrt{R^2 + (X_L - X_C)^2}$: $Z = \sqrt{(400\ \Omega)^2 + (500\ \Omega - 200\ \Omega)^2} = 500\ \Omega$

The rms current is $I_{rms} = \dfrac{\Delta V_{rms}}{Z} = \dfrac{70.7\text{ V}}{500\ \Omega} = 0.141\text{ A}$

The average power is $P_{av} = I^2_{rms}R = (0.141\text{ A})^2(400\ \Omega) = 8.00\text{ W}$ ■

Finalize: The power delivered to the resistor is less than 12.5 W, so our answer appears to be reasonable. As with other *RLC* circuits, the power would be maximized at the resonance frequency where $X_L = X_C$ so that $Z = R$. Then the average power will simply be the 12.5 W we calculated first.

---

39. In a certain series *RLC* circuit, $I_{rms} = 9.00$ A, $\Delta V_{rms} = 180$ V, and the current leads the voltage by 37.0°. (a) What is the total resistance of the circuit? (b) Calculate the reactance of the circuit $(X_L - X_C)$.

Solution

Conceptualize: From the values of voltage and current we can directly calculate the impedance as 20 Ω, but this is not the resistance. The resistance must be somewhat smaller, the x component of the 20 Ω in phase space. And the other component of Z will be the net reactance.

Categorize: The geometry of the resistance and reactance phasors will give us the answers.

Analyze: The power is $P_{av} = I_{rms}\Delta V_{rms}\cos\phi = I^2_{rms}R$.

(a) Therefore, $R = \dfrac{\Delta V_{\text{rms}} \cos\phi}{I_{\text{rms}}} = \dfrac{(180\text{ V})\cos(-37.0°)}{9.00\text{ A}} = 16.0\ \Omega$ ■

(b) $\tan\phi = \dfrac{X_L - X_C}{R}$:

$$X_L - X_C = R\tan\phi = (16.0\ \Omega)\tan(-37.0°) = -12.0\ \Omega$$ ■

Finalize: We write out the solution in equation terms, to show that it proceeds step by step. But the solution is easier by drawing a phasor diagram. The phasors for the numbers of ohms are defined to be in the same directions as the voltages associated with each impedance component. They form a 3-4-5 right triangle. 16 Ω is along the *x* axis to the right, 20 Ω is the hypotenuse, and 12 Ω points vertically downward. Draw the diagram yourself.

---

43. An *RLC* circuit is used in a radio to tune into an FM station broadcasting at f = 99.7 MHz. The resistance in the circuit is R = 12.0 Ω, and the inductance is L = 1.40 μH. What capacitance should be used?

Solution

Conceptualize: For a fairly high resonance frequency the capacitance should be fairly small, perhaps less than a nanofarad.

Categorize: We will not need the resistance value. The problem is just about how inductance and capacitance determine the resonance frequency.

Analyze: The circuit is to be in resonance when $\omega L = \dfrac{1}{\omega C}$.

Solving for the capacitance gives

$$C = \frac{1}{\omega^2 L} = \frac{1}{4\pi^2 f^2 L} = \frac{1}{4\pi^2(99.7\text{ MHz})^2(1.40\ \mu\text{V}\cdot\text{s/A})} = 1.82\text{ pF}$$ ■

Finalize: The capacitance is smaller than we might have guessed. We did not need the resistance value to get the answer, but we can use it to show that the circuit has a high-quality resonance. Its quality factor is

$$Q = \frac{\omega_0 L}{R} = \frac{99.7\times10^6 \times 1.40\times10^{-6}\ \Omega}{12.0\ \Omega} = 11.6$$

This is much larger than 1, so the radio can have good selectivity, responding to the signal from the transmitter you want to receive and not to signals of other frequencies.

---

49. The primary coil of a transformer has $N_1 = 350$ turns and the secondary coil has $N_2 =$ 2 000 turns. If the input voltage across the primary coil is $\Delta v = 170 \cos \omega t$, where Δv is in volts and t is in seconds, what rms voltage is developed across the secondary coil?

Solution

Conceptualize: This is a step-up transformer with a turns ratio of roughly 6, so the secondary voltage will be about $6 \times 120 \text{ V} \approx 700 \text{ V}$.

Categorize: We must remember that the rms voltage is predictably smaller than the amplitude. We assume the transformer is ideal, so that the turns ratio is the voltage step-up factor.

Analyze: The rms primary voltage is $\Delta V_{1,\text{rms}} = \dfrac{170 \text{ V}}{\sqrt{2}} = 120 \text{ V}$

The rms voltage across the bigger coil is

$$\Delta V_{2,\text{rms}} = \frac{N_2}{N_1}\Delta V_{1,\text{rms}} = \frac{2000}{350}(120 \text{ V}) = 687 \text{ V} \quad \blacksquare$$

Finalize: A transformer like this might be in the power supply for a laboratory electron-beam tube.

---

53. The *RC* high-pass filter shown in Figure P33.53 has a resistance $R = 0.500\ \Omega$ and a capacitance $C = 613\ \mu\text{F}$. What is the ratio of the amplitude of the output voltage to that of the input voltage for this filter for a source frequency of 600 Hz?

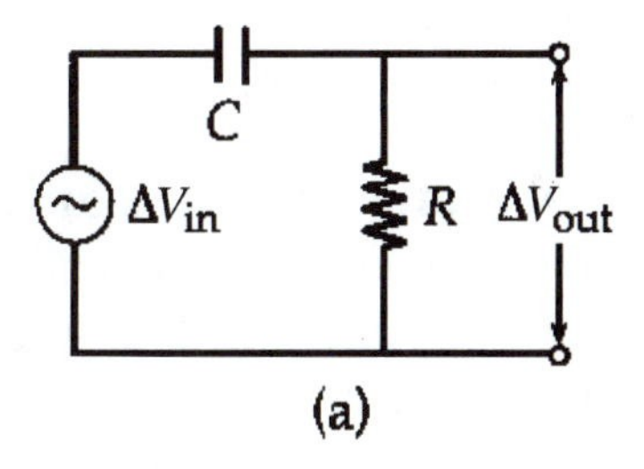

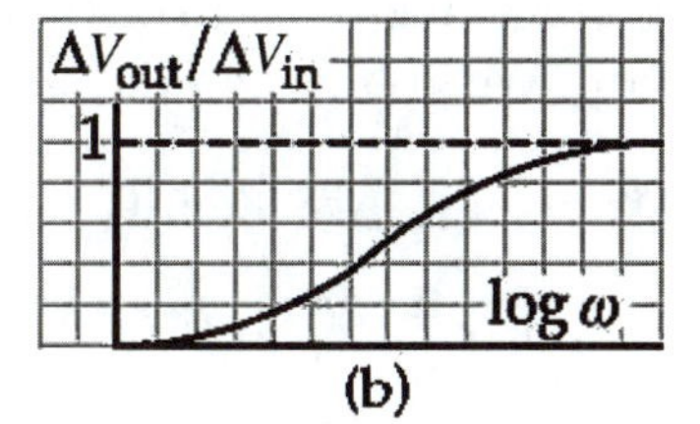

Figure P33.53

Solution

Conceptualize: 600 Hz is not a very high audio frequency, so we might guess that the voltage transfer ratio $\Delta V_{\text{out}}/\Delta V_{\text{in}}$ will be less than 0.5. This ratio is often called the "gain," even though for a passive filter it must always be less than one.

Categorize: The output voltage of this circuit is taken across the resistor, but the input sees the impedance of the resistor and the capacitor. Therefore, the gain will be the ratio of the resistance to the impedance.

Analyze: The voltage gain ratio is

$$\frac{\Delta V_{\text{out}}}{\Delta V_{\text{in}}} = \frac{\Delta V_R}{\Delta V_{\text{source}}} = \frac{I_{\text{rms}}R}{I_{\text{rms}}Z} = \frac{R}{Z} = \frac{R}{\sqrt{R^2 + X_C^2}} = \frac{R}{\sqrt{R^2 + (1/\omega C)^2}}$$

At 600 Hz, we have $\omega = (2\pi \text{ rad})(600 \text{ s}^{-1})$, so

$$\frac{\Delta V_{out}}{\Delta V_{in}} = \frac{0.500\ \Omega}{\sqrt{(0.500\ \Omega)^2 + (1/(1\,200\ \pi \text{ rad/s})(613\ \mu\text{F}))^2}} = 0.756$$ ■

Finalize: The gain is larger than we expected. With a relatively large value for the capacitance, it turns out that this circuit has gain 0.5 at 300 Hz. Our new version of the graph more transparently represents how this filter attenuates very low frequency signals. If this filter is used in an audio system, it will reduce very low frequency "hum" or "rumble" sounds while allowing high-pitch sounds to pass through. A low pass filter might also be needed to reduce high frequency "static" or "hiss" noise.

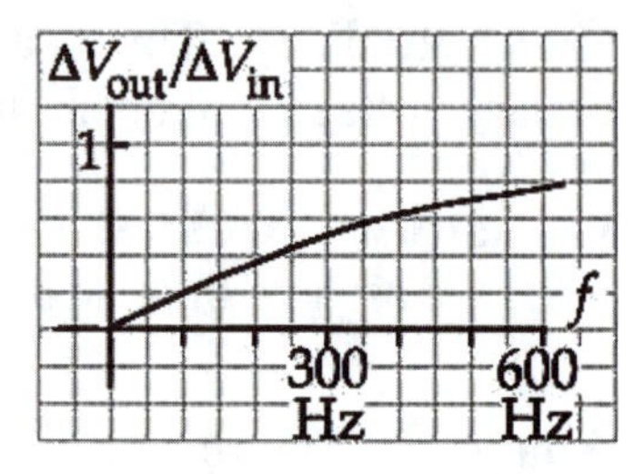

---

60. Consider a series *RLC* circuit having the following circuit parameters: $R = 200\ \Omega$, $L = 663$ mH, and $C = 26.5\ \mu$F. The applied voltage has an amplitude of 50.0 V and a frequency of 60.0 Hz. Find (a) the current I_{max} and its phase relative to the applied voltage Δv, (b) the maximum voltage ΔV_R across the resistor and its phase relative to the current, (c) the maximum voltage ΔV_C across the capacitor and its phase relative to the current, and (d) the maximum voltage ΔV_L across the inductor and its phase relative to the current.

Solution

Conceptualize: The impedance will be somewhat or a lot larger than 200 Ω, so the current will be somewhat or a lot smaller than 50 V/200 Ω = 250 mA. The voltage across each circuit element will be just the single current multiplied by the resistance or reactance of that element. The phase of each individual voltage is known in general . . .

Categorize: . . . to be (b) 0° for the resistor, (c) −90° for the capacitor, and (d) +90° for the coil. We go through the standard steps to find the total impedance and then the current.

Analyze: We identify that

$$R = 200\ \Omega,\ L = 663 \text{ mH},\ C = 26.5\ \mu F,\ \omega = 377 \text{ rad/s}, \quad \text{and} \quad \Delta V_{max} = 50.0 \text{ V}$$

So, $\omega L = 250\ \Omega$, and $1/\omega C = 100\ \Omega$

The impedance is

$$Z = \sqrt{R^2 + \left(\omega L - \frac{1}{\omega C}\right)^2} = \sqrt{(200\ \Omega)^2 + (250\ \Omega - 100\ \Omega)^2} = 250\ \Omega$$

(a) The amplitude of the current is

$$I_{max} = \frac{\Delta V_{max}}{Z} = \frac{50.0 \text{ V}}{250\ \Omega} = 0.200 \text{ A}$$ ■

The phase angle of the voltage relative to the current is

$$\phi = \tan^{-1}\left(\frac{X_L - X_C}{R}\right) = 36.8^\circ \text{ with } \Delta v \nearrow \text{ leading } i \rightarrow$$ ■

(b) $\Delta V_R = I_{max} R = (0.200\text{ A})(200\ \Omega) = 40.0\text{ V at } \phi = 0^\circ$ ■

(c) $\Delta V_C = I_{max} X_C = (0.200\text{ A})(100\ \Omega) = 20.0\text{ V at } \phi = -90.0^\circ$ ■

(d) $\Delta V_L = I_{max} X_L = (0.200\text{ A})(250\ \Omega) = 50.0\text{ V at } \phi = 90.0^\circ$ ■

Finalize: This is a good problem to do before a quiz. Notice how it asks about quantities that require several steps to compute, about quantities that are very easy to compute, and about quantities known without calculation. The phasor diagram of voltages is just like the phasor diagram of impedances, with each impedance multiplied by the same constant current.

---

67. A series *RLC* circuit consists of an 8.00-Ω resistor, a 5.00-μF capacitor, and a 50.0-mH inductor. A variable frequency source applies an emf of 400 V (rms) across the combination. Assuming the frequency is equal to one-half the resonance frequency, determine the power delivered to the circuit.

Solution

Conceptualize: Maximum power is delivered at the resonance frequency, and the power delivered at other frequencies depends on the quality factor Q. For the relatively small resistance in this circuit, we could expect a high $Q = \omega_0 L/R$. So at half the resonant frequency, the power should be a small fraction of the maximum power,

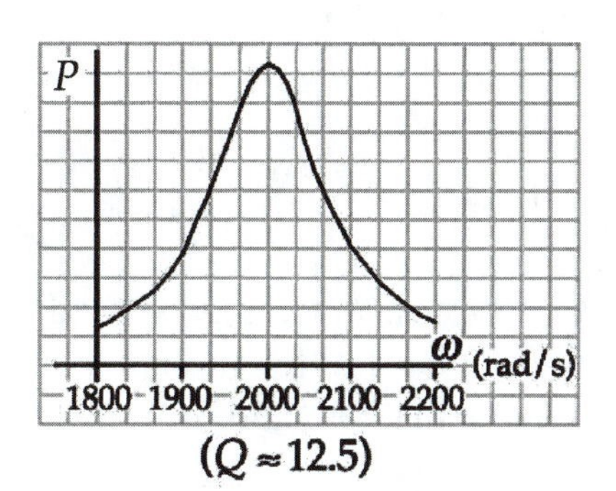

$$P_{av,\,max} = \Delta V_{rms}^2/R = (400\text{ V})^2/8\ \Omega = 20\text{ kW}$$

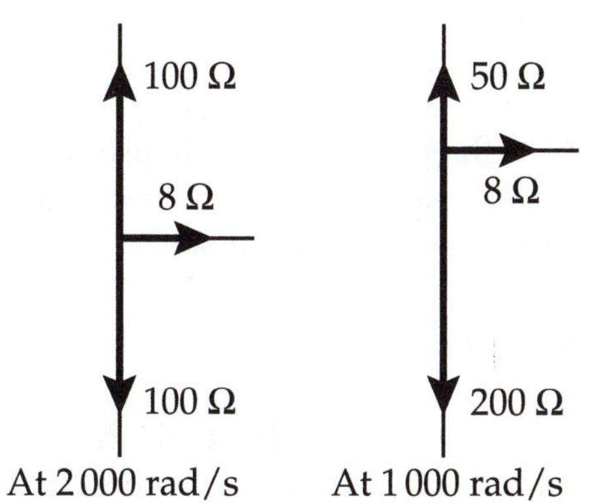

Categorize: We must first calculate the resonance frequency in order to find half this frequency. Then the power delivered by the source must equal the power taken out by the resistor. This power can be found from $P_{av} = I^2 R$, where $I_{rms} = \Delta V_{rms}/Z$.

Analyze: The resonance frequency is

$$f_0 = \frac{1}{2\pi\sqrt{LC}} = \frac{1}{2\pi\sqrt{(0.050\,0\text{ H})(5.00\times10^{-6}\text{ F})}} = 318\text{ Hz}$$

The operating frequency is $f = f_0/2 = 159$ Hz. We can calculate the impedance at this frequency. The inductive reactance is

$$X_L = 2\pi f L = 2\pi\,(159\text{ Hz})(0.050\,0\text{ H}) = 50.0\ \Omega$$

The capacitive reactance is $X_C = \dfrac{1}{2\pi f C} = \dfrac{1}{2\pi(159\text{ Hz})(5.00\times10^{-6}\text{ F})} = 200\ \Omega$

The impedance is

$$Z = \sqrt{R^2 + (X_L - X_C)^2} = \sqrt{8.00^2 + (50.0 - 200)^2}\ \Omega = 150\ \Omega$$

So the current is $I_{\text{rms}} = \dfrac{\Delta V_{\text{rms}}}{Z} = \dfrac{400\ V}{150\ \Omega} = 2.66\text{ A}$

The power delivered by the source is the power delivered to the resistor:

$$P_{\text{av}} = I_{\text{rms}}^2 R = (2.66\text{A})^2\,(8.00\ \Omega) = 56.7\text{ W}$$ ■

Finalize: This power is only about 0.3% of the 20 kW peak power delivered at the resonance frequency. The significant reduction in power for frequencies away from resonance is a consequence of the relatively high Q-factor of about 12.5 for this circuit. A high Q is beneficial if, for example, you want to listen to your favorite radio station that broadcasts at 101.5 MHz, and you do not want to receive the signal from another local station that broadcasts at 101.9 MHz.

---

73. A transformer may be used to provide maximum power transfer between two AC circuits that have different impedances Z_1 and Z_2. This process is called *impedance matching*. (a) Show that the ratio of turns N_1/N_2 for this transformer is

$$\frac{N_1}{N_2} = \sqrt{\frac{Z_1}{Z_2}}$$

(b) Suppose you want to use a transformer as an impedance-matching device between an audio amplifier that has an output impedance of 8.00 kΩ and a speaker that has an input impedance of 8.00 Ω. What should your N_1/N_2 ratio be?

Solution

Conceptualize: Example 28.2 showed that maximum power is transferred in a DC circuit when the internal resistance of the DC source is equal to the resistance of the load. This problem is about an extension of the idea of impedance matching to AC circuits. An ideal transformer steps up current if it steps down voltage, so the circuit connected to the input side does see quite a different impedance for the transformer, compared to the circuit connected to the output side.

Categorize: Assembling simple equations about the functioning of an ideal transformer and about the definition of impedance will solve the problem.

Analyze: The turns ratio is the factor of change in voltage:

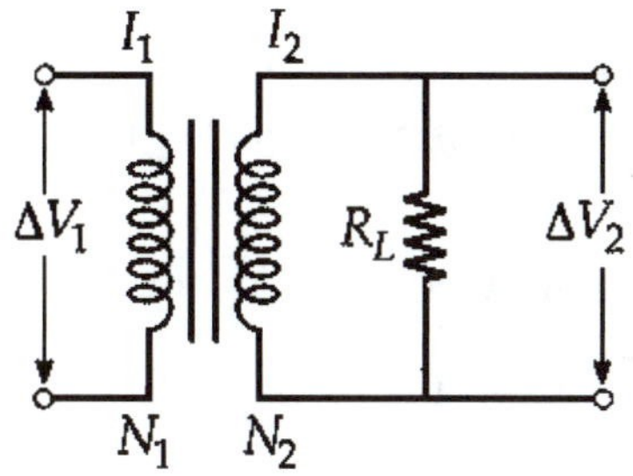

$$\frac{N_1}{N_2} = \frac{\Delta V_1}{\Delta V_2}$$

with $Z_1 = \dfrac{\Delta V_1}{I_1}$ and $Z_2 = \dfrac{\Delta V_2}{I_2}$

we have $\dfrac{N_1}{N_2} = \dfrac{Z_1 I_1}{Z_2 I_2}$

(a) Since $\dfrac{I_1}{I_2} = \dfrac{N_2}{N_1}$ we find $\dfrac{N_1}{N_2} = \dfrac{Z_1 N_2}{Z_2 N_1}$ so $\dfrac{N_1^2}{N_2^2} = \dfrac{Z_1}{Z_2}$ and $\dfrac{N_1}{N_2} = \sqrt{\dfrac{Z_1}{Z_2}}$ ■

(b) $\dfrac{N_1}{N_2} = \sqrt{\dfrac{8\,000\ \Omega}{8.00\ \Omega}} = 31.6$ ■

Finalize: This is a voltage step-down transformer to step up the current to drive the low-impedance speaker. Without the transformer, much less of the output power of the amplifier would get into the speaker. You can think of the transformer as like the drive train of a bicycle, that changes one product of large torque and low angular speed into a product of a small torque and high angular speed.

---

76. A series *RLC* circuit in which $R = 1.00\ \Omega$, $L = 1.00$ mH, and $C = 1.00$ nF is connected to an AC source delivering 1.00 V (rms). (a) Make a precise graph of the power delivered to the circuit as a function of the frequency and (b) verify that the full width of the resonance peak at half-maximum is $R/2\pi L$.

Solution

Conceptualize: We expect the graph to show a fairly sharp peak at the resonance frequency. We will read the frequencies above and below resonance at the half-power points and find the full width at half power Δf between them. The quality factor is $\omega_0/\Delta\omega = f_0/\Delta f$, so when we prove that Δf is $R/2\pi L$ we will be showing in this particular case that $Q = f_0/\Delta f = f_0/(R/2\pi L) = 2\pi f_0 L/R = \omega_0 L/R$, as the chapter text states without proof.

Categorize: We must choose a variety of frequencies below, at, and above resonance, find the impedance at each, then the current, and then the power. A spreadsheet will be useful.

Analyze: (a) At resonance,

$$\omega = \frac{1}{\sqrt{LC}} = \frac{1}{\sqrt{(1.00\times10^{-3}\text{ H})(1.00\times10^{-9}\text{ F})}} = 1.00\times10^{6}\text{ rad/s}$$

At that point $Z = R = 1.00\ \Omega$ and $I = \dfrac{1.00\text{ V}}{1.00\ \Omega} = 1.00\text{ A}$

The power is $I^2R = (1.00\text{ A})^2(1.00\ \Omega) = 1.00\text{ W}$

We compute the power at some other angular frequencies. Thus,

ω, 10^6 rad/s	ωL, Ω	$1/\omega C$, Ω	Z, Ω	I, A	$I^2R = P$, W
0.999 0	999	1 001	2.24	0.447	0.199 84
0.999 4	999.4	1 000.6	1.56	0.640	0.409 69
0.999 5	999.5	1 000.5	1.41	0.707	0.499 87
0.999 6	999.6	1 000.4	1.28	0.781	0.609 66
0.999 8	999.8	1 000.2	1.08	0.928	0.862 05
1	1 000	1 000	1	1	1
1.000 2	1 000.2	999.8	1.08	0.928	0.862 09
1.000 4	1 000.4	999.6	1.28	0.781	0.609 85
1.000 5	1 000.5	999.5	1.41	0.707	0.500 12
1.000 6	1 000.6	999.4	1.56	0.640	0.409 98
1.001	1 001	999	2.24	0.447	0.200 16

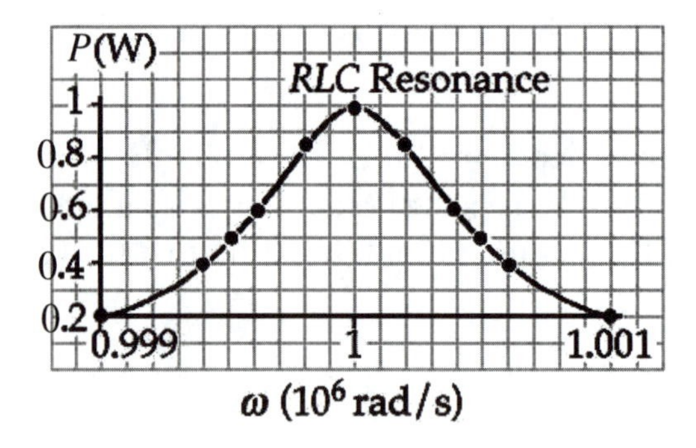

■

(b) The angular frequencies giving half the maximum power are

$$0.999\,5\times10^{6}\text{ rad/s} \quad\text{and}\quad 1.000\,5\times10^{6}\text{ rad/s}$$

so the full width at half the maximum is

$$\Delta\omega = (1.000\,5 - 0.999\,5)\times10^{6}\text{ rad/s}$$
$$\Delta\omega = 1.00\times10^{3}\text{ rad/s}$$

Since $\Delta\omega = 2\pi\Delta f$, $\Delta f = 159\text{ Hz}$

and for comparison $\frac{R}{2\pi L} = \frac{1.00\ \Omega}{2\pi\left(1.00\times10^{-3}\,\mathrm{H}\right)} = 159\ \mathrm{Hz}$

The two quantities agree. ■

Finalize: The actual quality factor of this circuit is $\omega_0/\Delta\omega = 10^6/10^3 = 1\,000$. It agrees with $\omega_0 L/R = 1\,000\ \Omega/1\ \Omega = 1\,000$. Note that a footnote in the chapter text says that a third way to think of the quality factor is in terms of the stored energy in the resonant oscillation divided by the energy input in each cycle. Still further, when the quality factor is high the resonance frequency will very nearly agree with the frequency of free oscillations when the input signal is taken away. From Equation 32.32, the angular frequency of damped free oscillations is

$$\omega_d = \left[\frac{1}{LC} - \left(\frac{R}{2L}\right)^2\right]^{1/2} = \left[\omega_0^2 - \left(\frac{\omega_0}{2Q}\right)^2\right]^{1/2} = \omega_0\left[1 - \frac{1}{4Q^2}\right]^{1/2}$$

The quality factor is useful for describing mechanical and acoustical resonance, as well as electrical resonance.

34

Electromagnetic Waves

EQUATIONS AND CONCEPTS

Maxwell's equations are the fundamental laws governing the behavior of electric and magnetic fields. Electromagnetic waves are a natural consequence of these laws. You should notice that the integrals in Equation 34.4 and Equation 34.5 are *surface integrals* in which the normal components of electric and magnetic fields are integrated over a *closed surface*. Equation 34.6 and Equation 34.7 involve *line integrals* in which the tangential components of electric and magnetic fields are integrated around a *closed path*.

$$\oint \vec{\mathbf{E}} \cdot d\vec{\mathbf{A}} = \frac{q}{\epsilon_0} \quad \text{(Gauss's law)} \qquad (34.4)$$

$$\oint \vec{\mathbf{B}} \cdot d\vec{\mathbf{A}} = 0 \quad \text{(Gauss's law in magnetism)} \qquad (34.5)$$

$$\oint \vec{\mathbf{E}} \cdot d\vec{\mathbf{s}} = -\frac{d\Phi_B}{dt} \quad \text{(Faraday's law)} \qquad (34.6)$$

$$\oint \vec{\mathbf{B}} \cdot d\vec{\mathbf{s}} = \mu_0 I + \mu_0 \epsilon_0 \frac{d\Phi_E}{dt}$$

(Ampère-Maxwell law) (34.7)

The **wave equations for electromagnetic waves in free space** (where $Q = 0$ and $I = 0$), as stated here, represent linearly polarized waves traveling with a speed c. *Both* $\vec{\mathbf{E}}$ *and* $\vec{\mathbf{B}}$ *satisfy a differential equation which has the form of the general wave equation.*

$$\frac{\partial^2 E}{\partial x^2} = \mu_0 \epsilon_0 \frac{\partial^2 E}{\partial t^2} \qquad (34.15)$$

$$\frac{\partial^2 B}{\partial x^2} = \mu_0 \epsilon_0 \frac{\partial^2 B}{\partial t^2} \qquad (34.16)$$

The **speed of electromagnetic waves** in vacuum is the same as the speed of light in vacuum.

$$c = \frac{1}{\sqrt{\mu_0 \epsilon_0}} \qquad (34.17)$$

The **electric and magnetic fields** of electromagnetic waves vary in position and time as sinusoidal transverse waves. *Their planes of vibration are perpendicular to each other and perpendicular to the direction of propagation.* In Equations 34.18 and 34.19 k is the angular wave number and ω is the angular frequency.

$$E = E_{max} \cos(kx - \omega t) \qquad (34.18)$$

$$B = B_{max} \cos(kx - \omega t) \qquad (34.19)$$

$$k = \frac{2\pi}{\lambda} \qquad \omega = 2\pi f$$

The **ratio of the magnitudes** of the electric and magnetic fields is constant and equal to the speed of light c.

$$\frac{E_{max}}{B_{max}} = \frac{E}{B} = c \qquad (34.21)$$

The **Poynting vector** ($\vec{\mathbf{S}}$) describes the energy flow associated with an electromagnetic wave. *The direction of $\vec{\mathbf{S}}$ is along the direction of propagation, and the magnitude of $\vec{\mathbf{S}}$ is the rate at which electromagnetic energy crosses a unit surface area perpendicular to the direction of $\vec{\mathbf{S}}$.*

$$\vec{\mathbf{S}} \equiv \frac{1}{\mu_0}\vec{\mathbf{E}} \times \vec{\mathbf{B}} \qquad (34.22)$$

The **wave intensity** is the time average of the magnitude of the Poynting vector; E_{max} and B_{max} are the maximum values of the field magnitudes.

$$I = S_{avg} = \frac{E_{max}^2}{2\mu_0 c} = \frac{cB_{max}^2}{2\mu_0} \qquad (34.24)$$

The **instantaneous energy densities** of the electric and magnetic fields are equal.

$$u_B = u_E = \tfrac{1}{2}\epsilon_0 E^2 = \frac{B^2}{2\mu_0}$$

The **total instantaneous energy density** (u) is equal to the sum of the energy densities associated with the electric and magnetic fields.

$$u = u_E + u_B = \epsilon_0 E^2 = \frac{B^2}{\mu_0}$$

The **total average energy density** is obtained by averaging the instantaneous energy density over one or more cycles.

$$u_{avg} = \tfrac{1}{2}\epsilon_0 E^2{}_{max} = \frac{B^2{}_{max}}{2\mu_0} \qquad (34.25)$$

The **intensity of an electromagnetic wave** is proportional to the average energy density.

$$I = S_{avg} = cu_{avg} \qquad (34.26)$$

The **linear momentum** (p) **delivered to an absorbing surface** by an electromagnetic wave at normal incidence depends on the fraction of the total incident energy (T_{ER}) which is absorbed.

$$p = \frac{T_{ER}}{c} \quad \left(\text{complete absorption}\right) \qquad (34.27)$$

$$p = \frac{2T_{ER}}{c} \quad \left(\text{complete reflection}\right) \qquad (34.29)$$

T_{ER} = total incident energy

Radiation pressure on an absorbing surface (at normal incidence) depends on the magnitude of the Poynting vector and the degree of absorption.

$$P = \frac{S}{c} \quad \left(\text{Perfectly absorbing surface}\right) \qquad (34.28)$$

$$P = \frac{2S}{c} \quad \left(\text{Perfectly reflecting surface}\right) \qquad (34.30)$$

The Spectrum of Electromagnetic Waves. All electromagnetic waves are produced by accelerating charges. Various types of electromagnetic waves can be characterized by a "typical" range of frequencies or wavelengths. The following are listed in order of increasing frequency.

- **Radio waves** ($\sim 10^4$ m > λ > ~ 0.1 m) are the result of electric charges accelerating through a conducting wire (antenna).
- **Microwaves** (~ 0.3 m > λ > $\sim 10^{-4}$ m) are generated by electronic devices.
- **Infrared waves** ($\sim 10^{-3}$ m > λ > $\sim 7 \times 10^{-7}$ m) are produced by high temperature objects and molecules.
- **Visible light** ($\sim 7 \times 10^{-7}$ m > λ > $\sim 4 \times 10^{-7}$ m) is produced by the rearrangement of electrons in atoms and molecules .
- **Ultraviolet (UV) light** ($\sim 4 \times 10^{-7}$ m > λ > $\sim 6 \times 10^{-10}$ m) is an important component of radiation from the Sun.
- **X-rays** ($\sim 10^{-8}$ m > λ > $\sim 10^{-12}$ m) are produced when high-energy electrons strike a target of high atomic number (e.g., metal or glass).
- **Gamma rays** ($\sim 10^{-10}$ m > λ > $\sim 10^{-14}$ m) are emitted by radioactive nuclei.

Remember, the wavelength ranges stated above are approximate. For example, on the long wavelength end, radio waves can be arbitrarily long; and on the short wavelength end, gamma rays can be arbitrarily short. Regions of the electromagnetic spectrum overlap in wavelength; see Figure 34.13 of the text.

REVIEW CHECKLIST

You should be able to:

- State Maxwell's equations and describe the essential features of the apparatus and procedure used by Hertz in his experiments leading to the discovery and understanding of the source and nature of electromagnetic waves. (Section 34.2)
- Write the equation for the magnetic field of an electromagnetic wave when given the equation of the corresponding electric field. (Section 34.3)
- Calculate the values for the Poynting vector (magnitude), wave intensity, and instantaneous and average energy densities in a plane electromagnetic wave. (Section 34.4)
- Calculate the radiation pressure on a surface and the linear momentum delivered to a surface by an electromagnetic wave. (Section 34.5)
- Describe the production of electromagnetic waves and radiation of energy by a half-wave (or dipole) antenna. Use a diagram to show the relative directions for $\vec{\mathbf{E}}$, $\vec{\mathbf{B}}$, and $\vec{\mathbf{S}}$,

and describe their space and time dependencies. Account for the intensity of the radiated wave at points near the dipole and at distant points. (Section 34.6)

- Place the various types of electromagnetic waves in the correct sequence in the electromagnetic spectrum and describe the basis of production particular to each of the wave types. (Section 34.7)

ANSWER TO AN OBJECTIVE QUESTION

5. Assume you charge a comb by running it through your hair and then hold the comb next to a bar magnet. Do the electric and magnetic fields produced constitute an electromagnetic wave? (a) Yes they do, necessarily. (b) Yes they do because charged particles are moving inside the bar magnet. (c) They can, but only if the electric field of the comb and the magnetic field of the magnet are perpendicular. (d) They can, but only if both the comb and the magnet are moving. (e) They can, if either the comb or the magnet or both are accelerating.

Answer (e). Charge on the comb creates an electric field and the bar magnet sets up a magnetic field. If these fields are constant or are steadily changing, they will not start to recreate each other and move as a wave. Only nonzero acceleration of the source charge or current radiates a wave.

□ □ □ □

ANSWERS TO SELECTED CONCEPTUAL QUESTIONS

3. Radio stations often advertise "instant news." If that means you can hear the news the instant the radio announcer speaks it, is the claim true? What approximate time interval is required for a message to travel from Maine to California by radio waves? (Assume the waves can be detected at this range.)

Answer The claim is untrue, but the transit time delay is usually too short to notice. Radio waves move at the speed of light. They can travel around the curved surface of the Earth, bouncing between the ground and the ionosphere, which has an altitude that is small when compared to the radius of the Earth. The distance across the lower forty-eight states is approximately 5 000 km, requiring a time interval of $(5 \times 10^6\text{ m})/(3 \times 10^8\text{ m/s}) \sim 10^{-2}\text{ s}$. To go halfway around the Earth takes only 0.07 s. In other words, a speech can be heard on the other side of the world before it is heard at the back of a large room.

□ □ □ □

10. What does a radio wave do to the charges in the receiving antenna to provide a signal for your car radio?

Answer Consider a typical metal rod antenna for a car radio. The rod detects the electric field portion of the carrier wave. Variations in the amplitude of the carrier wave cause the electrons in the rod to vibrate with amplitudes emulating those of the carrier wave. Likewise, for frequency modulation, the variations of the frequency of the carrier wave cause constant-amplitude vibrations of the electrons in the rod but at frequencies that imitate those of the carrier.

13. Suppose a creature from another planet has eyes that are sensitive to infrared radiation. Describe what the alien would see if it looked around your library. In particular, what would appear bright and what would appear dim?

Answer There is a lot of uniform dark gray, for the walls, the wooden tables, the metal shelves, the books, and the illustrated magazines and art books on display. The incandescent bulbs in the reading lamps are very bright, much brighter than the fluorescent tubes in the ceiling. It is the evening of a hot day, and a window air conditioner is making a large puddle of black on the floor in front of it. With a similar shape, a small puddle of brightness has spilled onto the tabletop from the side vent in a student's notebook computer. The computer screen shows more of the uniform dark gray. People of all races have skin with the same light gray color, but there are subtle variations for the woman checking her email. Her face gets a little darker when she is obligated to read something boring, and her face lights up a bit as soon as she opens a message from a good friend. She pulls a black cafeteria apple from an insulated lunch bag. Clothing is dark as a rule, but the seat of another student's pants glows like a baboon's bottom when he gets up from his chair, and he leaves behind a patch of the same blush on the chair. When he comes back from the restroom, you can tell that he has washed his hands with hot water. His face appears lit from within, like a jack-o-lantern. His nostrils and the openings of his ear canals are bright; brighter still are just the pupils of his eyes.

□ □ □ □

SOLUTIONS TO SELECTED PROBLEMS

1. A 0.100-A current is charging a capacitor that has square plates 5.00 cm on each side. The plate separation is 4.00 mm. Find (a) the time rate of change of electric flux between the plates and (b) the displacement current between the plates.

Solution

Conceptualize: The charge on the capacitor plates is changing, so the electric field between the plates is changing. This situation is described by a displacement current, …

Categorize: … which we can compute from its definition.

Analyze: The electric field in the space between the plates is $E = \dfrac{\sigma}{\epsilon_0} = \dfrac{Q}{\epsilon_0 A}$

The flux of this field is $\Phi_E = \vec{\mathbf{E}} \cdot \vec{\mathbf{A}} = \left(\dfrac{Q}{\epsilon_0 A}\right) A \cos 0^\circ = \dfrac{Q}{\epsilon_0}$

(a) The rate of change of flux is $\dfrac{d\Phi_E}{dt} = \dfrac{d}{dt}\dfrac{Q}{\epsilon_0} = \dfrac{1}{\epsilon_0}\dfrac{dQ}{dt} = \dfrac{I}{\epsilon_0}$

$$\frac{d\Phi_E}{dt} = \left(\frac{0.100\text{ A}}{8.85\times10^{-12}\text{ C}^2/\text{N}\cdot\text{m}^2}\right)(1\text{ C/A}\cdot\text{s}) = 1.13\times10^{10}\text{ N}\cdot\text{m}^2/\text{C}\cdot\text{s} \quad ■$$

(b) The displacement current is defined as

$$I_d = \epsilon_0\frac{d\Phi_E}{dt} = (8.85\times10^{-12}\text{ C}^2/\text{N}\cdot\text{m}^2)(1.13\times10^{10}\text{ N}\cdot\text{m}^2/\text{C}\cdot\text{s}) = 0.100\text{ A} \quad ■$$

Finalize: The changing electric field would create magnetic field in the surrounding space in the same way that 100 mA of electric current would. No electric current exists in the space between the capacitor plates, but Maxwell called his measure of the changing flux of electric field a displacement current because it has the units of current, creates magnetic field as a current does, and can even be included in a generalization of Kirchhoff's junction rule.

---

5. A proton moves through a region containing a uniform electric field given by $\vec{\mathbf{E}} = 50.0\hat{\mathbf{j}}$ V/m and a uniform magnetic field $\vec{\mathbf{B}} = (0.200\hat{\mathbf{i}} + 0.300\hat{\mathbf{j}} + 0.400\hat{\mathbf{k}})$ T. Determine the acceleration of the proton when it has a velocity $\vec{\mathbf{v}} = 200\hat{\mathbf{i}}$ m/s.

Solution

Conceptualize: The electric field is weak, the magnetic field strong, and the velocity small. Both fields will contribute to the net force on the proton, which will be a tiny fraction of a newton but will cause some huge acceleration. Gravity will be a negligible influence by comparison.

Categorize: We use the Lorentz force equation. It amounts to adding the electric force and the magnetic force as vectors. Then we use Newton's second law, or the particle under net force model.

Analyze: The net force on the proton is the Lorentz force, as described by

$$\sum\vec{\mathbf{F}} = m\vec{\mathbf{a}} = q\vec{\mathbf{E}} + q\vec{\mathbf{v}}\times\vec{\mathbf{B}} \quad \text{so that} \quad a = \frac{e}{m}\left[\vec{\mathbf{E}} + \vec{\mathbf{v}}\times\vec{\mathbf{B}}\right]$$

Taking the cross product of $\vec{\mathbf{v}}$ and $\vec{\mathbf{B}}$,

$$\vec{\mathbf{v}}\times\vec{\mathbf{B}} = \begin{vmatrix}\hat{\mathbf{i}} & \hat{\mathbf{j}} & \hat{\mathbf{k}}\\ 200 & 0 & 0\\ 0.200 & 0.300 & 0.400\end{vmatrix} = -200(0.400)\hat{\mathbf{j}} + 200(0.300)\hat{\mathbf{k}}$$

$$\vec{\mathbf{a}} = \left(\frac{1.60\times10^{-19}}{1.67\times10^{-27}}\right)\left[50.0\,\hat{\mathbf{j}} - 80.0\,\hat{\mathbf{j}} + 60.0\,\hat{\mathbf{k}}\right] = 9.58\times10^{7}[-30\hat{\mathbf{j}} + 60\,\hat{\mathbf{k}}]$$

$$\vec{\mathbf{a}} = (2.87\times10^{9})(-\hat{\mathbf{j}} + 2\hat{\mathbf{k}})\text{ m/s}^2 = \left(-2.87\times10^{9}\hat{\mathbf{j}} + 5.75\times10^{9}\hat{\mathbf{k}}\right)\text{ m/s}^2 \quad ■$$

Finalize: Don't forget that a proton has charge $+e$, with the same magnitude as the electron charge. It may seem obvious that a particle feels force $q\vec{\mathbf{E}} + q\vec{\mathbf{v}} \times \vec{\mathbf{B}}$ when it is in both an electric and a magnetic field, but the elucidation of this part of the relationship between electricity and magnetism was a real step that came comparatively late.

---

13. Figure P34.13 shows a plane electromagnetic sinusoidal wave propagating in the x direction. Suppose the wavelength is 50.0 m and the electric field vibrates in the xy plane with an amplitude of 22.0 V/m. Calculate (a) the frequency of the wave and (b) the magnetic field $\vec{\mathbf{B}}$ when the electric field has its maximum value in the negative y direction. (c) Write an expression for $\vec{\mathbf{B}}$ with the correct unit vector, with numerical values for B_{max}, k, and ω, and with its magnitude in the form $B = B_{max}\cos(kx - \omega t)$.

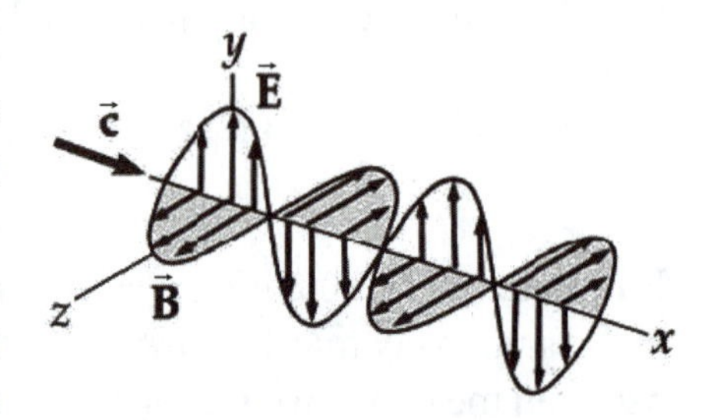

Figure P34.13

Solution

Conceptualize: With wavelength 50 m, this is a radio wave with a frequency in megahertz that an electronic timer could count. The electric field amplitude is weak, so the magnetic field amplitude should be a tiny fraction of a tesla. The wave function will contain x and t as variables, so that it can represent the moving graph at all points in space and time.

Categorize: We are reviewing the description of "a wave" from Chapter 16. Amplitude (here two vector amplitudes related by $E = Bc$), wavelength, frequency, speed, wave number, and angular frequency are all constants of the motion.

Analyze:

(a) $c = f\lambda$ gives the frequency as

$$f = \frac{c}{\lambda} = \frac{3.00 \times 10^8 \text{ m/s}}{50.0 \text{ m}} = 6.00 \times 10^6 \text{ Hz}$$ ■

(b) $c = E/B$ gives the magnetic field amplitude as

$$B = \frac{E}{c} = \frac{22.0 \text{ V/m}}{3.00 \times 10^8 \text{ m/s}} = 7.33 \times 10^{-8} \text{ T} = 73.3 \text{ nT}$$ ■

$\vec{\mathbf{B}}$ must be directed along **negative z direction** when $\vec{\mathbf{E}}$ is in the negative y direction, so that $\vec{\mathbf{S}} = \vec{\mathbf{E}} \times \vec{\mathbf{B}}/\mu_0$ will propagate in the direction $(-\hat{\mathbf{j}}) \times (-\hat{\mathbf{k}}) = +\hat{\mathbf{i}}$. ■

(c) In $B = B_{max}\cos(kx - \omega t)$ we require $k = \dfrac{2\pi}{\lambda} = \dfrac{2\pi}{50.0 \text{ m}} = 0.126 \text{ m}^{-1}$

and $\omega = 2\pi f = (2\pi \text{ rad})(6.00 \times 10^6 \text{ Hz}) = 3.77 \times 10^7 \text{ rad/s}$

Thus, $\vec{\mathbf{B}} = 73.3\cos\left[0.126x - 3.77 \times 10^7 t\right](-\hat{\mathbf{k}})$ where B is in nanoteslas, x is in meters, and t is in seconds ■

Finalize: In Figure P34.13, hold your right hand so that the curving fingers show the rotation from the electric field direction into the magnetic field direction in the diagram. Check that your extended thumb points in the direction of the wave's motion. Figure out where $c = f\lambda$ came from. It will be essential for the rest of the book.

---

14. In SI units, the electric field in an electromagnetic wave is described by

$$E_y = 100 \sin\left(1.00 \times 10^7 x - \omega t\right)$$

Find (a) the amplitude of the corresponding magnetic field oscillations, (b) the wavelength λ, and (c) the frequency f.

Solution

Conceptualize: The rules about electromagnetic wave propagation are so strict that two numbers describing a continuous wave, here the electric-field amplitude and the wave number, determine the required values for several other constants characterizing the wave.

Categorize: We identify the meaning of 100 N/C and 1.00×10^7 by comparison to a general sinusoidal wave function. Then using $E = Bc$ and $c = f\lambda$ tells us the rest. We assume the wave is moving in vacuum.

Analyze:

(a) 100 V/m is the amplitude of the electric field, so the amplitude of the magnetic field is

$$B_{\text{max}} = \frac{E_{\text{max}}}{c} = \frac{100 \text{ V/m}}{3.00 \times 10^8 \text{ m/s}} = 3.33 \times 10^{-7} \text{ T} \quad \blacksquare$$

(b) We compare the given wave function with $y = A \sin(kx - \omega t)$ to see that the wave number is $k = 1.00 \times 10^7 \text{ m}^{-1}$
With $k = 2\pi/\lambda$ we then have the wavelength as

$$\lambda = \frac{2\pi}{k} = \frac{2\pi}{1.00 \times 10^7 \text{ m}^{-1}} = 6.28 \times 10^{-7} \text{ m} \quad \blacksquare$$

(c) The frequency is $f = \dfrac{c}{\lambda} = \dfrac{3.00 \times 10^8 \text{ m/s}}{6.28 \times 10^{-7} \text{ m}} = 4.77 \times 10^{14} \text{ Hz}$ ■

Finalize: The magnetic field is very small. The frequency is too high for an electronic counter to keep up with. But this is a familiar wave: orangey-red visible light. The wavelength tells us that.

---

19. What is the average magnitude of the Poynting vector 5.00 mi from a radio transmitter broadcasting isotropically (equally in all directions) with an average power of 250 kW?

Solution

Conceptualize: As the distance from the source is increased, the power per unit area will decrease, so at a distance of 5 miles from the source, the power per unit area will be a small fraction of the Poynting vector near the source.

Categorize: The Poynting vector is the power per unit area, where A is the surface area of a sphere with a 5-mile radius.

Analyze: The Poynting vector is $S_{av} = \dfrac{Power}{A} = \dfrac{Power}{4\pi r^2}$

In meters, $r = (5.00 \text{ mi})(1\,609 \text{ m/mi}) = 8\,045 \text{ m}$

and the intensity of the wave is $S = \dfrac{250 \times 10^3 \text{ W}}{4\pi(8\,045 \text{ m})^2} = 3.07 \times 10^{-4} \text{ W/m}^2$ ■

Finalize: The magnitude of the Poynting vector ten meters from the source is 199 W/m^2, on the order of a million times larger than it is 5 miles away! It is surprising how little power is actually received by a radio. At the 5-mile distance, the signal would only be about 30 nW, assuming a receiving area of about 1 cm^2.

---

21. A community plans to build a facility to convert solar radiation to electrical power. The community requires 1.00 MW of power, and the system to be installed has an efficiency of 30.0% (that is, 30.0% of the solar energy incident on the surface is converted to useful energy that can power the community). If sunlight has a constant intensity of 1 000 W/m^2, what must be the effective area of a perfectly absorbing surface used in such an installation?

Solution

Conceptualize: We are not worrying about nights, cloudy days, orienting the collector to face the sun, or storing the energy. Then just some hundred-meter by hundred-meter area should be adequate, like the roof of a shopping mall.

Categorize: We use the definition of intensity S_{avg} as power transferred by the wave per perpendicular area.

Analyze: At 30.0% efficiency, the collected power is $Power = 0.300\,SA$.

Then the area is $A = \dfrac{Power}{0.300\,S} = \dfrac{1.00 \times 10^6 \text{ W}}{0.300\left(1\,000 \text{ W/m}^2\right)} = 3\,330 \text{ m}^2 \approx 0.8 \text{ acre.}$ ■

Finalize: Bad news: Storing the energy can be a serious concern. It does not make sense to use solar-voltaic electricity to run an electric laundry dryer—it is much more efficient to hang the clothes out on a line. Good news: A million watts can keep a lot of computers and telephones running. With efficient light-emitting diode battery lanterns, a lot of students can do their homework.

---

30. A radio wave transmits 25.0 W/m^2 of power per unit area. A flat surface of area A is perpendicular to the direction of propagation of the wave. Assuming the surface is a perfect absorber, calculate the radiation pressure on it.

Solution

Conceptualize: The radio wave delivers no mass, but it does carry energy, and it delivers momentum. We can find the radiation pressure…

Categorize: … from the expression given in the chapter text.

Analyze: For complete absorption,

$$P = \frac{S}{c} = \frac{25.0\ \text{W/m}^2}{3.00 \times 10^8\ \text{m/s}} = 8.33 \times 10^{-8}\ \text{N/m}^2 \qquad \blacksquare$$

Finalize: The pressure is so small as to be very challenging to measure. Note that for a stream of material particles running into a wall and stopping, the pressure is

$$P = F/A = (mv - 0)/tA = 2(0.5\ mv^2)/Atv = 2K/Atv = 2I/v$$

where I represents the intensity of kinetic energy transport. The expression for light pressure can be written as $P = I/v$. It is similar but different by a factor of 2. The energy of light is not kinetic energy.

---

33. A 15.0-mW helium–neon laser emits a beam of circular cross section with a diameter of 2.00 mm. (a) Find the maximum electric field in the beam. (b) What total energy is contained in a 1.00-m length of the beam? (c) Find the momentum carried by a 1.00-m length of the beam.

Solution

Conceptualize: We will see the reasonable size, in kV/m, of the electric field in a beam of bright light. The light moves so fast that only a tiny fraction of a joule is contained in a one-meter length, and only a very tiny fraction of a kilogram-meter-per-second of momentum. Note again that it carries energy and momentum without possessing any mass.

Categorize: The intensity is related to the wave amplitude by an equation given in the chapter. From the definition of intensity, we can relate it to the energy content of the beam. Then $p = energy/c$ will tell us the momentum in this chunk of light.

Analyze: The intensity of the light is the average magnitude of the Poynting vector:

$$I = \frac{Power}{\pi r^2} = \frac{E^2_{\text{max}}}{2\mu_0 c}$$

(a) Therefore, the maximum electric field is

$$E_{\max} = \sqrt{\frac{Power(2\mu_0 c)}{\pi r^2}}$$

$$= \sqrt{\frac{(0.015\,0\text{ W})(2)(4\pi \times 10^{-7}\text{T}\cdot\text{m/A})(3.00 \times 10^8\text{m/s})}{\pi(0.001\,00\text{ m})^2}} = 1.90 \times 10^3\text{N/C} \quad \blacksquare$$

(b) The power being 15.0 mW means that 15.0 mJ passes through a cross section of the beam in one second. This energy is uniformly spread through a beam length of 3.00×10^8 m, since that is how far the front end of the energy travels in one second. Thus, the energy in just a one-meter length is

$$T_{ER} = \left(\frac{15.0 \times 10^{-3}\text{ J/s}}{3.00 \times 10^8\text{ m/s}}\right)(1.00\text{ m}) = 5.00 \times 10^{-11}\text{ J} \quad \blacksquare$$

(c) The linear momentum carried by a 1.00-m length of the beam is the momentum that would be received by an absorbing surface, under complete absorption:

$$p = \frac{T_{ER}}{c} = \frac{5.00 \times 10^{-11}\text{ J}}{3.00 \times 10^8\text{ m/s}} = 1.67 \times 10^{-19}\text{ kg}\cdot\text{m/s} \quad \blacksquare$$

Finalize: We study light after studying waves and also electric and magnetic fields, so that you can think of it in terms that are as familiar as possible. The electric field here has a size such as a charged balloon might produce. But still the very high speed of an electromagnetic wave is remarkable. It shows up here in the small values for energy and especially momentum content.

---

41. Two vertical radio-transmitting antennas are separated by half the broadcast wavelength and are driven in phase with each other. In what horizontal directions are (a) the strongest and (b) the weakest signals radiated?

Solution

Conceptualize: The strength of the radiated signal will be a function of the location around the two antennas and will depend on the interference of the waves.

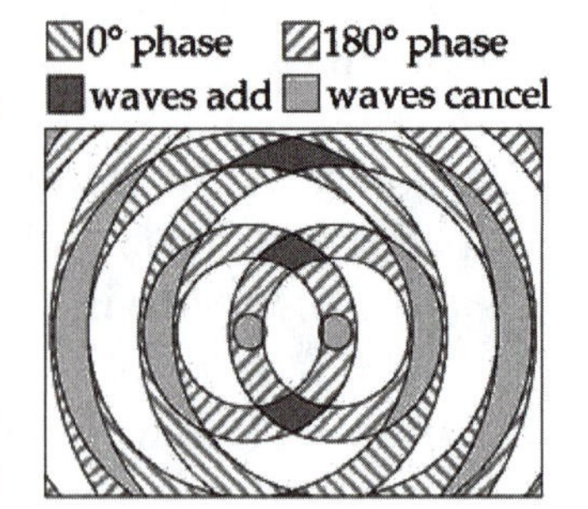

Categorize: A diagram helps to visualize this situation. The two antennas are driven in phase, which means that they both create maximum electric field strength at the same time, as shown in the diagram. The electromagnetic radio waves travel radially outwards from the antennas, and the received signal will be the vector sum of the two waves.

Analyze:

(a) Along the perpendicular bisector of the line joining the antennas, the distance is the same to both transmitting antennas. The transmitters oscillate in phase, so along this line the two signals will be received in phase, constructively interfering to produce a maximum signal strength that is twice the amplitude of one transmitter. ■

(b) Along the extended line joining the sources, the wave from the more distant antenna must travel one-half wavelength farther, so the waves are received 180° out of phase. They interfere destructively to produce the weakest signal in this direction. ■

Finalize: Radio stations may use an antenna array to direct the radiated signal toward a highly-populated region and reduce the signal strength delivered to a sparsely-populated area.

---

45. What are the wavelengths of electromagnetic waves in free space that have frequencies of (a) 5.00×10^{19} Hz and (b) 4.00×10^{9} Hz?

Solution

Conceptualize: Both frequencies are very high compared to frequencies for sound or mechanical waves we have studied. The spectrum chart shows that ~10^{20} Hz characterizes an x-ray or gamma ray with a picometer wavelength. And ~10^{10} Hz characterizes a very different microwave or radio wave with an ordinary-size wavelength in centimeters.

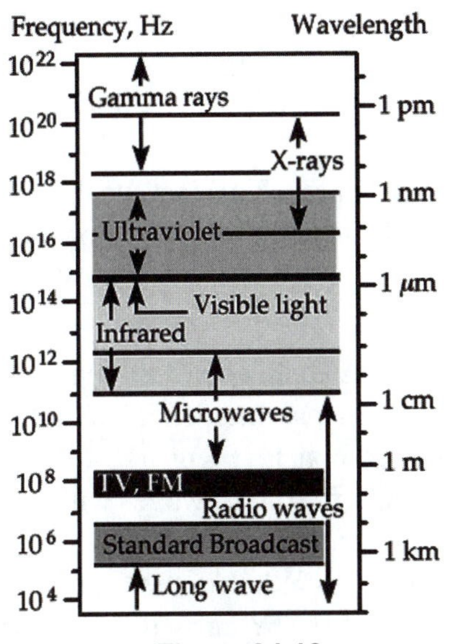

Figure 34.13

Categorize: We use just the $v = f\lambda$ relationship that applies to any continuous wave. For an electromagnetic wave in vacuum, the speed is c.

Analyze:

(a) $$\lambda = \frac{c}{f} = \frac{3.00 \times 10^{8}\ \text{m/s}}{5.00 \times 10^{19}\ \text{s}^{-1}} = 6.00\ \text{pm}$$ ■

(b) $$\lambda = \frac{c}{f} = \frac{3.00 \times 10^{8}\ \text{m/s}}{4.00 \times 10^{9}\ \text{s}^{-1}} = 7.50\ \text{cm}$$ ■

Finalize: The wave in (a) would be called an x-ray if it were emitted when an inner electron in an atom or an electron in a vacuum tube loses energy. It would be called a gamma ray if it were radiated by an atomic nucleus. By Figure 34.13, the finger-length wave in part (b) is called a radio wave or a microwave.

---

48. In 1965, Arno Penzias and Robert Wilson discovered the cosmic microwave radiation left over from the big bang expansion of the Universe. Suppose the energy density of this background radiation is 4.00×10^{-14} J/m^3. Determine the corresponding electric field amplitude.

Solution

Conceptualize: The primordial background radiation is moving past all points in all directions. Penzias and Wilson observed it as arriving in New Jersey from a sphere around us, about 14 billion light-years in radius. The problem does not quote an intensity, but we can still solve it...

Categorize: ...by using the relationship between energy density and electric field magnitude that we studied previously for a static field.

Analyze: The energy density can be written as $u = \frac{1}{2}\epsilon_0 E_{max}^2$

so $$E_{max} = \sqrt{\frac{2u}{\epsilon_0}} = \sqrt{\frac{2\left(4.00 \times 10^{-14} \text{ N}\cdot\text{m}^2\right)}{8.85 \times 10^{-12} \text{ C}^2/\text{N}\cdot\text{m}^2}} = 95.1 \text{ mV/m}$$ ■

Finalize: This is a weak field, but an antenna built for microwave relaying of telephone calls detected the radiation as a hum, isotropic and constant in time.

---

57. A dish antenna having a diameter of 20.0 m receives (at normal incidence) a radio signal from a distant source as shown in Figure P34.57. The radio signal is a continuous sinusoidal wave with amplitude $E_{max} = 0.200$ μV/m. Assume the antenna absorbs all the radiation that falls on the dish. (a) What is the amplitude of the magnetic field in this wave? (b) What is the intensity of the radiation received by this antenna? (c) What is the power received by the antenna? (d) What force is exerted by the radio waves on the antenna?

Figure P34.57

Solution

Conceptualize: Radio waves can be detected with remarkably low amplitudes. The force especially will be very small.

Categorize: Increasing any one of the electric-field amplitude, magnetic-field amplitude, intensity, or radiation pressure would make the others all increase in a predictable way. These quantities characterize the wave in itself. The energy received and the force are proportional to the area of the antenna as an absorbing surface intercepting the wave.

Analyze:

(a) The magnetic-field amplitude is

$$B_{max} = E_{max}/c = 6.67 \times 10^{-16} \text{ T}$$ ■

(b) The intensity is the Poynting vector averaged over one or more cycles, given by

$$S_{av} = E^2_{max}/2\mu_0 c = 5.31 \times 10^{-17}\ \text{W/m}^2 \qquad \blacksquare$$

(c) The power tells how fast the antenna receives energy. It is

$$Power_{av} = S_{av}A = S_{av}\pi r^2 = 1.67 \times 10^{-14}\ \text{W} \qquad \blacksquare$$

(d) The force tells how fast the antenna receives momentum. It is

$$F = PA = (S_{av}/c)A = 5.56 \times 10^{-23}\ \text{N} \qquad \blacksquare$$

Finalize: Do not confuse the power $Power = S_{av}A$ with the pressure $P = S_{av}/c$, which we used to find the force. The force is immeasurably small, like the weight of a few thousand hydrogen atoms, but the radio wave can be detected and so carries measurable power.

---

67. A linearly polarized microwave of wavelength 1.50 cm is directed along the positive x axis. The electric field vector has a maximum value of 175 V/m and vibrates in the xy plane. Assuming the magnetic field component of the wave can be written in the form $B = B_{max}\sin(kx - \omega t)$, give values for (a) B_{max}, (b) k, and (c) ω. (d) Determine in which plane the magnetic field vector vibrates. (e) Calculate the average value of the Poynting vector for this wave. (f) If this wave were directed at normal incidence onto a perfectly reflecting sheet, what radiation pressure would it exert? (g) What acceleration would be imparted to a 500-g sheet (perfectly reflecting and at normal incidence) with dimensions of 1.00 m × 0.750 m?

Solution

Conceptualize: This is a microwave that could be used for communication. It will have an intensity of some watts per square meter. But the other answers will be a small magnetic field, a small fraction of a pascal for the pressure, and a small acceleration for the sail.

Categorize: The given wavelength will tell us the values of wave number and angular frequency. The given electric-field amplitude will tell us the magnetic-field amplitude, intensity, and radiation pressure. The particle under net force model will tell us the acceleration of the sail.

Analyze:

(a) The magnetic field has amplitude

$$B_{max} = \frac{E_{max}}{c} = \frac{175\ \text{V/m}}{3.00 \times 10^8\ \text{m/s}} = 5.83 \times 10^{-7}\ \text{T} \qquad \blacksquare$$

(b) The wave number is $k = \dfrac{2\pi}{\lambda} = \dfrac{2\pi}{0.015\ \text{m}} = 419\ \text{m}^{-1}$ $\blacksquare$

(c) The angular frequency is

$$\omega = kc = (419\ \text{m}^{-1})(3.00 \times 10^8\ \text{m/s}) = 1.26 \times 10^{11}\ \text{rad/s} \qquad \blacksquare$$

(d) The magnetic field must be in the z direction so that $\vec{\mathbf{S}} = \vec{\mathbf{E}} \times \vec{\mathbf{B}}/\mu_0$ can be in the $\hat{\mathbf{j}} \times \hat{\mathbf{k}} = \hat{\mathbf{i}}$ direction. The magnetic field vibrates in the xz plane. ■

(e) The magnitude of the average Poynting vector is the wave intensity

$$S_{av} = \frac{E_{max}B_{max}}{2\mu_0} = \frac{(175\text{ V/m})(5.83\times10^{-7}\text{ T})}{2(4\pi\times10^{-7}\text{ N/A}^2)} = 40.6\text{ W/m}^2$$

The Poynting vector itself points in the direction of energy transport:

$$\vec{\mathbf{S}}_{av} = (40.6\text{ W/m}^2)\,\hat{\mathbf{i}}$$ ■

(f) For perfect reflection, the pressure is

$$P_r = \frac{2S}{c} = \frac{2(40.6\text{ W/m}^2)}{3.00\times10^8\text{ m/s}} = 2.71\times10^{-7}\text{ N/m}^2$$ ■

(g) We use Newton's second law. With magnitude

$$a = \frac{F}{m} = \frac{P_rA}{m} = \frac{(2.71\times10^{-7}\text{ N/m}^2)(0.750\text{ m}^2)}{0.500\text{ kg}} = 4.06\times10^{-7}\text{ m/s}^2$$

the acceleration is $\vec{\mathbf{a}} = (406\text{ nm/s}^2)\hat{\mathbf{i}}$ ■

Finalize: We have not forgotten that acceleration is a vector, like the magnetic field and the Poynting vector. Pressure and angular frequency are scalars. Wave number could be defined as a vector in the direction of wave velocity, but we treat it as a scalar.

---

69. An astronaut, stranded in space 10.0 m from her spacecraft and at rest relative to it, has a mass (including equipment) of 110 kg. Because she has a 100-W flashlight that forms a directed beam, she considers using the beam as a photon rocket to propel herself continuously toward the spacecraft. (a) Calculate the time interval required for her to reach the spacecraft by this method. (b) **What If?** Suppose she throws the 3.00-kg flashlight in the direction away from the spacecraft instead. After being thrown, the flashlight moves at 12.0 m/s relative to the recoiling astronaut. After what time interval will the astronaut reach the spacecraft?

Solution

Conceptualize: Based on our everyday experience, the force exerted by photons is too small to feel, so it may take a very long time (maybe days!) for the astronaut to travel 10 m with her "photon rocket." Using the momentum of the thrown lamp seems like a better solution, but it will still take a while (maybe a few minutes) for the astronaut to reach the spacecraft because her mass is so much larger than the mass of the flashlight.

Categorize: In part (a), the radiation pressure can be used to find the force that accelerates the astronaut toward the spacecraft. Then we will use the particle under a net force and particle under constant acceleration models to find the travel time interval. In part (b), the principle of conservation of momentum can be applied to the flashlight-astronaut system to find the time required to travel the 10 m.

Analyze:

(a) Light exerts on the astronaut a pressure $P = F/A = S/c$, and a force of

$$F = \frac{SA}{c} = \frac{Power}{c} = \frac{100\ \text{J/s}}{3.00 \times 10^8\ \text{m/s}} = 3.33 \times 10^{-7}\ \text{N}$$

By Newton's second law,

$$a = \frac{F}{m} = \frac{3.33 \times 10^{-7}\ \text{N}}{110\ \text{kg}} = 3.03 \times 10^{-9}\ \text{m/s}^2$$

This acceleration is constant, so the distance traveled is $x = \frac{1}{2}at^2$, and the amount of time she travels is

$$t = \sqrt{\frac{2x}{a}} = \sqrt{\frac{2(10.0\ \text{m})}{3.03 \times 10^{-9}\ \text{m/s}^2}} = 8.12 \times 10^4\ \text{s} = 22.6\ \text{h}$$ ■

(b) Because there are no external forces, the momentum of the astronaut-flashlight system before throwing the lamp is the same as afterwards, when the now 107-kg astronaut is moving at speed v towards the spacecraft and the flashlight is moving away from the spacecraft at $(12.0\ \text{m/s} - v)$.

Thus, $\sum \vec{\mathbf{p}}_i = \sum \vec{\mathbf{p}}_f$ gives $0 = (107\ \text{kg})\,v - (3.00\ \text{kg})(12.0\ \text{m/s} - v)$

$$0 = (107\ \text{kg})\,v - (36.0\ \text{kg} \cdot \text{m/s}) + (3.00\ \text{kg})v$$

$$v = \frac{36.0\ \text{kg} \cdot \text{m/s}}{110\ \text{kg}} = 0.327\ \text{m/s}.$$

Then the time interval for the astronaut's trip is

$$t = \frac{x}{v} = \frac{10.0\ \text{m}}{0.327\ \text{m/s}} = 30.6\ \text{s}$$ ■

Finalize: Throwing the light source away is certainly a quicker way to reach the spacecraft, but if you are a little off with the direction, you will miss the ship. You will not be able to retrieve the flashlight and try again unless it has a very long cord. How long would the cord need to be, and does its length depend on how hard the astronaut throws the lamp? (You can verify that the minimum cord length is 367 m, independent of the speed that the lamp is thrown.)

70. Review. In the absence of cable input or a satellite dish, a television set can use a dipole-receiving antenna for VHF channels and a loop antenna for UHF channels. In Figure CQ34.9, the "rabbit ears" form the VHF antenna and the smaller loop of wire is the UHF antenna. The UHF antenna produces an emf from the changing magnetic flux through the loop. The television station broadcasts a signal with a frequency f, and the signal has an electric-field amplitude E_{max} and a magnetic-field amplitude B_{max} at the location of the receiving antenna. (a) Using Faraday's law, derive an expression for the amplitude of the emf that appears in a single-turn, circular loop antenna with a radius r that is small compared with the wavelength of the wave. (b) If the electric field in the signal points vertically, what orientation of the loop gives the best reception?

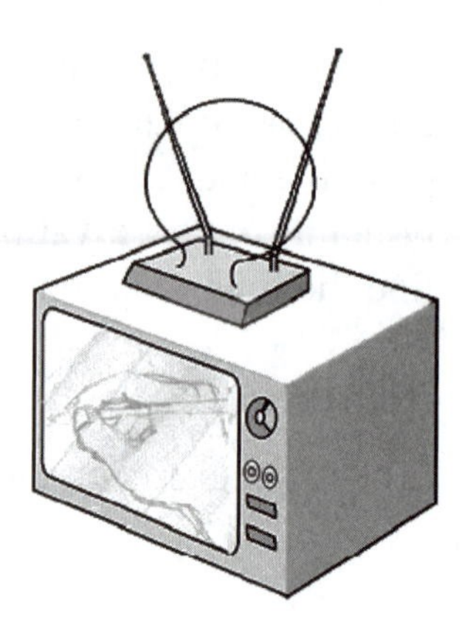

Figure CQ34.9

Solution

Conceptualize: The amplitude should be proportional to the amplitude of the magnetic field, the area of the loop, and the frequency.

Categorize: The magnetic flux through the loop changes as the wave moves past, so Faraday's law is appropriate for finding the induced voltage.

Analyze: We can approximate the magnetic field as uniform over the area of the loop while it oscillates in time as $B = B_{max} \cos\omega t$. The induced voltage is

$$\mathcal{E} = -\frac{d\Phi_B}{dt} = -\frac{d}{dt}(BA\cos\theta) = -A\frac{d}{dt}\left(B_{max}\cos\omega t\cos\theta\right)$$

$$\mathcal{E} = AB_{max}\omega\,(\sin\omega t\cos\theta)$$

(a) Since the angular frequency is $\omega = 2\pi f$, the amplitude of this emf is $\mathcal{E}_{max} = 2\pi^2 r^2 f B_{max}\cos\theta$ where θ is the angle between the magnetic field and the normal to the loop. ■

(b) If $\vec{\mathbf{E}}$ is vertical, then $\vec{\mathbf{B}}$ is horizontal, so the plane of the loop should be vertical, and the plane should contain the line of sight to the transmitter. This will make $\theta = 0°$, so $\cos\theta$ takes on its maximum value. ■

Finalize: The derived equation shows the predicted proportionalities to r^2, to B_{max}, and to f. We see also proportionality to $\cos\theta$, and that the proportionality constant is $2\pi^2$. The moral is that the laws set up to describe static or local or individual electric and magnetic fields also give a complete theory of electromagnetic waves. Maxwell was right when he said his theory was "great guns!"

35

The Nature of Light and the Principles of Ray Optics

EQUATIONS AND CONCEPTS

Reflection and refraction can occur when a light ray is incident obliquely on a smooth planar surface which forms the boundary between two transparent media of different optical densities. *As shown in the figure, the subscript 1 refers to parameters for light in the initial medium and subscript 2 refers to corresponding parameters in the new medium. A portion of the energy associated with each incoming ray will be reflected back into the original medium, while the remaining fraction will be transmitted into the second medium.*

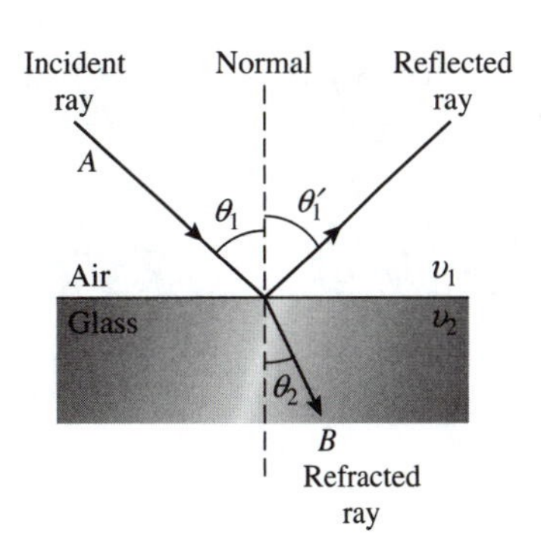

The **energy of a photon** is proportional to the frequency of the associated electromagnetic wave. The constant of proportionality, h, in Equation 35.1 is called Planck's constant.

$$E = hf \tag{35.1}$$
$$h = 6.63 \times 10^{-34}\ \text{J} \cdot \text{s}$$

The **law of reflection** states that the angle of incidence (the angle measured between the incident ray and the normal line) equals the angle of reflection (the angle measured between the reflected ray and the normal line). *The incident ray, reflected ray, and the normal line lie in the same plane.*

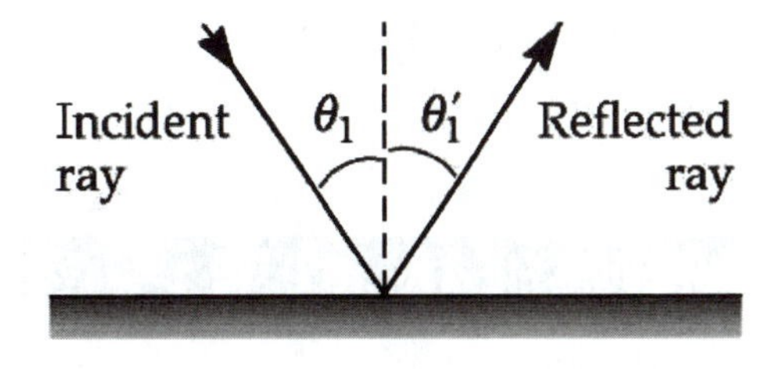

$$\theta_1' = \theta_1 \tag{35.2}$$

The **index of refraction** of a transparent medium equals the ratio of the speed of light in vacuum to the speed of light in the medium. The index of refraction of a given medium can also be expressed as the ratio of the wavelength of light in vacuum to the wavelength in that medium.

$$n \equiv \frac{\text{speed of light in vacuum}}{\text{speed of light in a medium}} = \frac{c}{v} \tag{35.4}$$

$$n = \frac{\lambda}{\lambda_n} \tag{35.7}$$

Snell's law, Equation 35.8, states that the degree of change in direction of propagation of a light ray which crosses the boundary between two materials depends on the relative values of the indices of refraction of the materials.

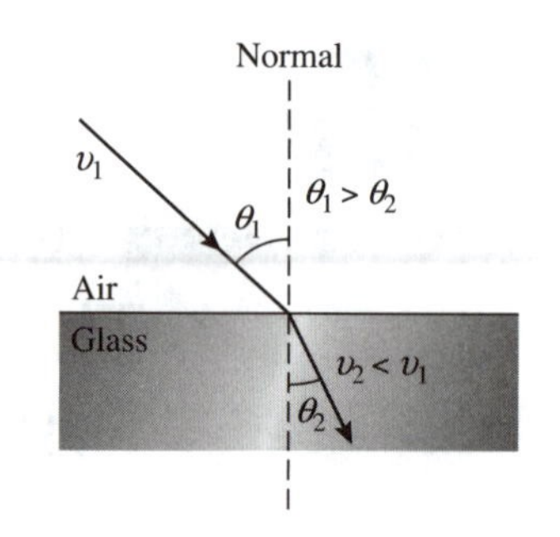

$$n_1 \sin \theta_1 = n_2 \sin \theta_2 \qquad (35.8)$$

***The frequency of a wave* is characteristic of the source**; *as light travels from one medium into another of different index of refraction, the frequency remains constant but the wavelength changes.*

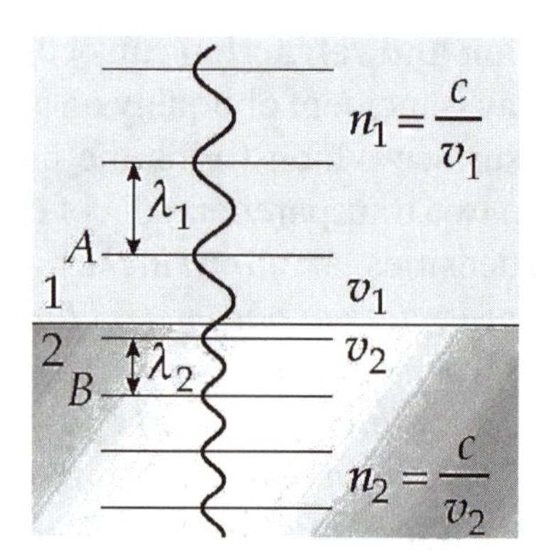

Total internal reflection occurs at the interface between two media when an incident ray is totally internally reflected back into the first medium. This effect is observed for angles of incidence, equal to or greater than the ***critical angle*** defined in Equation 35.10, *and is possible only when a light ray is directed from a medium of a given index of refraction into a medium of lower index of refraction.*

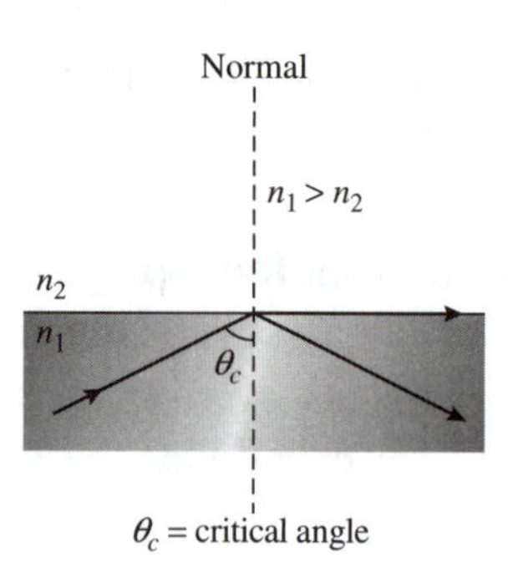

$$\sin \theta_c = \frac{n_2}{n_1} \quad (\text{for } n_1 > n_2) \qquad (35.10)$$

REVIEW CHECKLIST

You should be able to:

- Describe the methods used by Roemer and Fizeau for the measurement of c and make calculations using sets of typical values for the quantities involved. (Section 35.2)
- Make calculations using the law of reflection and the law of refraction (Snell's law). (Sections 35.4 and 35.5)
- Calculate the angle of deviation and the angular dispersion in a prism. (Section 35.7)
- Understand the conditions under which total internal reflection can occur in a medium and determine the critical angle for a given pair of adjacent media. (Section 35.8)

ANSWER TO AN OBJECTIVE QUESTION

12. Suppose you find experimentally that two colors of light, A and B, originally traveling in the same direction in air, are sent through a glass prism, and A changes direction more than B. Which travels more slowly in the prism, A or B? Alternatively, is there insufficient information to determine which moves more slowly?

Answer As the light slows down upon entering the prism, a light ray that is bent more suffers a greater loss of speed as it enters the new medium. Therefore, light ray A travels more slowly.

□ □ □ □

ANSWERS TO SELECTED CONCEPTUAL QUESTIONS

1. Why do astronomers looking at distant galaxies talk about looking backward in time?

Answer Light travels through a vacuum at a speed of 300 000 km per second. Thus, an image we see of a distant star or galaxy must have been generated some time ago.

For example, the star Altair is 16 light-years away; if we look at Altair today, we know only what was happening 16 years ago. A star does not ordinarily change much in 16 years, so the "look-back time" may not initially seem significant. But astronomers who study other galaxies can gain an idea of what galaxies looked like when they were significantly younger. Thus, it actually makes sense to speak of "looking backward in time."

□ □ □ □

8. Explain why a diamond sparkles more than a glass crystal of the same shape and size.

Answer Diamond has a larger index of refraction than glass, and consequently has a smaller critical angle for internal reflection. A brilliant-cut diamond is shaped to admit light from above, reflect it totally at the converging facets on the underside of the jewel, and let the light escape only at the top. Glass will have less light internally reflected.

□ □ □ □

SOLUTIONS TO SELECTED PROBLEMS

3. In an experiment to measure the speed of light using the apparatus of Armand H. L. Fizeau (see Fig. P35.2), the distance between light source and mirror was 11.45 km and the wheel had 720 notches. The experimentally determined value of c was 2.998×10^8 m/s when the outgoing light passed through one notch and then returned through the next notch. Calculate the minimum angular speed of the wheel for this experiment.

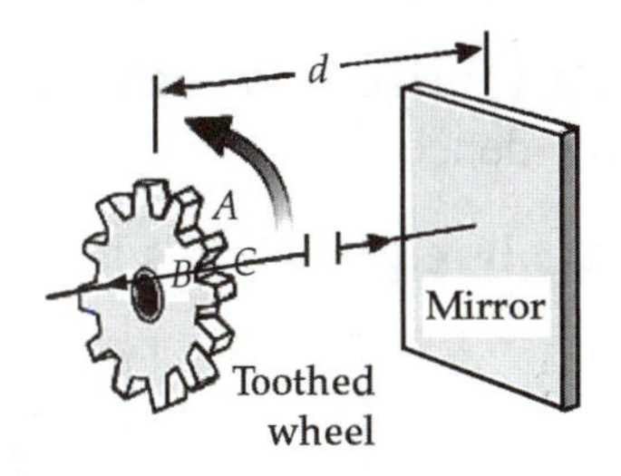

Figure P35.2

Solution

Conceptualize: If you have already measured speeds for mechanics carts and for sound, ask your instructor whether you can measure the speed of light. It is not too challenging for freshmen. You may use a modulated laser or a pulsed light-emitting diode. With Fizeau's apparatus, the toothed wheel chops a light beam into chunks. Visualize the advance of the front end of one of the chunks.

Categorize: We will use just the definitions of speed and angular speed.

Analyze: The distance down to the mirror and back is

$$2 \times 11.45 \times 10^3 \text{ m} = 2.290 \times 10^4 \text{ m}$$

The round-trip time is expected to be

$$t = \frac{x}{v} = \frac{2.290 \times 10^4 \text{ m}}{2.998 \times 10^8 \text{ m/s}} = 7.64 \times 10^{-5}\text{ s}$$

In this time the wheel should turn by (1/720) rev to move one notch into the place of the previous notch. Its angular speed should be

$$\omega = \frac{\theta}{t} = \left(\frac{1 \text{ rev}}{720}\right)\left(\frac{1}{7.64 \times 10^{-5} \text{ s}}\right) = 18.2 \text{ rev/s} = 114 \text{ rad/s} \quad \blacksquare$$

Finalize: Higher angular speeds must be available so that the experimenter can home in on the special maximum-brightness setting from both sides. Additional data, possibly with higher precision, can be taken at 9.09 rev/s, where one notch is replaced by a tooth while the light goes out and back. There the experimenter can zero in on the angular speed that gives darkness for the returning light beam.

---

6. An underwater scuba diver sees the Sun at an apparent angle of 45.0° above the horizontal. What is the actual elevation angle of the Sun above the horizontal?

Solution

Conceptualize: The sunlight refracts as it enters the water from the air. Because the water has a higher index of refraction, the light slows down and bends toward the vertical line that is normal to the interface. Therefore, the elevation angle of the Sun above the water will be less than 45°, as shown in the diagram, even though it appears to the diver that the sun is 45° above the horizon.

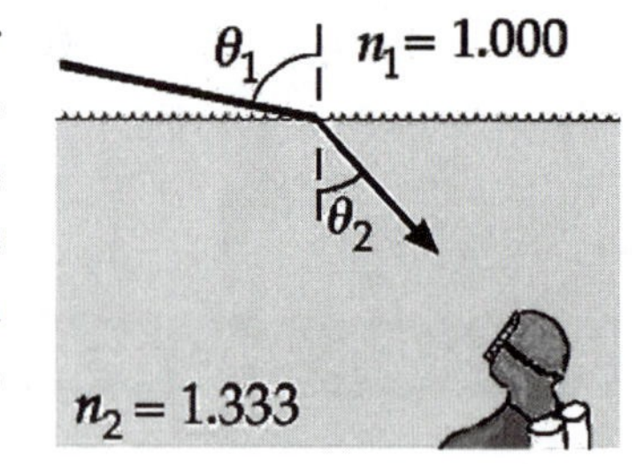

Categorize: We can use the wave under refraction model to find the precise angle of incidence.

Analyze: In Snell's law $n_1 \sin\theta_1 = n_2 \sin\theta_2$, we know n_1, n_2, and θ_2, so we find

$$\theta_1 = \sin^{-1}\left(\frac{n_2 \sin\theta_2}{n_1}\right) = \sin^{-1}\left(\frac{1.333 \sin 45.0°}{1.000}\right) = \sin^{-1} 0.943 = 70.5°$$

The sunlight impinges at $\theta_1 = 70.5°$ to the vertical, so the Sun is 19.5° above the horizon. ■

Finalize: The calculated result agrees with our prediction. When applying Snell's law, it is easy to mix up the index values and to confuse angles-with-the-normal and angles-with-the-surface. Making a sketch and a prediction as we did here helps avoid careless mistakes.

Bonus problem: The following extension will remind you that we are studying wave motion. Consider yellow light from the Sun with wavelength 580 nm in air. (The diver's face mask might block out other colors.) Find its frequency and speed in air and in water, and its wavelength in water.

Solution: We answer the questions in a convenient order. The speed of light in air is $v = c/n = (3.00 \times 10^8 \text{ m/s})/1.000 = 300$ Mm/s.

The frequency of this yellow light is $f = v/\lambda$. We take the speed in air and the wavelength in air to find $f = v/\lambda = (3.00 \times 10^8 \text{ m/s})/5.80 \times 10^{-7} \text{ m} = 517$ THz. The frequency in water has the same value, 517 THz, because every wave crest coming down to a point on the water surface immediately propagates as one wave crest into the water. The same number of wavefronts in air and in water pass a point on the interface in any particular time interval, notably in one second.

In water the speed is $v = c/n = (3.00 \times 10^8 \text{ m/s})/1.333 = 225$ Mm/s.

The wavelength in water is $\lambda = v/f = (2.25 \times 10^8 \text{ m/s})/(5.17 \times 10^{14}\text{/s}) = 435$ nm. This wavelength could also be computed as the vacuum wavelength divided by water's index of refraction: 580 nm/1.333 = 435 nm.

In vacuum, 435 nm would be the wavelength of blue light. The light still looks yellow to the diver when she is under water. The frequency controls what the light does when it is absorbed. It is conventional to classify different colors by their wavelengths in vacuum because wavelength can be measured (easily) and the $\sim 10^{15}$ Hz frequency cannot be measured directly. But it is the frequency that is set by the source and intrinsically carried by the light ray.

---

12. A ray of light strikes a flat block of glass ($n = 1.50$) of thickness 2.00 cm at an angle of 30.0° with the normal. Trace the light beam through the glass and find the angles of incidence and refraction at each surface.

Solution

Conceptualize: The ray changes direction sharply at the point of incidence on the block. Then it travels straight through the block to the opposite face and switches direction there again. Bending toward the normal at entry, it will bend away from the normal at exit.

Categorize: We use Snell's law about entry and separately about exit.

Analyze: At entry, the wave under refraction model, expressed as $n_1 \sin\theta_1 = n_2 \sin\theta_2$, gives

$$\theta_2 = \sin^{-1}\left(\frac{n_1 \sin\theta_1}{n_2}\right) = \sin^{-1}\left(\frac{1.000 \sin 30.0°}{1.50}\right) = \sin^{-1}\frac{0.500}{1.50} = 19.5° \qquad ■$$

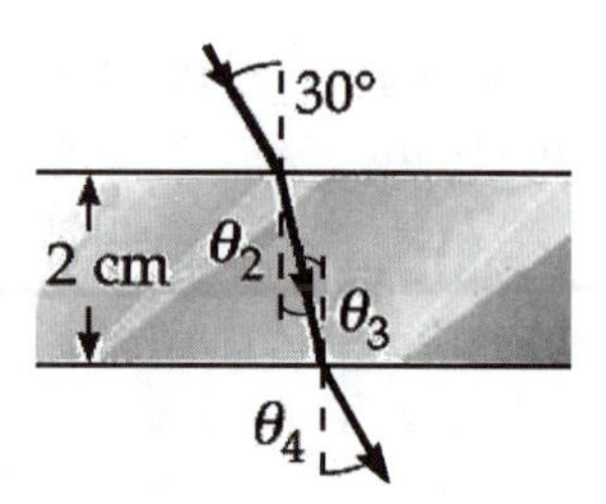

To do geometric optics, you must remember some geometry. The surfaces of entry and exit are parallel so their normals are parallel. Then angle θ_2 of refraction at entry and the angle θ_3 of incidence at exit are alternate interior angles formed by the ray as a transversal cutting parallel lines.

Therefore $\theta_3 = \theta_2 = 19.5°$ ■

At the exit point, $n_2 \sin\theta_3 = n_1 \sin\theta_4$ gives

$$\theta_4 = \sin^{-1}\left(\frac{n_2 \sin\theta_3}{n_1}\right) = \sin^{-1}\left(\frac{1.50 \sin 19.5°}{1.000}\right) = \sin^{-1} 0.333 = 30.0°$$ ■

Because θ_1 and θ_4 are equal, the departing ray in air is parallel to the original ray.

Finalize: This is a useful theorem, with practical applications. A car windshield of uniform thickness will not distort, but shows the driver the actual direction to every object outside. In the next chapter we will use the rule that a ray going through the center of a lens is unchanged in direction, because it encounters parallel surfaces of entry and exit.

---

13. A prism that has an apex angle of 50.0° is made of cubic zirconia. What is its minimum angle of deviation?

Solution

Conceptualize: Shining a beam of laser light through a D-shaped piece of glass or plastic is a good way to study Snell's law and the law of reflection too. After doing that, obtain any prism and shine the narrow beam through it. Turn the prism slowly clockwise about an axis perpendicular to the direction of the light. You will see the exiting beam swinging counterclockwise, slowing down, stopping, and reversing. At the point of stopping, the net change in direction of the light is a minimum. In experiments it is not hard to set up minimum deviation and it is easy to relate the measured deviation angle to the index of refraction of the material.

Categorize: We use the equation derived in the chapter for this situation, and the refractive index of the material from the table in the text.

Analyze: From Equation 35.9, $n = \dfrac{\sin\left(\dfrac{\Phi + \delta_{\min}}{2}\right)}{\sin\left(\dfrac{\Phi}{2}\right)}$

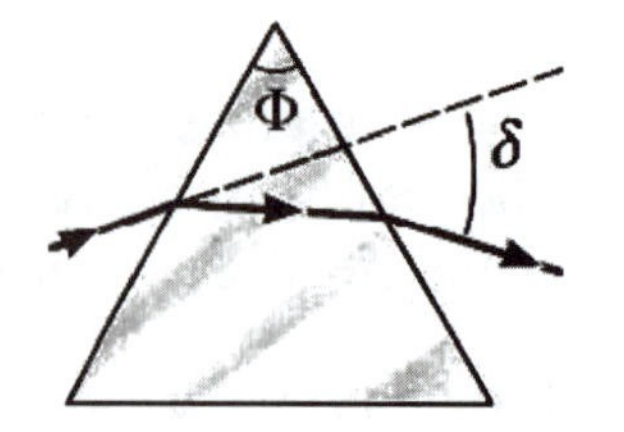

Solving for the angle of minimum deviation gives

$$\delta_{\min} = 2\sin^{-1}\left(n \sin\frac{\Phi}{2}\right) - \Phi$$

Then $\delta_{\min} = 2\sin^{-1}(2.20 \sin 25.0°) - 50.0° = 86.8°$ ■

Finalize: The prism need not be a triangular block. Φ refers to the angle between the surface of entry and the surface of exit, whatever the shape of the refracting object. So here is a prettier and perhaps more convenient demonstration of minimum deviation: Obtain a single crystal from a chandelier. Hang it by a string from the top of a sunny window and set it spinning. On the wall or floor opposite, spots of light will sweep toward the shadow of the crystal, stop at a certain radial distance, and sweep away again. The index of refraction depends on the wavelength, so each of these spots of light will show a full spectrum.

---

29. A beam of light both reflects and refracts at the surface between air and glass, as shown in Figure P35.29. If the refractive index of the glass is n_g, find the angle of incidence θ_1 in the air that would result in the reflected ray and the refracted ray being perpendicular to each other.

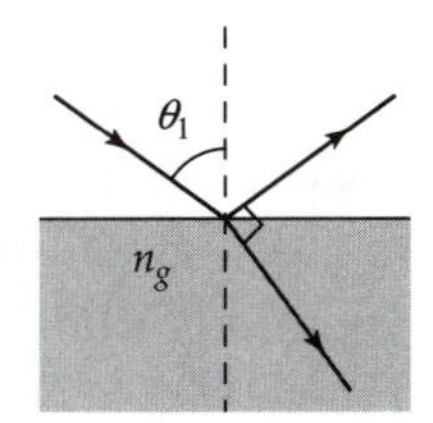

Figure P35.29

Solution

Conceptualize: When a wave encounters an interface where its speed changes sharply, some of its energy will reflect and some can pass into the new medium. In this book we do not study how much energy takes each path. We study the directions that the two outgoing waves take …

Categorize: … modeling them separately as a wave under reflection and a wave under refraction. It is most physical to think of the two phenomena as happening together, as we do in this problem.

Analyze: Think of a horizontal interface and the line normal (perpendicular) to it as the axes of a rectangular coordinate system, with its origin at the point where the light is incident on the surface. Think of the incident ray as lying in the second quadrant. This ray makes angle θ_1 with the normal to the interface. The reflected ray is then in the first quadrant, and there makes the same angle with the normal. The reflected ray makes angle $90° - \theta_1$ with the interface itself. The refracted ray is in the fourth quadrant and makes angle θ_2 with the same normal. The refracted ray makes angle $90° - \theta_2$ with the interface. To be perpendicular, the reflected and refracted rays must be in directions specified by angles that add up like this:

$$90° - \theta_1 + 90° - \theta_2 = 90°$$

This simplifies to $\theta_1 + \theta_2 = 90°$

We combine this condition with that specified by the wave under refraction model in $n_1 \sin\theta_1 = n_g \sin\theta_2$

by substitution: $1.000 \sin\theta_1 = n_g \sin(90° - \theta_1) = n_g \cos\theta_1$

Then $\dfrac{\sin\theta_1}{\cos\theta_1} = n_g = \tan\theta_1$ and $\theta_1 = \tan^{-1} n_g$ ■

Finalize: For a concrete example, if $n_g = 1.50$, then the special angle of incidence is 56.3°. We do not study the polarization of light until Chapter 38, but something special happens to the polarization of the reflected light at this angle of incidence, called Brewster's angle. Because reflection and refraction can often happen together, thinking of the incident, reflected, and refracted rays as in the second, first, and fourth quadrants of an interface-normal rectangular coordinate system can be useful in keeping track of things on exams and in laboratory. The remaining (third) quadrant is empty of energy. But note that reflection can happen without refraction, as when light falls on a smooth electrically conducting surface, and also when the angle of incidence is greater than the critical angle for total internal reflection.

---

33. The index of refraction for violet light in silica flint glass is 1.66 and that for red light is 1.62. What is the angular spread of visible light passing through a prism of apex angle 60.0° if the angle of incidence is 50.0°? See Figure P35.33.

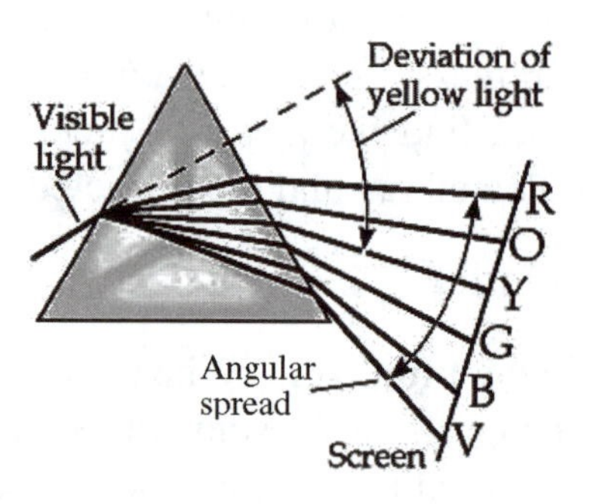

Figure P35.33

Solution

Conceptualize: The prism will produce a "rainbow," separating the colors of a narrow beam of incoming white light. Even if the angular dispersion is small, by placing a white card far away from the prism we can get a clear spectrum.

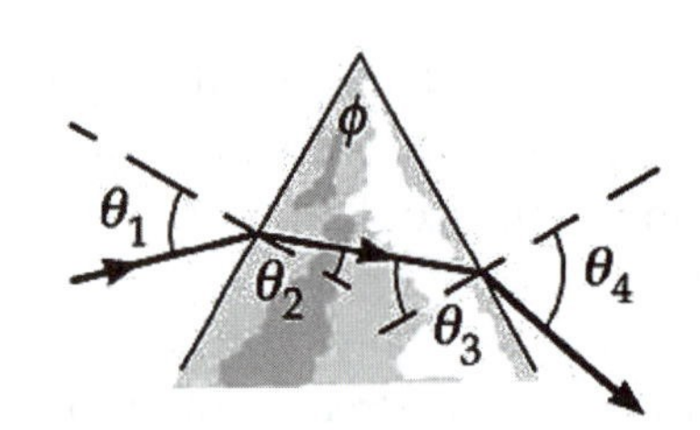

Categorize: We trace rays of violet light and red light through the prism, finding the direction of each as it exits. Then subtraction gives the angular width of the spectrum between them.

Analyze: Call the angles of incidence and refraction, at the surfaces of entry and exit, θ_1, θ_2, θ_3, and θ_4, in the order as shown. The apex angle ($\phi = 60.0°$) is the angle between the surfaces of entry and exit. The ray in the glass forms a triangle with these surfaces, in which the interior angles must add to 180°.

Thus,

$$(90° - \theta_2) + \phi + (90° - \theta_3) = 180°$$

and $\theta_3 = \phi - \theta_2$

This is a general rule for light going through prisms.

For the incoming ray, $\sin\theta_2 = \dfrac{\sin\theta_1}{n}$:

$$(\theta_2)_{\text{violet}} = \sin^{-1}\left(\frac{\sin 50.0°}{1.66}\right) = 27.48°$$

$$(\theta_2)_{\text{red}} = \sin^{-1}\left(\frac{\sin 50.0°}{1.62}\right) = 28.22°$$

For the ray exiting the glass, $\theta_3 = 60° - \theta_2$ and $\sin\theta_4 = n \sin\theta_3$

$$(\theta_4)_{\text{violet}} = \sin^{-1}\left[1.66 \sin 32.52°\right] = 63.17°$$

$$(\theta_4)_{\text{red}} = \sin^{-1}[1.62 \sin 31.78°] = 58.56°$$

The dispersion or angular spread is the difference between these two angles:

$$\Delta\theta_4 = 63.17° - 58.56° = 4.61°$$ ■

Finalize: Figure 35.22 in the textbook shows the effect nicely. Reflection cannot separate the colors of white light. Refraction does it, provided that the speeds of different wavelengths are different in the prism material. To help you understand lenses in the next chapter, you should learn that all colors of the light bend around the thicker part of the prism, away from the apex angle.

---

37. A triangular glass prism with apex angle $\Phi = 60.0°$ has an index of refraction $n = 1.50$ (Fig. P35.37). What is the smallest angle of incidence θ_1 for which a light ray can emerge from the other side?

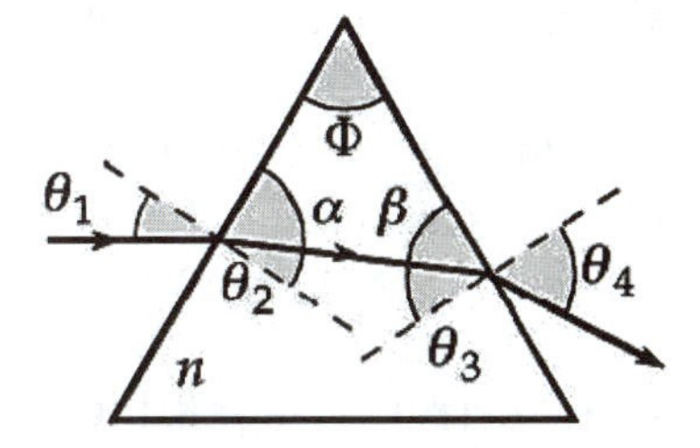

Figure P35.37 (modified)

Solution

Conceptualize: The alternative to emerging is undergoing total internal reflection. Figure 35.27 in the textbook shows the effect nicely. Going down through the incident rays in that figure, if θ_1 is too small θ_2 will be too small and θ_3 will be too large for refraction to take place at the second surface of the prism. Then the ray will be totally internally reflected there instead of emerging.

Categorize: We must trace a generic ray through the prism, finding how θ_2 and then θ_3 depend on θ_1. Then we apply the threshold condition for total internal reflection to θ_3 and evaluate the corresponding θ_1.

Analyze: Call the angles of incidence and refraction, at the surfaces of entry and exit, θ_1, θ_2, θ_3, and θ_4, in order as shown. The apex angle Φ is the angle between the surface of entry and the second surface. The ray in the glass forms a triangle with these surfaces, in which the interior angles must add to 180°. Thus, with $\Phi = 60.0°$,

we have $(90° - \theta_2) + 60° + (90° - \theta_3) = 180°$

so $\theta_2 + \theta_3 = 60.0°$ **[1]**

which exemplifies a general rule for light going through prisms. At the first refraction, Snell's law gives

$$\sin\theta_1 = 1.50 \sin\theta_2$$ **[2]**

At the second boundary, we want to almost reach the condition for total internal reflection:

$$1.50 \sin\theta_3 = 1.00 \sin 90° = 1.00$$

or $\theta_3 = \sin^{-1}(1.00/1.50) = 41.8°$

Now by Equation [1] above, $\theta_2 = 60.0° - 41.8° = 18.2°$

while by Equation [2], we find that $\theta_1 = \sin^{-1}(1.50 \sin 18.2°)$

So $\theta_1 = 27.9°$ ■

Finalize: Compare the solution of this problem to the solution of its symbolic version, Problem 38. The numeric solution is significantly less complicated. Whether a given problem is more conveniently solved analytically or numerically depends on the complexity of the problem.

---

38. A triangular glass prism with apex angle Φ has an index of refraction n. (Fig. P35.37.) What is the smallest angle of incidence θ_1 for which a light ray can emerge from the other side?

Solution

Conceptualize: The alternative to emerging is undergoing total internal reflection. Figure 35.27 in the textbook shows the effect nicely. Going down through the incident rays in that figure, if θ_1 is too small θ_2 will be too small and θ_3 will be too large for refraction to take place at the second surface of the prism. Then the ray will be totally internally reflected there instead of emerging.

Categorize: We must trace a generic ray through the prism, finding how θ_2 and then θ_3 depend on θ_1. Then we apply the threshold condition for total internal reflection to θ_3 and evaluate the corresponding θ_1.

Analyze: Call the angles of incidence and refraction, at the surfaces of entry and exit, θ_1, θ_2, θ_3, and θ_4, in order as shown. The apex angle Φ is the angle between the surface of entry and the second surface. The ray in the glass forms a triangle with these surfaces, in which the interior angles must add to 180°. Thus we have $(90° - \theta_2) + \Phi + (90° - \theta_3) = 180°$

so $\theta_2 + \theta_3 = \Phi$ **[1]**

which is a general rule for light going through prisms. At the first interface between the air and prism, Snell's law gives

$$\sin\theta_1 = n \sin\theta_2 \quad \textbf{[2]}$$

At the second interface, we want to almost reach the condition for total internal reflection:

$$n \sin\theta_3 = 1 \sin 90° = 1$$

or $\theta_3 = \sin^{-1}(1/n)$

Now by Equation [1] above, $\theta_2 = \Phi - \theta_3 = \Phi - \sin^{-1}(1/n)$

while by Equation [2], we find that

$$\theta_1 = \sin^{-1}\left[n \sin\left(\Phi - \sin^{-1}(1/n)\right)\right]$$

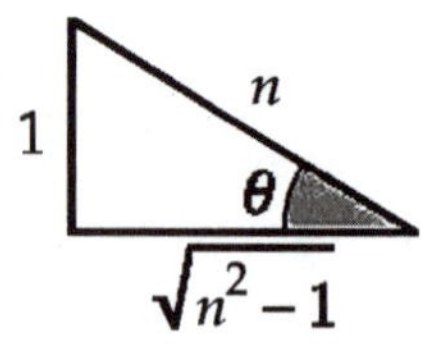

Remember the identity $\sin(\alpha - \beta) = \sin\alpha\cos\beta - \cos\alpha\sin\beta$. Looking at the drawing of a right triangle, prove to yourself that

$$\cos\left[\sin^{-1}(1/n)\right]=\frac{\sqrt{n^2-1}}{n}$$

Our expression for θ_1 therefore simplifies to

$$\theta_1=\sin^{-1}\left[\left(\sqrt{n^2-1}\right)\sin\Phi-\cos\Phi\right]$$ ■

Finalize: We can check this answer by substituting in n = 1.50 and Φ = 60.0°. Then we have $\theta_1=\sin^{-1}\left[\left(\sqrt{1.5^2-1}\right)\sin 60°-\cos 60°\right]=27.9°$, in agreement with the answer to Problem 37 above. The answer to this problem contains infinitely more information. For example, it shows that as n increases and also as Φ increases, θ_1 increases, to make it easier to produce total internal reflection. If the quantity $\left(\sqrt{n^2-1}\right)\sin\Phi-\cos\Phi$ is greater than 1, light can never emerge from the second surface, for any value of θ_1. If the quantity is negative, θ_1 is negative. Then the critical incident ray must be on the other side of the normal to the first surface from that shown, but the equation is still true.

---

41. Consider a common mirage formed by superheated air immediately above a roadway. A truck driver whose eyes are 2.00 m above the road, where n = 1.000 293, looks forward. She perceives the illusion of a patch of water ahead on the road. The road appears wet only beyond a point on the road at which her line of sight makes an angle of 1.20° below the horizontal. Find the index of refraction of the air immediately above the road surface.

Solution

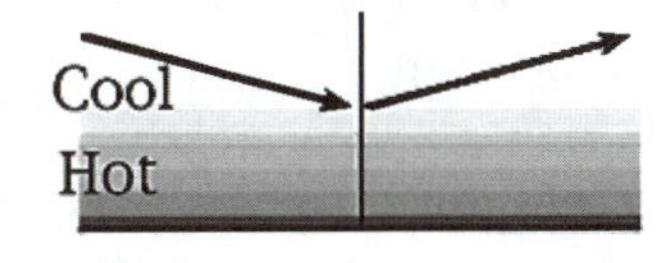

Conceptualize: A wet-road mirage is indeed common. Think of the air as in two discrete layers, the first medium being cooler air on top with n_1 = 1.000 293 and the second medium being hot air with a lower index, which reflects light from the sky by total internal reflection. Then reflection from the cool-air-hot-air interface will mimic reflection from a smooth water surface, at all points where the angle of the trucker's sightline is less than 1.20 degrees with the horizontal, which means at all points more than a certain distance away from the driver.

Categorize: We use the condition for total internal reflection, which is…

Analyze: … $n_1 \sin\theta_1 \geq n_2 \sin 90°$

At the edge of the mirage of wetness $1.000\,293 \sin 88.8° = n_2$

so $n_2 = 1.000\,073\,6$ ■

Finalize: Perhaps a more realistic model is that a nearly vertical wavefront continuously distorts as it moves through the nonuniform medium, with its lower portions moving faster than its upper portions. The direction of propagation (the ray) bends upward continuously as the wave first enters and then leaves sequentially hotter, lower-density layers closer to the road surface.

Do you think that the answer contains too many significant digits? O thou of little faith! A technique using a Michelson interferometer, which you will meet in Chapter 37, makes it possible to measure just the amount by which the index of refraction of a gas differs from the exact value 1. Let this difference be represented by $x = n - 1$. Snell's law of refraction can be put in terms of angles between rays and the interface (so-called grazing angles) instead of angles that rays make with the normal. If in this problem situation we define $\alpha_1 = 1.20°$ and $\alpha_2 = 0$ by $\alpha_1 = 90° - \theta_1$ and $\alpha_2 = 90° - \theta_2$, the condition for total internal reflection becomes

$$(1 + x_1)\cos \alpha_1 \geq (1 + x_2)1$$

We have next $(1 + x_1)(1 - \alpha_1^2/2 + \alpha_1^4/4 + \ldots) \geq 1 + x_2$ for α in radians.

For a small angle this gives simply $x_1 - \alpha_1^2/2 \geq x_2$ and the answer as stated above.

---

45. A small light fixture on the bottom of a swimming pool is 1.00 m below the surface. The light emerging from the still water forms a circle on the water surface. What is the diameter of this circle?

Solution

Conceptualize: Only the light that is directed upwards and hits the water's surface at less than the critical angle will be transmitted to the air so that someone outside can see it. The light that hits the surface farther from the center at an angle greater than θ_c will be totally reflected within the water, and cannot be seen from the outside. From the diagram, the diameter of this circle of light appears to be about 2 m.

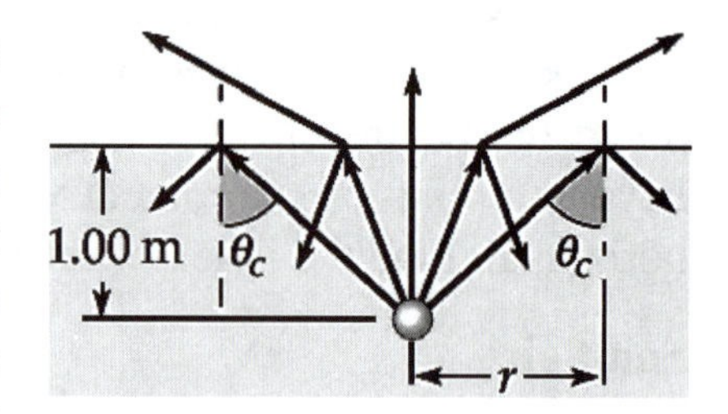

Categorize: We apply Snell's law to find the critical angle, and then find the diameter from geometry.

Analyze: The critical angle is found by imagining the refracted ray just grazing the surface ($\theta_2 = 90°$). The index of refraction of water is $n_1 = 1.333$, and $n_2 = 1.00$ for air, so $n_1 \sin\theta_c = n_2 \sin 90°$ gives $\theta_c = \sin^{-1}(1/1.333) = \sin^{-1}(0.750) = 48.6°$.

The radius then satisfies $\tan\theta_c = \dfrac{r}{1.00 \text{ m}}$

So the diameter is $d = 2r = 2(1.00 \text{ m})(\tan 48.6°) = 2.27 \text{ m}$ ■

Finalize: Only the light rays within a 97.2° cone above the lamp escape the water and can be seen by an outside observer. This angle does not depend on the depth of the light source. The path of a light ray is always reversible, so if a person were located beneath the water, they could see the whole hemisphere above the water surface within this cone; this is a good experiment to try the next time you go swimming!

---

51. A light ray enters the atmosphere of the Earth and descends vertically to the surface a distance $h = 100$ km below. The index of refraction where the light enters the atmosphere is precisely 1, and it increases linearly with distance to have the value $n = 1.000\ 293$ at the Earth's surface. (a) Over what time interval does the light traverse this path? (b) By what percentage is the time interval larger than that required in the absence of the Earth's atmosphere?

Solution

Conceptualize: The electromagnetic wave has its speed drop to lower and lower values. It gets bogged down by having to cause oscillation of charges in molecules of denser and denser air. If the air were absent, the time interval to traverse 100 km would be $h/c = 100$ km/(300 Mm/s) $= 333\ \mu$s.

Because the index of refraction of even the densest air is close to 1, the time for the light to come down through the atmosphere will be just a little longer, by a fraction of one percent.

Categorize: We used the particle under constant velocity model for our estimate in the Conceptualize step, but that will not work for the calculation itself, and neither will the particle under constant acceleration model, because the speed of light changes linearly with distance and not with time. We must drop back to the basic identification of speed as $v = dx/dt$ …

Analyze: … which lets us say that the incremental distance traveled in an incremental time interval is

$$dx = v\,dt = (c/n)dt$$

Then the time elapsing as the light moves down by distance dx is

$$dt = n\,dx/c \qquad \textbf{[1]}$$

Here n is a function of x while c is of course constant. We choose to let x represent the distance traversed downward from the top of the atmosphere. From the description in the problem we can identify the functional form of n as

$$n = 1.000\ 000 + (0.000\ 293/100\text{ km})\,x = 1 + (0.000\ 293/h)x$$

just because this is the linear function that fits the boundary conditions that n is 1 exactly at $x = 0$ and $n = 1.000\ 293$ at $x = 100$ km.

Now the whole time for the light to pass vertically through the atmosphere is given by integrating both sides of our equation [1] from the entry of the light at the top to its arrival at the bottom:

$$\Delta t = \int_0^t dt = \int_0^h n\,dx/c = \frac{1}{c}\int_0^{100\text{ km}} \left(1 + \frac{0.000\ 293}{h}x\right)dx$$

$$\Delta t = \frac{1}{c}\left(x + \frac{0.000\ 293}{h}\frac{x^2}{2}\right)\Bigg|_0^h = \frac{h}{c} + \frac{0.000\ 293}{ch}\frac{h^2}{2} = \frac{1.00\times10^5\text{ m}}{3.00\times10^8\text{ m/s}}\left(1 + \frac{0.000\ 293}{2}\right)$$

$$= 3.33\times10^{-4}\text{ s} + 4.88\times10^{-8}\text{ s}$$

So to three significant digits we have 3.33×10^{-4} s ■

(b) As noted in the Conceptualize step, the time interval without air would be h/c, so the fractional excess time attributable to the presence of the air is

$$\frac{\Delta t - h/c}{h/c} = \frac{\frac{h}{c} + \frac{0.000\,293\,h}{2\,c} - \frac{h}{c}}{\frac{h}{c}} = \frac{0.000\,293}{2} \times 100\% = 0.014\,6\%$$ ■

Finalize: For the actual atmosphere, the excess travel time might be hard to measure directly. But we will see in Chapter 37, about interference, that remarkably precise measurements can be made of the number of extra wavelengths traveled by light going through a layer of glass or even of air, compared to light going through vacuum.

---

57. The light beam in Figure P35.57 strikes surface 2 at the critical angle. Determine the angle of incidence θ_1.

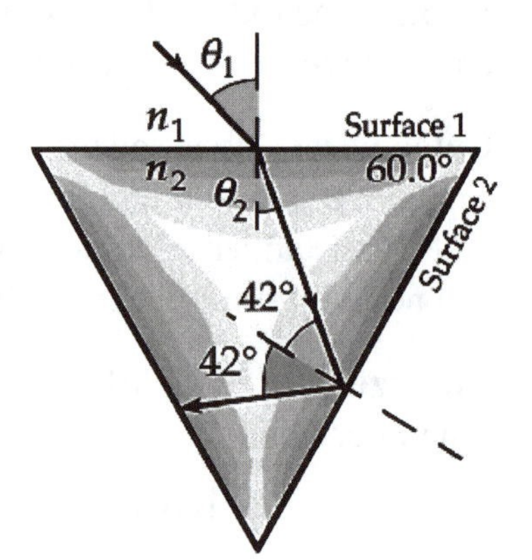

Figure P35.57

Solution

Conceptualize: From the diagram it appears that the angle of incidence is about 40°.

Categorize: We can find θ_1 by applying Snell's law at the first interface where the light is refracted. At surface 2, knowing that the 42.0° angle of reflection is the critical angle, we can work backwards to find θ_1.

Analyze: Define n_1 to be the index of refraction of the surrounding medium and n_2 to be that for the prism material. We can use the critical angle of 42.0° to find the ratio n_2/n_1:

$$n_2 \sin 42.0° = n_1 \sin 90.0°$$

So, $\frac{n_2}{n_1} = \frac{1}{\sin 42.0°} = 1.49$

(If we had started with the assumption $n_1 = 1.000$, we would have $n_2 = 1.49$, and the remainder of the solution would proceed with no real change, to give the same answer that we will obtain.)

Call the angle of refraction θ_2 at the surface 1. The ray inside the prism forms a triangle with surfaces 1 and 2, so the sum of the interior angles of this triangle must be 180°.

Thus, $(90.0° - \theta_2) + 60.0° + (90.0° - 42.0°) = 180°$

Therefore, $\theta_2 = 18.0°$

Applying Snell's law at surface 1, $n_1 \sin\theta_1 = n_2 \sin\theta_2$

$$\theta_1 = \sin^{-1}\left(\frac{n_2 \sin\theta_2}{n_1}\right) = \sin^{-1}(1.49 \sin 18.0°) = 27.5°$$ ■

Finalize: The result is a bit less than the 40.0° we expected, but this is probably because the figure is not drawn to scale. This problem was a bit tricky because it required four key concepts (refraction, reflection, critical angle, and geometry) in order to find the solution. One practical extension of this problem is to consider what would happen to the exiting light if the angle of incidence were varied slightly. Would all the light still be reflected off surface 2, or would some light be refracted and pass through this second surface? See Figure 35.27 in the textbook.

---

60. A light ray of wavelength 589 nm is incident at an angle θ on the top surface of a block of polystyrene as shown in Figure P35.60. (a) Find the maximum value of θ for which the refracted ray undergoes total internal reflection at the point *P* located at the left vertical face of the block. **What If?** Repeat the calculation for the case in which the polystyrene block is immersed in (b) water and (c) carbon disulfide. Explain your answers.

Solution

Conceptualize: If θ_1 is too large, θ_2 will be too large and θ_3 too small for total internal reflection to happen at the vertical surface. We might estimate an angle not far from 45°.

Categorize: We can find θ_3 from the condition for total internal reflection. Then we work backwards to find θ_2 and θ_1.

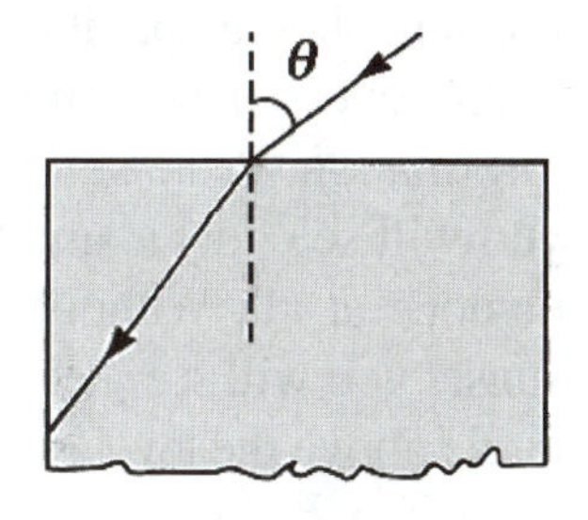

Figure P35.60

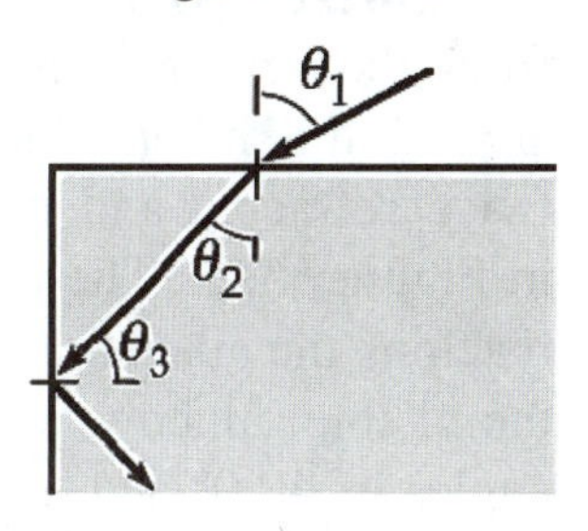

Analyze: We look up the index of refraction (for 589 nm light) for each material as listed here:

Air	1.00
Water	1.33
Polystyrene	1.49
Carbon disulfide	1.63

(a) For polystyrene **surrounded by air**, the critical angle for total internal reflection is $\theta_3 = \sin^{-1}(1/1.49) = 42.2°$

and then from geometry, $\theta_2 = 90.0° - \theta_3 = 47.8°$

From Snell's law, $\sin\theta_1 = 1.49 \sin 47.8° = 1.10$

This has no solution; thus, the real maximum value for θ_1 is 90.0°.

Total internal reflection always occurs. ■

(b) For polystyrene **surrounded by water**, we have $\theta_3 = \sin^{-1}\left(\dfrac{1.33}{1.49}\right) = 63.2°$

and $\theta_2 = 26.8°$

From Snell's law, $n_1 \sin\theta_1 = n_2 \sin\theta_2$: $1.33 \sin\theta_1 = 1.49 \sin 26.8°$

and $\theta_1 = 30.3°$ ■

(c) Total internal reflection at the second surface is not possible because the beam is initially traveling in polystyrene, a medium of lower index of refraction. When the block is surrounded by carbon disulfide, total internal reflection never happens. ■

Finalize: Note that $\sin\theta = 1.10$ does not describe an angle larger than 90°. It describes no angle whatsoever. We had to think to figure out the answer to part (a). It was not what we guessed. Thinking is a good habit in any case.

---

71. A hiker stands on an isolated mountain peak near sunset and observes a rainbow caused by water droplets in the air at a distance of 8.00 km along her line of sight to the most intense light from the rainbow. The valley is 2.00 km below the mountain peak and entirely flat. What fraction of the complete circular arc of the rainbow is visible to the hiker?

Solution

Conceptualize: The textbook's chapter-opening photograph, its Figures 35.23 and 35.24, and our diagram contain a lot of information. If the sun is high in the sky you will see only a small arc of rainbow, or none at all. Without the mountain, at sunset you will see just a 180° arc, a half-circle above the level ground.

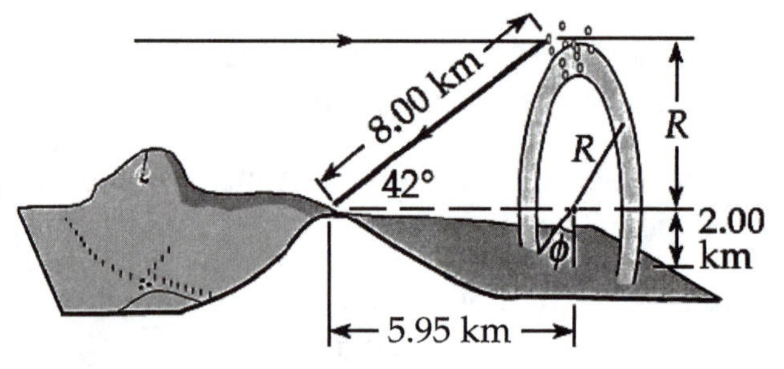

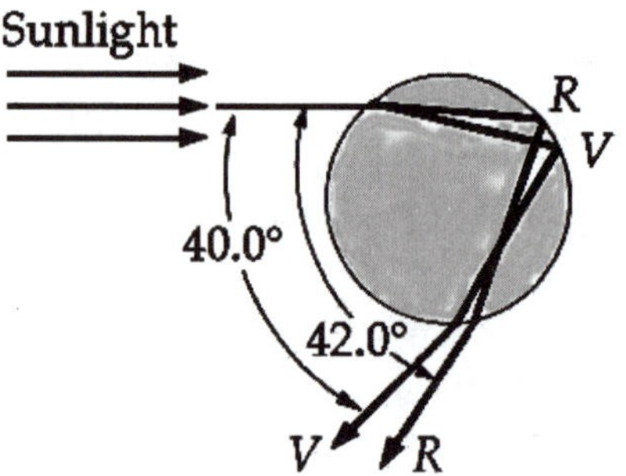

Figure 35.23

Categorize: Horizontal light rays from the setting Sun pass above and around the hiker. The light rays are twice refracted and once reflected, as in Figure 35.23. The most intense light reaching the hiker, that which represents the visible rainbow, is located between angles of 40.0° and 42.0° from the hiker's shadow. The hiker sees a greater percentage of the violet inner edge, so we consider the red outer edge. The radius R of the circle of droplets is ...

Analyze: $$R = (8.00\text{ km})\sin 42.0° = 5.35\text{ km}$$

Then the angle ϕ between the vertical and the radius where the bow touches the ground is given by

$$\cos\phi = \frac{2.00\text{ km}}{R} \quad \text{so} \quad \phi = \cos^{-1}\left(\frac{2.00\text{ km}}{5.35\text{ km}}\right) = \cos^{-1}0.374 = 68.1°$$

The angle filled by the visible bow is $360° - 2(68.1°) = 224°$, so the visible bow is

$$\frac{224°}{360°} = 62.2\% \text{ of a circle}$$ ■

Finalize: This striking view motivated Charles Wilson's 1906 invention of the cloud chamber, a standard tool of nuclear physics. The effect is mentioned in the Bible, Ezekiel 1:28. Look for a full circle of color around your shadow when you fly in an airplane. With a stepladder over a lawn sprinkler you can show children a full-circle rainbow when the summer sun is high in the sky. Do not let the wet children fall from the ladder.

---

75. Refer to Problem 74 for the statement of Fermat's principle of least time. Derive the law of reflection (Eq. 35.2) from Fermat's principle.

Solution

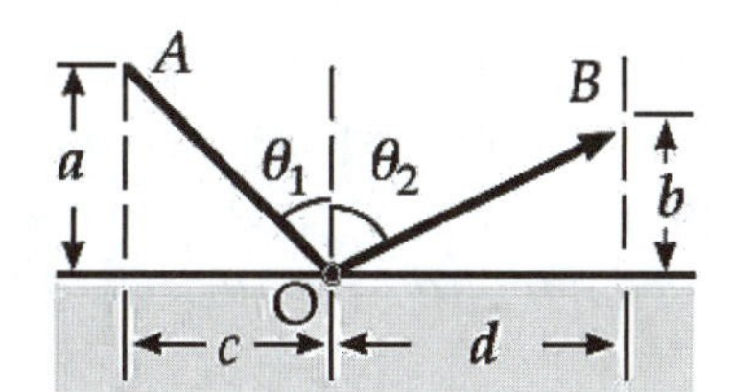

Conceptualize: For light getting from one point A to another point B by reflecting from a mirror along the way, we must prove that the path described by the law of reflection requires the least time for the light to travel. Since the light travels in a single medium at constant speed, we must prove this path is shorter than other paths that the light could follow if it could reflect following some rule different from $\theta_1 = \theta_2$. We can take the mirror as horizontal and the points A and B as above it. If the point of reflection were not along the line segment between the points on the mirror below A and B, the path length would surely be longer than necessary.

Categorize: Let point O be on the mirror surface as shown in the diagram. Let $c + d = R$, so that we can replace d with $R - c$. We want to minimize the distance of travel $D = \sqrt{a^2 + c^2} + \sqrt{b^2 + (R - c)^2}$. We can do this by differentiating with respect to c and setting the derivative equal to zero.

Analyze: From $$D = \sqrt{a^2 + c^2} + \sqrt{b^2 + (R - c)^2}$$

We require that $\frac{dD}{dc} = 0$: $$\frac{2c}{\sqrt{a^2 + c^2}} - \frac{2(R - c)}{\sqrt{b^2 + (R - c)^2}} = 0$$

By inspecting the triangles of the diagram, it follows that

$$\sin\theta_1 = \frac{c}{\sqrt{a^2 + c^2}}$$

and $$\sin\theta_2 = \frac{R - c}{\sqrt{b^2 + (R - c)^2}}$$

Substituting these values, $\sin\theta_1 = \sin\theta_2$

Therefore, $\theta_1 = \theta_2$ and we have derived the law of reflection. ■

Finalize: We think of the locations of points A and B relative to the reflecting surface as given, so we are allowed to think of their coordinates a, R, and b as constants when we take the derivative. After that, the derivation went through pretty quickly. Many physical laws can be put in terms of variational principles. Some people think of the variational principles as the most fundamental laws.

36

Image Formation

EQUATIONS AND CONCEPTS

In using the equations related to the image-forming properties of spherical mirrors, spherical refracting surfaces, and thin lenses, you must be very careful to use the correct algebraic sign for each physical quantity. *The sign conventions appropriate for the equation forms stated here are summarized in the Suggestions, Skills, and Strategies section.*

Lateral magnification is defined as the ratio of image height to the object height. *This ratio always has a value of +1 for a plane mirror since the erect image is always the same size as the object.*

$$M \equiv \frac{\text{image height}}{\text{object height}} = \frac{h'}{h} \quad (36.1)$$

The **lateral magnification of a spherical mirror** can be stated either as a ratio of image size to object size or in terms of the ratio of image distance to object distance. *A negative value of magnification indicates an inverted image.*

$$M = \frac{h'}{h} = -\frac{q}{p} \quad (36.2)$$

The **focal length of a spherical mirror** is the distance from the vertex of the mirror to the focal point, F, located midway between the center of curvature and the vertex of the mirror. *For a concave mirror, the focal point is in front of the mirror and f is positive. For a convex mirror, the focal point is back of the mirror and f is negative.*

$$f = \frac{R}{2} \quad (36.5)$$

R = radius of curvature

The **mirror equation** is used to locate the position of an image formed by reflection of paraxial rays.

$$\frac{1}{p} + \frac{1}{q} = \frac{1}{f} \quad \begin{pmatrix} p = \text{object distance} \\ q = \text{image distance} \end{pmatrix} \quad (36.6)$$

A **single spherical refracting surface** of radius R which separates two media whose indices of refraction are n_1 and n_2 will form an image of an object. *This equation is valid regardless of the relative values of the indices of refraction.*

$$\frac{n_1}{p} + \frac{n_2}{q} = \frac{n_2 - n_1}{R} \quad (36.8)$$

A **flat refracting surface** ($R = \infty$) forms a virtual image that is on the same side of the refracting surface as the object.

$$q = -\frac{n_2}{n_1}p \tag{36.9}$$

Several **combinations of lens parameters and image characteristics for a thin lens** are shown in the following equations.

Features of the **geometry of a thin lens** are shown in the figures at right. A lens can be considered "thin" when the lens thickness is much less that the radii of the surfaces (R_1 and R_2). A given radius is positive if the center of curvature is in back of the lens and negative if the center of curvature is in front of the lens. In the figure, R_1 is positive and R_2 is negative. A convex lens (bottom left) converges incident parallel light rays to the focal point (F_2). Parallel light rays entering a concave lens (bottom right) appear to diverge from the focal point (F_1).

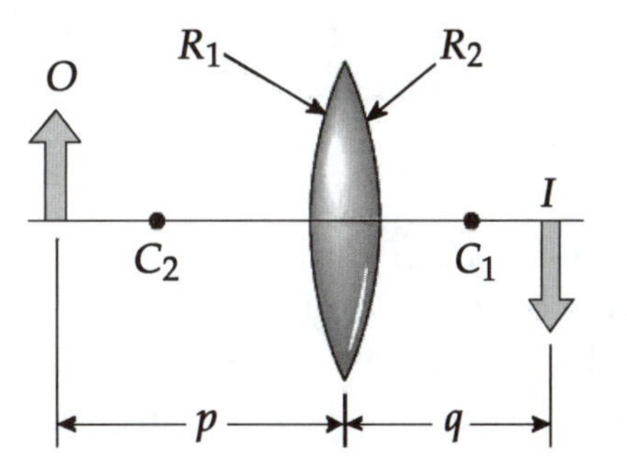

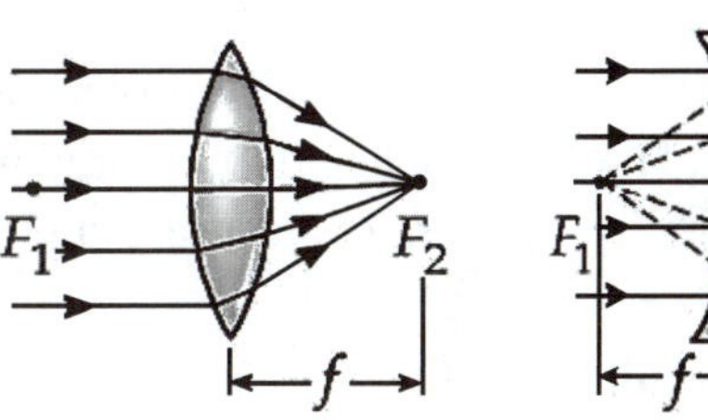

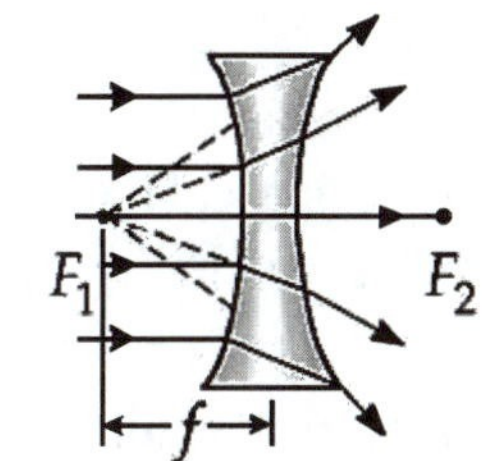

The **lens maker's equation** relates the focal length to the physical properties of the lens. *If the lens is surrounded by a medium other than air, the index of refraction given in Equation 36.14 and Equation 36.15 must be the ratio of the index of refraction of the lens to that of the surrounding medium. A hollow convex lens ("air lens"), if immersed in water, would have a negative focal length.*

$$\frac{1}{p} + \frac{1}{q} = (n-1)\left(\frac{1}{R_1} - \frac{1}{R_2}\right) \tag{36.14}$$

$$\frac{1}{f} = (n-1)\left(\frac{1}{R_1} - \frac{1}{R_2}\right) \tag{36.15}$$

Algebraic signs of R_1 and R_2 are determined as described in the paragraph above.

The **thin lens equation** can be used to find the image location when the focal length and the object distance are known.

$$\frac{1}{p} + \frac{1}{q} = \frac{1}{f} \tag{36.16}$$

The **lateral magnification** of a thin lens is given by the same equation as that of a spherical mirror. *The overall magnification of an image formed by two optical elements (e.g., lenses, mirrors, or lens-mirror combination) equals the product of the two individual magnifications.*

$$M = \frac{h'}{h} = -\frac{q}{p} \tag{36.17}$$

$$M = M_1 M_2 \tag{36.18}$$

Two thin lenses in contact are equivalent to a single lens with a focal length given by Equation 36.19. *If the two lenses are of the same type (converging or diverging), the equivalent focal length will be less than the focal length of either lens.*

$$\frac{1}{f} = \frac{1}{f_1} + \frac{1}{f_2} \tag{36.19}$$

Image forming properties of several optical instruments are described in the following paragraphs and equations.

Camera:
The **light intensity** (I) incident on a film is inversely proportional to the square of the ratio of the focal length of the lens to its diameter. This ratio is called the ***f*-number**.

$$f\text{-number} \equiv \frac{f}{D} \tag{36.20}$$

$$I \propto \frac{1}{(f\text{-number})^2} \tag{36.21}$$

Eye:
The **power of a lens**, measured in diopters, is the reciprocal of the focal length measured in meters (including the correct algebraic sign).

$$P = \frac{1}{f}$$

Simple magnifier:
Angular magnification is the ratio of the angle subtended by an object using a lens to the angle subtended by the object at the near point without a lens.

$$m \equiv \frac{\theta}{\theta_0} \tag{36.22}$$

When the **image is at the near point** (25 cm), the angular magnification is maximum.

$$m_{\text{max}} = 1 + \frac{25\text{ cm}}{f} \tag{36.24}$$

When the **image is at infinity** (most relaxed for the eye), the angular magnification is minimum.

$$m_{\text{min}} = \frac{25\text{ cm}}{f} \tag{36.25}$$

Compound microscope:
The **overall magnification** is the product of the lateral magnification (M_o) of the objective lens and the angular magnification (m_e) of the eyepiece.

$$M = M_o m_e = -\frac{L}{f_o}\left(\frac{25\text{ cm}}{f_e}\right) \tag{36.26}$$

L = distance between lenses

Astronomical telescope:
The **angular magnification** is equal to the ratio of the objective focal length to the eyepiece focal length. *The two converging lenses are separated by a distance equal to the sum of their focal lengths.*

$$m_e = -\frac{f_o}{f_e} \tag{36.27}$$

SUGGESTIONS, SKILLS, AND STRATEGIES

A major portion of this chapter is devoted to the development and presentation of equations which can be used to determine the location and nature of images formed by various optical components acting either singly or in combination. It is essential that these equations be used with the correct algebraic sign associated with each quantity involved. You must understand clearly the sign conventions for mirrors, refracting surfaces, and lenses. A review of equations, sign conventions, and illustrations of image characteristics for spherical mirrors, refracting surfaces, and thin lenses begins on the following page.

SIGN CONVENTIONS FOR SPHERICAL MIRRORS

Equations: $\frac{1}{p}+\frac{1}{q}=\frac{1}{f}=\frac{2}{R}$; $M=\frac{h'}{h}=-\frac{q}{p}$

The front side of the mirror is the region on which light rays are incident and reflected.

p is + if the object is in front of the mirror (real object).
p is – if the object is in back of the mirror (virtual object).

q is + if the image is in front of the mirror (real image).
q is – if the image is in back of the mirror (virtual image).

Both f and R are + if the center of curvature is in front (concave mirror).
Both f and R are – if the center of curvature is in back (convex mirror).

If M is positive, the image is upright.
If M is negative, the image is inverted.

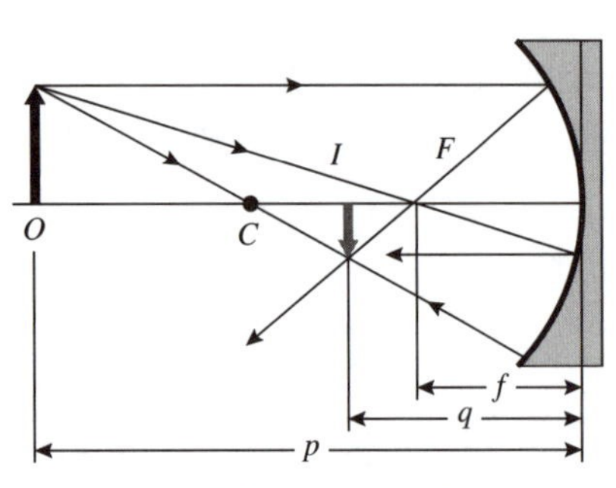

(a) Concave Mirror ($p > 2f$)
Image: real, inverted, diminished

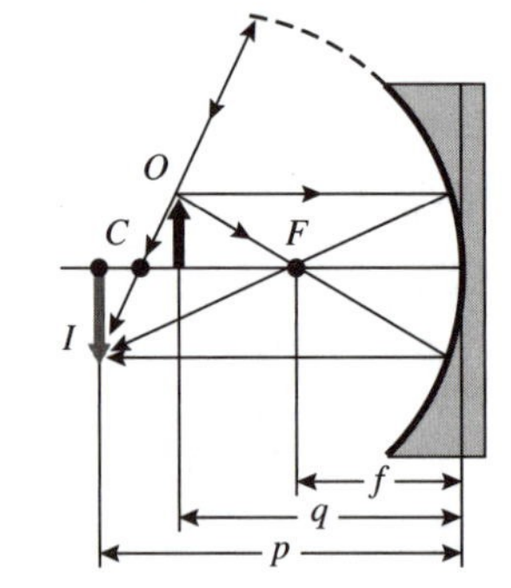

(b) Concave Mirror ($2f > p > f$)
Image: real, inverted, enlarged

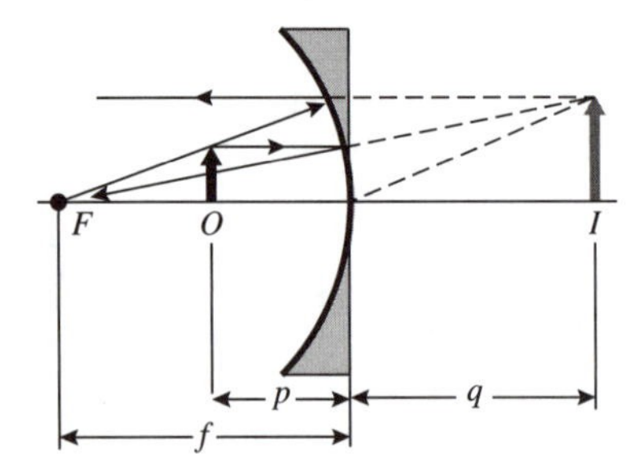

(c) Concave Mirror ($p < f$)
Image: virtual, upright, enlarged

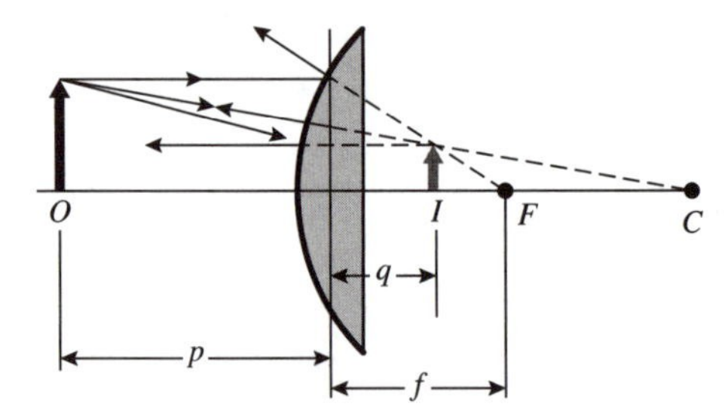

(d) Convex Mirror (any value of p)
Image: virtual, upright, diminished

SIGN CONVENTIONS FOR REFRACTING SURFACES

Equations: $\dfrac{n_1}{p}+\dfrac{n_2}{q}=\dfrac{n_2-n_1}{R}$; $M=\dfrac{h'}{h}=-\dfrac{n_1 q}{n_2 p}$

In the following table, the **front** side of the surface is the side **from which the light is incident**.

p is + if the object is in front of the surface (real object).
p is – if the object is in back of the surface (virtual object).

q is + if the image is in back of the surface (real image).
q is – if the image is in front of the surface (virtual image).

R is + if the center of curvature is in back of the surface.
R is – if the center of curvature is in front of the surface.

n_1 refers to the index of refraction of the first medium (before refraction).
n_2 is the index of refraction of the second medium (after refraction).

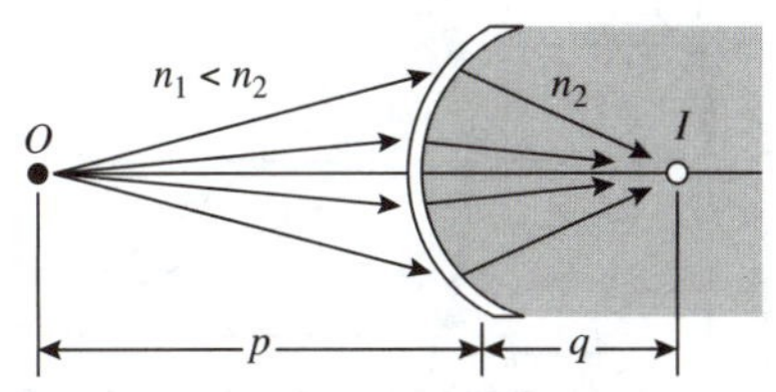

(a) Convex Refracting Surface
(object outside surface)
Image: real, inside surface

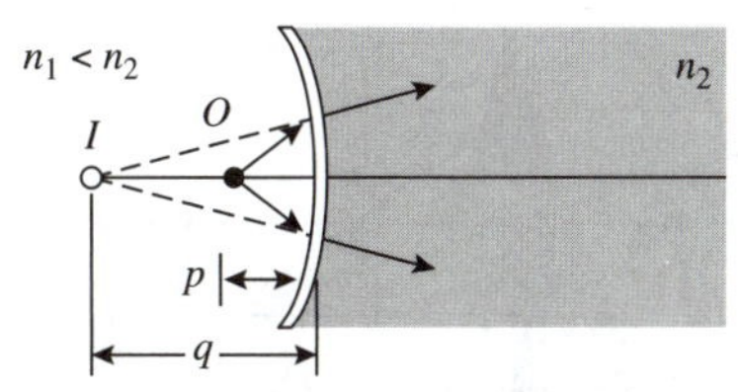

(b) Concave Refracting Surface
(object outside surface)
Image: virtual, outside surface

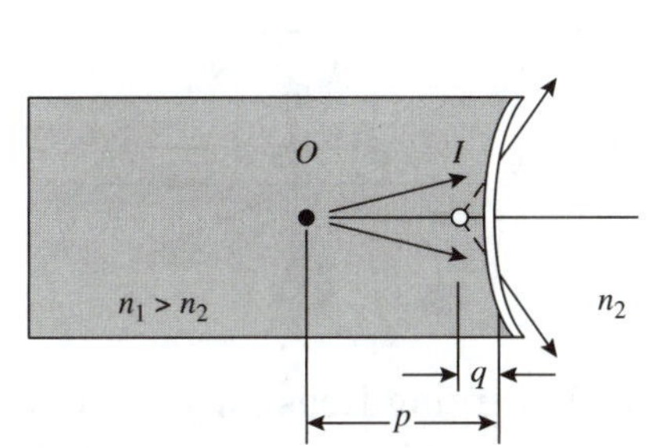

(c) Concave Refracting Surface
(object inside surface)
Image: virtual, inside surface

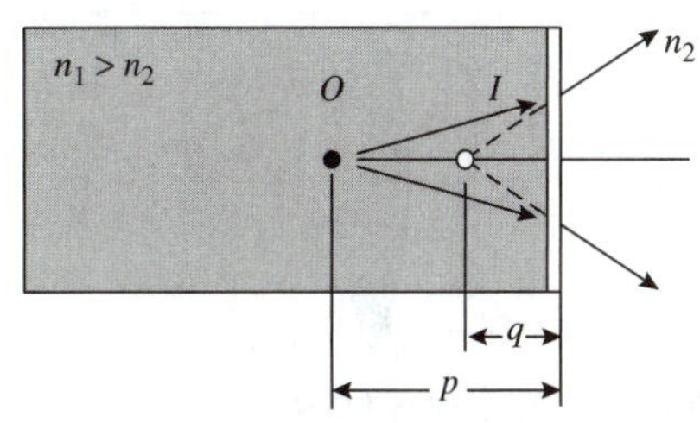

(d) Flat Refracting Surface
(object inside surface)
Image: virtual, inside surface

SIGN CONVENTIONS FOR THIN LENSES

Equations: $\dfrac{1}{p}+\dfrac{1}{q}=\dfrac{1}{f}=(n-1)\left(\dfrac{1}{R_1}-\dfrac{1}{R_2}\right);\quad M=\dfrac{h'}{h}=-\dfrac{q}{p}$

In the following table, the **front** of the lens is the **side from which the light is incident.**

p is + if the object is in front of the lens. p is – if the object is in back of the lens. q is + if the image is in back of the lens. q is – if the image is in front of the lens. f is + if the lens is thickest at the center. f is – if the lens is thickest at the edges. R_1 and R_2 are + if the center of curvature is in back of the lens. R_1 and R_2 are – if the center of curvature is in front of the lens.

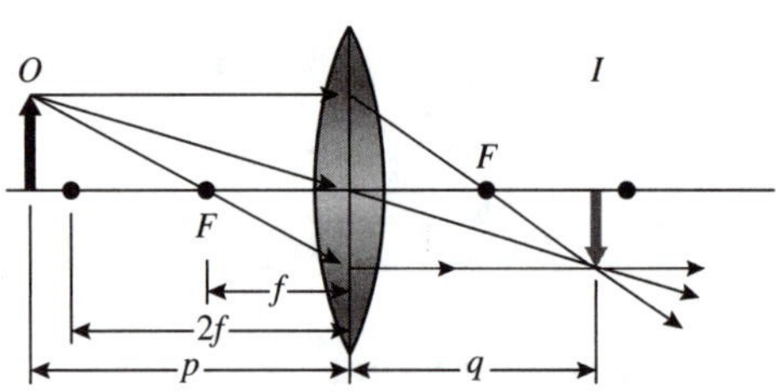

(a) Converging Lens ($p > 2f$)
Image: real, inverted, diminished

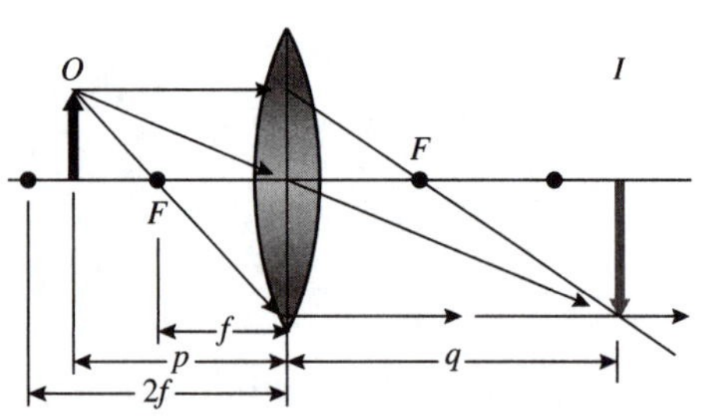

(b) Converging Lens ($2f > p > f$)
Image: real, inverted, enlarged

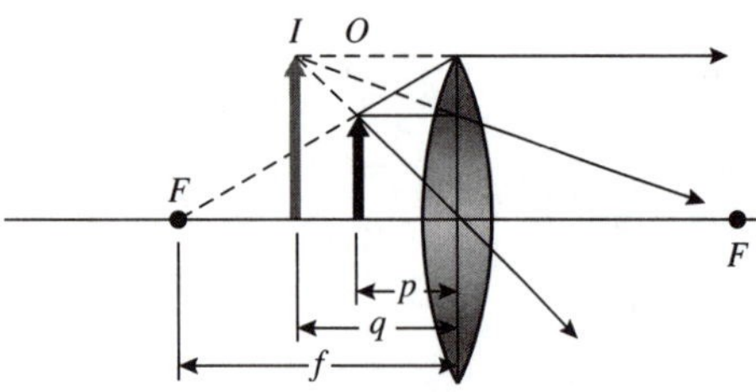

(c) Converging Lens ($p < f$)
Image: virtual, upright, enlarged

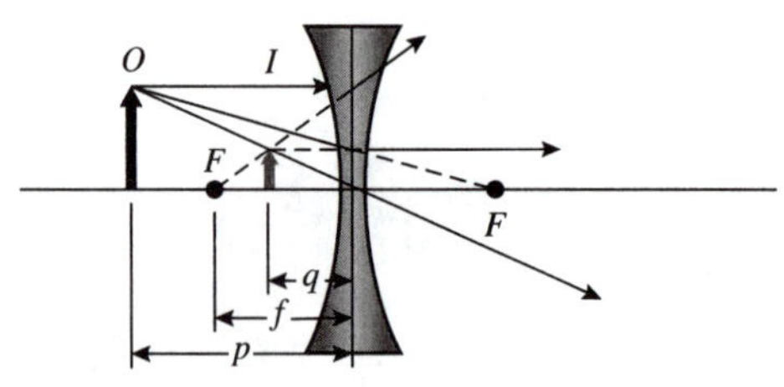

(d) Diverging Lens (any value of p)
Image: virtual, upright, diminished

REVIEW CHECKLIST

You should be able to:

- Correctly use the required equations and associated sign conventions to calculate the location of the image of a specified object as formed by a plane mirror, spherical mirror, plane refracting surface, spherical refracting surface, thin lens, or a combination of two or more of these devices. Determine the magnification and character of the image (real or virtual, upright or inverted, enlarged or diminished) in each case. (Sections 36.1, 36.2, 36.3, and 36.4)

- Construct ray diagrams to determine the location and nature of the image of a given object when the geometrical characteristics of the optical device (lens or mirror) are known. (Sections 36.2 and 36.4)

- Make calculations of magnification for a simple magnifier, compound microscope, and refracting telescope. (Sections 36.8, 36.9, and 36.10)

ANSWER TO AN OBJECTIVE QUESTION

8. Lulu looks at her image in a makeup mirror. It is enlarged when she is close to the mirror. As she backs away, the image becomes larger, then impossible to identify when she is 30.0 cm from the mirror, then upside down when she is beyond 30.0 cm, and finally small, clear, and upside down when she is much farther from the mirror. **(i)** Is the mirror (a) convex, (b) plane, or (c) concave? **(ii)** Is the magnitude of its focal length (a) 0, (b) 15.0 cm, (c) 30.0 cm, (d) 60.0 cm, or (e) ∞ ?

Answer **(i)** (c) A plane mirror always produces a rightside-up, actual-size image. A convex mirror always produces a rightside-up image of a real object, diminished in size. A concave mirror has the property of producing an enlarged, rightside-up image of an object closer than its focal point, and producing an inverted image of an object beyond its focal point. **(ii)** (c) For an object at the focal point, the image is at infinity, or the reflected rays originating from one point on the object are parallel, or no image is formed.

□ □ □ □

ANSWERS TO SELECTED CONCEPTUAL QUESTIONS

6. Explain why a fish in a spherical goldfish bowl appears larger than it really is.

Answer As in the diagram, let the center of curvature *C* of the fishbowl and the bottom point of the fish define the optical axis, intersecting the fishbowl at vertex *V*. A ray from the top point of the fish that reaches the bowl along a radial line through C has angle of incidence zero and angle of refraction zero. This ray exits the bowl unchanged in direction. A ray from the top of the fish to V is refracted to bend away from the normal. Its extension back inside the fishbowl determines the location of the image and the characteristics of the image. It is upright, virtual, and enlarged.

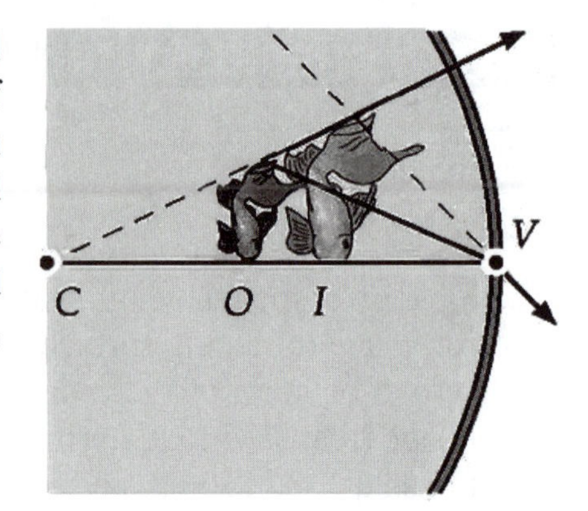

□ □ □ □

8. Lenses used in eyeglasses, whether converging or diverging, are always designed so that the middle of the lens curves away from the eye like the center lenses of Figures 36.25a and 36.25b. Why?

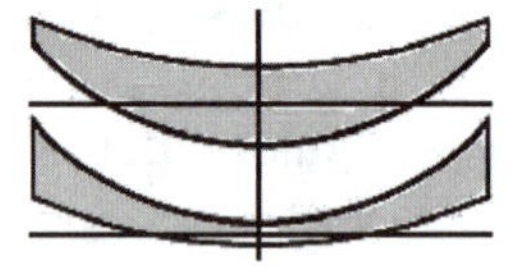

Answer With this so-called meniscus design, when you direct your gaze near the outer circumference of the lens you receive a ray that has passed through glass with more nearly parallel surfaces of entry and exit. Thus, the lens minimally distorts the direction to the object you are looking at. If you wear glasses, you can demonstrate this by turning them around and looking through them the wrong way, maximizing the distortion.

□ □ □ □

SOLUTIONS TO SELECTED PROBLEMS

3. Determine the minimum height of a vertical flat mirror in which a person 178 cm tall can see his or her full image. *Suggestion:* Drawing a ray diagram would be helpful.

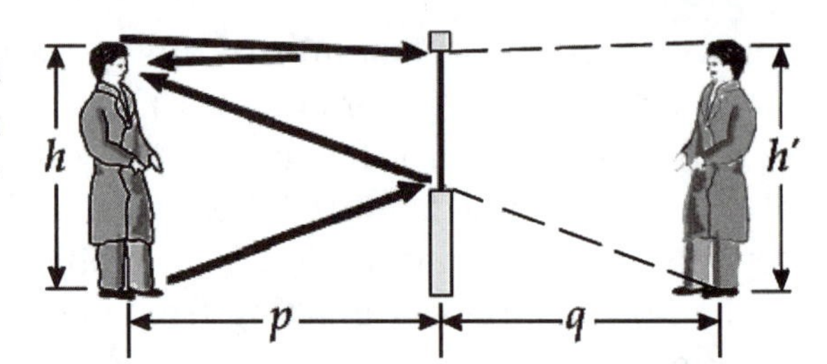

Solution

Conceptualize: Almost anyone looking for a full-length mirror would look for one around 2 meters tall. Many people think that their image lies on the mirror surface.

Categorize: We will use general techniques based on the mirror-lens equation, the magnification equation, and a diagram.

Analyze: The flatness of the mirror is described by $R = \infty$, $f = \infty$, and $1/f = 0$. By our general mirror and lens equation,

$$\frac{1}{p} + \frac{1}{q} = \frac{1}{f} \quad \text{we then have} \quad q = -p$$

Thus, the image is as far behind the mirror as the person is in front. The magnification is then

$$M = \frac{-q}{p} = 1 = \frac{h'}{h} \quad \text{so} \quad h' = h' = 178 \text{ cm}$$

The required height of the mirror is defined by the triangle from the person's eyes to the top and bottom of the image, as shown. From the geometry of the triangle, we see that the mirror height must be:

$$h'\left(\frac{p}{p-q}\right) = h'\left(\frac{p}{2p}\right) = \frac{h'}{2}$$

Thus, the mirror must be at least 89.0 cm high. ■

Finalize: On the bedroom or hallway wall you would really want a mirror a bit taller than 90 cm, to accommodate people whose eyes are different distances from the floor. But the mirror only has to be half as wide as your shoulders. Knowing some physics saves you some money.

Look back at the first line of the Analyze section. The flatness of the mirror is perfectly real, and the equation $R = \infty$ is a perfectly real and meaningful mathematical description of it. The variables p, q, h, h', R, f, and M range over an "augmented" number line, incorporating the usual real numbers and a single extra infinite value. An object at $p = \infty$, for example, is the source of incoming light rays that can be modeled as parallel. You do not need an ∞ key on your calculator, but recognize that $1/\infty = 0$.

---

7. A concave spherical mirror has a radius of curvature of magnitude 20.0 cm. (a) Find the location of the image for object distances of **(i)** 40.0 cm, **(ii)** 20.0 cm, and **(iii)** 10.0 cm. For each case state whether the image is (b) real or virtual and (c) upright or inverted. (d) Find the magnification in each case.

Solution

Conceptualize: It is a good idea to draw a principal-ray diagram as early as possible in any optics problem. This gives a qualitative sense of how the image appears relative to the object. From the ray diagrams below, we see that when the object is 40 cm from the mirror, the image is real, inverted, diminished, and located a bit more than 10 cm from the mirror. When the object is at 20 cm, the image is real, inverted, actual-size, and also at 20 cm. When the object is at 10 cm, the reflected rays do not intersect, so we can say that no image is formed.

Categorize: The mirror equation gives a precise value for the image distance. Then $M = -q/p$ gives a precise value for the magnification. Together they complement the ray diagrams.

Analyze: We apply the mirror equation using the sign conventions listed in *Suggestions, Skills, and Strategies.*

(i) (a) $1/p + 1/q = 2/R$ becomes

$$q = \frac{1}{2/R - 1/p} = \frac{1}{2/(20.0\text{ cm}) - 1/(40.0\text{ cm})} = 13.3\text{ cm}$$ ■

(b) The positive value for q indicates that the image is real. ■

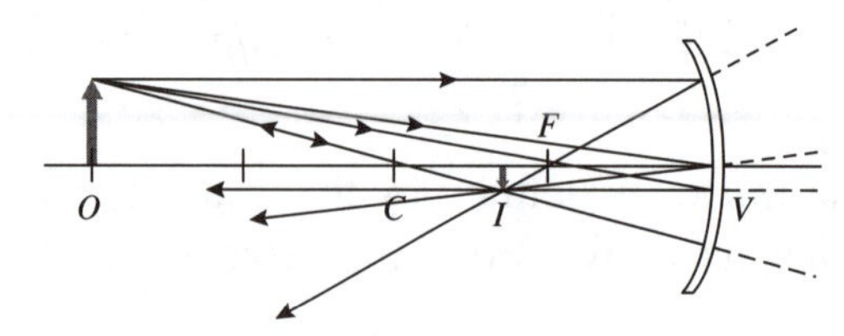

The ray diagram shows this identification more clearly, and (c) that the image is inverted. ■

(d) The magnification is $M = -\dfrac{q}{p} = -\dfrac{13.3\text{ cm}}{40.0\text{ cm}} = -0.333$ ■

Its value indicates that the image is inverted and one-third the height of the object.

(ii) (a) The object is now at the center of curvature. Following the same steps gives

$$q = \frac{1}{2/R - 1/p} = \frac{1}{2/(20.0\text{ cm}) - 1/(20.0\text{ cm})} = 20.0\text{ cm}$$ ■

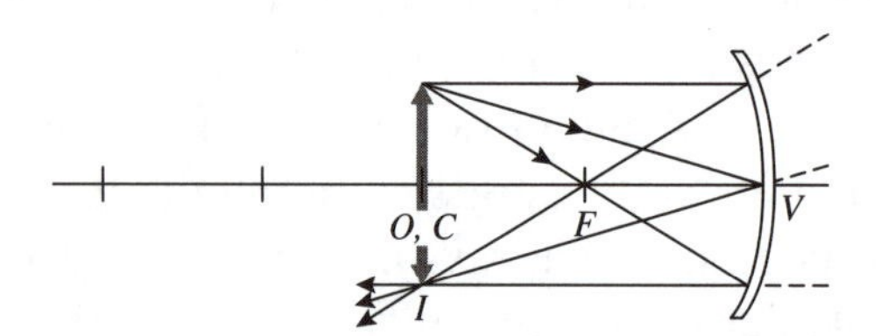

(b) The image is real, as shown by the ray diagram and by the positive value for q. ■

(c) The ray diagram shows that the image is inverted. ■

(d) The magnification is $M = -\dfrac{q}{p} = -\dfrac{20.0\text{ cm}}{20.0\text{ cm}} = -1.00$ ■

Its value indicates that the image is inverted and the same height as the object in this special case.

(iii) (a) The object is now at the focal point of the mirror. Following the same steps gives

$$q = \frac{1}{2/R - 1/p} = \frac{1}{2/(20.0\text{ cm}) - 1/(10.0\text{ cm})} = \frac{1}{0} = \infty$$

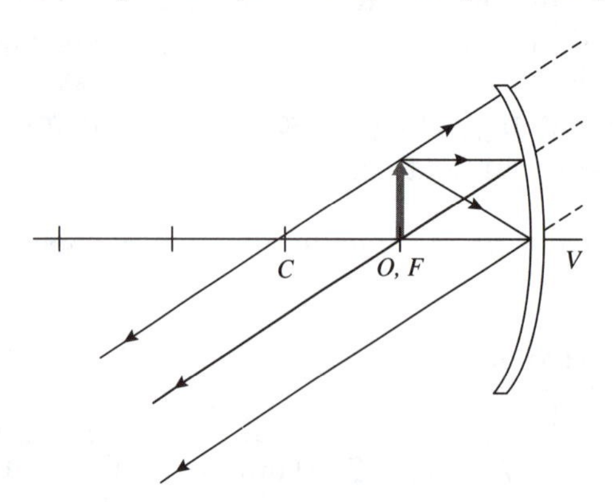

We can say that no image is formed, or that the image is at an infinite distance. ■

(b) In this special case the reflected rays do not intersect. We cannot classify the image as real or virtual. ■

(c) We cannot classify the image as upright or inverted. ■

A screen placed at a large distance in front of the mirror can intercept the reflected light energy, showing the appearance of an upside-down real image, but it is not sharp for any finite distance. You can look into the mirror to view the image as a rightside-up virtual image, with your eye focused on infinity. ■

(d) The magnification is $M = -\frac{q}{p} = -\frac{\infty}{20.0\text{ cm}} = \infty$ ■

In this special case, if we say no image is formed at a finite distance, it has no finite magnification. If we say the image is at infinity, then its height and its magnification are also infinite. There is no physical difference between $+\infty$ and $-\infty$.

Finalize: The calculations of image characteristics agree well with the conclusions from our ray diagrams. Especially in trial (iii), the ray diagram helps to explain the meaning of the calculated results. It is easy to miss a minus sign or to make a computational mistake in using the mirror-lens equation, so the characteristics and approximate values obtained from the ray diagrams are useful as a check on the calculated values.

Each of the three situations has a practical application. Situation (i) can be considered a model for a parabolic microphone, used for concentrating the sound of a bird on an electronic transducer. If the object distance were much greater than the radius of curvature, it would be clearly a model for a satellite dish, a reflecting telescope, and a solar-energy stove. Situation (ii) is used for hallway displays in physics buildings. The image of a lightbulb looks just like a lightbulb when your eyes intercept the reflected rays as they diverge from the image after converging to it; but you can move your finger through the image. Situation (iii) models a searchlight.

In the ray diagram for situation (ii), with the object at the center of curvature, it is not possible to draw a meaningful ray from O to C. In situation (iii) we could not draw a ray from O to F. So in all of the diagrams we supplement the rays mentioned in the textbook with a ray from the top of the object to point V, the mirror vertex or origin of coordinates. This ray reflects making an equal angle on the other side of the axis.

---

9. A convex spherical mirror has a radius of curvature of magnitude 40.0 cm. Determine the position of the virtual image and the magnification for object distances of (a) 30.0 cm and (b) 60.0 cm. (c) Are the images in parts (a) and (b) upright or inverted?

Solution

Conceptualize: Think of your reflection in a backyard gazing globe or, if it is more familiar, the back of a shiny spoon. We expect right side up images, diminished in size, behind the reflecting surface. As the object moves farther away from (a) to (b), will the image move closer to the reflecting surface or farther away?

Categorize: This problem is in standard cause-and-effect, locate-and-describe format. The lens-mirror equation will tell us the image location and $M = -q/p$ will tell us the magnification. A ray diagram is the best way to determine and understand the image description.

Analyze:

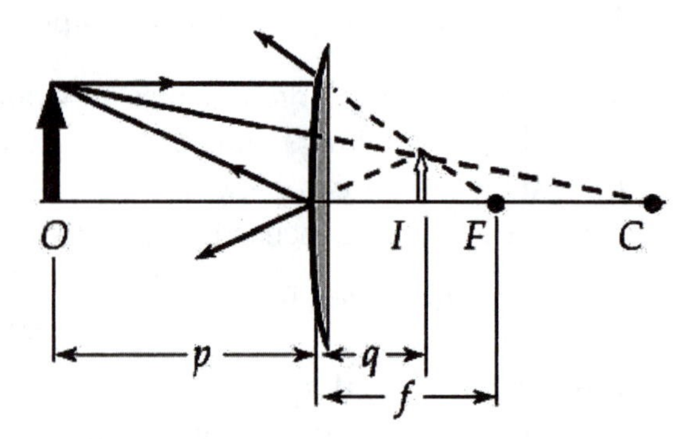

The convex mirror is described by

$$f = \frac{R}{2} = \frac{-40.0\text{ cm}}{2} = -20.0\text{ cm}$$

(a) Then $\frac{1}{p} + \frac{1}{q} = \frac{1}{f}$ gives

$$q = \frac{1}{1/f - 1/p} = \frac{1}{1/(-20.0\text{ cm}) - 1/(30.0\text{ cm})} = -12.0\text{ cm} \quad \blacksquare$$

the magnification factor is $M = -\frac{q}{p} = -\left(\frac{-12.0\text{ cm}}{30.0\text{ cm}}\right) = +0.400$ ■

(c) As shown by the figure above, the image is behind the mirror, upright, virtual, and diminished. ■

(b) Following the same steps, $q = \frac{1}{1/f - 1/p} = \frac{1}{1/(-20.0\text{ cm}) - 1/(60.0\text{ cm})} = -15.0\text{ cm}$

■

and $M = \frac{-q}{p} = -\left(\frac{-15.0\text{ cm}}{60.0\text{ cm}}\right) = +0.250$ ■

(c) The principal ray diagram is an essential complement to the numerical description of the image. Add rays on this diagram for a 60-cm object distance.

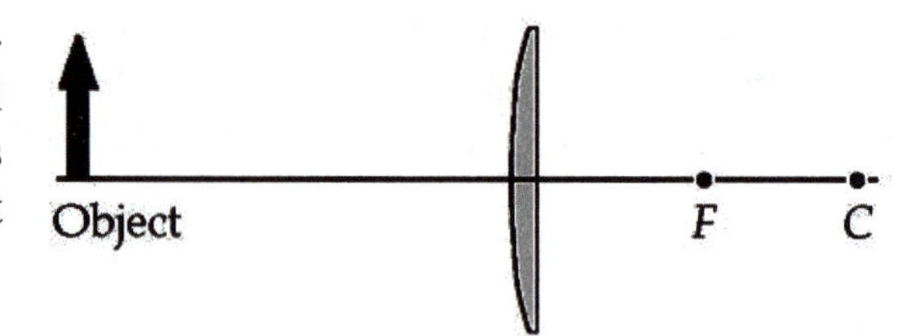

Use your diagram to confirm that the image is behind the mirror, upright, virtual, and diminished. ■

Finalize: A negative answer for q indicates that the image is behind the mirror and virtual. A positive answer for M indicates that the image is upright. A value of M less than 1 in absolute value indicates that the image is diminished. Drawing the diagram really is easier and clearer than trying to remember those different rules.

23. A spherical mirror is to be used to form an image 5.00 times the size of an object on a screen located 5.00 m from the object. (a) Is the mirror required concave or convex? (b) What is the required radius of curvature of the mirror? (c) Where should the mirror be positioned relative to the object?

Solution

Conceptualize: A mirror is not often used as a slide projector, but it can work well. The image must be real to be on a screen.

Categorize: We are careful not to assume that the image is right side up. The standard equations, supplemented by the diagram, will give us the answer.

Analyze: The object's distance from the mirror is p and the image's distance from the mirror is $q = p + 5.00$ m.

Further, from $M = -\dfrac{q}{p}$ we have $|M| = 5.00$

Since the image must be real, q must be positive, so M must be negative:

$$M = -5.00 \quad \text{and} \quad q = 5.00\,p$$

(c) Solving first for the distance of the mirror from the object,

$$p + 5.00 \text{ m} = 5.00p \qquad \text{and} \qquad p = 1.25 \text{ m} \qquad ■$$

Applying the lens-mirror equation

$$\frac{1}{f} = \frac{1}{p} + \frac{1}{q}$$

gives $$\frac{1}{f} = \frac{1}{1.25 \text{ m}} + \frac{1}{6.25 \text{ m}}$$

so that the focal length of the mirror is $f = 1.04$ m

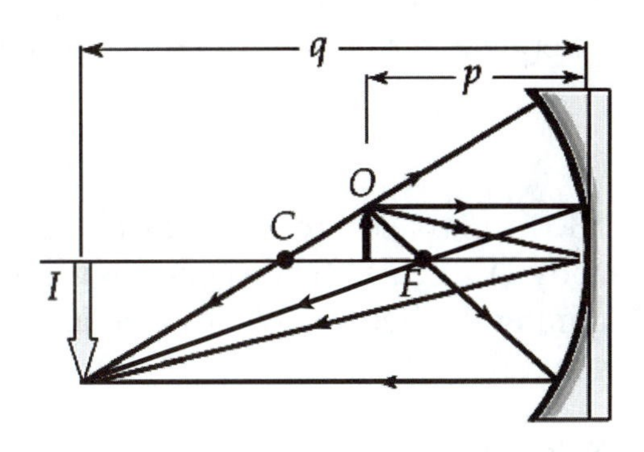

(a, b) Noting that the image is real, inverted, and enlarged, we can say that the mirror must be concave, and must have a radius of curvature of $R = 2f = 2.08$ m. ■

Finalize: With a shaving mirror or telescope mirror, a darkened room, and a small light bulb you can set up this situation to scale. Putting the light bulb precisely at the focal point makes a model of a searchlight. It is very instructive to put the light bulb at different locations in front of the mirror and find and observe the images.

---

28. A cubical block of ice 50.0 cm on a side is placed over a speck of dust on a level floor. Find the location of the image of the speck as viewed from above. The index of refraction of ice is 1.309.

Solution

Conceptualize: Light rays diverging from the dust are changed in direction when they leave the top surface of the ice. They can be traced back to a point from which they apparently diverge, and that is the image, somewhere (we think) inside the block.

Categorize: The upper surface of the block is a single refracting surface with zero curvature, and with infinite radius of curvature. We use the image-formation-by-a-single-refracting-surface equation.

Analyze:

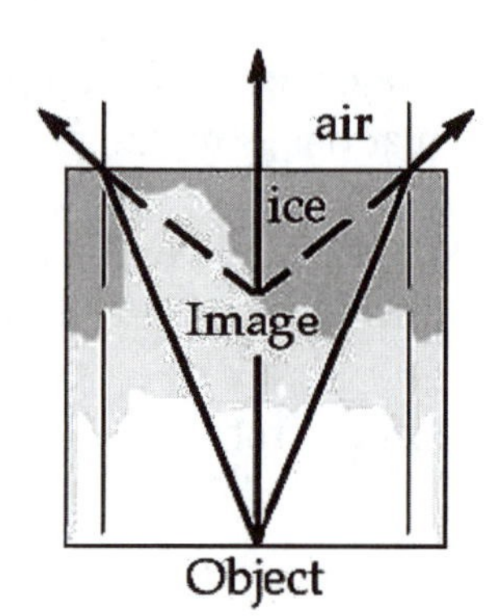

The equation is $\frac{n_1}{p}+\frac{n_2}{q}=\frac{n_2-n_1}{R}$ or

$$q=\frac{n_2}{(n_2-n_1)/R-n_1/p}=\frac{1.00}{(1.00-1.309)/\infty-1.309/(50.0\text{ cm})}$$

$$=\frac{-50.0\text{ cm}}{1.309}=-38.2\text{ cm}$$

The dust speck appears to be 38.2 cm below the upper surface of the ice. ■

Finalize: The image is virtual, upright, and actual size. A virtual image is a perfectly visible thing at a perfectly definite location. It is the intersection point of exiting rays. In optics an object is not necessarily a chunk of material. An object is an intersection point for rays entering a mirror, lens, or refracting surface.

---

33. A glass sphere ($n = 1.50$) with a radius of 15.0 cm has a tiny air bubble 5.00 cm above its center. The sphere is viewed looking down along the extended radius containing the bubble. What is the apparent depth of the bubble below the surface of the sphere?

Solution

Conceptualize: This problem is just like Problem 28, immediately above, except for different numbers and a curved rather than a flat refracting surface. We guess that the image will be quite close below the upper surface of the glass.

Categorize: The same methods as in Problem 28 should work.

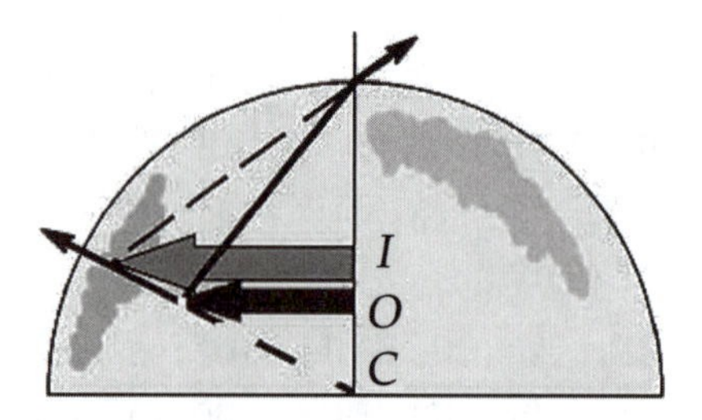

Analyze: From the equation about image formation by a single refracting surface,

$$\frac{n_1}{p}+\frac{n_2}{q}=\frac{n_2-n_1}{R}$$

We solve for q to find $q = \dfrac{n_2 R p}{p(n_2 - n_1) - n_1 R}$

In this case, $n_1 = 1.50$, $n_2 = 1.00$, $p = 10.0$ cm, and $R = -15.0$ cm.

So the image location is

$$q = \frac{(1.00)(-15.0\text{ cm})(10.0\text{ cm})}{(10.0\text{ cm})(1.00 - 1.50) - (1.50)(-15.0\text{ cm})} = -8.57\text{ cm}$$

The depth of the image is 8.57 cm. ■

Finalize: Oops! If the surface had been flat rather than curved, the image position would have been –10 cm/1.5 = –6.67 cm. This can be seen from our symbolic equation $q = \dfrac{n_2 R p}{p(n_2 - n_1) - n_1 R}$ by taking $R = \infty$. The curvature of the surface makes the image distance larger, not smaller as we guessed. To sketch a ray diagram we use the wave under refraction model, as shown. The image is virtual, upright, and enlarged.

---

36. An object is located 20.0 cm to the left of a diverging lens having a focal length $f = -32.0$ cm. Determine (a) the location and (b) the magnification of the image. (c) Construct a ray diagram for this arrangement.

Solution

Conceptualize: If you are nearsighted, your eyeglass lenses are diverging, likely with longer focal length that the lens considered here. We expect the image distance to be negative and smaller in absolute value than 20 cm. The magnification should be positive and less than 1. That is, the image should be virtual, upright, and diminished.

Categorize: The mirror-and-lens equation and the magnification equation should do the job for precise answers.

Analyze: The mirror-and-lens equation $\dfrac{1}{p} + \dfrac{1}{q} = \dfrac{1}{f}$

becomes here $q = \dfrac{1}{1/f - 1/p}$

(a) So $q = -\left(\dfrac{1}{20.0\text{ cm}} + \dfrac{1}{32.0\text{ cm}}\right)^{-1} = -12.3\text{ cm}$ ■

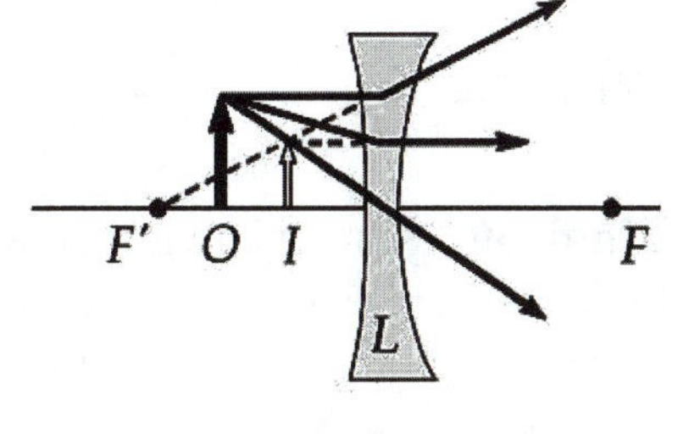

(b) The magnification is

$$M = -\frac{q}{p} = -\frac{(-12.3\text{ cm})}{20.0\text{ cm}} = 0.615$$ ■

(c) The image is virtual, upright, and diminished; the ray diagram is shown. ■

Finalize: Another practical application of a diverging lens is a wide-angle lens in a peephole in a door.

In drawing the ray diagram, the focal points might give you trouble. Review how light goes through a prism, in the textbook's Figure 35.16 or the drawing accompanying Problem 35.13 in this manual. In the diagram for this problem, draw first the ray from the top of the object through the center of the lens, continuing on straight. Draw next the ray from the top of the object parallel to the axis. It bends "around the thicker part of the prism," up in the diagram, to go off as if it were coming from F'. Then F has not yet been used. A ray from the top of the object toward F bends around the thicker part of the "prism" it encounters to go off parallel to the axis. Extending the outgoing rays backwards determines the image by their intersection.

---

43. The nickel's image in Figure P36.43 has twice the diameter of the nickel and is 2.84 cm from the lens. Determine the focal length of the lens.

Solution

Conceptualize: The lens is being used as a magnifying glass. We expect a positive focal length of a few centimeters.

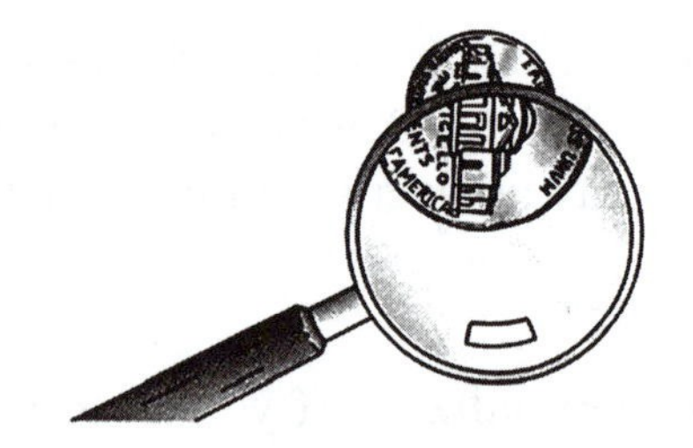

Figure P36.43

Categorize: The picture shows us that the image is right side up. We have $M = +2$ and not -2. Using this with the magnification equations and the mirror-lens equation will let us find p first and then f.

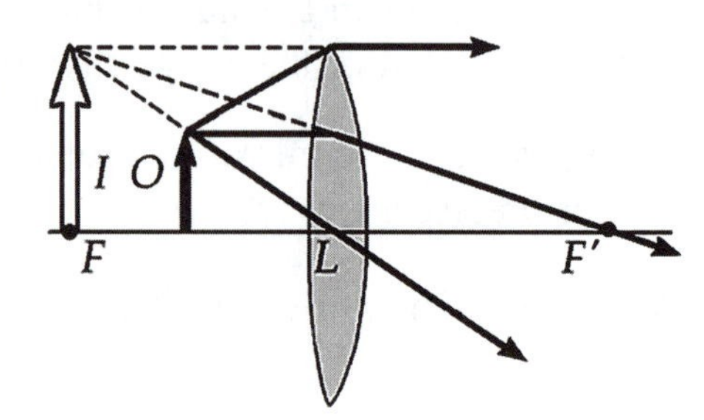

Analyze: Looking through the lens, you see the image beyond the lens. Therefore, the image is virtual, with $q = -2.84$ cm.

Now, $M = \dfrac{h'}{h} = 2 = -\dfrac{q}{p}$ so $p = -\dfrac{q}{2} = 1.42$ cm

A check is that p is positive, as it must be for a real object.

Thus, $f = \left(\dfrac{1}{p} + \dfrac{1}{q}\right)^{-1} = \left[\dfrac{1}{1.42\text{ cm}} + \dfrac{1}{(-2.84\text{ cm})}\right]^{-1} = 2.84$ cm ■

Finalize: The image is virtual, upright, and enlarged. A virtual image is a useful thing, because your eyes and brain are adapted to accept diverging rays and trace them back, always in straight lines, to find where they intersect. The pattern of light divergence from a virtual image is just like the pattern of light divergence from a real object. If you do not put a screen in the way, the rays forming a real image will pass through the image plane and then diverge. If your eyes intercept these rays, your brain will do the same processing as for a virtual image, and you will see the real image floating in space at the image plane.

---

45. The left face of a biconvex lens has a radius of curvature of magnitude 12.0 cm, and the right face has a radius of curvature of magnitude 18.0 cm. The index of refraction of the glass is 1.44. (a) Calculate the focal length of the lens for light incident from the left. (b) **What If?** After the lens is turned around to interchange the radii of curvature of the two faces, calculate the focal length of the lens for light incident from the left.

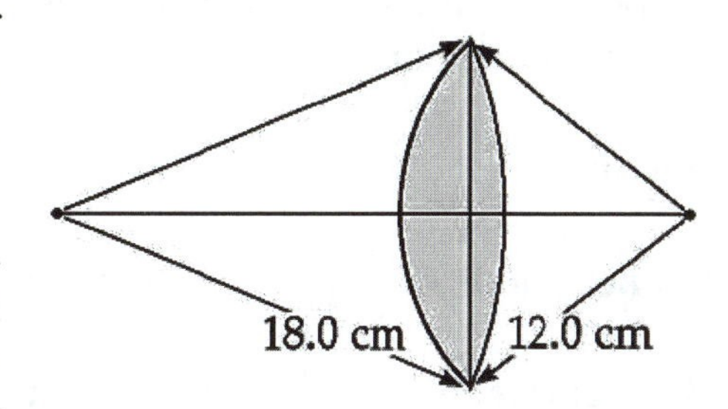

Solution

Conceptualize: Since this is a biconvex lens, the center is thicker than the edges, and the lens will tend to converge incident light rays. Therefore, it has a positive focal length. Exchanging the radii of curvature amounts to turning the lens around so the light enters the opposite side first. However, this does not change the fact that the center of the lens is thicker than the edges, so we should not expect the focal length of the lens to be different.

Categorize: The lens makers' equation can be used to find the focal length of this lens.

Analyze: The centers of curvature of the lens surfaces are on opposite sides, so the second surface has a negative radius:

(a) $\frac{1}{f} = (n-1)\left(\frac{1}{R_1} - \frac{1}{R_2}\right)$ gives

$$f = \frac{1}{(n-1)(1/R_1 - 1/R_2)} = \frac{1}{(1.44-1)\left(1/(12.0\text{ cm}) - 1/(-18.0\text{ cm})\right)} = 16.4\text{ cm} \quad \blacksquare$$

(b) Similarly,

$$f = \frac{1}{(n-1)(1/R_1 - 1/R_2)} = \frac{1}{(0.440)\left(1/(18.0\text{ cm}) - 1/(-12.0\text{ cm})\right)} = 16.4\text{ cm} \quad \blacksquare$$

Finalize: As expected, reversing the orientation of the lens does not change what it does to the light, as long as the lens is relatively thin (variations may be noticed with a thick lens). If you can find a converging lens that is flat on one side, set it up to make an image of a distant object on a card, and ask your lab partner or sibling what will happen when you flip the lens so that the other surface is facing the object. They may think that the image will turn to be right side up. An explanation based on how the lens alters an incoming flat wavefront may be best for them.

---

52. A nearsighted person cannot see objects clearly beyond 25.0 cm (her far point). If she has no astigmatism and contact lenses are prescribed for her, what (a) power and (b) type of lens are required to correct her vision?

Solution

Conceptualize: We expect a strong (short-focal-length) diverging lens.

Categorize: We use the mirror-lens equation for a very distant object. If the patient can see this object clearly as an image at her far point and has normal accommodation, she will be able to see closer objects clearly as well.

Analyze: The lens should take parallel light rays from a very distant object ($p = \infty$) and make them diverge from a virtual image at the woman's far point, which is 25.0 cm beyond the lens, at $q = -25.0$ cm.

Thus,
$$\frac{1}{f} = \frac{1}{p} + \frac{1}{q} = \frac{1}{\infty} + \frac{1}{-25.0\text{ cm}}$$

(a) Hence, the power of the lens is $P = \frac{1}{f} = -\frac{1}{0.250\text{ m}} = -4.00$ diopters ■

(b) With negative focal length, this is a diverging lens. ■

Finalize: A goal of vision correction is always to provide a clear view of objects at infinity. A car down the highway and a player across the football field are so far away, compared to the size of the eye, that you receive essentially parallel rays from them. Thus they are well described as at infinity.

---

56. The refracting telescope at the Yerkes Observatory has a 1.00-m diameter objective lens of focal length 20.0 m. Assume it is used with an eyepiece of focal length 2.50 cm. (a) Determine the magnification of Mars as seen through this telescope. (b) Are the Martian polar caps right side up or upside down?

Solution

Conceptualize: We expect a magnification of some hundreds of times, large compared to ten times for typical binoculars. To minimize light loss, astronomical telescopes are made as simple as possible (If there is a tube, there is nothing inside it!) and the image is inverted.

Categorize: The equation for angular magnification of a telescope, derived in the chapter text, will tell us the answer.

Analyze:

(a) The angular magnification is

$$m = -\frac{f_o}{f_e} = -\frac{20.0\text{ m}}{0.0250\text{ m}} = -800 \qquad \blacksquare$$

(b) The minus sign means the image is inverted relative to the object. ■

Finalize: This is angular magnification *m*, not linear magnification *M*. Do not think of the telescope as making things *bigger*—an image of a star larger than the star would be of no use because the intercepted light would be spread too thin. A telescope you look through makes things look *closer*. The telescope maker constructs basically just the objective: for this telescope, just the one-meter aperture lens. The astronomer can place a photographic plate or a sheet of light detecting pixels (charge-coupled detectors) at the focal plane of the objective to record an image. Alternatively, she can pop in an eyepiece for use as a magnifying glass to inspect the same image. She chooses the focal length of the eyepiece to get an angular magnification convenient for inspecting the interesting part of the image. Larger magnification can always be obtained with an eyepiece of shorter focal length, but higher magnification is not better once a convenient choice has been made. An eyepiece cannot reveal details that are not already present in the image made by the objective. What is always better is a larger objective diameter. With an objective mirror of larger diameter, ability to see details can improve and the telescope will definitely intercept more light power. Then the exposure time interval for each picture can be shorter and more things can be imaged in a night.

---

70. A parallel beam of light enters a glass hemisphere perpendicular to the flat face as shown in Figure P36.70. The magnitude of the radius of the hemisphere is $R = 6.00$ cm, and its index of refraction is $n = 1.560$. Assuming paraxial rays, determine the point at which the beam is focused.

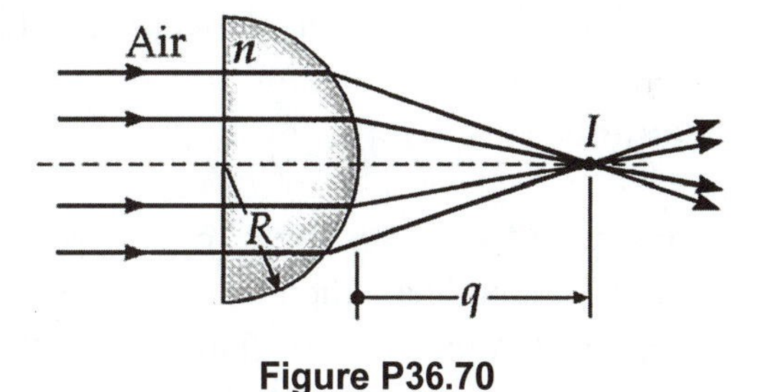

Figure P36.70

Solution

Conceptualize: "Focused" is the right word because the object is at infinity. We must calculate the image distance.

Categorize: A hemisphere is too thick to be described as a thin lens. The light is undeviated on entry into the flat face. Having noted this, we consider the light's exit from the second surface, for which $R = -6.00$ cm. We use the equation about image formation by a single refracting surface.

Analyze: The incident rays are parallel, so $p = \infty$.

Then $\dfrac{n_1}{p} + \dfrac{n_2}{q} = \dfrac{n_2 - n_1}{R}$

becomes $q = \dfrac{n_2}{(n_2 - n_1)/R - n_1/p} = \dfrac{1.00}{(1.00 - 1.560)/(-6.00\text{ cm}) - 0} = 10.7\text{ cm}$ ■

Finalize: The image is real, diminished, and inverted. You can tell it is inverted by noticing that the image point would move down in the picture if the object rays came in from the upper right.

What if you did try to use the thin-lens model? You would compute a focal length for the hemisphere from

$$\frac{1}{f} = (n-1)\left(\frac{1}{R_1} - \frac{1}{R_2}\right) \text{ giving}$$

$$f = \frac{1}{(n-1)\left(1/R_1 - 1/R_2\right)} = \frac{1}{(1.560-1)\left(1/\infty - 1/(-6.00\text{ cm})\right)} = \frac{1}{0.093\,3/\text{cm}} = 10.7\text{ cm}$$

You would likely interpret this as a distance from the center of mass of the hemisphere, whereas really the focal point is 16.7 cm from the center of the flat face and 10.7 cm from the point where the axis passes through the curved face. Further, if the object were at a finite distance, the thin-lens approximation would give an unambiguously wrong answer.

---

73. An object is placed 12.0 cm to the left of a diverging lens of focal length –6.00 cm. A converging lens of focal length 12.0 cm is placed a distance d to the right of the diverging lens. Find the distance d so that the final image is infinitely far away to the right.

Solution

Conceptualize: We must think separately about how the first lens affects the light and how the second lens then again changes how rays intersect. We expect the first lens to produce a virtual image less than 6 cm to the left of it. This should be at the focal point of the second lens so that the second lens can render the rays parallel. So we estimate 9 cm for distance d.

Categorize: We use the mirror-lens equation to locate the image formed by the first lens. This virtual image will be a real object for the second lens at a distance from it that we can represent in an expression involving d. Then the mirror-lens equation applied to the second lens will let us evaluate the object distance and so d.

Analyze: From the mirror-and-lens equation $1/p + 1/q = 1/f$

we have $q_1 = \dfrac{f_1 p_1}{p_1 - f_1} = \dfrac{(-6.00\text{ cm})(12.0\text{ cm})}{12.0\text{ cm} - (-6.00\text{ cm})} = -4.00\text{ cm}$

The first lens forms an image 4.00 cm to its left. The rays between the lenses diverge from this image, so the second lens receives diverging light. It sees a real object at distance $p_2 = d - (-4.00\text{ cm}) = d + 4.00\text{ cm}$.

For the second lens, when we require that $\quad q_2 \to \infty$

The mirror-lens equation becomes $p_2 = f_2 = 12.0\text{ cm}$.

Since the object for the converging lens must be 12.0 cm to its left, and since this object is the image for the diverging lens, which is 4.00 cm to **its** left, the two lenses must be separated by 8.00 cm.

Mathematically, $\quad f_2 = 12.0\text{ cm} = p_2 = d + 4.00\text{ cm} \quad \text{and} \quad d = 8.00\text{ cm}$ ■

We could draw separate ray diagrams for the two lenses, but we choose to combine them here.

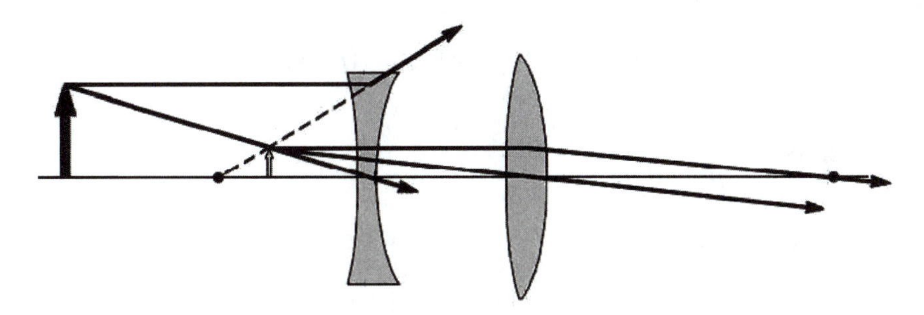

Finalize: We could say that no final image is formed, because the outgoing parallel rays do not intersect. Using the idea of an augmented number line including a point at infinity, we could say that the final image is at infinity on the right, real, enlarged, and inverted. The arrangement could be a useful invention in that it takes light from a fairly large source, concentrates it, and turns it into a fairly narrow searchlight beam. Looking into the outgoing rays, an observer will see a virtual image at infinity to the left, upright and magnified—this is an equally accurate description.

---

76. In a darkened room, a burning candle is placed 1.50 m from a white wall. A lens is placed between the candle and the wall at a location that causes a larger, inverted image to form on the wall. When the lens is in this position, the object distance is p_1. When the lens is moved 90.0 cm toward the wall, another image of the candle is formed on the wall. From this information, we wish to find p_1 and the focal length of the lens. (a) From the lens equation for the first position of the lens, write an equation relating the focal length f of the lens to the object distance p_1, with no other variables in the equation. (b) From the lens equation for the second position of the lens, write another equation relating the focal length f of the lens to the object distance p_1. (c) Solve the equations in (a) and (b) simultaneously to find p_1. (d) Use the value in (c) to find the focal length f of the lens.

Solution

Conceptualize: Light rays are always reversible. We guess that the object distance for the second image is just the same as the image distance for the first and the image distance the same as the first object distance. This symmetry implies that when the lens has moved

45 of the 90 cm it must be halfway between candle and wall. The pair of distances must be 75 cm – 45 cm and 75 cm + 45 cm.

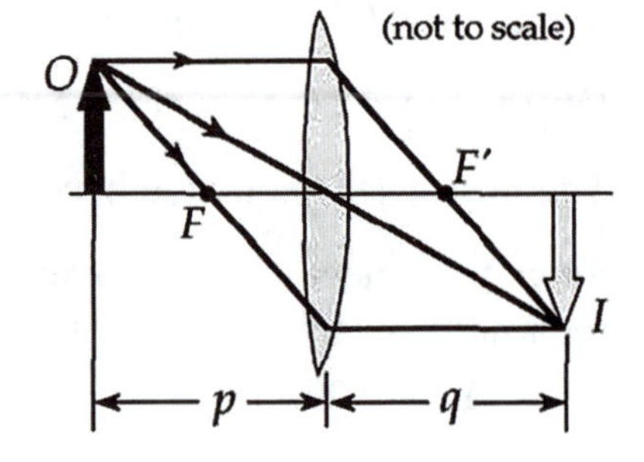

Categorize: The mirror-lens equation will be useful, together with the recognition that $p + q = 1.5$ m. We note that p_1 and q_1 must both be positive because the image must be real to be formed on a wall.

Analyze: The ray diagram is not to scale, but it shows correctly that originally and finally

$$p_1 + q_1 = 1.50 \text{ m} = p_2 + q_2$$

In the final situation, $p_2 = p_1 + 0.900$ m

And $q_2 = q_1 - 0.900 \text{ m} = (1.50 \text{ m} - p_1) - 0.900 \text{ m} = 0.600 \text{ m} - p_1$

Our lens equation is $\dfrac{1}{p_1} + \dfrac{1}{q_1} = \dfrac{1}{f} = \dfrac{1}{p_2} + \dfrac{1}{q_2}$

(a) We read the first answer from the first equality sign,

$$\frac{1}{p_1} + \frac{1}{1.50 \text{ m} - p_1} = \frac{1}{f}$$ ■

(b) and the second answer from the second,

$$\frac{1}{p_1 + 0.900 \text{ m}} + \frac{1}{0.600 \text{ m} - p_1} = \frac{1}{f}$$ ■

(c) The two expressions for the lens power $1/f$ must be equal, and so we have an equation with only one unknown:

$$\frac{1}{p_1} + \frac{1}{1.50 \text{ m} - p_1} = \frac{1}{p_1 + 0.900} + \frac{1}{0.600 - p_1}$$

Adding the fractions, $\dfrac{1.50 \text{ m} - p_1 + p_1}{p_1(1.50 \text{ m} - p_1)} = \dfrac{0.600 - p_1 + p_1 + 0.900}{(p_1 + 0.900)(0.600 - p_1)}$

Simplified, this becomes $p_1(1.50 \text{ m} - p_1) = (p_1 + 0.900)(0.600 - p_1)$

Thus, $p_1 = \dfrac{0.540}{1.80} \text{m} = 0.300 \text{ m}$ ■

and $p_2 = p_1 + 0.900 \text{ m} = 1.20 \text{ m}$

(d) $\dfrac{1}{f} = \dfrac{1}{0.300 \text{ m}} + \dfrac{1}{1.50 \text{ m} - 0.300 \text{ m}}$

and $f = 0.240$ m ■

The second image is real, inverted, and diminished, with

$$M = -\frac{q_2}{p_2} = -\frac{0.600\text{ m} - 0.300\text{ m}}{1.20\text{ m}} = -0.250$$ ■

Finalize: Do this as a demonstration. Obtain any convenient converging lens. Measure its focal length by making an image of a distant object. Set up the light source in a darkened room, distant from the wall by more than four times the focal length. Form the first image on the wall. This arrangement is a model for a movie projector. Now move the lens toward the wall until you get the second image. It is like the first in being real and inverted, but different in being diminished. This arrangement models a camera or your eye.

In effect, the problem required solving the four simultaneous equations

$$p_1 + q_1 = 1.50\text{ m} \qquad p_2 + q_2 = 1.50\text{ m} \qquad p_2 = p_1 + 0.900\text{ m}$$

and $\frac{1}{p_1} + \frac{1}{q_1} = \frac{1}{p_2} + \frac{1}{q_2}$ for p_1. We did it by substitution. If someone tries to tell you to solve by typing it all into a calculator, or by determinants or by matrix inversion, do not believe them. If you wish you could believe them, try it and find out what is inconvenient and what doesn't work at all.

---

77. The disk of the Sun subtends an angle of 0.533° at the Earth. What are (a) the position and (b) the diameter of the solar image formed by a concave spherical mirror with a radius of curvature of magnitude 3.00 m?

Solution

Conceptualize: The Sun is so far away that the image will be very nearly at the focal point, 1.5 m in front of the collector. For the diameter we guess something small, perhaps a millimeter.

Categorize: We can take the distance to our star as known. The mirror-lens equation and the magnification equation $h'/h = -q/p$ will give us the answers, but we may need to watch our approximations.

Analyze: For the mirror, the focal length is $f = R/2 = +1.50$ m. In the mirror equation, because the distance to the Sun is so much larger than f and q, we can take $p = \infty$. Alternatively, we could look up the distance to the Sun and substitute it in. In either case, the mirror equation, $\frac{1}{p} + \frac{1}{q} = \frac{1}{f}$ then gives $q = f = 1.50$ m. ■

Now, in $M = -q/p = h'/h$, the magnification is nearly zero, but we can be more precise: the definition of radian measure means that h/p is the angular diameter of the object. Thus the image size is

$$h' = -\frac{hq}{p} = (-0.533^\circ)\left(\frac{\pi}{180}\text{rad/deg}\right)(1.50\text{ m}) = -0.0140\text{ m} = -1.40\text{ cm}$$

The negative sign refers to the image being inverted. Anyone looking at the image would say its diameter is 1.40 cm. ■

Finalize: We could take $p = \infty$ in one equation but not in the other. In the magnification equation, we could have substituted in the diameter of the Sun in kilometers and its distance away, to obtain the same result. This method would in effect verify that the angular diameter of the Sun is 0.533°. The image diameter is bigger than we guessed. It is quite suitable for toasting a marshmallow, and this could be done quite quickly if the face diameter of the mirror is large. Especially for large mirror areas on sunny days, the arrangement described here is dangerous. It might set things on fire, produce burns, and very quickly produce eye damage. It is the face diameter of the mirror, its "objective diameter," that controls how much solar power is collected. Both the objective diameter and the radius of curvature affect the temperature attained at the image.

37

Wave Optics

EQUATIONS AND CONCEPTS

In **Young's double-slit experiment**, two slits, $\mathbf{S}_1$ and $\mathbf{S}_2$, separated by a distance d, serve as *monochromatic coherent sources*. The light intensity at any point on the screen is the resultant of light reaching the screen from both slits. *As illustrated in the figure, a point P on the screen can be identified by the angle θ or by the distance y from the center of the screen.*

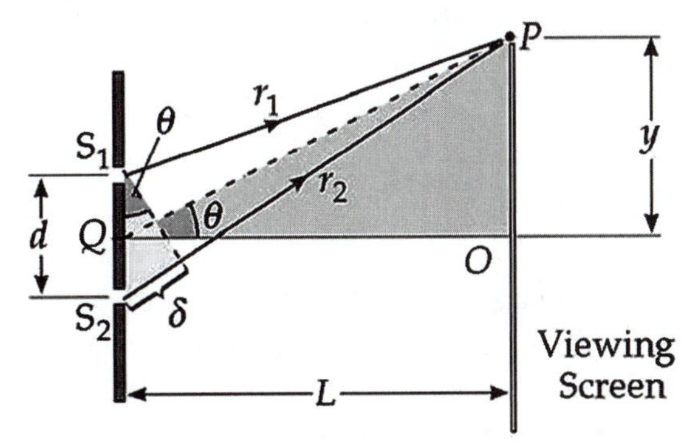

$$\tan\theta = \frac{y}{L} \tag{37.4}$$

A **path difference** (δ) is due to the unequal distances traveled by waves from $\mathbf{S}_1$ and $\mathbf{S}_2$ to any point on the screen (except the center point). *The value of δ determines whether the waves from the two slits arrive in phase or out of phase.*

$$\delta = r_2 - r_1 = d\sin\theta \tag{37.1}$$

Constructive interference (bright fringes) will appear at points on the screen for which the path difference is equal to an integer multiple of the wavelength. The positions of bright fringes can also be located by calculating their vertical distance from the center of the screen (y). *In each case, the number m is called the order number of the fringe. The central bright fringe ($\theta = 0$, $m = 0$) is called the zeroth-order maximum.*

$$d\sin\theta_{\text{bright}} = m\lambda \quad (m = 0, \pm1, \pm2, \ldots) \tag{37.2}$$

$$y_{\text{bright}} = L\frac{m\lambda}{d} \quad (\textit{small angles}) \quad (m = 0, \pm1, \pm2, \ldots) \tag{37.7}$$

Destructive interference (dark fringes) will occur at points on the screen which correspond to path differences of an odd multiple of half wavelengths. *For these points, waves which leave the two slits in phase arrive at the screen 180° out of phase.*

$$d\sin\theta_{\text{dark}} = \left(m + \tfrac{1}{2}\right)\lambda \quad (m = 0, \pm1, \pm2, \ldots) \tag{37.3}$$

$$y_{\text{dark}} = L\left(\frac{m + \frac{1}{2}}{d}\right)\lambda; \quad (\textit{small angles}) \quad (m = 0, \pm1, \pm2, \ldots) \tag{37.8}$$

The **phase difference** (ϕ) between the two waves at any point on the screen depends on the path difference at that point.

$$\phi = \frac{2\pi}{\lambda}\delta = \frac{2\pi}{\lambda} d \sin\theta \qquad (37.10)$$

The **resultant electric field** (due to the superposition of two waves initially in phase and of equal amplitude, E_0) at some point on the screen has a magnitude which depends on the phase difference. *The frequency of the resultant wave has the same frequency as that of the two coherent sources; the amplitude, $2E_0\cos(\phi/2)$, is multiplied by a factor that depends on the phase difference.*

$$E_p = 2E_0 \cos\left(\frac{\phi}{2}\right)\sin\left(\omega t + \frac{\phi}{2}\right) \qquad (37.12)$$

The **average light intensity** (I) at any point P on the screen is proportional to the square of the amplitude of the resultant wave. The average intensity can be expressed

- as a function of phase difference ϕ

$$I = I_{\max}\cos^2\left(\frac{\phi}{2}\right) \qquad (37.13)$$

- as a function of $\sin\theta$, where θ is the angle that locates a point on the screen

$$I = I_{\max}\cos^2\left(\frac{\pi d \sin\theta}{\lambda}\right) \qquad (37.14)$$

- as a function of the vertical distance y from the center of the screen

$$I \approx I_{\max}\cos^2\left(\frac{\pi d}{\lambda L} y\right) \qquad (37.15)$$

Increasing the number of equally spaced slits *will increase the number of secondary maxima; the principal maxima will become narrower but remain fixed in position. See Figure 37.7 of the textbook.*

The **wavelength of light** in a medium having an index of refraction n is smaller than the wavelength in vacuum.

$$\lambda_n = \frac{\lambda}{n}$$

Interference in thin films depends on wavelength, film thickness, and the indices of refraction of the film and surrounding media. *Differences in phase may be due to path difference or phase change upon reflection.* There are two general cases as described below:

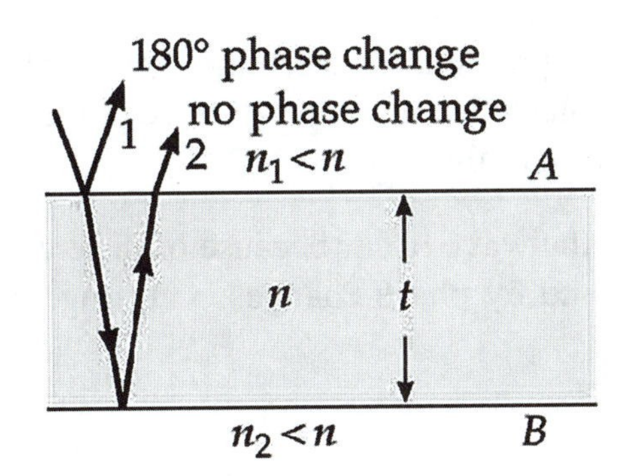

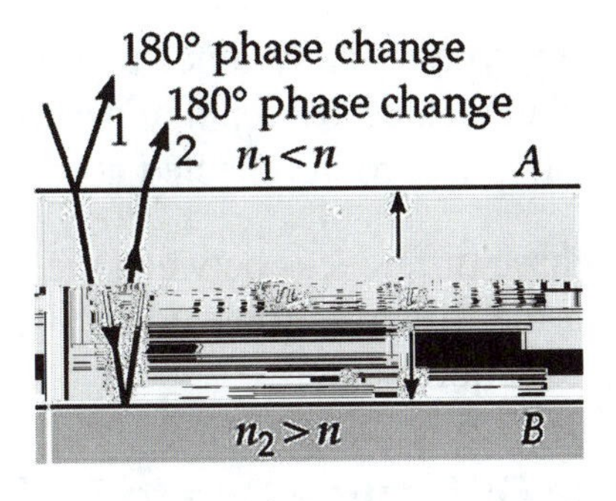

Case (1) Phase change at only one film surface.

The indices of refraction of the media on both sides of the film are less than that of the film ($n_1 < n$ and $n_2 < n$), as shown in the figure above left; or the indices on both sides of film are greater than that of the film (not shown).

constructive interference:

$$2nt = \left(m + \tfrac{1}{2}\right)\lambda \quad (m = 0, 1, 2, \ldots) \quad (37.17)$$

destructive interference:

$$2nt = m\lambda \quad (m = 0, 1, 2, \ldots) \quad (37.18)$$

Case (2) Phase changes at both or neither film surface.

The film is between two media, either of which has an index of refraction greater than that of the film and the other a smaller index; ($n_1 < n < n_2$) as shown the figure above right; or ($n_1 > n > n_2$).

constructive interference:

$$2nt = m\lambda \quad (m = 0, 1, 2, \ldots) \quad (37.18)$$

destructive interference:

$$2nt = \left(m + \tfrac{1}{2}\right)\lambda \quad (m = 0, 1, 2, \ldots) \quad (37.17)$$

Note that the roles of the equations in Case (1) and Case (2) are reversed for constructive and destructive interference.

SUGGESTIONS, SKILLS, AND STRATEGIES

THIN-FILM INTERFERENCE PROBLEMS

- Identify the thin film from which interference effects are being observed.
- The type of interference that occurs in a specific problem is determined by the phase relationship between that portion of the wave reflected at the upper surface of the film and that portion reflected at the lower surface of the film.
- Phase differences between the two portions of the wave occur because of differences in the distances traveled by the two portions and by phase changes occurring upon reflection.

Phase Difference Due to Path Difference

The wave reflected from the lower surface of the film has to travel a distance equal to twice the thickness of the film before it returns to the upper surface of the film, where it interferes with that portion of the wave reflected at the upper surface.

Phase Change Due to Reflection

When a wave traveling in a particular medium reflects off a surface having a higher index of refraction than the one it is in, a 180° phase shift occurs. This has the same effect as if the wave lost $\frac{1}{2}\lambda$. This effect must be considered in addition to the phase difference related to the greater distance traveled by one of the waves.

- When distance and phase changes upon reflection are both taken into account:

  Constructive interference will occur when the phase difference is an integral multiple of λ (zero, λ, 2λ, 3λ, . . .).

  Destructive interference will occur when the phase difference is an odd number of half wavelengths ($\frac{1}{2}\lambda, \frac{3}{2}\lambda, \frac{5}{2}\lambda, \ldots$).

REVIEW CHECKLIST

You should be able to:

- Describe Young's double-slit experiment to demonstrate the wave nature of light. Account for the phase difference between light waves from the two sources as they arrive at a given point on the screen. State the conditions for constructive and destructive interference in terms of each of the following: path difference δ, phase difference ϕ, distance from the center of the screen y, and angle subtended by the observation point at the source mid-point θ. (Sections 37.1 and 37.2)
- Calculate the ratio of average intensities at two points in a double-slit interference pattern. (Section 37.3)

- State the conditions, and write the corresponding equations, for constructive and destructive interference in thin films considering both path difference and any expected phase changes due to reflection. Calculate the minimum film thickness to produce constructive/destructive interference in a film between media of known indices of refraction. (Section 37.5)

ANSWER TO AN OBJECTIVE QUESTION

1. A plane monochromatic light wave is incident on a double slit as illustrated in Active Figure 37.1. **(i)** As the viewing screen is moved away from the double slit, what happens to the separation between the interference fringes on the screen? (a) It increases. (b) It decreases. (c) It remains the same. (d) It may increase or decrease, depending on the wavelength of the light. (e) More information is required. **(ii)** As the slit separation increases, what happens to the separation between the interference fringes on the screen? Select from the same choices.

Answer **(i)** (a) The equations $d \sin\theta_1 = 1\lambda$ and $y_1 = L \tan\theta_1$ describe the location of the first bright fringe to the side of the central bright fringe. As the distance L from the slits to the screen increases, θ_1 is constant and the distance y_1 on the screen gets larger in direct proportion to L. **(ii)** (b) In the same equations think of d getting larger. Then θ_1 gets smaller and all of the fringes get closer together.

□ □ □ □

ANSWERS TO SELECTED CONCEPTUAL QUESTIONS

1. What is the necessary condition on the path length difference between two waves that interfere (a) constructively and (b) destructively?

Answer (a) Two waves interfere constructively if their path difference is either zero or some integral multiple of the wavelength, that is, if the path difference equals $m\lambda$. (b) Two waves interfere destructively if their path difference is an odd multiple of one-half of a wavelength, that is, if the path difference equals $(m + \frac{1}{2})\lambda$.

□ □ □ □

6. (a) In Young's double-slit experiment, why do we use monochromatic light? (b) If white light is used, how would the pattern change?

Answer (a) Suppose we use a mixture of colors. Then each color produces its own pattern, with a spacing between the maxima that is characteristic of the wavelength. With several colors, the patterns are superimposed, and it can become difficult to pick out a single maximum. Using monochromatic light eliminates this problem.

(b) With white light, the central maximum is white. The first side maximum is a full spectrum, with violet on the inside and red on the outside. The second side maximum is a full spectrum also, but red in the second maximum overlaps the violet in the third maximum. At larger angles, the light soon starts mixing to white again, but it may be so faint that you would call it gray.

□ □ □ □

SOLUTIONS TO SELECTED PROBLEMS

5. Young's double-slit experiment is performed with 589-nm light and a distance of 2.00 m between the slits and the screen. The tenth interference minimum is observed 7.26 mm from the central maximum. Determine the spacing of the slits.

Solution

Conceptualize: For the situation described, the observed interference pattern is very narrow (the minima are less than 1 mm apart when the screen is 2 m away). In fact, the minima and maxima are so close together that it would probably be difficult to resolve adjacent maxima, so the pattern might look like a uniform blur to the naked eye. Since the angular spacing of the pattern is inversely proportional to the slit width, we should expect that for this narrow pattern, the space between the slits will be larger than the typical fraction of a millimeter, and certainly much greater than the wavelength of the light ($d >> \lambda = 589$ nm).

Categorize: Since we are given the location of the tenth minimum for this interference pattern, we should use the equation for **destructive interference** from a double slit. The figure for Problem 7 shows the critical variables for this problem.

Analyze: In the equation $d \sin\theta = \left[m + \frac{1}{2}\right]\lambda$, the first minimum is described by $m = 0$ and the tenth by $m = 9$. So $d \sin\theta = \lambda\left[9 + \frac{1}{2}\right]$

Also, $\tan \theta = y/L$, but for small θ, $\sin\theta \approx \tan\theta$. Thus, the distance between the slits is

$$d = \frac{9.5\lambda}{\sin\theta} = \frac{9.5\lambda L}{y} = \frac{9.5(589 \times 10^{-9}\text{ m})(2.00\text{ m})}{7.26 \times 10^{-3}\text{ m}}$$

$$= 1.54 \times 10^{-3}\text{ m} = 1.54\text{ mm}$$ ■

Finalize: The spacing between the slits is relatively large, as we expected (about 3 000 times greater than the wavelength of the light). In order to more clearly distinguish between maxima and minima, the pattern could be expanded by increasing the distance to the screen. However, as L is increased, the overall pattern would be less bright as the light expands over a larger area, so that beyond some distance, the light would be too dim to see.

---

7. A pair of narrow, parallel slits separated by 0.250 mm are illuminated by green light ($\lambda = 546.1$ nm). The interference pattern is observed on a screen 1.20 m away from the plane of the parallel slits. Calculate the distance (a) from the central maximum to the first bright region on either side of the central maximum and (b) between the first and second dark bands in the interference pattern.

Solution

Conceptualize: The spacing between adjacent maxima and minima should be fairly uniform across the pattern as long as the width of the pattern is much less than the distance to

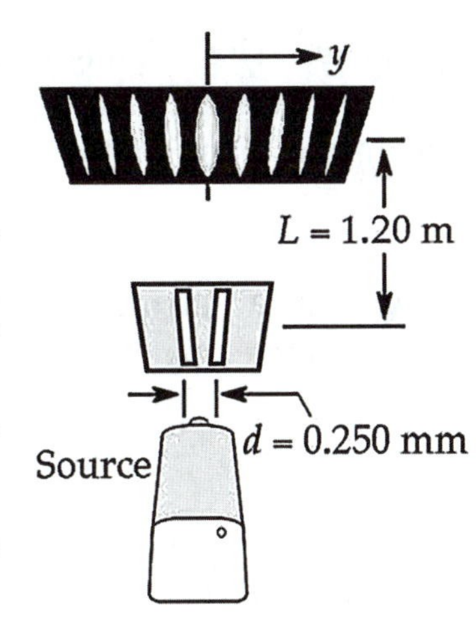

the screen (so that the small angle approximation is valid). The separation between fringes should be at least a millimeter if the pattern can be easily observed with a naked eye.

Categorize: The bright regions are areas of constructive interference and the dark bands are destructive interference, so the corresponding double-slit equations will be used to find the y distances.

It can be confusing to keep track of four different symbols for distances. Three are shown in the drawing. Note that:

d is the distance between the adjacent wave sources, here the slits.

L is the distance from the sources to the screen where interference fringes are observed.

y is the unknown distance from the bright central maximum ($m = 0$) to another maximum or minimum on either side of the center of the interference pattern.

λ is the wavelength of the light, determined by the source. The interference maxima are *not* wave crests. The stationary bright fringes are separated by a distance y of a few millimeters while the wave crests are individually unobservable, always moving, and separated by the distance $\lambda = 546.1$ nm.

Analyze:

(a) For any size θ we have $\tan\theta = y/L$

For **very small** θ we have $\sin\theta \approx \tan\theta$ and the equation for constructive interference $d\sin\theta = m\lambda$ becomes $dy/L \approx m\lambda$ or $y_{\text{bright}} \approx (\lambda L/d)m$

Substituting values, $$y_{\text{bright}} = \frac{(546.1\times10^{-9}\text{ m})(1.20\text{ m})}{0.250\times10^{-3}\text{ m}}(1) = 2.62\text{ mm}$$ ■

(b) If you have trouble remembering whether the equation with $m\lambda$ or the equation with $(m+\frac{1}{2})\lambda$ applies to a particular situation, you can remember that a zero-order bright band is in the center, and dark bands are halfway between bright bands. Thus, the made-up equation $d\sin\theta = (count)\lambda$ describes them all, with *count* = 0, 1, 2, . . . for bright bands, and with *count* = 0.5, 1.5, 2.5, . . . for dark bands.

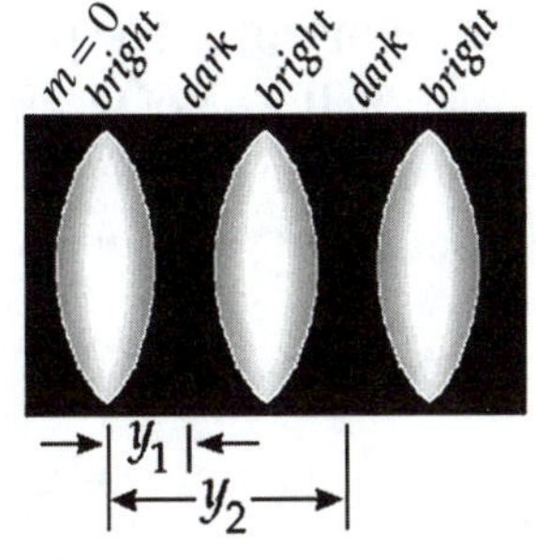

The dark band version of the equation $y_{\text{bright}} \approx (\lambda L/d)m$ is simply

$$y_{\text{dark}} = \frac{\lambda L}{d}\left[m+\tfrac{1}{2}\right]$$

Then the distance requested in the problem is

$$\Delta y_{\text{dark}} = \frac{\lambda L}{d}\left[1+\tfrac{1}{2}\right] - \frac{\lambda L}{d}\left[0+\tfrac{1}{2}\right] = \frac{\lambda L}{d} = 2.62\text{ mm}$$ ■

Finalize: This spacing is large enough for easy resolution of adjacent fringes. The distance between minima is the same as the distance between maxima. We expected this equality since the angles are small:

$$\theta = (2.62 \text{ mm})/(1.20 \text{ m}) = 0.002\ 18 \text{ rad} = 0.125°$$

Only when the angular spacing exceeds about 3° does $\sin\theta$ differ from $\tan\theta$ when they are written to three significant figures.

---

11. Two radio antennas separated by $d = 300$ m as shown in Figure P37.11 simultaneously broadcast identical signals at the same wavelength. A car travels due north along a straight line at position $x = 1\ 000$ m from the center point between the antennas, and its radio receives the signals. (a) If the car is at the position of the second maximum after that at point O when it has traveled a distance $y = 400$ m northward, what is the wavelength of the signals? (b) How much farther must the car travel to encounter the next minimum in reception? *Note:* Do not use the small-angle approximation in this problem.

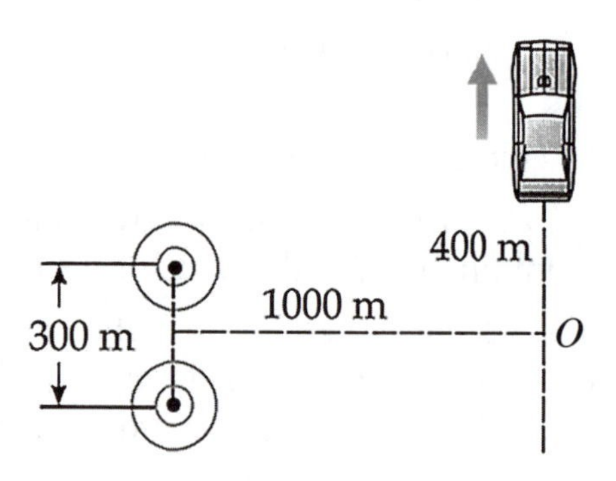

Figure P37.11

Solution

Conceptualize: The car is farther from the south transmitter than from the north transmitter, by two wavelengths. We estimate around 100 m for the wavelength. If the small-angle approximation were good, we would have maxima a 0, at 200 m, at 400 m, and at 600 m along the northward road. The next minimum would be at 500 m, which is 100 m of extra travel. Probably the accurate answer is somewhat more than 100 m.

Categorize: The problem suggests that with the conditions given, the small angle approximation does not work well. That is, $\sin\theta$, $\tan\theta$, and θ are significantly different. We use the Fraunhofer interference model, treating the waves from the two sources as moving along essentially parallel rays.

Analyze:

(a) At the $m = 2$ maximum, from $\tan\theta = y/L$ we have

$$\theta = \tan^{-1}\frac{y}{L} = \tan^{-1}\frac{400 \text{ m}}{1\ 000 \text{ m}} = 21.8°$$

so $d\sin\theta = m\lambda$

gives $\lambda = \dfrac{d\sin\theta}{m} = \dfrac{(300 \text{ m})(\sin 21.8°)}{2} = 55.7 \text{ m}$ ■

(b) The next minimum encountered is the $m = 2$ minimum, which is the third minimum away from the central maximum.

In that direction $d \sin\theta = \left[m + \frac{1}{2}\right]\lambda$: $d \sin\theta = \frac{5}{2}\lambda$

so $\theta = \sin^{-1}\dfrac{5\lambda}{2d} = \sin^{-1}\dfrac{5(55.7 \text{ m})}{2(300 \text{ m})} = \sin^{-1} 0.464 = 27.7°$

and $y = L\tan\theta = (1\,000 \text{ m})\tan 27.7° = 524 \text{ m}$

Therefore, the car must travel an additional 124 m. ■

Finalize: Our estimates were good in order-of-magnitude terms. If we considered Fresnel interference, we would more precisely find

(a) $\lambda = \frac{1}{2}\left(\sqrt{550^2 + 1\,000^2} - \sqrt{250^2 + 1\,000^2}\right) = 55.2 \text{ m}$

(b) $\Delta y = 123 \text{ m}$

Note well that interference of waves from side-by-side sources works in the same way for all kinds of waves. Observing how signals from side-by-side sources add together can be thought of as a definitive test for whether the signals are waves or classical particles.

---

16. In Figure P37.16 (not to scale), let L = 1.20 m and d = 0.120 mm and assume the slit system is illuminated with monochromatic 500-nm light. Calculate the phase difference between the two wave fronts arriving at P when (a) $\theta = 0.500°$ and (b) $y = 5.00$ mm. (c) What is the value of θ for which the phase difference is 0.333 rad? (d) What is the value of θ for which the path difference is $\lambda/4$?

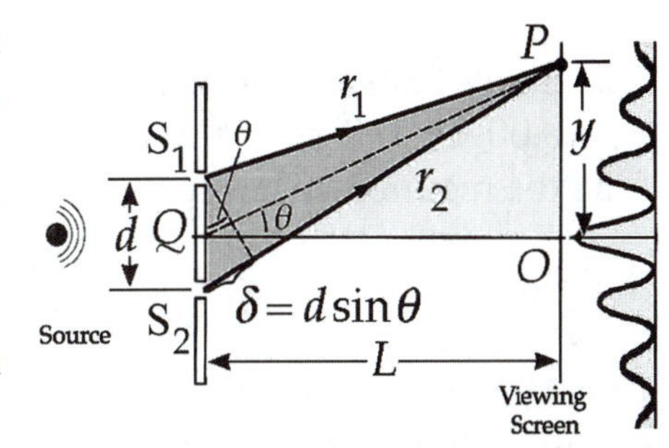

Figure P37.16 (modified)

Solution

Conceptualize: In parts (a) and (b), the phase differences will likely be between 0 and 4π. In parts (c) and (d), the angles will likely be between 0 and 2°. After doing some calculations, we will be able to sort the four cases into order as describing points P from the one closest to O to the farthest.

Categorize: Instead of the usual $d \sin\theta = m\lambda$, we will use more basic equations describing Young's experiment, from which $d \sin\theta = m\lambda$ is derived.

Analyze:

(a) The path difference is

$$\delta = d \sin \theta = (0.120 \times 10^{-3} \text{ m})(\sin 0.500°) = 1.05 \times 10^{-6} \text{ m}$$

The phase difference is $\phi = \dfrac{2\pi\delta}{\lambda} = \dfrac{2\pi\left(1.05 \times 10^{-6} \text{ m}\right)}{500 \times 10^{-9} \text{ m}} = 13.2 \text{ rad}$ ■

This produces the same effect as $\phi = 13.159 - 4\pi = 0.593 \text{ rad} = 34.0°$. ■

(b) $\tan\theta = \dfrac{y}{L} = \dfrac{5.00\times10^{-3}\text{ m}}{1.20\text{ m}} \approx \sin\theta$

$$\phi = \frac{2\pi d\sin\theta}{\lambda} = \frac{2\pi\left(0.120\times10^{-3}\text{ m}\right)\left(4.17\times10^{-3}\right)}{500\times10^{-9}\text{ m}} = 2\pi\text{ rad} = 0 \quad ■$$

(c) $\phi = \dfrac{2\pi d\sin\theta}{\lambda}$: $\theta = \sin^{-1}\left[\dfrac{\lambda\phi}{2\pi d}\right] = \sin^{-1}\dfrac{\left(500\times10^{-9}\text{ m}\right)(0.333)}{2\pi\left(1.20\times10^{-4}\text{ m}\right)} = 0.012\,7°$ ■

(d) $\dfrac{\lambda}{4} = d\sin\theta$: $\theta = \sin^{-1}\left[\dfrac{\lambda}{4d}\right] = \sin^{-1}\left[\dfrac{\left(500\times10^{-9}\text{ m}\right)}{4\left(1.20\times10^{-4}\text{ m}\right)}\right] = 0.059\,7°$ ■

Finalize: In order of distance away from point *O*, the points considered in this problem are c, d, b, and a. Points c and d are between the zero-order maximum and the first minimum. Point b is at the first side maximum, and point a is a bit beyond the second side maximum.

---

22. Show that the two waves with wave functions given by $E_1 = 6.00\sin(100\,\pi t)$ and $E_2 = 8.00\sin(100\,\pi t + \pi/2)$ add to give a wave with the wave function $E_R\sin(100\pi t + \phi)$. Find the required values for E_R and ϕ.

Solution

Conceptualize: We are to prove that the sum of two sinusoidal oscillations is a sinusoidal oscillation. We are to find the amplitude and phase constant of the resultant. These two oscillations are only 90° out of phase so the amplitude of their sum should be a bit larger than 8 and the phase constant somewhat smaller than 90°.

Categorize: We will use trigonometric identities.

Analyze: We write

$$E_1 + E_2 = 6.00\sin(100\,\pi t) + 8.00\sin(100\,\pi t + \pi/2)$$

$$= 6.00\sin(100\,\pi t) + 8.00\sin(100\,\pi t)\cos(\pi/2) + 8.00\cos(100\,\pi t)\sin(\pi/2)$$

$$E_1 + E_2 = 6.00\sin(100\,\pi t) + 8.00\cos(100\,\pi t)$$

Also, $E_1 + E_2 = E_R\sin(100\,\pi t + \phi) = E_R\sin(100\,\pi t)\cos\phi + E_R\cos(100\,\pi t)\sin\phi$

Our two expressions for $E_1 + E_2$,

$$6.00\sin(100\,\pi t) + 8.00\cos(100\,\pi t) = E_R\sin(100\,\pi t)\cos\phi + E_R\cos(100\,\pi t)\sin\phi$$

are equal if we require both $6.00 = E_R\cos\phi$ and $8.00 = E_R\sin\phi$

Then $6.00^2 + 8.00^2 = E_R^2(\cos^2\phi + \sin^2\phi) = E_R^2$ so $E_R = 10.0$ ■

and $\tan\phi = E_R \sin\phi / E_R \cos\phi = 8.00/6.00 = 1.33$ so $\phi = 53.1°$ ■

Finalize: The textbook describes interference between waves of equal amplitude. Here we have demonstrated interference of waves of different amplitude.

The calculation is more direct using phasors, introduced in Chapter 33. Let the x axis lie along $\vec{\mathbf{E}}_1$ in phase space.

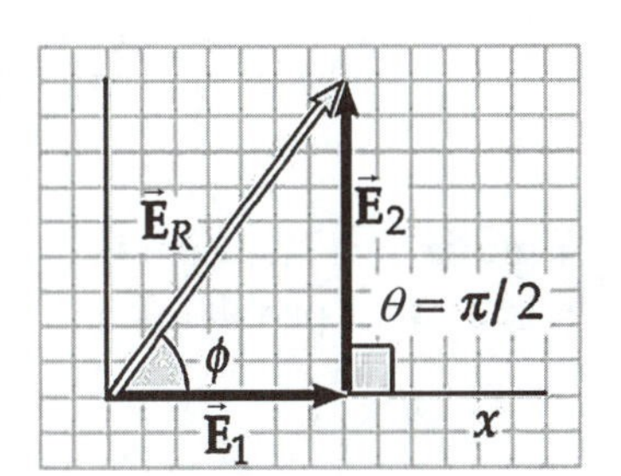

Its component form is then $\vec{\mathbf{E}}_1 = 6.00\,\hat{\mathbf{i}} + 0\,\hat{\mathbf{j}}$

The components of $\vec{\mathbf{E}}_2$ are

$$\vec{\mathbf{E}}_2 = 8.00\cos\left(\frac{\pi}{2}\right)\hat{\mathbf{i}} + 8.00\sin\left(\frac{\pi}{2}\right)\hat{\mathbf{j}} = 8.00\,\hat{\mathbf{j}}$$

The resultant is $\vec{\mathbf{E}}_R = \vec{\mathbf{E}}_1 + \vec{\mathbf{E}}_2 = 6.00\hat{\mathbf{i}} + 8.00\hat{\mathbf{j}}$

with amplitude $\sqrt{6.00^2 + 8.00^2} = 10.0$

and phase $\phi = \tan^{-1}\left(\dfrac{8.00}{6.00}\right) = 0.927$ rad

Thus, $E_R = 10.0\sin(100\pi t + 0.927)$

---

23. In Figure P37.16, let L = 120 cm and d = 0.250 cm. The slits are illuminated with coherent 600-nm light. Calculate the distance y above the central maximum for which the average intensity on the screen is 75.0% of the maximum.

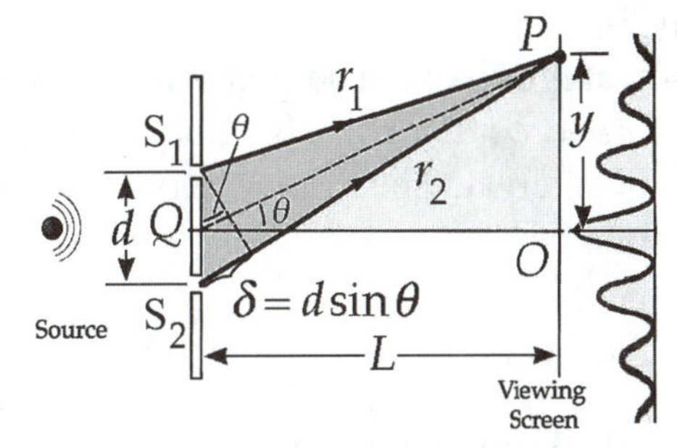

Figure P37.16 (modified)

Solution

Conceptualize: Visually, this point on the screen would be judged as part of the central bright fringe. We estimate a distance on the order of a centimeter.

Categorize: We will use the result stated in the chapter for intensity in a double-slit interference pattern,…

Analyze: …which is $I = I_{\text{max}}\cos^2\left[\dfrac{\pi d\sin\theta}{\lambda}\right]$

For small θ, from the drawing, $\sin\theta \approx \dfrac{y}{L}$

Substituting and solving gives $y = \dfrac{\lambda L}{\pi d}\cos^{-1}\sqrt{\dfrac{I}{I_{max}}}$

Next, with $I = 0.750 I_{max}$, we can substitute a value for each variable:

$$y = \frac{\left(6.00 \times 10^{-7}\text{ m}\right)(1.20\text{ m})}{\pi\left(2.50 \times 10^{-3}\text{ m}\right)}\cos^{-1}\sqrt{0.750} = 0.0480\text{ mm}$$ ■

Finalize: The distance is smaller than we estimated. The slit spacing is coarser than in some other problems and the distance to the screen is not large.

---

27. Two narrow, parallel slits separated by 0.850 mm are illuminated by 600-nm light, and the viewing screen is 2.80 m away from the slits. (a) What is the phase difference between the two interfering waves on a screen at a point 2.50 mm from the central bright fringe? (b) What is the ratio of the intensity at this point to the intensity at the center of a bright fringe?

Solution

Conceptualize: We cannot accurately predict the relative intensity at the point of interest without actually doing the calculation. The waves from each slit could meet in phase ($\phi = 0$) to produce a bright spot of **constructive interference**. They could add out of phase ($\phi = 180°$) to produce a dark region of **destructive interference**. They could most likely combine with the phase difference somewhere between these extremes, $0 < \phi < 180°$, so that the relative intensity will be described by $0 < I/I_{max} < 1$.

Categorize: The phase angle depends on the path difference δ of the waves according to $\phi = 2\pi\delta/\lambda$. This phase difference is used to find the average intensity at the point of interest. Then the relative intensity is simply this intensity divided by the maximum intensity.

Analyze:

(a) Using the variables shown in the diagram for Problem 7 we have, since, $y << L$,

$$\phi = \frac{2\pi d}{\lambda}\sin\theta = \frac{2\pi d}{\lambda}\left(\frac{y}{\sqrt{y^2 + L^2}}\right) \approx \frac{2\pi\, yd}{\lambda L}$$

$$\phi = \frac{2\pi\left(0.850 \times 10^{-3}\text{ m}\right)(0.00250\text{ m})}{\left(600 \times 10^{-9}\text{ m}\right)(2.80\text{ m})} = 7.95\text{ rad}$$

This is larger than 2π rad, so this point is beyond the first side maximum. As far as the intensity is concerned, the phase difference amounts to

$$7.95 - 2\pi = 1.66\text{ rad} = 95.4°$$ ■

(b) $$\frac{I}{I_{\max}} = \frac{\cos^2\left(\frac{\pi d}{\lambda}\sin\theta\right)}{\cos^2\left(\frac{\pi d}{\lambda}\sin\theta_{\max}\right)} = \frac{\cos^2\left(\frac{\phi}{2}\right)}{\cos^2(m\pi)} = \cos^2\left(\frac{\phi}{2}\right)$$

$$\frac{I}{I_{\max}} = \cos^2\left(\frac{95.4^\circ}{2}\right) = 0.453$$ ■

Finalize: At this point, the waves show **partial interference** so that the combination is about half the brightness found at the central maximum. We should remember that the equations used in this solution do not account for the diffraction caused by the finite width of each slit. This diffraction effect creates an "envelope" that diminishes in intensity away from the central maximum, as shown by the dotted lines in textbook Figure 37.7 and Figure P37.56. Therefore, the relative intensity at y = 2.50 mm will actually be slightly less than 0.453.

---

34. An oil film (n = 1.45) floating on water is illuminated by white light at normal incidence. The film is 280 nm thick. Find (a) the wavelength and color of the light in the visible spectrum most strongly reflected and (b) the wavelength and color of the light in the spectrum most strongly transmitted. Explain your reasoning.

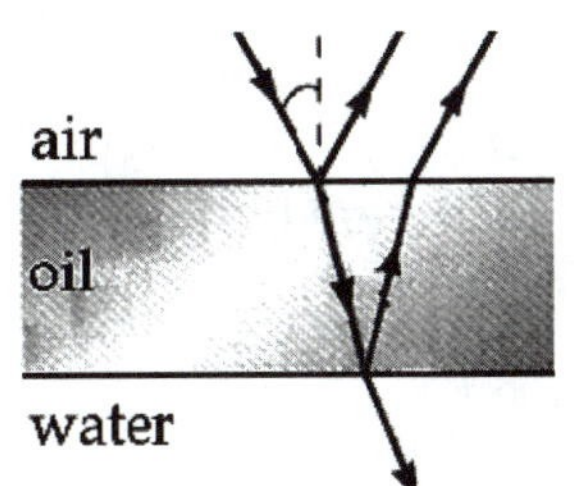

Solution

Conceptualize: We cannot guess the answers. The answers to (a) and (b) need not be colors that an artist would think of as opposites, but their wavelengths should be related by a simple fraction like 5/4. Light of one color must go through a half-integer number of oscillations in the same geometrical space that light of the other color goes through an integer number of oscillations.

Categorize: For thin-film interference, there are too many cases and variations to memorize equations for all of them. We will puzzle it through, thinking of all of the phase shifts of light reflected from the bottom of the film versus light reflected from the top. Table 34.1 in the textbook lists names for colors of various wavelengths.

Analyze: The light reflected from the top of the oil film undergoes phase reversal. Since 1.45 > 1.33, the light reflected from the bottom undergoes no reversal. For constructive interference of reflected light, we then have

$$2nt = \left[m + \tfrac{1}{2}\right]\lambda \quad \text{or} \quad \lambda_m = \frac{2nt}{m + \frac{1}{2}} = \frac{2(1.45)(280\text{ nm})}{m + \frac{1}{2}}$$

(a) Substituting for m, we have $m = 0$: $\lambda_0 = 1\,624$ nm (infrared)

$m = 1$: $\lambda_1 = 541$ nm (green) ■

$m = 2$: $\lambda_2 = 325$ nm (ultraviolet)

Both infrared and ultraviolet light are invisible to the human eye, so the dominant color is green. ■

(b) Any light that is not reflected is transmitted. To find the color transmitted most strongly, we find the wavelengths reflected least strongly. According to the condition for destructive interference, $2nt = m\lambda$.

Therefore, $\lambda = \dfrac{2nt}{m} = \dfrac{2(1.45)(280\text{ nm})}{m}$

For $m = 1$, $\lambda = 812$ nm (near infrared)

$m = 2$, $\lambda = 406$ nm (violet) ■

$m = 3$, $\lambda = 271$ nm (ultraviolet)

Thus, violet is the only visible color not attenuated by reflection, and the dominant color in the transmitted light. ■

Finalize: The ratio of the strongly and weakly reflected wavelengths is 541 nm/406 nm = 1.33 = 4/3, a ratio of integers as predicted. In case you did not recognize it, $2nt$ is the *optical path length* down and back through the film. Because we include the index of refraction here, every wavelength we write down is a wavelength in vacuum.

---

35. An air wedge is formed between two glass plates separated at one edge by a very fine wire as shown in Figure P37.35. When the wedge is illuminated from above by 600-nm light and viewed from above, 30 dark fringes are observed. Calculate the diameter d of the wire.

Figure P37.35

Solution

Conceptualize: The diameter of the wire is probably less than 0.1 mm since it is described as a "very fine wire."

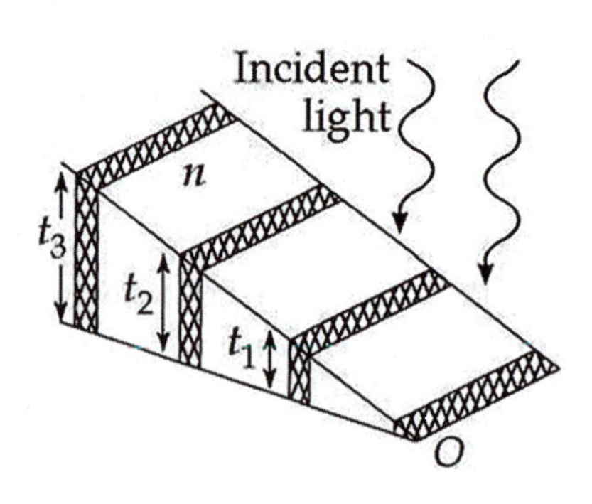

Categorize: Light reflecting from the bottom surface of the top plate undergoes no phase shift, while light reflecting from the top surface of the bottom plate is shifted by π, and also has to travel an extra distance $2t$, where t is the local thickness of the air wedge. Then…

Analyze: …for destructive interference, $2t = m\lambda \quad (m = 0, 1, 2, 3, \ldots)$.

The first dark fringe appears where $m = 0$ at the line of contact between the plates. The 30th dark fringe gives for the diameter of the wire $2d = 29\lambda$, and $d = 14.5\lambda$.

The diameter of the wire is then

$$d = 14.5\lambda = 14.5\left(600 \times 10^{-9}\text{m}\right) = 8.70\ \mu\text{m}$$ ■

Finalize: This wire is not only less than 0.1 mm thick; it is much thinner than a typical human hair (~50 μm).

---

40. Mirror M_1 in Active Figure 37.13 is moved through a displacement ΔL. During this displacement, 250 fringe reversals (formation of successive dark or bright bands) are counted. The light being used has a wavelength of 632.8 nm. Calculate the displacement ΔL.

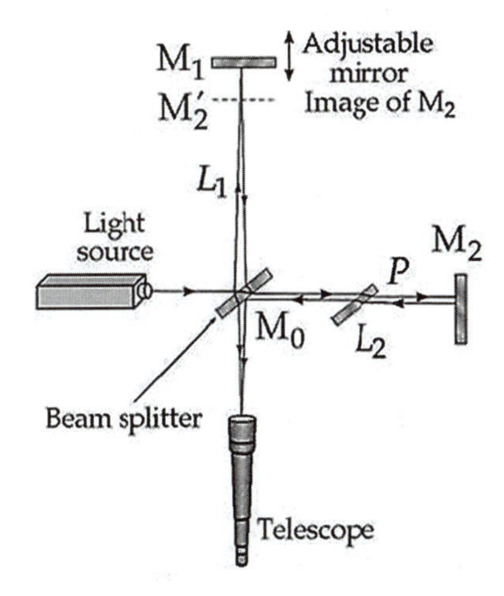

Figure 37.13

Solution

Conceptualize: An interferometer can be used for very precise distance measurements. Here the displacement should be much less than a millimeter.

Categorize: The basic idea is just that waves with a relative shift of an integer number of wavelengths add crest on crest to interfere constructively, and an extra half-wavelength makes them add crest on trough to interfere destructively.

Analyze: When the mirror on one arm is displaced by ΔL, the path difference increases by $2\Delta L$. A shift resulting in the reversal from dark to bright or bright to dark requires a path length change of one-half wavelength. Therefore, since in this case $m = 250$,

$$2\Delta L = \frac{m\lambda}{2} \quad \text{and} \quad \Delta L = \frac{m\lambda}{4} = \frac{250\left(6.328 \times 10^{-7}\text{m}\right)}{4} = 3.955 \times 10^{-5}\text{m}$$ ■

Finalize: One situation where optical interferometry is used for distance measurement is in calibrating machinists' gauge blocks.

---

54. Measurements are made of the intensity distribution within the central bright fringe in a Young's interference pattern (see Fig. 37.6). At a particular value of y, it is found that $I/I_{max} = 0.810$ when 600-nm light is used. What wavelength of light should be used to reduce the relative intensity at the same location to 64.0% of the maximum intensity?

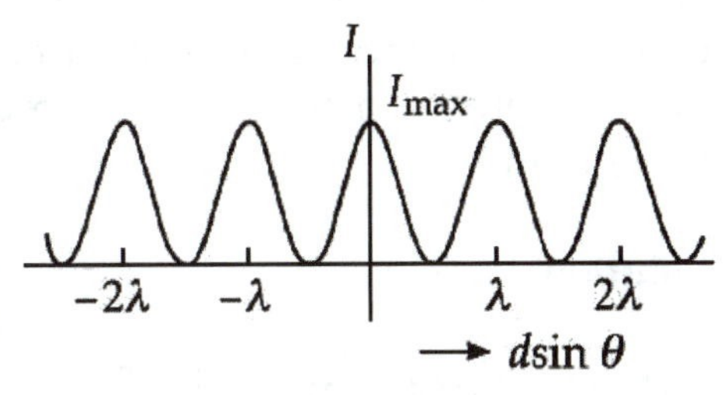

Figure 37.6

Solution

Conceptualize: Instead of being at the 81%-of-maximum point, we want to be farther down the curve, at 64%. The central maximum should be narrower on the screen, and so all the other fringes should be closer together. To make the pattern fold up a bit in $d\sin\theta = m\lambda$, we want a shorter wavelength. It will not be just (64/81)600 nm = 474 nm, because the graph of intensity versus screen position curves; but that is a rough estimate.

Categorize: We will use the intensity distribution in a double-slit interference pattern.

Analyze: From Equation 37.14, $\dfrac{I_1}{I_{\text{max}}} = \cos^2\left(\dfrac{\pi d\sin\theta}{\lambda_1}\right) = 0.810$

Therefore,

$$\pi d\sin\theta = \lambda_1 \cos^{-1}\sqrt{0.810} = (600\text{ nm})\cos^{-1}(0.900) = 271\text{ nm}$$

For the second wavelength, $\dfrac{I_2}{I_{\text{max}}} = \cos^2\left(\dfrac{\pi d\sin\theta}{\lambda_2}\right) = 0.640$

and $\lambda_2 = \dfrac{\pi d\sin\theta}{\cos^{-1}\sqrt{0.640}}$

But this is simply $\lambda_2 = \dfrac{271\text{ nm}}{\cos^{-1}\sqrt{0.640}} = 421\text{ nm}$ ■

Finalize: The light changes from orange to violet as the interference pattern folds up and the central maximum gets narrower. Can you show it to yourself and others in laboratory? Young's experiment can be thought of as a convenient way to measure the wavelengths of visible light, small as they are.

---

56. Consider the double-slit arrangement shown in Figure P37.56, where the slit separation is d and the distance from the slit to the screen is L. A sheet of transparent plastic having an index of refraction n and thickness t is placed over the upper slit. As a result, the central maximum of the interference pattern moves upward a distance y'. Find y'.

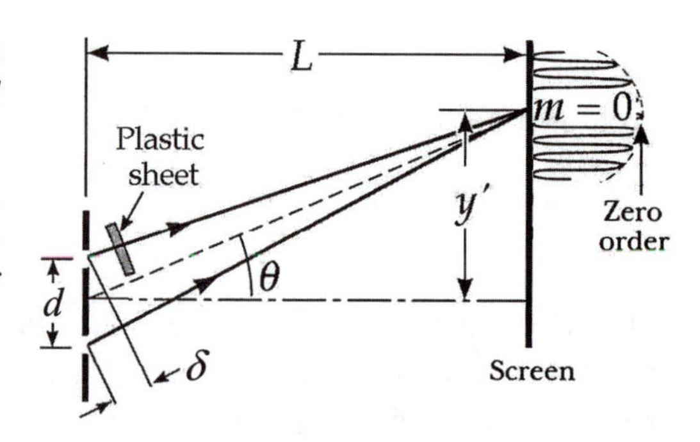

Figure P37.56

Solution

Conceptualize: Since the film shifts the pattern upward, we could expect y' to be proportional to n, t, and L.

Categorize: The film increases the optical path length of the light passing through the upper slit, so the physical distance of this path must be shorter for the waves to meet in phase ($\phi = 0$) to produce the central maximum. Thus, the added distance δ traveled by the light from the lower slit must introduce a phase difference equal to that introduced by the plastic film.

Analyze: First calculate the additional phase difference due to the plastic. Recall that the relation between phase difference and path difference is $\phi = 2\pi\delta/\lambda$. The presence of plastic affects this by changing the wavelength of the light, so that the phase changes of the light in air and plastic, as it travels over the thickness t, are $\phi_{air} = \dfrac{2\pi t}{\lambda_{air}}$ and $\phi_{plastic} = \dfrac{2\pi t}{\lambda_{air}/n}$

Thus, plastic causes an additional phase change of $\Delta\phi = \dfrac{2\pi t}{\lambda_{air}}(n-1)$

Next, in order to interfere constructively, we must calculate the additional distance that the light from the bottom slit must travel.

$$\delta = \frac{\Delta\phi\,\lambda_{air}}{2\pi} = t(n-1)$$

In the small angle approximation we can write $\delta \approx y'd/L$,

so

$$y' \approx \frac{tL(n-1)}{d} \quad \blacksquare$$

If we choose not to make the small-angle approximation, we have

$\delta = d\sin\theta$ where $\tan\theta = y'/L$

giving $y' = L\tan\theta = L\tan\left[\sin^{-1}\left(\dfrac{\delta}{d}\right)\right] = L\tan\left[\sin^{-1}\left(\dfrac{(n-1)t}{d}\right)\right]$ $\blacksquare$

This in turn can be written as $y' = \dfrac{L(n-1)t}{\sqrt{d^2-(n-1)^2t^2}}$

Finalize: As expected, y' is proportional to t and L. It increases with increasing n, being proportional to $(n-1)$. It could not be proportional to n itself, as we first guessed, because in the case $n = 1$ the film becomes just like air and the offset distance y' must go to zero. We see y' is also inversely proportional to the slit separation d, which makes sense since slits that are closer together make a wider interference pattern.

---

65. Astronomers observe a 60.0-MHz radio source both directly and by reflection from the sea as shown in Figure P37.15. If the receiving dish is 20.0 m above sea level, what is the angle of the radio source above the horizon at first maximum?

Solution

Conceptualize: Interference of radio waves follows the same rules as interference of visible light. This arrangement is a bit like Lloyd's mirror, mentioned in the chapter text. We might guess an angle of a few degrees or less, just because sources of coherent visible light show interference most conveniently when they are close together.

Categorize: In our diagram, one radio wave reaches the receiver R directly from the distant source at an angle θ above the horizontal. The other wave undergoes phase reversal as it reflects from the water at P. The line midway between P and R' is a wavefront, along which the phase would be constant if not for the reflection. This line lets us label the extra distance that the reflecting wave has to travel as δ.

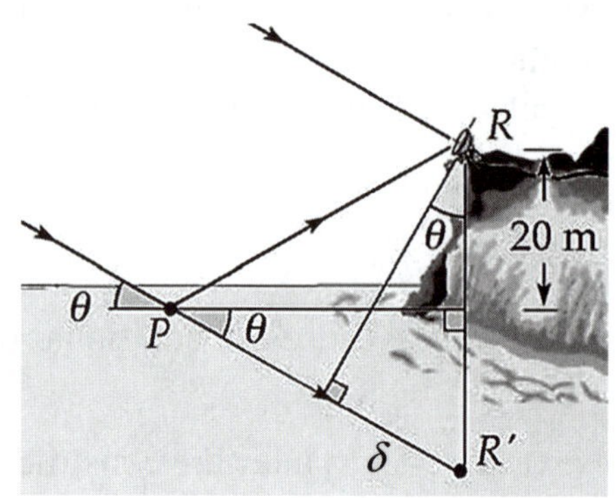

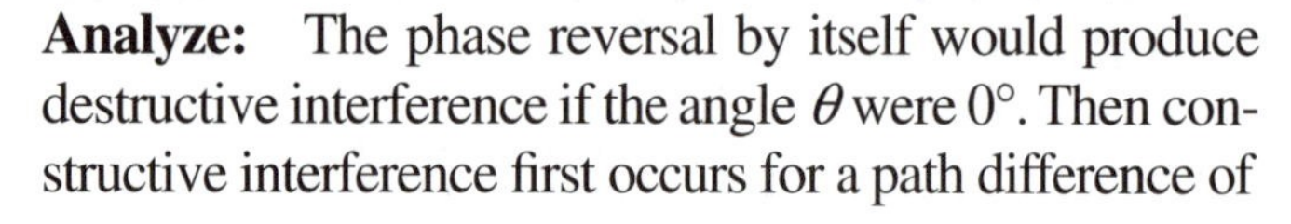

Analyze: The phase reversal by itself would produce destructive interference if the angle θ were 0°. Then constructive interference first occurs for a path difference of

$$\delta = \lambda/2 \qquad [1]$$

The telescope is 20 m vertically above the water surface and the telescope's mirror image R' is 20 m below the water surface. The angles θ in the diagram are equal because their sides are perpendicular, right side to right side and left side to left side.

So the path difference is $\delta = 2(20.0\text{ m})\sin\theta = (40.0\text{ m})\sin\theta$

The wavelength is $\lambda = \dfrac{c}{f} = \dfrac{3.00\times10^8\text{ m/s}}{60.0\times10^6\text{ Hz}} = 5.00\text{ m}$

Substituting for δ and λ in Eq. [1], $(40.0\text{ m})\sin\theta = \dfrac{5.00\text{ m}}{2}$

Solving for the angle θ gives $\theta = \sin^{-1}\left(\dfrac{5.00\text{ m}}{80.0\text{ m}}\right) = 3.58°$ ■

Finalize: Drawing and carefully analyzing the diagram were keys to the solution. The dish of the radio telescope probably accepts waves over only a limited range of angle, so alternating minima and maxima at larger angles may be unobservable.

---

68. The condition for constructive interference by reflection from a thin film in air as developed in Section 37.5 assumes nearly normal incidence. **What If?** Suppose the light is incident on the film at a nonzero angle θ_1 (relative to the normal). The index of refraction of the film is n and the film is surrounded by vacuum. Find the condition for constructive interference that relates the thickness t of the film, the index of refraction n of the film, the wavelength λ of the light, and the angle of incidence θ_1.

Solution

Conceptualize: We need to analyze the textbook's Figure 37.10 without the simplifying assumption of normal incidence.

Categorize: We draw the diagram as shown. The ray reflected from the top of the film undergoes phase reversal. The ray reflected from the bottom surface has no phase change on reflection. We must figure out geometrically the extra distance it travels relative to the upper ray.

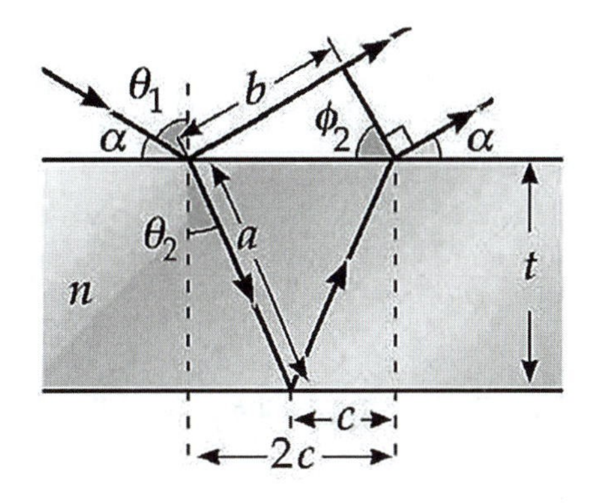

Analyze: Because refraction out of the film is the opposite of refraction into the film, the ray reflected from the bottom surface leaves the film parallel to the ray from the top surface. Then, since the angles marked α are equal, $\phi_2 + \alpha + 90° = 180°$

From the angle of incidence, $\theta_1 + \alpha = 90°$

so $\phi_2 = \theta_1$

Before they head off together along parallel paths, one beam travels a distance b and undergoes phase reversal on reflection; the other beam travels a distance $2a$ inside the film, where its wavelength is λ/n. The number of cycles it completes along this path is

$$\frac{2a}{\lambda/n} = \frac{2na}{\lambda}$$

The optical path length in the film is $2na$. Then the shift between the two outgoing rays is

$$\delta = 2na - b - \frac{\lambda}{2}$$

where a and b are as shown in the ray diagram, n is the index of refraction, and the term $\lambda/2$ is due to phase reversal at the top surface. For constructive interference, $\delta = m\lambda$ where m has integer values.

This condition becomes $2na - b = \left[m + \frac{1}{2}\right]\lambda$ **[1]**

From the figure's geometry, $a = \dfrac{t}{\cos\theta_2}$

and $c = a\sin\theta_2 = \dfrac{t\sin\theta_2}{\cos\theta_2}$

Further, $b = 2c\sin\theta_1 = \dfrac{2t\sin\theta_2}{\cos\theta_2}\sin\theta_1$

Also, from Snell's law, $\sin\theta_1 = n\sin\theta_2$

so $b = \dfrac{2nt\sin^2\theta_2}{\cos\theta_2}$

With these results, the condition for constructive interference given in Equation [1] becomes:

$$2n\frac{t}{\cos\theta_2} - \frac{2nt\sin^2\theta_2}{\cos\theta_2} = \frac{2nt}{\cos\theta_2}\left(1 - \sin^2\theta_2\right) = \left[m + \tfrac{1}{2}\right]\lambda$$

or $2nt\cos\theta_2 = \left[m + \tfrac{1}{2}\right]\lambda$

Finally, we apply the law of refraction at entry, $\sin\theta_1 = n\sin\theta_2$, like this:

$$2nt\cos\theta_2 = 2nt\sqrt{1 - \sin^2\theta_2} = 2nt\sqrt{1 - \left(\sin^2\theta_1\right)/n^2}$$

so the reflected light will show constructive interference provided that

$$2nt\sqrt{1 - \left(\sin^2\theta_1\right)/n^2} = \left(m + \tfrac{1}{2}\right)\lambda \qquad \text{where } m = 0,\ 1,\ 2, \ldots$$ ■

Finalize: If the light comes in perpendicular to the film, $\theta_1 = 0$ and the equation we proved reduces to $2nt = (m + 1/2)\lambda$, in agreement with Equation 37.17. As you turn the film, the color you see at one spot changes. The equation describes this as λ for constructive interference changing when θ_1 changes. This iridescence effect, characteristic of interference colors, is very different from the way pigment colors behave. Abalone shells show beautiful iridescence of thin sheets.

Was this problem too tough for you? If so, do Problem 67, where numbers can automatically keep track of what is known and what is unknown. But then come back to this problem again and do it.

38

Diffraction Patterns and Polarization

EQUATIONS AND CONCEPTS

In **single-slit Fraunhofer diffraction**, the total phase difference between waves from the top and bottom portions of the slit will depend on the angle θ, which determines the direction to a point on the screen. The diffraction pattern consists of a broad central bright band and a series of less intense narrow side bands.

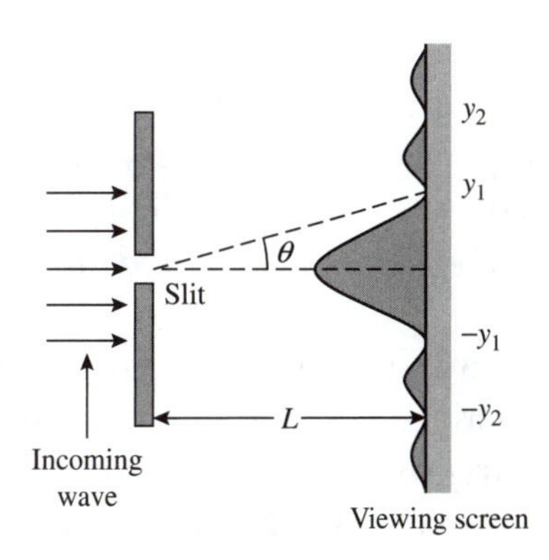

The **general condition for destructive interference in single slit diffraction** requires that $\sin\theta$ equal an integer multiple of (λ/a).

$$\sin\theta_{dark} = m\frac{\lambda}{a} \quad \text{(destructive interference)} \tag{38.1}$$

$$m = \pm 1, \pm 2, \pm 3, \ldots$$

The **intensity** at any point on the screen is given relative to the intensity of the central maximum (I_{max}).

$$I = I_{max}\left[\frac{\sin(\pi a \sin\theta/\lambda)}{\pi a \sin\theta/\lambda}\right]^2 \tag{38.2}$$

Rayleigh's criterion states the limiting condition for the resolution of the images due to two nearby sources. *Two images are at the limit of resolution when the central maximum of one image is located at the first minimum of the other as illustrated in the figure at right.*

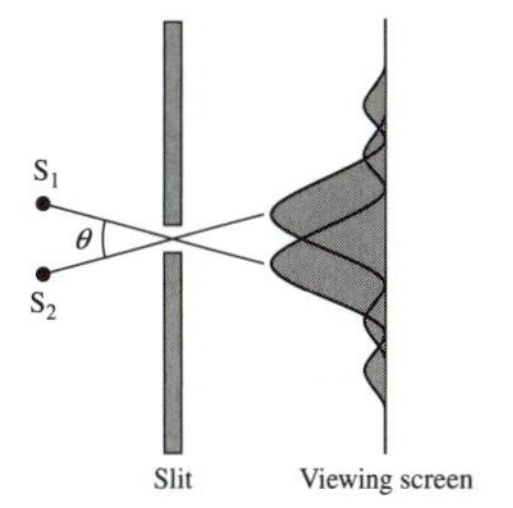

For a **slit**, the angular separation between the sources must be greater than the ratio of the wavelength to slit width.

$$\theta_{min} = \frac{\lambda}{a} \quad \text{(slit)} \tag{38.5}$$

For a **circular aperture** (or lens), the minimum angular separation depends on the diameter of the aperture (D).

$$\theta_{min} = 1.22\frac{\lambda}{D} \quad \text{(circular aperture)} \tag{38.6}$$

A **diffraction grating** (an array of a large number of parallel slits, each separated by a distance d) will produce an interference pattern in which there is a series of maxima for each wavelength. *Maxima due to wavelengths of different values comprise a spectral order denoted by an order number* (m). The figure at right shows the zeroth, first, and second order maxima in the diffraction grating intensity pattern of a monochromatic source.

$$d \sin\theta_{bright} = m\lambda \qquad m = 0, \pm 1, \pm 2, \pm 3, \ldots \tag{38.7}$$

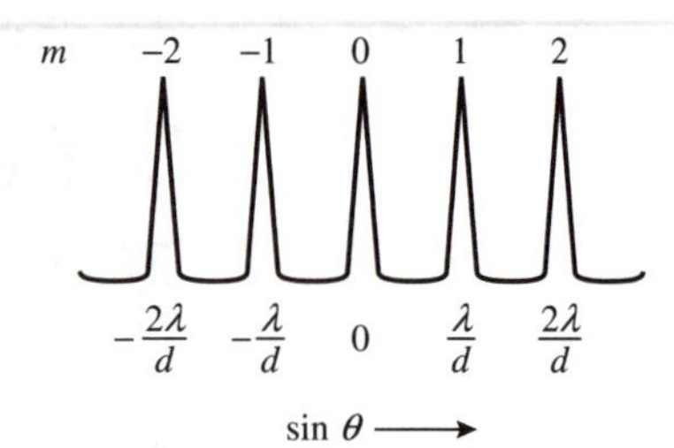

Bragg's law gives the conditions for constructive interference of x-rays reflected from the parallel planes of a crystalline solid separated by a distance d. *θ is the angle between the incident beam and the surface.*

$$2d \sin\theta = m\lambda \qquad m = 1, 2, 3, \ldots \tag{38.8}$$

Malus's law states that the fraction of initially polarized light (from a polarizer) that will be transmitted by a second sheet of polarizing material (the analyzer) depends on the square of the cosine of the angle between the transmission axis of the polarizer and that of the analyzer.

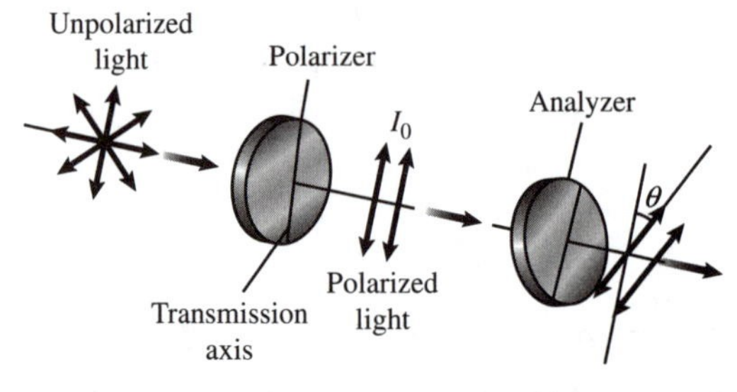

$$I = I_{max} \cos^2\theta \tag{38.9}$$

Brewster's law gives the angle of incidence (the polarizing angle, θ_p) for which the reflected beam will be completely polarized.

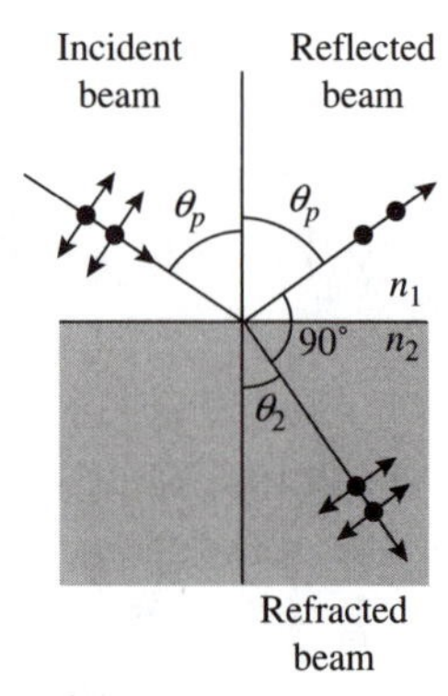

$$\tan\theta_p = \frac{n_2}{n_1} \tag{38.10}$$

REVIEW CHECKLIST

You should be able to:

- Determine the positions of the maxima and minima in a single-slit diffraction pattern and calculate the intensities of the secondary maxima relative to the intensity of the central maximum. (Sections 38.1 and 38.2)

- Calculate the intensities of interference maxima due to a double slit, expressed as a fraction of the intensity at the center of the pattern. (Section 38.2)
- Determine whether or not two sources under a given set of conditions are resolvable as defined by Rayleigh's criterion. (Section 38.3)
- Determine the positions of the principal maxima in the diffraction pattern of a diffraction grating. (Section 38.4)
- Describe the technique of x-ray diffraction and make calculations of the lattice spacing using Bragg's law. (Section 38.5)
- Describe how the state of polarization of a light beam can be determined by use of a polarizer-analyzer combination. Describe qualitatively the polarization of light by selective absorption, reflection, scattering, and double refraction. Make appropriate calculations using Malus's law and Brewster's law. (Section 38.6)

ANSWER TO AN OBJECTIVE QUESTION

5. Consider a wave passing through a single slit. What happens to the width of the central maximum of its diffraction pattern as the slit is made half as wide? (a) It becomes one-fourth as wide. (b) It becomes one-half as wide. (c) Its width does not change. (d) It becomes twice as wide. (e) It becomes four times as wide.

Answer The best answer is (d). Equation 38.1 describes the angles at which you get destructive interference; from it, we can obtain the width of the central maximum. For small angles, the equation can be rewritten as

$$\theta_m = \sin^{-1}(m\lambda/a) \approx m\lambda/a$$

Thus, as the width of the slit a is cut in half, the angle of the first destructive interference θ_1 doubles, and the width of the central maximum doubles as well.

□ □ □ □

ANSWER TO A CONCEPTUAL QUESTION

1. Why can you hear around corners, but not see around corners?

Answer Audible sound has wavelengths on the order of meters or centimeters, while visible light has wavelengths on the order of half a micron. In this world of breadbox-size objects, λ is comparable to the object size a for sound, and sound diffracts around walls and through doorways. But λ/a is much smaller for visible light passing ordinary-size objects or apertures, so light diffracts only through very small angles.

Another way of answering this question would be as follows. We can see by a small angle around a small obstacle or around the edge of a small opening. The side fringes in Figure 38.1 and the Arago spot in the center of Figure 38.3 (in the textbook) show this diffraction.

Conversely, we cannot always hear around corners. Out-of-doors, away from reflecting surfaces, have someone a few meters distant face away from you and whisper. The high-frequency, short-wavelength, information-carrying components of the sound do not diffract around his head enough for you to understand his words.

Suppose an opera singer loses the tempo and cannot immediately get it from the orchestra conductor. Then the prompter may make rhythmic kissing noises with her lips and teeth. Try it—you will sound like a birdwatcher trying to lure out a curious bird. This sound is clear on the stage but does not diffract around the prompter's box enough for the audience to notice it.

SOLUTIONS TO SELECTED PROBLEMS

7. A screen is placed 50.0 cm from a single slit, which is illuminated with light of wavelength 690 nm. If the distance between the first and third minima in the diffraction pattern is 3.00 mm, what is the width of the slit?

Solution

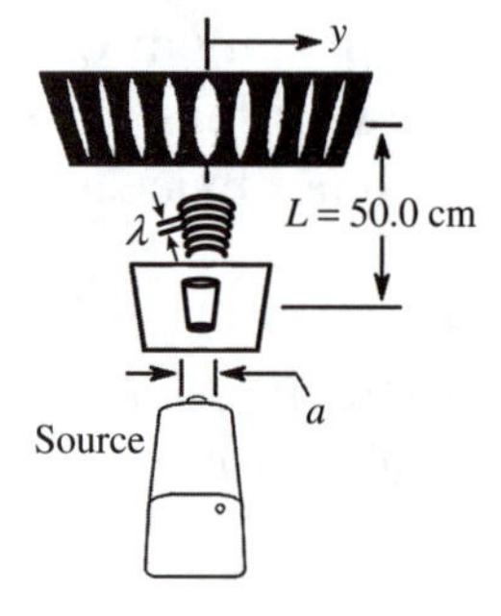

Conceptualize: We estimate between 0.1 mm and 1 mm. The width of the slit will be much greater than the wavelength of the light, but small enough to reveal that light is not a stream of classical particles.

Categorize: The textbook's analysis of single-slit diffraction will give us the answer. We will have to apply it both to the first and to the third minimum.

Analyze: In the equation for single-slit diffraction minima at small angles

$$\frac{y}{L} \approx \sin\theta_{\text{dark}} = \frac{m\lambda}{a}$$

we take differences between the first and third dark fringes, to see that

$$\frac{\Delta y}{L} = \frac{\Delta m\lambda}{a} \quad \text{with} \quad \Delta y = 3.00 \times 10^{-3}\text{ m} \quad \text{and} \quad \Delta m = 3 - 1 = 2$$

The width of the slit is then $a = \dfrac{\lambda L \Delta m}{\Delta y} = \dfrac{(690 \times 10^{-9}\text{ m})(0.500\text{ m})(2)}{3.00 \times 10^{-3}\text{ m}}$

$$a = 2.30 \times 10^{-4}\text{ m}$$ ■

Finalize: It is 0.230 mm; our estimate was good. It is the light that goes through the opening unimpeded that diffracts. We are not studying reflection from the edges of the slit. In sharp contrast to Newton's first law, the light does not travel just in a straight line. Diffraction can be thought of as an aspect of the basic dynamic of wave motion.

11. A diffraction pattern is formed on a screen 120 cm away from a 0.400-mm-wide slit. Monochromatic 546.1-nm light is used. Calculate the fractional intensity I/I_{max} at a point on the screen 4.10 mm from the center of the principal maximum.

Solution

Conceptualize: In an interference pattern of two side-by-side sources, the intensity can return to its maximum value at the center of each bright fringe. In the situation here, the fractional intensity must be less than 100%. The diffraction pattern is nowhere else so bright as at the center. If we are outside the central maximum, the intensity will be small.

Categorize: We use the intensity distribution in a single-slit diffraction pattern.

Analyze: First we find where we are. The angle to the side is small so

$$\sin\theta \approx \tan\theta = \frac{y}{L} = \frac{4.10\times 10^{-3}\text{ m}}{1.20\text{ m}} = 3.417\times 10^{-3}$$

The parameter controlling the intensity is

$$\frac{\pi a\sin\theta}{\lambda} = \frac{\pi\left(4.00\times 10^{-4}\text{ m}\right)\left(3.417\times 10^{-3}\right)}{546.1\times 10^{-9}\text{ m}} = 7.862\text{ rad}$$

This is between 2π and 3π, so the point analyzed is off in the second side fringe. The fractional intensity is

$$\frac{I}{I_{max}} = \left[\frac{\sin\left(\pi a\sin\theta/\lambda\right)}{\pi a\sin\theta/\lambda}\right]^2 = \left[\frac{\sin(7.862\text{ rad})}{7.862\text{ rad}}\right]^2 = 1.62\times 10^{-2}$$ ■

Finalize: In fact, 7.862 rad = 2.50π, so we have found the intensity very nearly at the center of the second side bright fringe. As we see, it is only a small percentage of the intensity at the center of the pattern.

---

19. A helium–neon laser emits light that has a wavelength of 632.8 nm. The circular aperture through which the beam emerges has a diameter of 0.500 cm. Estimate the diameter of the beam 10.0 km from the laser.

Solution

Conceptualize: A typical laser pointer makes a spot about 5 cm in diameter at 100 m, so the spot size at 10 km would be about 100 times bigger, or about 5 m across. Assuming that this HeNe laser is similar, we could expect a comparable beam diameter.

Categorize: We assume that the light is parallel and not diverging before it passes through and fills the circular aperture. However, after the light passes through the circular aperture, it will spread from diffraction according to Equation 38.6.

Analyze: The beam spreads into a cone of half-angle

$$\theta_{\text{min}} = 1.22\frac{\lambda}{D} = 1.22\left(\frac{632.8 \times 10^{-9}\text{ m}}{0.005\,00\text{ m}}\right) = 1.54 \times 10^{-4}\text{ rad}$$

The radius of the beam ten kilometers away is, from the definition of radian measure,

$$r_{\text{beam}} = \theta_{\text{min}}\left(1.00 \times 10^{4}\text{ m}\right) = 1.54\text{ m}$$

and its diameter is $d_{\text{beam}} = 2r_{\text{beam}} = 3.09\text{ m}$. ■

Finalize: The beam is several meters across as expected, and is about 600 times larger than the laser aperture. Since most HeNe lasers are low power units in the mW range, the beam at this range would be so spread out that it would be too dim to see on a screen.

---

21. Impressionist painter Georges Seurat created paintings with an enormous number of dots of pure pigment, each of which was approximately 2.00 mm in diameter. The idea was to have colors such as red and green next to each other to form a scintillating canvas, such as in his masterpiece, *A Sunday Afternoon on the Island of La Grande Jatte* (Fig. P38.21). Assume λ = 500 nm and a pupil diameter of 5.00 mm. Beyond what distance would a viewer be unable to discern individual dots on the canvas?

Solution

Conceptualize: If you are too far away from the painting, your eyes will not be able to resolve the dots of color. It is not too hard to get an experimental answer, as by seeing how far away you can see the separateness of millimeter marks on a ruler, or how far away you can read printed text. We estimate 5 m.

Categorize: When the pupil is open wide, the resolving power of human vision is probably limited by the coarseness of the light sensors on the retina. But we use Rayleigh's criterion as a handy indicator of how good our vision might be.

Analyze: We will assume that the dots are just touching and do not overlap, so that the distance between their centers is 2.00 mm. By Rayleigh's criterion, two dots separated center to center by 2.00 mm would be seen to overlap when

$$\theta_{\text{min}} = \frac{d}{L} = 1.22\frac{\lambda}{D} \quad \text{with} \quad d = 2.00\text{ mm},\ \lambda = 500\text{ nm}, \quad \text{and} \quad D = 5.00\text{ mm}$$

Thus, $$L = \frac{Dd}{1.22\lambda} = \frac{\left(5.00 \times 10^{-3}\text{ m}\right)\left(2.00 \times 10^{-3}\text{ m}\right)}{1.22\left(500 \times 10^{-9}\text{ m}\right)} = 16.4\text{ m}$$ ■

Finalize: An eagle's eye or any bionic eye, still with 5 mm pupil diameter, would be unable to resolve Seurat's dots beyond 16.4 m. Prof. Steven McCauley has made observations and reports that a viewer with normal vision cannot resolve pointillist dots more than about 2 m away.

---

24. White light is spread out into its spectral components by a diffraction grating. If the grating has 2 000 grooves per centimeter, at what angle does red light of wavelength 640 nm appear in first order?

Solution

Conceptualize: We expect an angle on the order of ten degrees, easily measurable on a tabletop and very precisely measurable with a spectrometer.

Categorize: We use the grating equation. Note that it is identical in form to the equation describing constructive interference with two side-by-side sources.

Analyze: The ruling engine that cut the diffraction grating (or the aluminum plate from which the gelatin or plastic was cast) sliced each centimeter into two thousand divisions. So the grating spacing is

$$d = \frac{1.00 \times 10^{-2}\text{ m}}{2\ 000} = 5.00 \times 10^{-6}\text{ m}$$

The light is deflected according to $d\sin\theta = m\lambda$

$$\theta = \sin^{-1}\frac{m\lambda}{d} = \sin^{-1}\frac{1(640\times10^{-9}\text{m})}{5.00\times10^{-6}\text{m}} = \sin^{-1}0.128 = 7.35° \qquad \blacksquare$$

Finalize: The textbook does not need to put a box around the equation stating that the grating spacing is the reciprocal of the number of grooves per unit width. Just think about a pie being cut into eighths for an eight-person family. A spotlight or the beam from a slide projector shining through a diffraction grating, onto a classroom projection screen, makes a beautiful display. An individual student can see the beauty of the visible spectrum by using a spectrometer with an ordinary light bulb.

---

28. A grating with 250 grooves/mm is used with an incandescent light source. Assume the visible spectrum to range in wavelength from 400 nm to 700 nm. In how many orders can one see (a) the entire visible spectrum and (b) the short-wavelength region of the visible spectrum?

Solution

Conceptualize: Compare this problem with Problem 30, immediately following this one. With 250 grooves per millimeter instead of 450 grooves per millimeter, the grating is coarser and the angles for constructive interference will be smaller. The red will be visible (if it is bright enough) in more than three orders and the blue in more than five orders.

Categorize: We use the grating equation, this time thinking with some care about the order of interference m as the unknown.

Analyze: The grating spacing is $d = \dfrac{1.00\text{ mm}}{250} = 4.00 \times 10^{-6}\text{ m}$.

(a) In each order of interference m, red light diffracts at a larger angle than the other colors with shorter wavelengths. We find the largest integer m satisfying

$$d\sin\theta = m\lambda \qquad \text{with} \qquad \lambda = 700\text{ nm}$$

With $\sin\theta$ having its largest possible value, we have

$$m = \frac{d\sin\theta}{\lambda} = \frac{(4.00\times10^{-6}\text{m})\sin 90°}{700\times10^{-9}\text{m}} = 5.71$$

Thus, the red light cannot be seen in the sixth order, and the full visible spectrum appears in only five orders. ■

(b) Now consider light at the boundary between violet and ultraviolet.

$$m = \frac{d\sin\theta}{\lambda} = \frac{(4.00\times10^{-6}\text{m})\sin 90°}{400\times10^{-9}\text{m}} = 10.0$$

and $m = 10$ ■

Finalize: Seeing more orders is not an advantage. It is natural to want a grating with closely spaced grooves so that the angles to be measured are nice and large for good precision. But note that the grating only works if the distance between grooves is larger than the wavelength.

"Order of interference" has a physical meaning: When you are looking at a particular color in third order, you are just three wavelengths farther from one grating groove than from the adjacent groove.

---

30. The hydrogen spectrum includes a red line at 656 nm and a blue-violet line at 434 nm. What are the angular separations between these two spectral lines for all visible orders obtained with a diffraction grating that has 4 500 grooves/cm?

Solution

Conceptualize: Many diffraction gratings yield several spectral orders within the viewing range of 90° to either side. So the angle between red and blue lines is probably 10° to 30°. (We call the color of the 434-nm line blue, but note that it is not the blue-green line of hydrogen at 486.1 nm.)

Categorize: The angular separation is the difference between the angles corresponding to the red and blue wavelengths for each visible spectral order according to the diffraction grating equation, $d\sin\theta = m\lambda$.

Analyze: The grating spacing is

$$d = 1.00 \times 10^{-2}\ \text{m}/4\ 500\ \text{grooves} = 2.22 \times 10^{-6}\ \text{m}$$

In the first-order spectrum ($m = 1$), the angles of diffraction are given by $\sin\theta = \lambda/d$. We have

$$\theta_{1r} = \sin^{-1}\frac{\lambda_r}{d} = \sin^{-1}\frac{656\times10^{-9}\ \text{m}}{2.22\times10^{-6}\ \text{m}} = \sin^{-1}0.295 = 17.17°$$

and for the light in the blue spectral line

$$\theta_{1b} = \sin^{-1}\frac{\lambda_b}{d} = \sin^{-1}\frac{434\times10^{-9}\ \text{m}}{2.22\times10^{-6}\ \text{m}} = \sin^{-1}0.195 = 11.26°$$

The angular separation is $\Delta\theta = \theta_{1r} - \theta_{1b} = 17.17° - 11.26° = 5.91°$. ■

In the second order of interference, with $m = 2$, similarly,

$$\Delta\theta_2 = \sin^{-1}\left(\frac{2\lambda_r}{d}\right) - \sin^{-1}\left(\frac{2\lambda_b}{d}\right) = 13.2°$$ ■

In the third order ($m = 3$), $\Delta\theta_3 = \sin^{-1}\left(\frac{3\lambda_r}{d}\right) - \sin^{-1}\left(\frac{3\lambda_b}{d}\right) = 26.5°$ ■

Examining the fourth order, we find the red line is not visible:

$$\theta_{4r} = \sin^{-1}\left(\frac{4\lambda_r}{d}\right) = \sin^{-1}(1.18)$$

does not exist, so the answer is complete with the angular separations in the first three orders. ■

Finalize: The full spectrum is visible in the first 3 orders with this diffraction grating, and the fourth is partially visible. We can also see that the pattern is dispersed more for higher spectral orders so that the angular separation between the red and blue lines increases as m increases. It is also worth noting that the spectra of different orders can overlap, as the graphical display indicates. For example, the red line in second order is at a larger angle than the blue line in third order. This effect can make the pattern look confusing if you do not know what you are looking for.

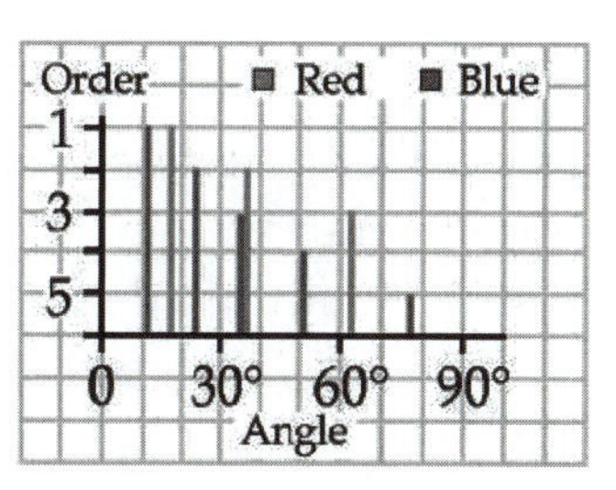

---

33. Light of wavelength 500 nm is incident normally on a diffraction grating. If the third-order maximum of the diffraction pattern is observed at 32.0°, (a) what is the number of rulings per centimeter for the grating? (b) Determine the total number of primary maxima that can be observed in this situation.

Solution

Conceptualize: The diffraction pattern described in this problem seems to be similar to previous problems about diffraction gratings with 2 000 to 5 000 grooves/cm. With the third-order maximum at 32°, there are probably 5 or 6 maxima on each side of the central bright fringe, for a total of 11 or 13 primary maxima.

Categorize: The diffraction grating equation can be used to find the grating spacing and the angles of the other maxima that should be visible within the viewing range of 90° on both sides of the central maximum.

Analyze:

(a) We use the grating equation $d\sin\theta = m\lambda$:

$$d = \frac{m\lambda}{\sin\theta} = \frac{3\left(5.00\times10^{-7}\text{ m}\right)}{\sin 32.0°} = 2.83\times10^{-6}\text{ m}$$

Thus the grating gauge is $\frac{1}{d} = 3.53\times10^{5}$ grooves/m $= 3\,530$ grooves/cm. ■

(b) For any interference maximum for this light going through this grating,

$$\sin\theta = m\left(\frac{\lambda}{d}\right) = \frac{m\left(5.00\times10^{-7}\text{ m}\right)}{2.83\times10^{-6}\text{ m}} = m(0.177)$$

For $\sin\theta \le 1$, we require that $m(0.177) \le 1$ or $m \le 5.65$. Because m must be an integer, its maximum value is really 5. Therefore, the total number of maxima is $2m + 1 = 11$. ■

Finalize: The results agree with our predictions, and there are 5 maxima on either side of the central maximum. If more maxima were desired, a coarser grating with **fewer** grooves/cm would be required; however, this could impede our ability to resolve the angles between exiting light beams that appear close together.

---

36. If the spacing between planes of atoms in an NaCl crystal is 0.281 nm, what is the predicted angle at which 0.140-nm x-rays are diffracted in a first-order maximum?

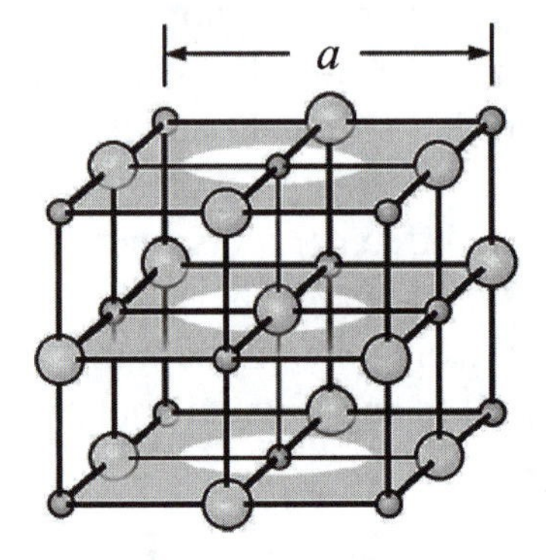

Figure 38.22

Solution

Conceptualize: X-rays have very short wavelength compared to visible light. It is convenient that a crystal forms a natural diffraction grating to prove that an x-ray moves as a wave, and to make its wavelength measurable. We expect an angle on the order of ten degrees, easily measurable on a divided circle.

Categorize: The atomic planes in this crystal are shown in Figure 38.22 of the text. The diffraction they produce is described by Bragg's law, …

Analyze: … which is $2d\sin\theta = m\lambda$

Solving for the angle gives

$$\theta = \sin^{-1}\frac{m\lambda}{2d} = \sin^{-1}\frac{1(0.140\times10^{-9}\text{ m})}{2(0.281\times10^{-9}\text{ m})} = \sin^{-1}0.249 = 14.4° \quad ■$$

Finalize: Note that this is a grazing angle, an angle between the incident beam and the surface. The angle of incidence is $90° - 14.4°$.

In the same way that regularly stacked atoms in a crystal diffract x-rays according to Bragg's law, soldiers standing in ranks and files on a parade ground can diffract sound. On a bed of nails, regularly spaced nails can diffract microwaves. Think up some more examples yourself.

---

43. Plane-polarized light is incident on a single polarizing disk with the direction of $\vec{\mathbf{E}}_0$ parallel to the direction of the transmission axis. Through what angle should the disk be rotated so that the intensity in the transmitted beam is reduced by a factor of (a) 3.00, (b) 5.00, and (c) 10.0?

Solution

Conceptualize: The disk should be called an analyzer. Maximum intensity is transmitted when its axis is at 0° to the incident light's polarization direction. Turning the disk more and more, approaching 90°, will cut down the transmitted intensity by larger and larger factors.

Categorize: We use Malus's law.

Analyze: We define the initial angle, at which all the light is transmitted, to be $\theta = 0$. Turning the disk to another angle will then reduce the transmitted light by an intensity factor as described by $I = I_{max}\cos^2\theta$.

Then $\theta = \cos^{-1}(I/I_{max})^{1/2}$

(a) For $I = I_{max}/3.00$, $\theta = \cos^{-1}\left(\dfrac{I}{I_{max}}\right)^{1/2} = \cos^{-1}\dfrac{1}{\sqrt{3.00}} = \cos^{-1}0.577 = 54.7°$ ■

(b) Now $\theta = \cos^{-1}\left(\dfrac{I}{I_{max}}\right)^{1/2} = \cos^{-1}\dfrac{1}{\sqrt{5.00}} = \cos^{-1}0.447 = 63.4°$ ■

(c) The largest factor of intensity reduction requires the largest crossing angle,

$$\theta = \cos^{-1}\left(\frac{I}{I_{max}}\right)^{1/2} = \cos^{-1}\frac{1}{\sqrt{10.0}} = \cos^{-1}0.316 = 71.6° \quad ■$$

Finalize: You can think of the cosine in Malus's law as taking the component of the incident electric field along the direction of the polarizing filter's transmission axis. It is squared because the energy of any oscillator is proportional to the square of its amplitude and the intensity of any traveling wave is proportional to the square of its amplitude.

---

45. The critical angle for total internal reflection for sapphire surrounded by air is 34.4°. Calculate the polarizing angle for sapphire.

Solution

Conceptualize: Total internal reflection depends on the index of refraction of the optically dense material. (We suppose it is larger than, say, 1.7 for a gem like sapphire.) The polarizing angle also depends on the index of refraction, so we can reason through the index to relate the two angles.

Categorize: We use the relationships between the two angles and the index of refraction of the medium.

Analyze: For the polarizing angle,

$$n_{sapphire}/n_{air} = \tan\theta_p \qquad \text{and} \qquad \theta_p = \tan^{-1}\left(n_{sapphire}/1.00\right)$$

For the critical angle for total internal reflection,

$$n_{sapphire}\sin\theta_c = n_{air}\sin 90^\circ = 1.00 \qquad \text{so} \qquad n_{sapphire} = \frac{1}{\sin\theta_c}$$

Therefore, $\theta_p = \tan^{-1}\left(\dfrac{1}{\sin\theta_c}\right) = \tan^{-1}\left(\dfrac{1}{\sin 34.4^\circ}\right) = 60.5^\circ$ ■

Finalize: Total internal reflection is really a condition drawn from the law of refraction—we think of the incident light as within the sapphire. The polarizing angle is drawn from a special case of reflection—we think of the incident light as in air. Just for curiosity's sake, we can compute the refractive index of sapphire as 1/sin 34.4° = 1.77, a comparatively large value.

---

49. In a single-slit diffraction pattern, assuming each side maximum is halfway between the adjacent minima, find the ratios of the intensity of the (a) first-order side maximum and (b) the second-order side maximum to the intensity of the central maximum.

Solution

Conceptualize: In an interference pattern produced by two (or more!) very fine slits, it is a good approximation to think of all the maxima as equal in intensity. On the other hand, consider the intensity-versus-sinθ graph for a single-slit diffraction pattern, as in the textbook's Figure 38.6a and reproduced here. It shows the central maximum much higher than the side maxima. In this problem we find how much higher the central bright fringe is in intensity.

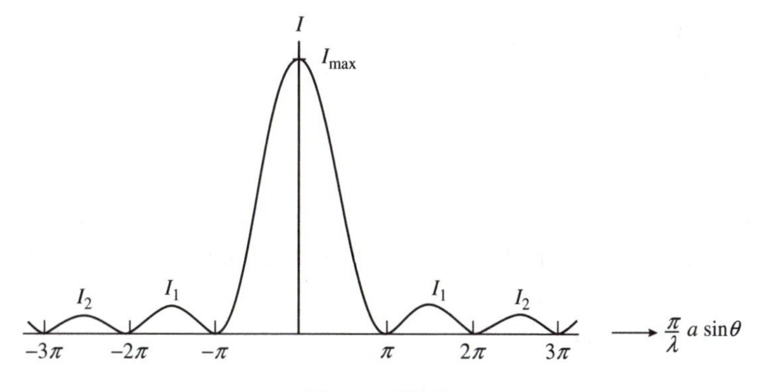

Figure 38.6a

Categorize: We will decode the answers from the intensity equation

$$I = I_{max}\left[\frac{\sin(\pi a \sin\theta / \lambda)}{\pi a \sin\theta / \lambda}\right]^2$$

Analyze: The central maximum is at $\theta = 0$. You learned in calculus class that $\lim_{\alpha\to 0}\frac{\sin\alpha}{\alpha} = 1$, so at the center of the pattern the intensity is correctly identified as $I = I_{max}[1]^2 = I_{max}$.

(a) The first side minimum is at $a\sin\theta = \lambda$, giving $\pi a \sin\theta/\lambda = \pi$.

The second side minimum is at $\pi a\sin\theta/\lambda = 2\pi$. Because $\sin(\pi\,\text{rad}) = \sin(2\pi) = 0$, both of these locations are completely dark.

The problem instructs us to assume that the first side maximum is close to halfway between these dark fringes, namely at $\pi a\sin\theta/\lambda = 1.5\pi$. Here then the intensity is

$$I_1 = I_{max}\left[\frac{\sin(\pi a\sin\theta/\lambda\,)}{\pi a\sin\theta/\lambda}\right]^2 = I_{max}\left[\frac{\sin(1.5\pi)}{1.5\pi}\right]^2 = \frac{I_{max}(-1)^2}{2.25\pi^2}$$

Then the intensity ratio is $\frac{I_1}{I_{max}} = \frac{1}{2.25\pi^2} = 0.045\,0$ ■

(b) Again, the third dark fringe is centered at the location described by $\pi a\sin\theta/\lambda = 3\pi$.

We assume the second bright fringe is at $\pi a\sin\theta/\lambda = 2.5\pi$.

Here the intensity ratio is $\frac{I_2}{I_{max}} = \left[\frac{\sin(2.5\pi)}{2.5\pi}\right]^2 = \frac{1}{6.25\pi^2} = 0.016\,2$ ■

Finalize: By coincidence, part (b) of this problem ends up being essentially like Problem 11, which is included in this manual.

In simple terms, the first side maximum has five percent of the intensity of the central bright fringe, and the second side maximum is down to two percent. Recall that the side bright fringes are only half as wide as the central bright fringe. The farther-out maxima get fainter still, fast enough so that the total power in all of them together is less than ten percent of that in the central maximum—Problem 72 asks for a mathematical argument for this identification.

In contrast to a two-slit interference pattern, in a single-slit diffraction pattern the side maxima are not just halfway between the minima that bracket them. Problem 65 more accurately finds the location of the first side maximum as $\pi a \sin\theta / \lambda = 1.430\ 3\pi$, rather than 1.50π. The second side maximum is really at $\pi a \sin\theta / \lambda = 2.459\ 0\pi$, rather than 2.50π. Then more accurate values for the fractional intensities at these two bright fringes are, respectively, $\frac{I_1}{I_{max}} = \left[\frac{\sin(1.430\ 3\pi)}{1.430\ 3\pi}\right]^2 = 0.047\ 19$

and $\frac{I_2}{I_{max}} = \left[\frac{\sin(2.459\ 0\pi)}{2.459\ 0\pi}\right]^2 = 0.016\ 48$

Observe that these answers agree fairly well (within 9.7 %) with the answers computed from the simplifying assumption of equally-spaced dark and bright fringes.

---

65. The intensity of light in the diffraction pattern of a single slit is described by the equation

$$I_\theta = I_{max} \frac{\sin^2 \phi}{\phi^2}$$

where $\phi = (\pi a \sin\theta)/\lambda$. The central maximum is at $\phi = 0$, and the side maxima are *approximately* at $\phi = (m + \frac{1}{2})\pi$ for $m = 1, 2, 3, \ldots$. Determine more precisely (a) the location of the first side maximum, where $m = 1$, and (b) the location of the second side maximum. *Suggestion*: Observe in Figure 38.6a that the graph of intensity versus ϕ has a horizontal tangent at maxima and also at minima.

Solution

Conceptualize: In Chapter 37 we studied the interference pattern of light from two fine slits. There the intensity as a function of angle was given by $I_\theta = I_{max} \cos^2\left(\frac{\pi d \sin\theta}{\lambda}\right)$, which we can write simply as $I_\theta = I_{max} \cos^2 \phi$. In that case, the maxima are all equally high and at locations given by $\phi = 0, \pi, 2\pi, \ldots, m\pi$ for $m = 0, 1, 2, 3, \ldots$. The minima are all zero and at locations halfway between the maxima, given by $\phi = (m + ½)\lambda$ for $m = 0, 1, 2, 3, \ldots$. The given function of the intensity in a single-slit diffraction pattern is more complicated, but the sine function gives it alternating maxima and minima. We find their locations in this problem.

Categorize: We use the standard calculus method of finding maxima and minima of a continuous function: We set its derivative equal to zero and solve for the values of the argument ϕ that produce the horizontal tangents.

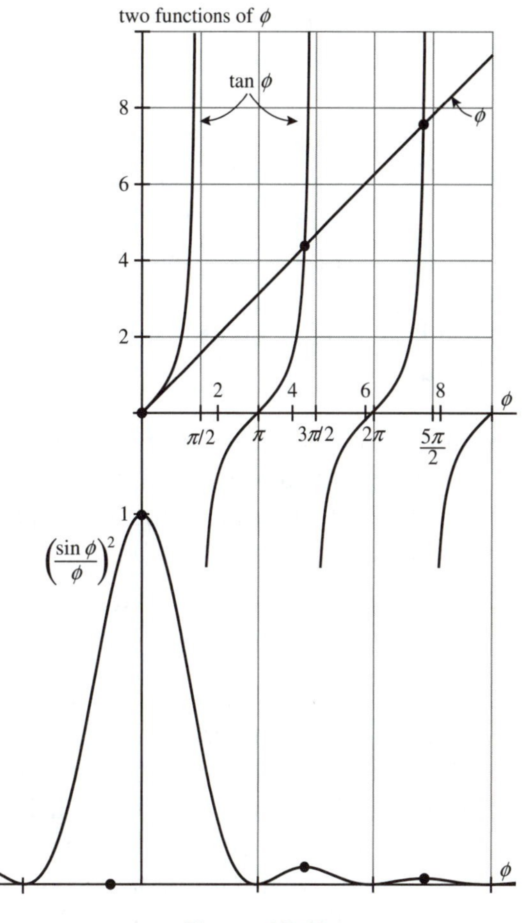

Figure 38.65

Analyze: From $I = I_{\max}\left(\frac{\sin\phi}{\phi}\right)^2$ we find

$$\frac{dI}{d\phi} = I_{\max}2\left(\frac{\sin\phi}{\phi}\right)\left(\frac{\phi\cos\phi - [\sin\phi]1}{\phi^2}\right)$$ and we require that this expression be equal to zero.

The possibility $I_{\max} = 0$ indicates that the graph would be flat if no light were sent through the slit. The possibility $\phi \to \infty$ refers to the diffraction pattern becoming uniform (uniformly dark) at large angles away from the center. The possibility $\sin\phi = 0$ gives the locations of the central maximum and of all of the minima, according to $\phi = 0, \pi, 2\pi, \ldots, m\pi$. This statement is true because $\frac{\pi a \sin\theta}{\lambda} = m\pi$ implies either $\theta = 0$ (at the central bright fringe) or $a\sin\theta = m\lambda$ for $m = 1, 2, 3, \ldots$ (at the dark fringes). Then the side maxima that the problem asks about are described by the possibility $\phi\cos\phi - \sin\phi = 0$ or $\phi = \sin\phi/\cos\phi = \tan\phi$ where ϕ is to be interpreted as in radians. The solution $\phi = 0$ describes again the central maximum of the function. Your calculator may be able to find the next two solutions of $\tan\phi = \phi$ if you use a "solve" function in the right way; but it is good to know a way to do it yourself. The graph of the two functions $\tan\phi$ and ϕ versus ϕ shows equality—intersections—at a few points, circled in the diagram. We should find solutions for ϕ a bit less than $3\pi/2 = 4.71$ and for ϕ a bit less than $5\pi/2 = 7.85$. We can home in on a solution by evaluating the function $\tan\phi - \phi$ for various trial values. This function can switch from positive to negative by running off to infinity as well as by crossing zero, so your calculator might get very confused, but we can avoid confusion by making a table of values:

ϕ	4	4.7	7	7.8
$\tan\phi - \phi$	−2.8	+76	−6.1	+11

(a) The first solution of $\tan\phi - \phi = 0$ we expect to find closer to 4 than to 4.7, so we try in sequence

ϕ	4.2	4.4	4.5	4.48	4.49	4.493	4.494	4.4933	4.4936	4.4934	4.4935
$\tan\phi - \phi$	−2.4	−1.3	+0.14	−0.25	−0.06	−0.008	+0.012	−0.002	+0.004	−0.0002	+0.0018

So we know to five digits that the root is $\phi = 4.493\,4 = 1.430\pi$ corresponding to $a \sin\theta = 1.430\lambda$. ■

(b) Similarly, for the second bright fringe on the side, we try

ϕ	7.5	7.6	7.7	7.75	7.72	7.73	7.725	7.726	7.7253	7.7252	7.72525
$\tan\phi - \phi$	−4.8	−3.7	−1.2	+1.8	−0.3	+0.29	−0.015	+0.045	+0.003	−0.003	−0.0001

So we know to five digits that the root is $\phi = 7.725\,3 = 2.459\pi$ corresponding to $a \sin\theta = 2.459\lambda$. ■

Finalize: It would take infinitely many steps to determine infinitely many digits of each solution, but that task is not of physical interest anyway. The brightest locations in the first two side bright fringes in a diffraction pattern are within 5% and 2% of being just halfway between the dark fringes that bracket them.

---

73. Suppose the single slit in Active Figure 38.4 is 6.00 cm wide and in front of a microwave source operating at 7.50 GHz. (a) Calculate the angle for the first minimum in the diffraction pattern. (b) What is the relative intensity I/I_{max} at $\theta = 15.0°$? (c) Assume two such sources, separated laterally by 20.0 cm, are behind the slit. What must be the maximum distance between the plane of the sources and the slit if the diffraction patterns are to be resolved? In this case, the approximation $\sin\theta \approx \tan\theta$ is not valid because of the relatively small value of a/λ.

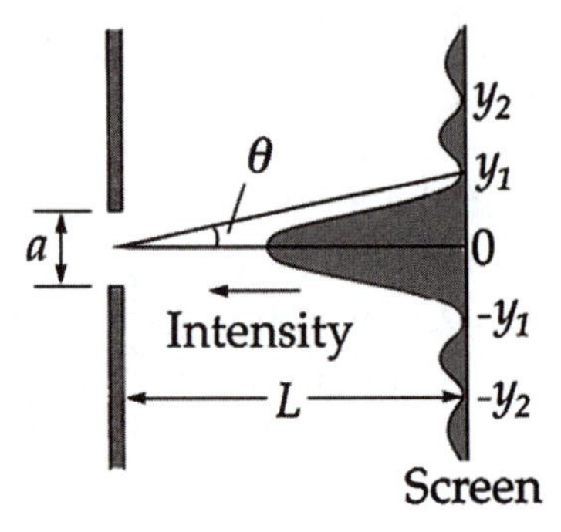

Figure 38.4a

Solution

Conceptualize: Microwaves have much larger wavelength than visible light, so the ideas applied to visible light going through a single narrow slit will describe microwaves going through this slot that a puppy could squirm through.

Categorize: We use ideas of single-slit diffraction and Rayleigh's criterion.

Analyze:

(a) From Equation 38.1, $\theta = \sin^{-1}\left(\dfrac{m\lambda}{a}\right)$

In this case, $m = 1$, and $\lambda = \dfrac{c}{f} = \dfrac{3.00 \times 10^8\ \text{m/s}}{7.50 \times 10^9\ \text{s}^{-1}} = 0.040\,0\ \text{m}$

so $\theta = \sin^{-1}\left(\dfrac{1(0.040\,0\ \text{m})}{0.060\,0\ \text{m}}\right) = \sin^{-1}(0.667) = 41.8°$ ■

(b) From Equation 38.2, $\dfrac{I}{I_{\max}} = \left(\dfrac{\sin\phi}{\phi}\right)^2$ where $\phi = \dfrac{\pi a \sin\theta}{\lambda}$

When $\theta = 15.0°$, $\phi = \pi(6.00 \text{ cm})(\sin 15.0°)/(4.00 \text{ cm}) = 1.22 \text{ rad}$

and $\dfrac{I}{I_{\max}} = \left(\dfrac{\sin(1.22 \text{ rad})}{1.22}\right)^2 = 0.593$ ■

(c) Let L' be the maximum distance between the plane of the two sources and the slit. So that the central maximum of one diffraction pattern will fall on the first minimum of the other, the angle subtended by the two sources at the slit should be $\theta = 41.8°$, and the half-angle between the sources is $\alpha = \theta/2 = 20.9°$.

Then from the diagram, $\tan\alpha = 10.0 \text{ cm}/L'$

so $L' = \dfrac{10.0 \text{ cm}}{\tan 20.9°} = 26.2 \text{ cm}$ ■

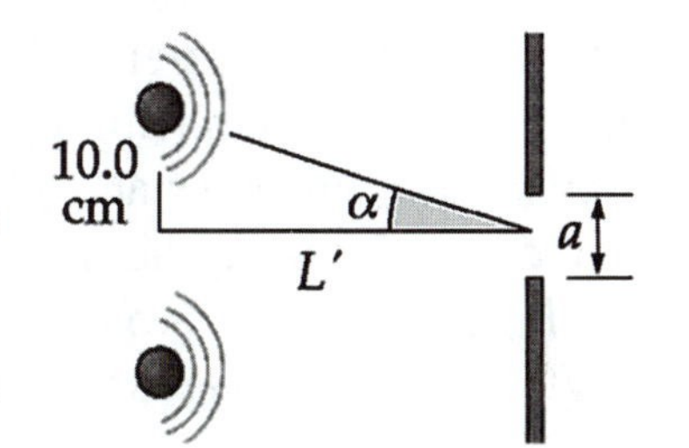

Finalize: The 15° direction considered in part (b) is inside the central diffraction maximum. This situation makes a nice demonstration. It could be done with sound of wavelength 4 cm going through a slot between blocks of plastic foam, as an alternative to microwaves.

39

Relativity

EQUATIONS AND CONCEPTS

Galilean space-time transformation equations transform the location and time of an event in one inertial frame of reference, S, with space-time coordinates (x, y, z, t) to a second frame of reference, S′, with coordinates (x', y', z', t') moving along common x and x' axes with constant velocity of magnitude v relative to the first. *Observers in both frames measure the same time interval between two successive events.*

$$\left.\begin{aligned} x' &= x - vt \\ y' &= y \\ z' &= z \\ t' &= t \end{aligned}\right\} \quad (39.1)$$

The **Galilean velocity transformation equation** relates the velocity of a particle measured in a rest frame (u_x) relative to the velocity (u_x') measured in a frame moving in the x direction with velocity of magnitude v relative to the rest frame.

$$u_x' = u_x - v \quad (39.2)$$

Basic postulates of the special theory of relativity:

- The laws of physics are the same in all inertial frames of reference.
- The speed of light has the same value in all inertial frames.

Important consequences of the theory of special relativity:

- **Simultaneity** — events observed as simultaneous in one frame of reference are not necessarily observed as simultaneous in a second frame moving relative to the first.
- **Time dilation** — A time interval Δt measured by an *observer moving with respect to a clock* is longer than the time interval Δt_p (the proper time)

$$\Delta t = \frac{\Delta t_p}{\sqrt{1 - \dfrac{v^2}{c^2}}} = \gamma \Delta t_p \quad (39.7)$$

measured by an observer *at rest with respect to the clock. Moving clocks run slower than clocks at rest with respect to an observer.*

$$\gamma = \frac{1}{\sqrt{1-\frac{v^2}{c^2}}} \tag{39.8}$$

- **Length contraction.** If an object has a length L_p (the proper length) when measured by an observer *at rest* with respect to the object, the length L measured by an observer *moving* with respect to the object will be less than L_p. *Length contraction occurs only along the dimension parallel to the direction of motion.*

$$L = \frac{L_p}{\gamma} = L_p\sqrt{1-\frac{v^2}{c^2}} \tag{39.9}$$

- **Relativistic Doppler effect.** When a light source and observer approach each other, the observed frequency is greater than the frequency of the source. *When the light and observer recede from each other, negative values are used for v in the equation.*

$$f' = \frac{\sqrt{1+v/c}}{\sqrt{1-v/c}}f \tag{39.10}$$

The **Lorentz transformation equations** transform space and time coordinates from the rest frame of reference (S) into the frame (S′) moving along a common axis with relative velocity of magnitude v. *In order to transform coordinates in S′ to coordinates in S, replace v by $-v$ and interchange primed and unprimed coordinates in Equation 39.11.*

$$\begin{aligned} x' &= \gamma(x - vt) \\ y' &= y \\ z' &= z \\ t' &= \gamma\left(t - \frac{v}{c^2}x\right) \end{aligned} \tag{39.11}$$

The **Lorentz velocity transformation equations** relate the velocity of a moving object as measured by an observer in a fixed frame (S) to the measurement of the velocity by an observer in a frame (S′) moving with velocity of magnitude v with respect to (S) along a common x axis. Equations 39.16 and 39.17 allow for transformation of velocity values along each of the coordinate axes.

$$u'_x = \frac{u_x - v}{1-\frac{u_x v}{c^2}} \tag{39.16}$$

$$u'_y = \frac{u_y}{\gamma\left(1-\frac{u_x v}{c^2}\right)} \tag{39.17}$$

$$u'_z = \frac{u_z}{\gamma\left(1-\frac{u_x v}{c^2}\right)}$$

The **relativistic linear momentum** of a particle with mass m and moving with velocity $\vec{\mathbf{u}}$ satisfies the following conditions:

(i) Momentum is conserved in all collisions, and

(ii) The relativistic value of momentum approaches the classical value ($m\vec{\mathbf{u}}$) as $\vec{\mathbf{u}}$ approaches zero.

$$\vec{\mathbf{p}} \equiv \frac{m\vec{\mathbf{u}}}{\sqrt{1-\dfrac{u^2}{c^2}}} = \gamma m\vec{\mathbf{u}} \qquad (39.19)$$

The **relativistic kinetic energy** of a particle of mass m moving with a speed u includes the **rest energy** term mc^2.

$$K = \frac{mc^2}{\sqrt{1-\dfrac{u^2}{c^2}}} - mc^2 \qquad (39.23)$$

$$= \gamma mc^2 - mc^2 = (\gamma - 1)mc^2$$

The **rest energy** of a particle is independent of the speed of the particle. The mass m must have the same value in all inertial frames.

$$E_R = mc^2 \qquad (39.24)$$

The **total energy** of a **relativistic particle** is the sum of the kinetic energy and the rest energy. *This expression shows that mass is a form of energy.*

$$E = \frac{mc^2}{\sqrt{1-\dfrac{u^2}{c^2}}} = \gamma mc^2 \qquad (39.26)$$

The **energy-momentum relationship for a relativistic particle** is useful when the momentum or energy of the particle is known (rather than the speed).

$$E^2 = p^2c^2 + (mc^2)^2 \qquad (39.27)$$

Photons ($m = 0$) travel with the speed of light. *Equation 39.28 is an exact expression relating energy and momentum for particles that have zero mass.*

$$E = pc \qquad (39.28)$$

The **electron volt** (eV) is a convenient energy unit in which to express the energies of electrons and other subatomic particles.

$$1 \text{ eV} = 1.60 \times 10^{-19} \text{ J}$$

REVIEW CHECKLIST

You should be able to:

- Explain the Michelson-Morley experiment, its objectives, results, and the significance of its outcome. (Section 39.2)
- State Einstein's two postulates of the special theory of relativity. (Section 39.3)
- Make calculations using the equations for time dilation, length contraction, and relativistic Doppler effect. (Section 39.4)
- Make calculations using the Lorentz transformation equations and the Lorentz velocity transformation equations. (Sections 39.5 and 39.6)
- Make calculations using relativistic expressions for momentum, kinetic energy, and total energy of a particle. (Sections 39.7, 39.8, and 39.9)

ANSWER TO AN OBJECTIVE QUESTION

7. Two identical clocks are set side by side and synchronized. One remains on the Earth. The other is put into orbit around the Earth moving rapidly toward the east. **(i)** As measured by an observer on the Earth, does the orbiting clock (a) run faster than the Earth-based clock, (b) run at the same rate, or (c) run slower? **(ii)** The orbiting clock is returned to its original location and brought to rest relative to the Earth-based clock. Thereafter, what happens? (a) Its reading lags farther and farther behind the Earth-based clock. (b) It lags behind the Earth-based clock by a constant amount. (c) It is synchronous with the Earth-based clock. (d) It is ahead of the Earth-based clock by a constant amount. (e) It gets farther and farther ahead of the Earth-based clock.

Answer

(i) If the first clock is at the equator, the Earth's daily rotation carries it toward the east at 40 000 km per 24 h, relative to the Sun. The question gives enough information to ensure that the second clock is moving more rapidly. Then the Earth-based experimenter observes the orbiting clock ticking more slowly, as described by time dilation. This is after the observer corrects his measurements for the Doppler effect. The answer is (c).

(ii) After the clocks are side by side and again at rest relative to each other, time dilation is no longer observed. The clocks are measured to tick at the same rate. But, as Hafele and Keating directly observed, the accumulated effect of ticking more slowly during its trip makes the second clock lag behind the one stuck to the mantelpiece by a constant amount, as stated in answer (b).

□ □ □ □

ANSWERS TO SELECTED CONCEPTUAL QUESTIONS

2. Explain why, when defining the length of a rod, it is necessary to specify that the positions of the ends of the rod are to be measured simultaneously.

Answer Suppose a railroad train is moving past you. One way to measure its length is this: You mark the tracks at the front of the moving engine at 9:00:00 AM, while your assistant marks the tracks at the back of the caboose at the same time. Then you find the distance between the marks on the tracks with a tape measure. You and your assistant must make the marks simultaneously (in your reference frame), for otherwise the motion of the train would make its length different from the distance between marks.

□ □ □ □

4. List three ways our day-to-day lives would change if the speed of light were only 50 m/s.

Answer For a wonderful fictional exploration of this question, get a "Mr. Tompkins" book by George Gamow. All of the relativity effects, such as time dilation and length contraction, would be obvious in our lives, so you can make many choices of three. Driving home in a hurry, you push on the gas pedal not to increase your speed very much, but to make the blocks shorter. Big Doppler shifts in wave frequencies make red lights look green as you approach, and make car horns and radios useless. High-speed transportation is both very expensive, requiring huge fuel purchases, as well as dangerous, since a speeding car can knock down a building. Having had breakfast at home, when you arrive home hungry for lunch, you find that you have missed dinner. There is a five-day delay in transit when you watch the Olympics in Australia on live TV. We are unable to see the Milky Way, since the fireball of the Big Bang surrounds us at the distance of Rigel or Deneb.

□ □ □ □

9. Give a physical argument that shows it is impossible to accelerate an object of mass m to the speed of light, even with a continuous force acting on it.

Answer As an object approaches the speed of light, its energy approaches infinity. Hence, it would take an infinite amount of work to accelerate the object to the speed of light under the action of a constant force, or it would take an infinitely large force.

□ □ □ □

SOLUTIONS TO SELECTED PROBLEMS

2. In a laboratory frame of reference, an observer notes that Newton's second law is valid. Assume forces and masses are measured to be the same in any reference frame for speeds small compared with the speed of light. (a) Show that Newton's second law is also valid for an observer moving at a constant speed, small compared with the speed of light, relative to the laboratory frame. (b) Show that Newton's second law is *not* valid in a reference frame moving past the laboratory frame with a constant acceleration.

Solution

Conceptualize: We are to demonstrate theoretically part of the principle of Galilean relativity. This principle says that all reference frames moving with constant velocity, with respect to any one unaccelerated reference frame, are equally good for describing reality.

Categorize: The first observer watches some object accelerate under applied forces. Call the instantaneous velocity of the object $\vec{\mathbf{u}}$. The first observer confirms that $\sum \vec{\mathbf{F}} = m\dfrac{d\vec{\mathbf{u}}}{dt}$. With the term we introduced in Chapter 5, he confirms that his is an inertial reference frame. The second observer is moving with velocity $\vec{\mathbf{v}}$ relative to the first.

Analyze: She measures the object to have velocity $\vec{\mathbf{u}} \ = \vec{\mathbf{u}} - \vec{\mathbf{v}}$.

(a) With the relative velocity of the reference frames constant, we differentiate to find that the object's acceleration is

$$\frac{d\vec{\mathbf{u}}'}{dt} = \frac{d\vec{\mathbf{u}}}{dt} - 0$$

This is the same as measured by both observers. In this nonrelativistic case they measure also the same mass and forces; so the second observer also confirms that

$$\sum \vec{\mathbf{F}} = m\frac{d\vec{\mathbf{u}}'}{dt} \qquad \blacksquare$$

(b) Now suppose for contrast that the second reference frame has a changing velocity relative to the first, according to $d\vec{\mathbf{v}}/dt \neq 0$. The second observer measures the object's acceleration to be

$$\frac{d\vec{\mathbf{u}}'}{dt} = \frac{d\vec{\mathbf{u}}}{dt} - \frac{d\vec{\mathbf{v}}}{dt}$$

Then she computes $m\dfrac{d\vec{\mathbf{u}}'}{dt} = m\dfrac{d\vec{\mathbf{u}}}{dt} - m\dfrac{d\vec{\mathbf{v}}}{dt} = \sum \vec{\mathbf{F}} - m\dfrac{d\vec{\mathbf{v}}}{dt} \neq \sum \vec{\mathbf{F}}$

so Newton's second law is not true in this noninertial reference frame. ■

Finalize: The principle is called *Galilean* relativity because Galileo gave an eloquent description of it, beginning "Shut yourself up below decks in some great ship." He wrote that any observation, such as a broad jump, free fall, the flight of insects, or projectile motion, has the same results if the ship is at rest or if it is moving with constant velocity. It was Einstein who generalized this principle to include electromagnetism.

---

15. A supertrain with a proper length of 100 m travels at a speed of $0.950c$ as it passes through a tunnel having a proper length of 50.0 m. As seen by a trackside observer, is the train ever completely within the tunnel? If so, by how much do the train's ends clear the ends of the tunnel?

Solution

Conceptualize: The train is moving at such a large fraction of the speed of light that length contraction will be a large effect. We think that it will contract to much less than 100 m in length, and even to less than 50 m in length.

Categorize: The trackside observer sees the proper length of the tunnel, 50.0 m, but sees the train Lorentz-contracted…

Analyze: … to a length of

$$L = L_p\sqrt{1 - v^2/c^2} = (100\text{ m})\sqrt{1 - (0.950)^2} = 31.2\text{ m}$$

This is shorter than the tunnel by 18.8 m, so it is completely within the tunnel. ■

Finalize: An observer on the train sees the train as 100 m long and the tunnel as only 15.6 m long, so the train will definitely not fit. This can be a self-consistent conclusion because the passenger and the trackside observer do not agree on the simultaneity of each other's observations of where the front and back ends of the train are, or where the entrance and exit of the tunnel are.

---

17. A spacecraft with a proper length of 300 m passes by an observer on the Earth. According to this observer, it takes 0.750 μs to pass a fixed point. Determine the speed of the spacecraft as measured by the Earth-based observer.

Solution

Conceptualize: We should first determine if the spaceship is traveling at a relativistic speed: classically, $v = (300\text{ m})/(0.750\ \mu\text{s}) = 4.00 \times 10^8$ m/s, which is faster than the speed of light and so impossible. Quite clearly, the relativistic correction must be used to find the correct speed of the spaceship, which we can guess will be close to the speed of light.

Categorize: We can use the contracted length equation to find the speed of the spaceship in terms of the proper length and the time interval. The time of 0.750 μs is the **proper time interval** measured by the Earth observer, because it is the time interval between two events that she sees as happening at the same point in space. The two events are the passage of the front end of the spaceship over her stopwatch, and the passage of the back end of the ship.

Analyze: The contracted length of the ship is $L = L_p/\gamma$, with $L = v\Delta t$.

So we have $$v\Delta t = L_p\left(1 - v^2/c^2\right)^{1/2}$$

Squaring both sides, $v^2\Delta t^2 = L_p^2\left(1 - v^2/c^2\right)$

or $v^2c^2 = L_p^2c^2/\Delta t^2 - v^2L_p^2/\Delta t^2$

Solving for the velocity,

$$v = \frac{cL_p/\Delta t}{\sqrt{c^2 + L_p^2/\Delta t^2}} = \frac{\left(3.00 \times 10^8 \text{ m/s}\right)(300 \text{ m})/\left(0.750 \times 10^{-6} \text{ s}\right)}{\sqrt{\left(3.00 \times 10^8 \text{ m/s}\right)^2 + (300 \text{ m})^2/\left(0.750 \times 10^{-6} \text{ s}\right)^2}}$$

So $v = 2.40 \times 10^8$ m/s ■

Finalize: The spaceship is traveling at $0.8c$. We can also verify that the general equation for the speed reduces to the classical relation $v = L_p/\Delta t$ when the time interval is relatively large.

---

19. An atomic clock moves at 1 000 km/h for 1.00 h as measured by an identical clock on the Earth. At the end of the 1.00-h interval, how many nanoseconds slow will the moving clock be compared with the Earth-based clock?

Solution

Conceptualize: This is not a considerable fraction of the speed of light, so time dilation will be a small-fraction effect. From the statement of the problem, we might expect a few nanoseconds rather than microseconds or picoseconds.

Categorize: This problem is slightly more difficult than most, for the simple reason that your calculator probably cannot hold enough decimal places to yield an accurate answer. However, we can bypass the difficulty by noting the approximation

$$\sqrt{1 - \frac{v^2}{c^2}} \approx 1 - \frac{v^2}{2c^2}$$

Squaring both sides shows that when v/c is small, these two terms are equivalent.

Analyze: We evaluate $\dfrac{v}{c} = \left(\dfrac{1000 \times 10^3 \text{ m/h}}{3.00 \times 10^8 \text{ m/s}}\right)\left(\dfrac{1 \text{ h}}{3600 \text{ s}}\right) = 9.26 \times 10^{-7}$

From Equation 39.7, the dilated time interval is

$$\Delta t = \gamma \Delta t_p = \frac{\Delta t_p}{\sqrt{1 - \dfrac{v^2}{c^2}}}$$

Rearranging, our approximation yields

$$\Delta t_p = \left(\sqrt{1 - \frac{v^2}{c^2}}\right)\Delta t \approx \left(1 - \frac{v^2}{2c^2}\right)\Delta t$$

and $\Delta t - \Delta t_p = \dfrac{v^2}{2c^2}\Delta t$

Substituting, $\Delta t - \Delta t_p = \dfrac{\left(9.26 \times 10^{-7}\right)^2}{2}(3\,600\text{ s})$

Thus, the time lag of the moving clock is $\Delta t - \Delta t_p = 1.54 \times 10^{-9}\text{ s} = 1.54\text{ ns}$. ■

Finalize: The Earth clock measures the proper time interval, because it is in an unaccelerated frame and the start and end of the hour timing occur at the same location on Earth. The traveling clock measures a longer time interval, as we have shown. As the textbook notes, just such a time dilation effect was measured by Hafele and Keating.

---

28. Figure P39.28 shows a jet of material (at the upper right) being ejected by galaxy M87 (at the lower left). Such jets are believed to be evidence of a supermassive black hole at the center of a galaxy. Suppose two jets of material from the center of a galaxy are ejected in opposite directions. Both jets move at $0.750c$ relative to the galaxy center. Determine the speed of one jet relative to the other.

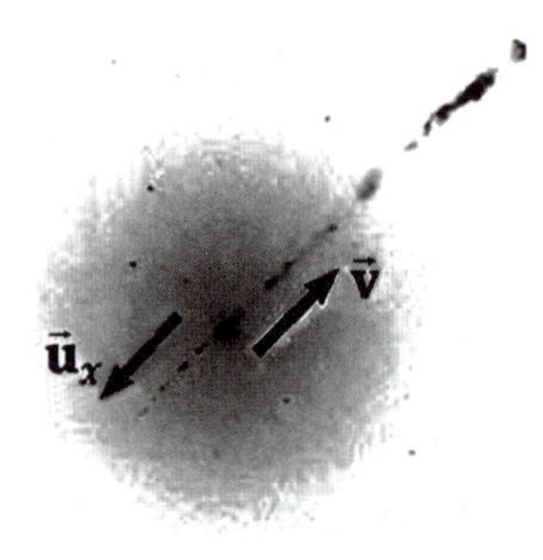

Solution

Conceptualize: In our frame of reference, the gap between one batch of material and the other opens at 1.5(300 Mm/s). An observer riding on a southbound glob of plasma would see the Earth moving north at $0.750c$. She would see material in the other jet moving north at a speed close to but less than the speed of light.

Categorize: We use the relativistic velocity-addition results.

Analyze: Take the galaxy as the unmoving frame. Arbitrarily define the jet moving upward to be the object, and the jet moving downward to be the "moving" frame:

$$u_x = 0.750c \qquad v = -0.750c$$

Then the speed of the upward-moving jet as measured from the downward-moving jet is

$$u'_x = \frac{u_x - v}{1 - u_x v/c^2} = \frac{0.750c - (-0.750c)}{1 - (0.750c)(-0.750c)/c^2} = \frac{1.50c}{1 + 0.750^2} = 0.960c$$ ■

Finalize: No energy and no signal can travel faster than light travels in a vacuum. Matter must always travel less rapidly, and these prohibitions include velocities arrived at by addition.

---

35. An unstable particle at rest spontaneously breaks into two fragments of unequal mass. The mass of the first fragment is 2.50×10^{-28} kg, and that of the other is 1.67×10^{-27} kg. If the lighter fragment has a speed of $0.893c$ after the breakup, what is the speed of the heavier fragment?

Solution

Conceptualize: The heavier fragment should have a speed less than that of the lighter piece since the momentum of the system must be conserved. However, due to the relativistic factor, the ratio of the speeds will not equal the simple ratio of the particle masses, which would give a speed of $0.134c$ for the heavier particle.

Categorize: Relativistic momentum of the system must be conserved. For the total momentum to be zero after the decay, as it was before, $\vec{\mathbf{p}}_1 + \vec{\mathbf{p}}_2 = 0$, where we will refer to the lighter particle with the subscript "1," and to the heavier particle with the subscript "2."

Analyze: From the expression for relativistic momentum we have

$$\gamma_2 m_2 u_2 + \gamma_1 m_1 u_1 = 0: \qquad \gamma_2 m_2 u_2 + \left(\frac{2.50 \times 10^{-28}\text{ kg}}{\sqrt{1 - 0.893^2}}\right)(0.893c) = 0$$

Rearranging, $$\left(\frac{1.67 \times 10^{-27}\text{ kg}}{\sqrt{1 - u_2^2/c^2}}\right)\frac{u_2}{c} = -4.96 \times 10^{-28}\text{ kg}$$

Squaring both sides, $$\left(2.79 \times 10^{-54}\text{ kg}^2\right)\left(\frac{u_2}{c}\right)^2 = \left(2.46 \times 10^{-55}\text{ kg}^2\right)\left(1 - \frac{u_2^2}{c^2}\right)$$

and $$\frac{u_2}{c} = \left(\frac{2.46}{27.9 + 2.46}\right)^{1/2} = -0.285$$

We choose the negative sign only to mean that the two particles must move in opposite directions. The speed, then, is $|u_2| = 0.285c$. ■

Finalize: The speed of the heavier particle is less than the lighter particle, as expected. We can also see that for this situation, the relativistic speed of the heavier particle is about twice as great as was predicted by a simple non-relativistic calculation.

---

37. A proton moves at $0.950c$. Calculate its (a) rest energy, (b) total energy, and (c) kinetic energy.

Solution

Conceptualize: The textbook's front endpaper lists the mass of any proton as 938 MeV/c^2, so its rest energy must be 938 MeV. We will calculate the rest energy from its definition. This proton is moving so fast, close to the speed of light, that its kinetic energy will be several thousand MeV's and its total energy larger by just the 938 MeV.

Categorize: We use relativistic expressions for rest energy and total energy.

Analyze: At $u = 0.950c$, it will be useful to know the gamma factor:

$$\gamma = \frac{1}{\sqrt{1 - u^2/c^2}} = \frac{1}{\sqrt{1 - 0.950^2}} = 3.20$$

(a) The rest energy is

$$E_R = mc^2 = \left(1.67 \times 10^{-27} \text{ kg}\right)\left(2.998 \times 10^8 \text{ m/s}\right)^2 = 1.50 \times 10^{-10} \text{ J}$$

$$= 1.50 \times 10^{-10} \text{ J}\left(\frac{1 \text{ eV}}{1.60 \times 10^{-19} \text{ J}}\right) = 938 \text{ MeV} \quad \blacksquare$$

(We use a value for c accurate to four digits so that we can be sure to get an answer accurate to three digits. Through the rest of the book we will use values for physical constants accurate to four digits or to three, whichever we like. We will still quote answers to three digits, and you can still think of the last digit as uncertain.)

(b) The total energy is

$$E = \gamma mc^2 = \gamma E_R = (3.20)(938 \text{ MeV}) = 3.00 \text{ GeV} \quad \blacksquare$$

(c) The kinetic energy is

$$K = E - E_R = 3.00 \text{ GeV} - 938 \text{ MeV} = 2.07 \text{ GeV} \quad \blacksquare$$

Finalize: It is a good habit to find the value of gamma as early in the solution as possible. Resist any temptation to find kinetic energy first or before finding total energy. In $K = (\gamma - 1)mc^2$, it is natural to find γ and mc^2 and γmc^2 first. There is nothing wrong with quoting speeds in meters per second and energies in femtojoules, but we save writing by quoting speed as a fraction of c and energies in MeV's.

---

39. A proton in a high-energy accelerator moves with a speed of $c/2$. Use the work–kinetic energy theorem to find the work required to increase its speed to (a) $0.750c$ and (b) $0.995c$.

Solution

Conceptualize: Since particle accelerators have typical maximum energies on the order of a GeV ($1 \text{ eV} = 1.60 \times 10^{-19}$ J), we could expect the work to be $\sim 10^{-10}$ J.

Categorize: The work–kinetic energy theorem is $W = \Delta K = K_f - K_i$, which for relativistic speeds (u comparable to c) is:

$$W = \left(\frac{1}{\sqrt{1 - u_f^{\,2}/c^2}} - 1\right)mc^2 - \left(\frac{1}{\sqrt{1 - u_i^{\,2}/c^2}} - 1\right)mc^2$$

or, simplified,

$$W = \left(1/\sqrt{1 - u_f^{\,2}/c^2} - 1/\sqrt{1 - u_i^{\,2}/c^2}\right)mc^2$$

Analyze: From our specialized equation,

$$\text{(a)}\quad W = \left(\frac{1}{\sqrt{1-0.750^2}} - \frac{1}{\sqrt{1-0.500^2}} \right)\left(1.67\times10^{-27}\ \text{kg}\right)\left(3.00\times10^{8}\ \text{m/s}\right)^2$$

$$W = (1.512 - 1.155)\left(1.50\times10^{-10}\ \text{J}\right) = 5.37\times10^{-11}\ \text{J} = 336\ \text{MeV} \quad ■$$

$$\text{(b)}\quad W = \left(\frac{1}{\sqrt{1-0.995^2}} - \frac{1}{\sqrt{1-0.500^2}} \right)\left(1.67\times10^{-27}\ \text{kg}\right)\left(3.00\times10^{8}\ \text{m/s}\right)^2$$

$$W = (10.01 - 1.155)\left(1.50\times10^{-10}\ \text{J}\right) = 1.33\times10^{-9}\ \text{J} = 8.32\ \text{GeV} \quad ■$$

Finalize: Even though these energies are small numbers, we must remember that the proton has very small mass, so these input energies are comparable to the rest energy of the proton (1.50×10^{-10} J). To produce a speed higher by 33%, the answer to part (b) is 25 times larger than the answer to part (a). Even with arbitrarily large accelerating energies, the particle will never reach or exceed the speed of light.

---

41. The total energy of a proton is twice its rest energy. Find the momentum of the proton in MeV/c units.

Solution

Conceptualize: The textbook's front endpaper lists the mass of a proton as 938 MeV/c^2. Then the rest energy of any proton is 938 MeV. We are supposed to think of a proton with this same amount of kinetic energy, so its speed will be a considerable fraction of c and its momentum on the order of one thousand MeV/c.

Categorize: We choose to do this problem in three steps. First, we compute the rest energy of the proton; then we use that in the equation for total energy to find the speed of the proton. Last, we substitute the speed and rest energy into the relativistic momentum equation to obtain the momentum.

Analyze: For a proton,

$$mc^2 = \left(1.67\times10^{-27}\ \text{kg}\right)\left(2.998\times10^{8}\ \text{m/s}\right)^2 = 1.50\times10^{-10}\ \text{J}$$

In MeV, $$mc^2 = \left(1.50\times10^{-10}\ \text{J}\right)\left(\frac{1\ \text{eV}}{1.60\times10^{-19}\ \text{kg}\cdot\text{m}^2/\text{s}^2}\right) = 938\ \text{MeV}$$

Total energy is $\gamma mc^2 = 2mc^2$ so that $\gamma = 2$

From $\dfrac{1}{\sqrt{1-u^2/c^2}} = \gamma$ we have $\dfrac{u}{c} = \dfrac{\sqrt{\gamma^2 - 1}}{\gamma} = \dfrac{\sqrt{2^2-1}}{2} = 0.866$

(We choose to represent the speed of the particle by u.)

Momentum is $$p = \gamma m u = \frac{\gamma mc^2}{c}(u/c)$$

So in this case $$p = \frac{2(938\text{ MeV})}{c}(0.866) = 1620\text{ MeV}/c$$ ■

Finalize: Compare problems 37, 39, and 41, solved in sequence here. Each asks about relating different combinations of speed, kinetic energy, total energy, and momentum for relativistic protons. By using Equation 39.27, $E^2 = p^2c^2 + (mc^2)^2$, we could calculate the momentum without first finding the speed. Try it and see how quick it is. Get used to representing the speed of a particle by u so that v is available to stand for something else. The textbook continues this symbol choice in Chapters 39, 40, and 41.

---

44. Show that the energy–momentum relationship in Equation 39.27, $E^2 = p^2c^2 + (mc^2)^2$, follows from the expressions $E = \gamma mc^2$ and $p = \gamma mu$.

Solution

Conceptualize: With zero momentum, the equation to be proved reduces to $E = mc^2$, and that is a correct statement about rest energy. As an object moves faster, its momentum p increases and the equation correctly says that its total energy E also increases. The equation even correctly describes a particle like a photon with no rest energy. For it, we have $E^2 = p^2c^2 + 0$ and $E = pc$.

Categorize: The statement of the problem tells us precisely what to do, but we need to get through the algebra. We could think about our goal as evaluating E^2. The equations we start with contain the speed in the factor gamma. We must eliminate that speed or that γ.

Analyze: We are told to start from $E = \gamma mc^2$ and $p = \gamma mu$.

Squaring both equations gives $E^2 = \left(\gamma mc^2\right)^2$ and $p^2 = (\gamma mu)^2$

We choose to multiply the second equation by c^2 and subtract it from the first:

$$E^2 - p^2c^2 = \left(\gamma mc^2\right)^2 - (\gamma mu)^2c^2$$

We factor to obtain $$E^2 - p^2c^2 = \gamma^2\left(\left(mc^2\right)\left(mc^2\right) - \left(mc^2\right)\left(mu^2\right)\right)$$

Extracting the (mc^2) factors gives $$E^2 - p^2c^2 = \gamma^2\left(mc^2\right)^2\left(1 - \frac{u^2}{c^2}\right)$$

We substitute the definition of γ: $$E^2 - p^2c^2 = \left(1 - \frac{u^2}{c^2}\right)^{-1}\left(mc^2\right)^2\left(1 - \frac{u^2}{c^2}\right)$$

The γ^2 factors divide out, leaving $E^2 - p^2c^2 = (mc^2)^2$ ■

Finalize: This equation is analogous to the nonrelativistic $K = p^2/2m$ in that it gives an energy–momentum relationship without reference to speed. In fact, the relativistic equation should reduce to the classical one, like this: Total energy is kinetic plus rest energy, so $(K + mc^2)^2 = p^2c^2 + (mc^2)^2$.

Calculating, $K^2 + 2Kmc^2 + (mc^2)^2 = p^2c^2 + (mc^2)^2$

Simplifying, $K^2/mc^2 + 2K = p^2c^2/mc^2$

In the low-speed limit, K is small compared to mc^2, so the first term in this equation is negligible compared to the others. Then we have nonrelativistically $2K = p^2/m$ and $K = p^2/2m$, just as it should be.

---

47. A pion at rest ($m_\pi = 273m_e$) decays to a muon ($m_\mu = 207m_e$) and an antineutrino $(m_{\bar{\nu}} \approx 0)$. The reaction is written $\pi^- \rightarrow \mu^- + \bar{\nu}$. Find (a) the kinetic energy of the muon and (b) the energy of the antineutrino in electron volts.

Solution

Conceptualize: The original rest energy is greater than the final rest energy. (Both masses are quoted for brevity as multiples of the mass of an electron.) The decrease in rest energy becomes the kinetic energy of both decay products. You can think of this as sliding downhill in energy, as a child on a sled gains kinetic energy by falling into a more stable state. The two energies we find will add up to $(273 - 207)m_ec^2 = 66(0.511\text{ MeV})$.

Categorize: We use, together, both the energy version and the momentum version of the isolated system model.

Analyze: By conservation of system energy, $m_\pi c^2 = \gamma m_\mu c^2 + |p_{\bar{\nu}}|c$

By conservation of system momentum $p_{\bar{\nu}} = -p_\mu = -\gamma m_\mu u$

Substituting the second equation into the first, $m_\pi c^2 = \gamma m_\mu c^2 + \gamma m_\mu uc$

Simplified, this equation then reads $m_\pi = m_\mu(\gamma + \gamma u/c)$

Substituting the masses, $273m_e = (207m_e)(\gamma + \gamma u/c)$

where the rest energy of an electron is $m_ec^2 = 0.511\text{ MeV}$

Numerically,

$$\frac{273m_e}{207m_e} = \frac{1 + u/c}{\sqrt{1 - (u/c)^2}} = \sqrt{\frac{1 + u/c}{1 - u/c}}$$

Solving for the muon speed, $\frac{u}{c}=\frac{273^2-207^2}{273^2+207^2}=0.270$

Therefore, $\gamma=\frac{1}{\sqrt{1-u^2/c^2}}=1.038\,5$

(a) and the muon's kinetic energy is

$$K_\mu=(0.038\,5)(207\times 0.511\text{ MeV})=4.08\text{ MeV} \qquad \blacksquare$$

(b) The energy of the antineutrino is

$$K_{\bar{\nu}}=(273\times 0.511\text{ MeV})-(207\times 0.511\text{ MeV}+4.08\text{ MeV})=29.6\text{ MeV} \qquad \blacksquare$$

Finalize: You may see these elementary particles (pion, muon, antineutrino) again in Chapter 46, but energy and momentum conservation are the same ideas whatever they are applied to. It turned out to be easiest to solve for the muon speed on the way to finding the energies. Note that we used at one point $1-u^2/c^2=(1+u/c)(1-u/c)$.

---

53. The power output of the Sun is 3.85×10^{26} W. By how much does the mass of the Sun decrease each second?

Solution

Conceptualize: Anything losing energy loses what we measure as mass at the same time. It will be plenty of kilograms, but a very small fraction of the mass of the Sun.

Categorize: We use Einstein's famous mass-energy relation.

Analyze: From $E_R=mc^2$, for one second of the Sun's radiation we have

$$m=\frac{E_R}{c^2}=\frac{3.85\times 10^{26}\text{ J}}{\left(3.00\times 10^8\text{ m/s}\right)^2}=4.28\times 10^9\text{ kg} \qquad \blacksquare$$

Finalize: In each nuclear reaction of hydrogen nuclei fusing to form a helium nucleus, about 0.7% of the original rest mass disappears. (This is demonstrated in Problem 57 below.) Enough reactions go on every second for the vanishing mass to amount to these billions of kilograms. The Sun has a finite life.

---

57. The net nuclear fusion reaction inside the Sun can be written as $4\,^1\text{H}\rightarrow\,^4\text{He}+E$. The rest energy of each hydrogen atom is 938.78 MeV, and the rest energy of the helium-4 atom is 3 728.4 MeV. Calculate the percentage of the starting mass that is transformed to other forms of energy.

Solution

Conceptualize: It is a reaction of the nuclei of atoms. Everything below the visible surface of the Sun is so hot that there are no neutral atoms—everything is an ionized plasma. But it is conventional to quote masses of neutral atoms, and every cubic millimeter of the Sun contains enough electrons as well as nuclei for electrical neutrality.

Categorize: We add up the original mass and the final mass and look for the difference.

Analyze: The original rest energy of four protons is

$$E_R = 4(938.78 \text{ MeV}) = 3\ 755.12 \text{ MeV}$$

The energy given off is $|\Delta E| = (3\ 755.12 - 3\ 728.4) \text{ MeV} = 26.7 \text{ MeV}$

The fractional energy released is $\dfrac{|\Delta E|}{E_R} = \dfrac{26.7 \text{ MeV}}{3\ 755 \text{ MeV}} \times 100\% = 0.712\%$ ■

Finalize: A nuclear fusion reaction is colossally energetic compared to any mechanical process or chemical reaction. It is more energetic per unit mass than the fission of a heavy nucleus. But only less than one percent of the mass gets converted into energy. In $E = mc^2$, if E represents the energy output, then m is not the whole original mass, but the bit of mass that is annihilated.

---

61. The cosmic rays of highest energy are protons that have kinetic energy on the order of 10^{13} MeV. (a) As measured in the proton's frame, what time interval would a proton of this energy require to travel across the Milky Way galaxy, which has proper diameter ~ 10^5 ly? (b) From the point of view of the proton, how many kilometers across is the galaxy?

Solution

Conceptualize: We can guess that the energetic cosmic rays will be traveling close to the speed of light, so the time it takes a proton to traverse the Milky Way will be much less in the proton's frame than the 10^5 years that it takes in our frame. The galaxy will also appear smaller to the high-speed protons than the galaxy's proper diameter of 10^5 light years.

Categorize: The kinetic energy of the protons can be used to determine the relativistic γ-factor, which can then be applied to the time dilation and length contraction equations to find the time interval and distance in the proton's frame of reference.

Analyze: The relativistic kinetic energy of such a proton is

$$K = (\gamma - 1)mc^2 = 10^{13} \text{ MeV}$$

Its rest energy is

$$mc^2 = (1.67 \times 10^{-27} \text{ kg})(2.998 \times 10^8 \text{ m/s})^2 \left(\frac{1 \text{ eV}}{1.60 \times 10^{-19} \text{ kg} \cdot \text{m}^2/\text{s}^2} \right) = 938 \text{ MeV}$$

So $\quad 10^{13}\text{ MeV} = (\gamma - 1)(938\text{ MeV}),\quad$ and therefore $\quad \gamma = 1.07 \times 10^{10}$

The proton's speed in the galaxy's reference frame can be found from

$$\gamma = 1/\sqrt{1 - u^2/c^2} \qquad \text{so} \qquad 1 - u^2/c^2 = 8.80 \times 10^{-21}$$

and $\quad u = c\sqrt{1 - 8.80 \times 10^{-21}} = \left(1 - 4.40 \times 10^{-21}\right)c \approx 3.00 \times 10^8\text{ m/s}$

The proton's speed is nearly as large as the speed of light. In the galaxy frame, the traversal time is

$$\Delta t = x/u = 10^5\text{ light years}/c = 10^5\text{ years}$$

(a) This is dilated from the proper time measured in the proton's frame. The proper time interval is found from $\Delta t = \gamma \Delta t_p$:

$$\Delta t_p = \Delta t/\gamma = 10^5\text{ yr}/1.07 \times 10^{10} = 9.38 \times 10^{-6}\text{ years} = 296\text{ s}$$

$$\Delta t \sim \text{a few hundred seconds}$$ ■

(b) The proton sees the galaxy moving by at a speed nearly equal to c, passing in 296 s:

$$\Delta L_{proton\ frame} = u\Delta t_p = \left(3.00 \times 10^8\text{ m/s}\right)(296\text{ s}) = 8.88 \times 10^7\text{ km} \sim 10^8\text{ km}$$ ■

$$\Delta L_{proton\ frame} = \left(8.88 \times 10^{10}\text{ m}\right)\left(\frac{1\text{ ly}}{9.46 \times 10^{15}\text{ m}}\right) = 9.39 \times 10^{-6}\text{ ly} \sim 10^{-5}\text{ ly}$$

Finalize: The results agree with our predictions, although we may not have guessed that the protons would be traveling quite so close to the speed of light! The calculated results should be rounded to zero significant figures since we were given order of magnitude data. We should also note that the relative speed of motion u and the value of γ are the same in both the proton and galaxy reference frames.

---

64. Spacecraft I, containing students taking a physics exam, approaches the Earth with a speed of $0.600c$ (relative to the Earth), while spacecraft II, containing professors proctoring the exam, moves at $0.280c$ (relative to the Earth) directly toward the students. If the professors stop the exam after 50.0 min have passed on their clock, for what time interval does the exam last as measured by (a) the students and (b) an observer on the Earth?

Solution

Conceptualize: Think of two events as the professors sending out a "start" signal and the professors sending out a "stop" signal. The 50 minutes on a clock stationary in their spacecraft is a proper time interval. Everyone else, especially the students, sees this clock as moving and so running slow, so for us on Earth and especially for the students the exam lasts longer.

Categorize: This is a problem about time dilation. We are equipped to think only about two reference frames at a time.

Analyze: Suppose that in the Earth frame, the students are moving to the right at $u_x = 0.600c$ and the professors are moving to the left, with velocity component $v = -0.280c$ In the professors' frame, the Earth moves to the right at $u'_e = 0.280c$ and the students move to the right at

$$u'_x = \frac{u_x - v}{1 - u_x v/c^2}: \qquad u'_x = \frac{0.600c - (-0.280c)}{1 - (0.600c)(-0.280c)/c^2} = 0.753c$$

The professors measure 50 minutes on a clock at rest in their frame: they measure proper time and everyone else sees longer, dilated time intervals.

(a) For the students, $\Delta t = \gamma \Delta t_p = \dfrac{50.0 \text{ min}}{\sqrt{1-(0.753)^2}} = 76.0 \text{ min}$ ■

(b) On Earth, $\Delta t = \gamma \Delta t_p = \dfrac{50.0 \text{ min}}{\sqrt{1-(0.280)^2}} = 52 \text{ min}, 5 \text{ sec}$ ■

Finalize: Einstein's relativity is so fascinating to nonscientists (and scientists) because of surprising conclusions about everyday-life situations. We remind you that it is a scientific theory, fully confirmed by experiment and in accord with other theories. The natural clocks of muons undergoing radioactive decay show just the sort of time dilation considered here.

---

74. A particle with electric charge q moves along a straight line in a uniform electric field $\vec{\mathbf{E}}$ with speed u. The electric force exerted on the charge is $q\vec{\mathbf{E}}$. The velocity of the particle and the electric field are both in the x direction. (a) Show that the acceleration of the particle in the x direction is given by

$$a = \frac{du}{dt} = \frac{qE}{m}\left(1 - \frac{u^2}{c^2}\right)^{3/2}$$

(b) Discuss the significance of the dependence of the acceleration on the speed. (c) **What If?** If the particle starts from rest at $x = 0$ at $t = 0$, how would you proceed to find the speed of the particle and its position at time t?

Solution

Conceptualize: Relativistically, a constant force does not produce a constant acceleration. It makes the momentum increase steadily, but momentum is γmu instead of just mu. In the equation we derive, as u increases the acceleration will decrease. In this way the speed will approach a finite limit of c as time goes on.

Categorize: The constant force exerted on a charged particle in an electric field is given by $F = qE$. Further, at any speed, the momentum of the particle is given by

$$p = \gamma mu = \frac{mu}{\sqrt{1 - u^2/c^2}}$$

We will use the momentum version of the nonisolated system model.

Analyze:

(a) With Newton's law expressed as $F = qE = \dfrac{dp}{dt}$,

we have $qE = \dfrac{d}{dt}\left(mu\left(1 - u^2/c^2\right)^{-1/2}\right)$

We take the derivative:

$$qE = m\left(1 - \frac{u^2}{c^2}\right)^{-1/2}\frac{du}{dt} + \tfrac{1}{2}mu\left(1 - \frac{u^2}{c^2}\right)^{-3/2}\left(\frac{2u}{c^2}\right)\frac{du}{dt}$$

Simplifying, we find that

$$\frac{qE}{m} = \frac{du}{dt}\left(1 - \frac{u^2}{c^2}\right)^{-3/2} \quad \text{and} \quad a = \frac{du}{dt} = \frac{qE}{m}\left(1 - \frac{u^2}{c^2}\right)^{3/2} \quad \blacksquare$$

(b) As $u \to c$, we see that $a \to 0$. The particle thus never attains the speed of light. If $u \ll c$, the equation turns into $a \approx qE/m$, in agreement with the nonrelativistic account. ■

(c) Taking the acceleration equation, isolating the velocity terms, and integrating, we have

$$\int_0^u \left(1 - \frac{u^2}{c^2}\right)^{-3/2} du = \int_0^t \frac{qE}{m}\,dt \quad \text{then} \quad u = \frac{qEct}{\sqrt{m^2c^2 + q^2E^2t^2}} = \frac{dx}{dt} \quad \blacksquare$$

We can find the position by direct integration:

$$x = \int_0^x dx = qEc\int_0^t \frac{t\,dt}{\sqrt{m^2c^2 + q^2E^2t^2}} = \frac{c}{qE}\left(\sqrt{m^2c^2 + q^2E^2t^2} - mc\right) \quad \blacksquare$$

Finalize: In the velocity equation for small t we have $u = qEct/mc = qEt/m$, as we should expect classically. In the same equation for large t we have $u \to qEct/qEt = c$. With unlimited energy input the particle can approach arbitrarily close to the speed of light.

40

Introduction to Quantum Physics

EQUATIONS AND CONCEPTS

Stefan's law states: The total power emitted by a body as thermal radiation depends on the fourth power of the Kelvin temperature. The parameter σ is the Stefan–Boltzmann constant and e is the emissivity of the surface of area A. *For an ideal black body, $e = 1$.*

$$P = \sigma A e T^4 \quad (40.1)$$

$$\sigma = 5.670 \times 10^{-8}\ \text{W/m}^2 \cdot \text{K}^4$$

According to the **Wien displacement law**, as the temperature of a black body increases, the radiation intensity increases and the peak of the distribution shifts to shorter wavelengths. *The total radiation emitted is the area under the Intensity vs. Wavelength curve.*

$$\lambda_{max} T = 2.898 \times 10^{-3}\ \text{m} \cdot \text{K} \quad (40.2)$$

Discrete energy values of an atomic oscillator are determined by a quantum number, n. Each discrete energy value corresponds to a quantum state.

$$E_n = nhf \quad (40.4)$$

$$(n = 1, 2, 3, \ldots)$$

The **energy of a quantum** or photon corresponds to the energy difference between initial and final quantum states. *An oscillator emits or absorbs energy only when there is a transition between quantum states.*

$$E = hf \quad (40.5)$$

Planck's constant is a fundamental constant of nature.

$$h = 6.626 \times 10^{-34}\ \text{J} \cdot \text{s} \quad (40.7)$$

Photoelectrons have a maximum kinetic energy which depends on the energy of the incident photon and the work function (ϕ), of the metal which is typically a few eV.

$$K_{max} = hf - \phi \quad (40.9)$$

The cutoff wavelength (and corresponding frequency) depends on the value of the work function of a specific surface; for wavelengths greater than λ_c the photoelectric effect will not be observed.

$$\lambda_c = \frac{hc}{\phi} \quad (40.10)$$

$$hc = 1\,240 \text{ eV} \cdot \text{nm}$$

The **Compton shift** is the change in wavelength of a photon when scattered from an electron. *The scattered photon makes an angle θ with the direction of the incident photon.*

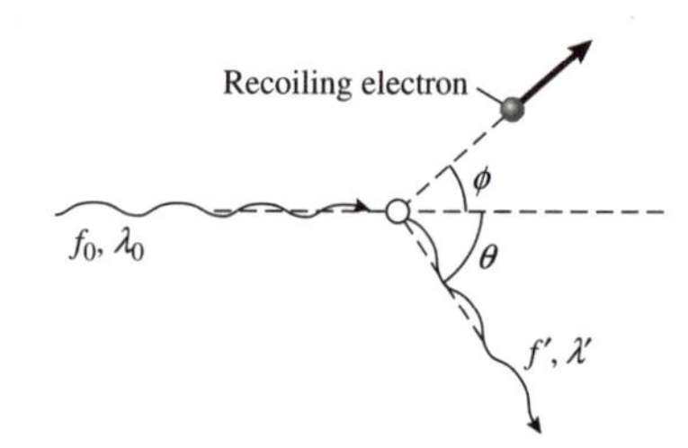

$$\lambda' - \lambda_0 = \frac{h}{m_e c}(1 - \cos\theta) \quad (40.11)$$

λ_C is called the Compton wavelength of the electron.

$$\lambda_C = \frac{h}{m_e c} = 0.002\,43 \text{ nn}$$

The **de Broglie wavelength** of a particle is inversely proportional to the momentum of the particle. In Chapters 39, 40, and 41 u represents the speed of a particle with mass and v is used when no rest energy is involved.

$$\lambda = \frac{h}{p} = \frac{h}{mu} \quad (40.15)$$

Phase speed refers to the speed of a crest on a single wave in a wave packet.

$$v_{phase} = \frac{\omega}{k} \quad (40.18)$$

$$(\omega = 2\pi f \text{ and } k = 2\pi/\lambda)$$

Group speed refers to the speed of a packet (envelope) or group of waves.

$$v_g = \frac{d\omega}{dk} \quad (40.19)$$

The **Heisenberg uncertainty principle** can be stated in two forms:

- **Simultaneous measurements of position and momentum** with respective uncertainties Δx and Δp_x.

$$\Delta x \, \Delta p_x \geq \frac{\hbar}{2} \quad (40.23)$$

- **Simultaneous measurements of energy and lifetime** with uncertainties ΔE and Δt.

$$\Delta E \, \Delta t \geq \frac{\hbar}{2} \quad (40.24)$$

REVIEW CHECKLIST

You should be able to:

- Describe the formula for blackbody radiation proposed by Planck, the assumption made in deriving this formula, and the related experimental results. (Section 40.1)
- Describe the Einstein model for the photoelectric effect, including the important experimental results. Make calculations using the photoelectric equation. (Section 40.2)
- Describe the Compton effect (the scattering of x-rays by electrons) and make calculations using the equation for the Compton shift. (Section 40.3)
- Make calculations using the de Broglie wavelength equation. (Section 40.5)
- Calculate the phase speed and group speed associated with a quantum particle. (Section 40.6)
- Make calculations using both forms of the Heisenberg uncertainty principle. (Section 40.8)

ANSWERS TO SELECTED OBJECTIVE QUESTIONS

4. In a certain experiment, a filament in an evacuated lightbulb carries a current I_1 and you measure the spectrum of light emitted by the filament, which behaves as a black body at temperature T_1. The wavelength emitted with highest intensity (symbolized by λ_{max}) has the value λ_1. You then increase the potential difference across the filament by a factor of 8, and the current increases by a factor of 2. **(i)** After this change, what is the new value of the temperature of the filament? (a) $16T_1$ (b) $8T_1$ (c) $4T_1$ (d) $2T_1$ (e) still T_1 **(ii)** What is the new value of the wavelength emitted with highest intensity? (a) $4\lambda_1$ (b) $2\lambda_1$ (c) λ_1 (d) $\frac{1}{2}\lambda_1$ (e) $\frac{1}{4}\lambda_1$

Answer The temperature T_1 is no doubt well above room temperature, and the temperature T_2 with higher voltage will be higher still. So we might expect the resistance of the filament to increase, and that is why the current may increase only by a factor of 2 as the potential difference becomes 8 times larger. Taking the product, we find the electric power $P = I\Delta V$ delivered to the filament is 16 times higher at the final setting. With a steady-state temperature in a vacuum, the filament must radiate all of this power as electromagnetic waves, according to $P = 1A\sigma T^4$. Thermal expansion in surface area has a negligible effect, so the final temperature is then given by $P_2/P_1 = (T_2/T_1)^4$ so $T_2 = T_1(P_2/P_1)^{1/4} = T_1(16)^{1/4} = 2T_1$. For question **(i)** this is answer (d). Now Wien's displacement law, that the wavelength of strongest emission is inversely proportional to the absolute temperature, says that **(ii)** the new wavelength where the spectrum intensity peaks is $\lambda_1/2$, answer (d).

□ □ □ □

8. An x-ray photon is scattered by an originally stationary electron. Relative to the frequency of the incident photon, is the frequency of the scattered photon (a) lower, (b) higher, or (c) unchanged?

Answer (a) The x-ray photon transfers some of its energy to the electron. Thus, its energy, and therefore its frequency, must be decreased.

□ □ □ □

ANSWERS TO SELECTED CONCEPTUAL QUESTIONS

4. If the photoelectric effect is observed for one metal, can you conclude that the effect will also be observed for another metal under the same conditions? Explain.

Answer No. Suppose that the incident light frequency at which you first observed the photoelectric effect is above the cutoff frequency of the first metal, but less than the cutoff frequency of the second metal. In that case, the photoelectric effect would not be observed at all in the second metal.

□ □ □ □

6. Why does the existence of a cutoff frequency in the photoelectric effect favor a particle theory for light over a wave theory?

Answer A theory modeling light as a classical wave would predict that the photoelectric effect should occur at any frequency, provided that the light intensity is high enough. As is implied by the question, this is in contradiction to experimental results. Light with a frequency below the cutoff cannot knock electrons out of a particular target at all.

□ □ □ □

13. If matter has a wave nature, why is this wave-like characteristic not observable in our daily experiences?

Answer For any object that we can perceive directly, the de Broglie wavelength $\lambda = h/mu$ is too small to be measured by any means; therefore, no wavelike characteristics can be observed. The object will not diffract noticeably when it goes through an aperture. It will not show resolvable interference maxima and minima when it goes through two openings. It will not show resolvable nodes and antinodes if it is in resonance.

□ □ □ □

SOLUTIONS TO SELECTED PROBLEMS

1. The human eye is most sensitive to 560-nm (green) light. What is the temperature of a black body that would radiate most intensely at this wavelength?

Solution

Conceptualize: The color could be called yellow as well as green. A red-hot object might be at a couple of thousand kelvins. To have this shorter wavelength as its wavelength of maximum emission, the object will need to be at somewhat higher temperature.

Categorize: We use Wien's law.

Analyze: It is expressed by $\lambda_{max}T = 2.898 \times 10^{-3}$ m · K.

$$T = \frac{2.90 \text{ mm} \cdot \text{K}}{560 \times 10^{-6} \text{ mm}} = 5\,180 \text{ K} \quad \blacksquare$$

Finalize: This is close to the temperature of the surface of the Sun (which acts as a pretty good black body). Living things on Earth evolved to be sensitive to electromagnetic waves near this wavelength because there is such a lot of it bouncing around, carrying information.

---

6. **(i)** Calculate the energy, in electron volts, of a photon whose frequency is (a) 620 THz, (b) 3.10 GHz, and (c) 46.0 MHz. **(ii)** Determine the corresponding wavelengths for the photons listed in part (i) and **(iii)** state the classification of each on the electromagnetic spectrum.

Solution

Conceptualize: From part (a) through (b) to (c) the frequency of the electromagnetic wave under discussion goes down by on the order of ten million times. So the photon energy, starting from a few electronvolts, will go down by ~10^7 times to a fraction of a microelectronvolt. The wavelength will go up by this factor, from something small compared to your height to something large.

Categorize: We use just Planck's equation for the energy of a photon and the relationship between wavelength and frequency for any continuous wave. 1 eV = 1.60×10^{-19} J is a unit of energy we can use for anything if we want to.

Analyze: **(i)** Planck's equation is $E = hf$. The photon energies are

(a) $$E = \left(6.63 \times 10^{-34} \text{ J} \cdot \text{s}\right)\left(6.20 \times 10^{14} \text{ Hz}\right) = 4.11 \times 10^{-19} \text{ J} = 2.57 \text{ eV} \quad \blacksquare$$

(b) $$E = \left(6.63 \times 10^{-34} \text{ J} \cdot \text{s}\right)\left(3.10 \times 10^{9} \text{ Hz}\right) = 2.06 \times 10^{-24} \text{ J} = 12.8\ \mu\text{eV} \quad \blacksquare$$

(c) $$E = \left(6.63 \times 10^{-34} \text{ J} \cdot \text{s}\right)\left(46.0 \times 10^{6} \text{ Hz}\right) = 3.05 \times 10^{-26} \text{ J} = 1.91 \times 10^{-7} \text{ eV} \quad \blacksquare$$

(ii) The wavelengths are

(a) $$\lambda = \frac{c}{f} = \frac{3.00 \times 10^{8} \text{ m/s}}{6.20 \times 10^{14} \text{ s}^{-1}} = 4.84 \times 10^{-7} \text{ m} = 484 \text{ nm} \quad \blacksquare$$

(b) $$\lambda = \frac{3.00 \times 10^{8} \text{ m/s}}{3.10 \times 10^{9} \text{ s}^{-1}} = 0.096\,8 \text{ m} = 9.68 \text{ cm} \quad \blacksquare$$

(c) $$\lambda = \frac{3.00 \times 10^{8} \text{ m/s}}{46.0 \times 10^{6} \text{ s}^{-1}} = 6.52 \text{ m} \quad \blacksquare$$

(iii) These wavelengths correspond to (a) blue light, (b) microwave radiation, and (c) radio waves in the public LO band. ■

Finalize: Make sure you never confuse a photon with a proton. A photon is a quantum particle with zero mass, a parcel of electromagnetic radiation that cannot be subdivided in absorption or emission. It carries energy like a classical particle, with the amount given by $E = hf$; and it moves as a continuous wave, described by $\lambda = v/f$, just as classical waves do. It is neither a classical particle nor a classical wave, but a quantum particle.

---

8. An FM radio transmitter has a power output of 150 kW and operates at a frequency of 99.7 MHz. How many photons per second does the transmitter emit?

Solution

Conceptualize: Each photon is so small an energy bundle that the number must be colossally large.

Categorize: We use Planck's equation and the definition of power.

Analyze: Each photon has an energy

$$E = hf = \left(6.63 \times 10^{-34}\ \text{J}\cdot\text{s}\right)\left(99.7 \times 10^{6}\ \text{s}^{-1}\right) = 6.61 \times 10^{-26}\ \text{J}$$

The number of photons per second is the power divided by the energy per photon:

$$R = \frac{P}{E} = \frac{150 \times 10^{3}\ \text{J/s}}{6.61 \times 10^{-26}\ \text{J}} = 2.27 \times 10^{30}\ \text{photons/s}$$ ■

Finalize: The photoelectric effect and other phenomena demonstrate the graininess of energy carried by visible light. For radio waves the photon energy is so small that the quantization of energy cannot be directly observed—it is like trying to observe from Earth the separateness of grains of dust on the Moon.

---

17. Two light sources are used in a photoelectric experiment to determine the work function for a particular metal surface. When green light from a mercury lamp ($\lambda = 546.1$ nm) is used, a stopping potential of 0.376 V reduces the photocurrent to zero. (a) Based on this measurement, what is the work function for this metal? (b) What stopping potential would be observed when using the yellow light from a helium discharge tube ($\lambda = 587.5$ nm)?

Solution

Conceptualize: According to Table 40.1, the work function for most metals is on the order of a few eV, so this metal is probably similar. We can expect the stopping potential for the yellow light to be slightly lower than 0.376 V since the yellow light has a longer wavelength (lower frequency) and therefore less energy per photon than the green light.

Categorize: In this photoelectric experiment, the green light has sufficient energy hf to overcome the work function of the metal ϕ so that the ejected electrons have a maximum kinetic energy of 0.376 eV. With this information, we can use the photoelectric effect equation to find the work function, which can then be used to find the stopping potential for the less energetic yellow light.

Analyze:

(a) Einstein's photoelectric effect equation is $K_{max} = hf - \phi$ and the energy required to raise an electron through a 1-V potential is 1 eV, so that

$$K_{max} = e\Delta V_s = 0.376 \text{ eV}$$

The energy of a photon from the mercury lamp is:

$$hf = \frac{hc}{\lambda} = \frac{(6.626\times10^{-34}\text{ J}\cdot\text{s})(2.998\times10^{8}\text{m/s})}{546.1\times10^{-9}\text{m}}\left(\frac{1\text{ eV}}{1.602\times10^{-19}\text{J}}\right) = \frac{1240\text{ eV}\cdot\text{nm}}{546.1\text{ nm}} = 2.27\text{ eV}$$

Therefore, the work function for this metal is:

$$\phi = hf - K_{max} = 2.27\text{ eV} - 0.376\text{ eV} = 1.89\text{ eV}$$ ■

(b) For the yellow light, $\lambda = 587.5$ nm and the photon energy is

$$hf = \frac{hc}{\lambda} = \frac{1240\text{ eV}\cdot\text{nm}}{587.5\times10^{-9}\text{ m}} = 2.11\text{ eV}$$

Therefore the maximum energy that can be given to an ejected electron is

$$K_{max} = hf - \phi = 2.11\text{ eV} - 1.89\text{ eV} = 0.216\text{ eV}$$

so the stopping voltage is $\Delta V_s = 0.216$ V ■

Finalize: The work function for this metal is lower than we expected and does not correspond with any of the values in Table 40.1. Further examination in the *CRC Handbook of Chemistry and Physics* reveals that all of the metal elements have work functions between 2 and 6 eV. However, a single metal's work function may vary by about 1 eV depending on impurities in the metal, so it is just barely possible that a metal might have a work function of 1.89 eV.

The stopping potential for the yellow light is indeed lower than for the green light as we expected. An interesting calculation is to find the wavelength for the lowest energy light that will eject electrons from this metal. That threshold wavelength for $K_{max} = 0$ is 654 nm, which is red light in the visible portion of the electromagnetic spectrum.

25. A 0.001 60-nm photon scatters from a free electron. For what (photon) scattering angle does the recoiling electron have kinetic energy equal to the energy of the scattered photon?

Solution

Conceptualize: Scattering angles can lie between 0° and 180°. If photon and electron end up with equal energies in this Compton scattering process, the photon must give just half its original energy to the originally stationary electron, and the photon's wavelength must double. It is not clear that this is allowed …

Categorize: … by Compton's equation but, if it is, that equation will relate the change in wavelength to the scattering angle of the x-ray photon.

Analyze: The energy of the incoming photon is

$$E_0 = \frac{hc}{\lambda_0} = \frac{(6.63 \times 10^{-34} \text{ J}\cdot\text{s})(3.00 \times 10^8 \text{ m/s})}{0.001\,60 \times 10^{-9} \text{ m}} = 1.24 \times 10^{-13} \text{ J}$$

The outgoing photon and the electron share equally in this energy. The kinetic energy of the electron and the energy of the scattered photon are each one-half of E_0.

$$E' = 6.22 \times 10^{-14} \text{ J} \quad \text{and} \quad \lambda' = \frac{hc}{E'} = 3.20 \times 10^{-12} \text{ m}$$

The shift in wavelength is $\Delta\lambda = \lambda' - \lambda_0 = 1.60 \times 10^{-12}$ m

But by Equation 40.11, $\Delta\lambda = \lambda_C(1 - \cos\theta)$ where λ_C is the Compton wavelength.

Then $\cos\theta = 1 - \frac{\Delta\lambda}{\lambda_C}$ implies

$$\theta = \cos^{-1}\left(1 - \frac{\Delta\lambda}{\lambda_C}\right) = \cos^{-1}\left(1 - \frac{1.60 \times 10^{-12}\text{m}}{2.43 \times 10^{-12}\text{m}}\right) = \cos^{-1} 0.342 = 70.0° \quad \blacksquare$$

Finalize: Compton's equation says that the smallest possible change in photon wavelength is zero, occurring if the photon barely grazes the electron and its scattering angle is zero. The largest possible change in wavelength is $2\lambda_C = 4.86$ pm, if the photon recoils straight back from a head-on collision. So the 1.60-pm wavelength change in this problem is allowed.

---

35. Review. A helium–neon laser produces a beam of diameter 1.75 mm, delivering 2.00×10^{18} photons/s. Each photon has a wavelength of 633 nm. Calculate the amplitudes of (a) the electric field and (b) the magnetic field inside the beam. (c) If the beam shines perpendicularly onto a perfectly reflecting surface, what force does it exert on the surface? (d) If the beam is absorbed by a block of ice at 0°C for 1.50 h, what mass of ice is melted?

Solution

Conceptualize: We are reviewing several properties of light. It is a wave of electric field and magnetic field; we expect the amplitudes for this bright light to be several newtons per coulomb and a small fraction of a tesla. It carries momentum, and will exert a tiny force on the mirror. It carries energy and will melt a bit of ice, perhaps a few grams.

Categorize: The photon stream is the light beam. We will find its intensity. We have studied how the intensity is related to both of the field amplitudes, to the radiation pressure, and to the energy transported in a certain time interval.

Analyze: The energy of one photon is

$E = \frac{hc}{\lambda} = \frac{6.63\times10^{-34}\,\text{J}\cdot\text{s}\left(3\times10^{8}\text{ m/s}\right)}{633\times10^{-9}\text{m}} = 3.14\times10^{-19}\,\text{J}$. The power carried by the beam is $P = \left(2\times10^{18}\text{photons/s}\right)\left(3.14\times10^{-19}\text{ J/photon}\right) = 0.628\text{ W}$. Its intensity is the average Poynting vector

$$I = S_{\text{av}} = \frac{P}{A} = \frac{P}{\pi r^2} = \frac{0.628\text{ W}(4)}{\pi\left(1.75\times10^{-3}\text{m}\right)^2} = 2.61\times10^{5}\text{ W/m}^2$$

From Chapter 34, the intensity is related to the field amplitudes by

$$S_{\text{av}} = \left|\frac{1}{\mu_0}\vec{\mathbf{E}}\times\vec{\mathbf{B}}\right|_{\text{av}} = \frac{1}{\mu_0}E_{\text{rms}}B_{\text{rms}}\sin 90^\circ = \frac{1}{\mu_0}\frac{E_{\text{max}}}{\sqrt{2}}\frac{B_{\text{max}}}{\sqrt{2}}$$

We also have $E_{\text{max}} = B_{\text{max}}c$.

(a) So $S_{\text{av}} = \frac{E_{\text{max}}^2}{2\mu_0 c}$ and the electric field amplitude is

$$E_{\text{max}} = \left(2\mu_0 c S_{\text{av}}\right)^{1/2} = \left(2\left(4\pi\times10^{-7}\text{T}\cdot\text{m/A}\right)\left(3\times10^{8}\text{m/s}\right)\left(2.61\times10^{5}\text{W/m}^2\right)\right)^{1/2}$$
$$= 1.40\times10^{4}\,\text{N/C} \quad \blacksquare$$

(b) The magnetic field amplitude is $B_{\text{max}} = \frac{1.40\times10^{4}\,\text{N/C}}{3\times10^{8}\,\text{m/s}} = 4.68\times10^{-5}\,\text{T}$ ■

(c) Each photon carries momentum $\frac{E}{c}$. The beam transports momentum at the rate $\frac{P}{c}$. It would impart momentum to an absorbing surface at this rate, and it bounces back to give momentum to a perfectly reflecting surface at the rate

$$\frac{2P}{c} = \text{force} = \frac{2(0.628\text{ W})}{3\times10^{8}\,\text{m/s}} = 4.19\times10^{-9}\,\text{N} \quad \blacksquare$$

(d) The block of ice absorbs energy $L\Delta m = P\Delta t$, melting the mass

$$\Delta m = \frac{P\Delta t}{L} = \frac{0.628\text{ W}\left(1.5\times3\,600\text{ s}\right)}{3.33\times10^{5}\,\text{J/kg}} = 1.02\times10^{-2}\,\text{kg} \quad \blacksquare$$

Finalize: This problem involves lots of review. You might think of it as a fine point, but it is useful to know that the rms value is $1/\sqrt{2}$ times the maximum value for a sinusoidally varying signal, and it is the rms value that tells most directly about energy transport. It is a very fundamental idea that force is the time rate of momentum change. The laser in this problem is more powerful than a typical student-laboratory laser, at 628 milliwatts instead of on the order of one milliwatt. But it would be challenging to make direct measurements of the field amplitudes and the force on the mirror. It would be challenging to keep the ice block insulated well enough to see clearly that the light melts ten grams in ninety minutes. And it would be especially challenging to demonstrate the photon-granularity of the energy transport.

---

37. The resolving power of a microscope depends on the wavelength used. If you wanted to "see" an atom, a wavelength of approximately 1.00×10^{-11} m would be required. (a) If electrons are used (in an electron microscope), what minimum kinetic energy is required for the electrons? (b) **What If?** If photons are used, what minimum photon energy is needed to obtain the required resolution?

Solution

Conceptualize: Higher energy goes with shorter wavelength, both for photons and for electrons, but with quite different patterns. An electron with a few thousand eV's of kinetic energy may have wavelength 10^{-11} m. It will take a hard x-ray with much higher energy to get an electromagnetic wave with the same wavelength.

Categorize: Think of this as two separate problems. Quite different equations describe the wavelength of a quantum particle with mass (de Broglie's equation) and the wavelength of a photon.

Analyze:

(a) Since the de Broglie wavelength is $\lambda = h/p$, the electron momentum is

$$p_e = \frac{h}{\lambda} = \frac{6.63 \times 10^{-34}\ \text{J}\cdot\text{s}}{1.00 \times 10^{-11}\ \text{m}} = 6.63 \times 10^{-23}\ \text{kg}\cdot\text{m/s}$$

and its kinetic energy is

$$K = \frac{p_e^{\,2}}{2m_e} = \frac{\left(6.63 \times 10^{-23}\ \text{kg}\cdot\text{m/s}\right)^2}{2\left(9.11 \times 10^{-31}\ \text{kg}\right)} = 2.41 \times 10^{-15}\ \text{J} = 15.1\ \text{keV} \qquad \blacksquare$$

For better accuracy, you can use the relativistic equation

$$\left(m_e c^2 + K\right)^2 = p_e^{\,2} c^2 + m_e^{\,2} c^4 \quad \text{to find} \quad K = 14.9\ \text{keV} \qquad \blacksquare$$

(b) For photons the energy is

$$E = hf = \frac{hc}{\lambda} = \frac{\left(6.63 \times 10^{-34}\ \text{J}\cdot\text{s}\right)\left(3.00 \times 10^{8}\ \text{m/s}\right)}{1.00 \times 10^{-11}\ \text{m}} = 1.99 \times 10^{-14}\ \text{J} = 124\ \text{keV} \qquad \blacksquare$$

For the photon, this wavelength $\lambda = 10$ pm is in the x-ray range of the electromagnetic spectrum.

Finalize: The photon has on the order of ten times more energy than an electron with the same wavelength.

Never try to use $\lambda = h/mu$ for a photon. A photon has no mass. (But $\lambda = h/p$ does apply to photons, where $p = E/c$ is the momentum a photon transports, as evidenced by its radiation pressure.) Good advice is not to use $v = f\lambda$ for electrons or other quantum particles with mass. The v in this equation could not be interpreted as the speed of the particle. On the other hand, 1 eV is a convenient unit for energies of photons, protons, and other quantum particles and atomic systems, not just for electrons.

---

40. The nucleus of an atom is on the order of 10^{-14} m in diameter. For an electron to be confined to a nucleus, its de Broglie wavelength would have to be on this order of magnitude or smaller. (a) What would be the kinetic energy of an electron confined to this region? (b) Make an order-of-magnitude estimate of the electric potential energy of a system of an electron inside an atomic nucleus. (c) Would you expect to find an electron in a nucleus? Explain.

Solution

Conceptualize: The de Broglie wavelength of a normal ground-state orbiting electron is on the order of 10^{-10} m (the diameter of a hydrogen atom) so with a shorter wavelength, the electron would have more kinetic energy if confined inside the nucleus. If the kinetic energy is much greater than the potential energy characterizing its attraction with the positive nucleus, then the electron will escape from its electrostatic potential well.

Categorize: If we try to calculate the velocity of the electron from the de Broglie wavelength as

$$u = \frac{h}{m_e\lambda} = \frac{6.63 \times 10^{-34}\ \text{J}\cdot\text{s}}{\left(9.11 \times 10^{-31}\ \text{kg}\right)\left(10^{-14}\ \text{m}\right)} = 7.27 \times 10^{10}\ \text{m/s}$$

we find a value which is not possible since it exceeds the speed of light. Therefore, we must use the relativistic energy expression to find the kinetic energy of this fast-moving electron.

Analyze:

(a) We find the momentum of the particle:

$$p = \frac{h}{\lambda} = \frac{6.63 \times 10^{-34}\ \text{J}\cdot\text{s}}{10^{-14}\ \text{m}} = 6.63 \times 10^{-20}\ \text{N}\cdot\text{s}$$

We find the particle's relativistic total energy from $E^2 = (pc)^2 + \left(mc^2\right)^2$ as

$$E = \sqrt{\left(1.99 \times 10^{-11}\ \text{J}\right)^2 + \left(8.19 \times 10^{-14}\ \text{J}\right)^2} = 1.99 \times 10^{-11}\ \text{J}$$

Its relativistic kinetic energy is $K = E - mc^2$.

$$K = \frac{1.99 \times 10^{-11}\ \text{J} - 8.19 \times 10^{-14}\ \text{J}}{1.60 \times 10^{-19}\ \text{J/eV}} = 124\ \text{MeV} \sim 100\ \text{MeV} \qquad \blacksquare$$

(b) The electrostatic potential energy of an electron-proton system with a separation of 10^{-14} m is:

$$U = -\frac{k_e e^2}{r} = -\frac{\left(8.99 \times 10^9\ \text{N}\cdot\text{m}^2/\text{C}^2\right)\left(1.60 \times 10^{-19}\ \text{C}\right)^2}{10^{-14}\ \text{m}}$$
$$= -2.30 \times 10^{-14}\ \text{J} \sim -0.1\ \text{MeV} \qquad \blacksquare$$

(c) Since the kinetic energy is nearly 1 000 times greater than the magnitude of the potential energy, the electron would immediately escape the proton's attraction and would not be confined to the nucleus. ■

Finalize: It is also interesting to notice in the above calculations that the rest energy of the electron is negligible compared to the momentum contribution to the total energy.

---

46. Consider a freely moving quantum particle with mass m and speed u. Its energy is $E = K = \frac{1}{2}mu^2$. (a) Determine the phase speed of the quantum wave representing the particle and (b) show that it is different from the speed at which the particle transports mass and energy.

Solution

Conceptualize: Section 40.6 in the textbook demonstrates how the group speed of a wave packet is the speed u of the quantum particle with mass that the wave packet represents. A point of constant phase, such as a crest of the wave within the packet, can move at a different speed. Its speed, called the phase speed v_{phase}, can in principle be larger or smaller than the group speed; this problem identifies the phase speed.

Categorize: Note that Chapters 39, 40, and 41 use u to represent the speed of a particle with mass, so that v is available for the speeds of waves and reference frames. We are to determine the phase speed $v_{\text{phase}} = \omega/k$ of the wave representing the particle. We can use the Planck and de Broglie equations $E = hf$ and $\lambda = h/p$ that relate the energy and momentum of the quantum particle to the wavelength and frequency of the wave that represents it.

Analyze: The particle is freely moving, so we attribute no potential energy to it. Its energy is $E = K = \frac{1}{2}mu^2 = hf = (h/2\pi)(2\pi f) = \hbar\omega$.

For its momentum we have $p = mu = h/\lambda = (h/2\pi)(2\pi/\lambda) = \hbar k$

Thus $\omega = K/\hbar$ and $k = p/\hbar$

Then (a) the phase speed is

$$v_{\text{phase}} = f\lambda = (2\pi f)(\lambda/2\pi) = \omega/k = (K/\hbar)(\hbar/p) = K/p = (\tfrac{1}{2}mu^2/mu) = u/2$$

(b) We see that the phase speed is only one-half of the experimentally measurable speed u at which the quantum particle transports mass, energy, and momentum. In the textbook's Active Figure 40.20, individual wave crests would move forward more slowly than their envelope moves forward, so individual crests would appear to move backward relative to the packet containing them.

Finalize: By contrast, a photon is a quantum particle with no mass. It transports energy and momentum, related by $E = pc$. Its energy is not kinetic energy, but energy in electric and magnetic fields. For a photon in vacuum the same procedure identifies the phase speed as

$$v_{\text{phase}} = f\lambda = (2\pi f)(\lambda/2\pi) = \omega/k = (E/\hbar)(\hbar/p) = E/p = pc/p = c$$

Its phase speed is the same as its group speed, the speed of light.

---

49. Neutrons traveling at 0.400 m/s are directed through a pair of slits separated by 1.00 mm. An array of detectors is placed 10.0 m from the slits. (a) What is the de Broglie wavelength of the neutrons? (b) How far off axis is the first zero-intensity point on the detector array? (c) When a neutron reaches a detector, can we say which slit the neutron passed through? Explain.

Solution

Conceptualize: The momentum of each neutron is small, but Planck's constant is so very small that the de Broglie wavelength of the neutrons will be small compared to a millimeter. Still the two-slit diffraction pattern may be observable, with the answer to (b) being on the order of a millimeter.

Categorize: We use de Broglie's equation and the waves in interference model.

Analyze:

(a) The wavelength of the neutrons is

$$\lambda = \frac{h}{mu} = \frac{6.63\times10^{-34}\ \text{J}\cdot\text{s}}{\left(1.67\times10^{-27}\ \text{kg}\right)(0.400\ \text{m/s})} = 9.93\times10^{-7}\ \text{m} \quad \blacksquare$$

(b) The condition for destructive interference in a multiple-slit experiment is $d\sin\theta = \left(m+\frac{1}{2}\right)\lambda$ with $m = 0$ for the first minimum.

Then $\theta = \sin^{-1}\left(\dfrac{\lambda}{2d}\right) = \sin^{-1}\left(\dfrac{9.93\times10^{-7}\text{m}}{2\times10^{-3}\text{m}}\right) = \sin^{-1}\left(4.96\times10^{-4}\right) = 0.028\,4°$

And the geometry of a double-slit experiment implies that the distance to the first minimum on the screen is given by

$$\frac{y}{L} = \tan\theta: \quad y = L\tan\theta = (10.0\text{ m})\tan(0.028\ 4^\circ) = 4.96\text{ mm}$$ ■

(c) We cannot say the neutron passed through one slit. We can only say it passed through the pair of slits, as a water wave does to produce an interference pattern. ■

Finalize: Experimentally it would be hard to set up a monoenergetic beam of neutrons moving as slowly as 40 cm/s, but essentially this experiment has been done at higher energies with finer-grained diffracting targets, and the results agree with the theory of waves in interference. If we set up any kind of detector to see which slit a particular neutron passed through, the interference pattern would disappear.

---

53. An electron and a 0.020 0 kg bullet each have a velocity of magnitude 500 m/s, accurate to within 0.010 0%. Within what lower limit could we determine the position of each object along the direction of the velocity?

Solution

Conceptualize: It seems reasonable that a tiny particle like an electron could be located within a narrower region than a big object like a bullet, but we may find that the realm of the very small does not obey common sense.

Categorize: Heisenberg's uncertainty principle can be used to find the uncertainty in position from the uncertainty in the momentum.

Analyze: The uncertainty principle states $\Delta x \Delta p_x \geq \frac{\hbar}{2}$

where $\Delta p_x = m\Delta u$ and $\hbar \equiv h/2\pi$

Both the electron and bullet have a velocity uncertainty

$$\Delta u = (0.000\ 100)(500\text{ m/s}) = 0.050\ 0\text{ m/s}$$

For the electron, the minimum uncertainty in position is

$$\Delta x = \frac{h}{4\pi m\Delta u} = \frac{6.63\times10^{-34}\text{ J}\cdot\text{s}}{4\pi\left(9.11\times10^{-31}\text{ kg}\right)(0.050\ 0\text{ m/s})} = 1.16\text{ mm}$$ ■

For the bullet,

$$\Delta x = \frac{h}{4\pi m\Delta u} = \frac{6.63\times10^{-34}\text{ J}\cdot\text{s}}{4\pi(0.020\ 0\text{ kg})(0.050\ 0\text{ m/s})} = 5.28\times10^{-32}\text{ m}$$ ■

Finalize: Our intuition did not serve us well here, since the position of the center of the larger bullet can be determined much more precisely than the electron. Quantum mechanics describes all objects, but the quantum fuzziness in position is too small to observe for the bullet. It is large for the small-mass electron.

---

58. The accompanying table shows data obtained in a photoelectric experiment. (a) Using these data, make a graph similar to Active Figure 40.11 that plots as a straight line. From the graph, determine (b) an experimental value for Planck's constant (in joule-seconds) and (c) the work function (in electron volts) for the surface. (Two significant figures for each answer are sufficient.)

Wavelength (nm)	Maximum Kinetic Energy of Photoelectrons (eV)
588	0.67
505	0.98
445	1.35
399	1.63

Solution

Conceptualize: The table shows that the energy goes up as the wavelength goes down. We must test for the linearity predicted by Einstein's model …

Categorize: … by finding the frequency of each kind of photon. From Einstein's equation for the photoelectric effect, $K_{\max} = hf - \phi$, a graph of $K_{\max}$ versus f will have slope h and vertical-axis intercept $-\phi$.

Analyze: From each wavelength we find the corresponding frequency using the relation $\lambda f = c$, where c is the speed of light:

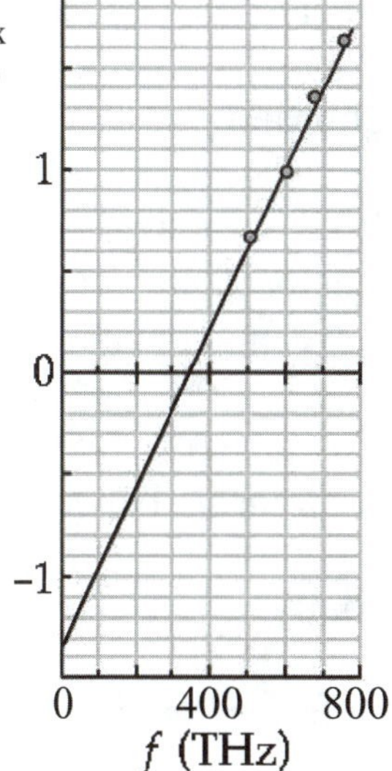

For $\lambda_1 = 588 \times 10^{-9}$ m $\qquad f_1 = \dfrac{c}{\lambda_1} = 5.10 \times 10^{14}$ Hz

$\lambda_2 = 505 \times 10^{-9}$ m $\qquad f_2 = 5.94 \times 10^{14}$ Hz

$\lambda_3 = 445 \times 10^{-9}$ m $\qquad f_3 = 6.74 \times 10^{14}$ Hz

$\lambda_4 = 399 \times 10^{-9}$ m $\qquad f_4 = 7.52 \times 10^{14}$ Hz

(a) We plot each point on an energy versus frequency graph, as shown.

We extend a straight line through the set of 4 points, as far as the negative y intercept.

(b) Our basic equation is $K_{\max} = hf - \phi$. Therefore, an experimental value for Planck's constant is the slope of the K-f graph, which can be found from a least-squares fit or from reading the graph as:

$$h_{\text{exp}} = \frac{\text{Rise}}{\text{Run}} = \frac{1.25 \text{ eV} - 0.25 \text{ eV}}{6.5 \times 10^{14} \text{ Hz} - 4.0 \times 10^{14} \text{ Hz}} = 4.0 \times 10^{-15} \text{ eV} \cdot \text{s} = 6.4 \times 10^{-34} \text{ J} \cdot \text{s} \quad \blacksquare$$

From the scatter of the data points on the graph, we estimate the uncertainty of the slope to be about 3%. Thus we choose to show two significant figures in writing the experimental value of Planck's constant.

(c) Again from the linear equation $K_{max} = hf - \phi$, the work function for the metal surface is the negative of the y-intercept of the graph, so $\phi_{exp} = -(-1.4 \text{ eV}) = 1.4 \text{ eV}$ ■

Based on the range of slopes that appear to fit the data, the estimated uncertainty of the work function is 5%.

Finalize: In a sense the most important feature of the graph is not its slope or its intercept, but its shape. By being a straight line it demonstrates Planck's direct proportionality of photon energy to frequency. The x intercept of the line is the frequency of light having the threshold wavelength for ejecting photoelectrons from this surface. It is the cutoff frequency.

---

67. Show that the ratio of the Compton wavelength λ_C to the de Broglie wavelength $\lambda = h/p$ for a relativistic electron is

$$\frac{\lambda_C}{\lambda} = \left[\left(\frac{E}{m_e c^2}\right)^2 - 1\right]^{1/2}$$

where E is the total energy of the electron and m_e is its mass.

Solution

Conceptualize: The Compton wavelength is a constant used in describing how the wavelength of an x-ray changes when it scatters from an electron. A moving electron has a wavelength in the same sense that a water wave does, and this is its de Broglie wavelength.

Categorize: We use the definitions of the Compton and de Broglie wavelengths and the relationship between energy and momentum for any particle with mass.

Analyze: The definition of the Compton wavelength is $\lambda_C = h/m_e c$.

The de Broglie wavelength is $\lambda = h/p$.

We take the ratio of the Compton wavelength to the de Broglie wavelength, and square it:

$$\frac{\lambda_C^2}{\lambda^2} = \frac{p^2}{(m_e c)^2}$$

From Equation 39.27, the momentum for a slowly-moving or rapidly-moving object is described by

$$p^2 = \frac{E^2 - m_e^2 c^4}{c^2}$$

Substituting and simplifying,

$$\frac{\lambda_C^2}{\lambda^2} = \frac{\left(E^2 - m_e^2 c^4\right)}{\left(m_e c^2\right)^2} = \left(\frac{E}{m_e c^2}\right)^2 - 1$$

and $\dfrac{\lambda_C}{\lambda} = \sqrt{\left(\dfrac{E}{m_e c^2}\right)^2 - 1}$ ■

Finalize: The result says that as the energy increases the ratio of the constant Compton wavelength to the variable de Broglie wavelength increases. This means that the de Broglie wavelength decreases, in qualitative agreement with $\lambda = h/p$. The relativistic total energy E includes the rest energy in $E = m_e c^2 + K$. We can confirm that this general result for wavelength gives the right limit in the nonrelativistic limit:

$$\frac{\lambda_C}{\lambda} = \left[\left(\frac{m_e c^2 + K}{m_e c^2}\right)^2 - 1\right]^{1/2} = \left[\left(1 + \frac{m_e u^2}{2m_e c^2}\right)^2 - 1\right]^{1/2} = \left[\frac{2u^2}{2c^2} + \frac{u^4}{4c^4}\right]^{1/2}$$

For u much less than c this equation simplifies to $\dfrac{h}{m_e c\lambda} = \dfrac{u}{c}$

so $\lambda = h/m_e u$ as it should be.

---

71. The total power per unit area radiated by a black body at a temperature T is the area under the $I(\lambda, T)$-versus-λ curve as shown in Active Figure 40.3. (a) Show that this power per unit area is

$$\int_0^\infty I(\lambda, T)\,d\lambda = \sigma T^4$$

where $I(\lambda, T)$ is given by Planck's radiation law and σ is a constant independent of T. This result is known as Stefan's law. (See Section 20.7.) To carry out the integration, you should make the change of variable $x = hc/\lambda k_B T$ and use

$$\int_0^\infty \frac{x^3\,dx}{e^x - 1} = \frac{\pi^4}{15}$$

(b) Show that the Stefan–Boltzmann constant σ has the value

$$\sigma = \frac{2\pi^5 k_B^4}{15c^2 h^3} = 5.67 \times 10^{-8}\ \text{W/m}^2\cdot\text{K}^4$$

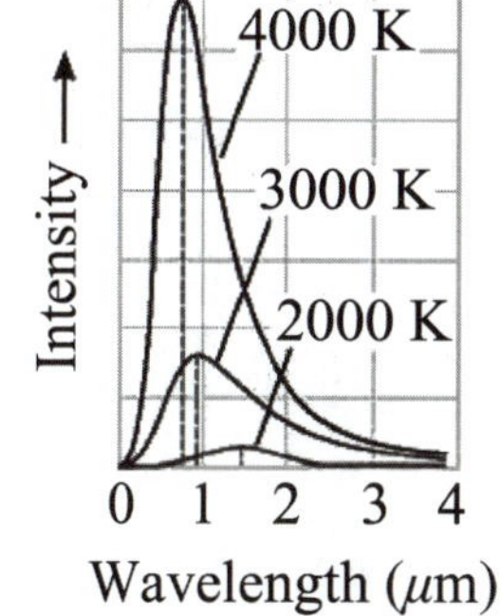

Figure 40.3

Solution

Conceptualize: Stefan's law and Wien's law were purely experimental results before Planck developed his theory of blackbody radiation. From Planck's radiation law they can be derived theoretically, as we show here for Stefan's law.

Categorize: The problem tells us just what to do, namely to calculate

$$\int_0^{\infty} I(\lambda, T)\, d\lambda = \int_0^{\infty} \frac{2\pi hc^2}{\lambda^5\left(e^{hc/\lambda k_B T} - 1\right)}\, d\lambda$$

Analyze: In order to make the suggested substitution, we find λ and $d\lambda$.

From $x = \dfrac{hc}{\lambda k_B T}$ we have $\lambda = \dfrac{hc}{x k_B T}$ and $d\lambda = -\dfrac{hc\, dx}{x^2 k_B T}$

We also note that the limits of integration change from $\lambda = (0, \infty)$ to $x = (\infty, 0)$. Substituting these variables into the integral, the intensity of the blackbody radiation is:

$$\int_0^{\infty} I(\lambda, T)\, d\lambda = \int_0^{\infty} \frac{2\pi hc^2}{\lambda^5\left(e^{hc/\lambda k_B T} - 1\right)}\, d\lambda = \int_{\infty}^{0} -\left(\frac{2\pi hc^2}{e^x - 1}\right)\left(\frac{x^5 k_B{}^5 T^5}{h^5 c^5}\right)\left(\frac{hc\, dx}{x^2 k_B T}\right)$$

Simplifying, $$\int_0^{\infty} I(\lambda, T)\, d\lambda = \frac{2\pi k_B{}^4 T^4}{h^3 c^2}\int_{\infty}^{0} -\frac{x^3}{e^x - 1}\, dx = \frac{2\pi k_B{}^4 T^4}{h^3 c^2}\int_0^{\infty} \frac{x^3}{e^x - 1}\, dx$$

Now the integral is $\pi^4/15$, so $\displaystyle\int_0^{\infty} I(\lambda, T)\, d\lambda = \frac{2\pi^5 k_B{}^4 T^4}{15 h^3 c^2} = \sigma T^4$ ■

with

$$\sigma = \frac{2\pi^5 k_B{}^4}{15 c^2 h^3} = \frac{2\pi^5\left(1.3807 \times 10^{-23}\ \text{J/K}\right)^4}{15\left(2.998 \times 10^8\ \text{m/s}\right)^2\left(6.626 \times 10^{-34}\ \text{J}\cdot\text{s}\right)^3} = 5.67 \times 10^{-8}\ \text{W/m}^2\cdot\text{K}^4$$ ■

Finalize: We have proved that the intensity of radiation is proportional to the fourth power of the temperature, a truly remarkable result. And we have found the proportionality constant in terms of more fundamental constants. Before Planck, theories of matter and electromagnetic waves were separate, but Planck unified them.

41

Quantum Mechanics

EQUATIONS AND CONCEPTS

The **wave function of a free particle** moving along the x axis is a sinusoidal wave. *The wave representing the particle has a constant amplitude A and angular wave number k.*

$$\psi(x) = Ae^{ikx} \quad (41.4)$$

The probability of finding a particle within an arbitrary interval equals the area under the curve of $|\psi|^2$ vs. x, between the end points of the interval. *The **probability density**, $|\psi|^2$, is the relative probability of finding a particle at a given point along the interval. The sum of all probabilities over all possible values of x must equal 1.*

$$P_{ab} = \int_a^b |\psi|^2\, dx \quad (41.6)$$

$$\int_{-\infty}^{+\infty} |\psi|^2\, dx = 1 \quad (41.7)$$

(normalization condition)

The **expectation value for the position** is the expected average value of the coordinate. *This is the average of the values of position if calculated from the known wave function.* To find the expectation value of any function $f(x)$, replace x in Equation 41.8 with $f(x)$.

$$\langle x \rangle \equiv \int_{-\infty}^{\infty} \psi^* x\, \psi\, dx \quad (41.8)$$

$$\langle f(x) \rangle \equiv \int_{-\infty}^{\infty} \psi^* f(x) \psi\, dx \quad (41.9)$$

For a **particle in a box of width L** in one dimensional motion:

- The **wave function** can be represented as a sinusoidal function.

$$\psi(x) = A \sin\left(\frac{n\pi x}{L}\right) \quad (41.12)$$

$$n = 1, 2, 3, \ldots$$

- The **normalized wave function** for a particle in a box of dimension L is stated in Equation 41.13.

$$\psi_n(x) = \sqrt{\frac{2}{L}} \sin\left(\frac{n\pi x}{L}\right) \quad (41.13)$$

$$n = 1, 2, 3, \ldots$$

- The **energy of the particle** is quantized. The ground state corresponds to $n = 1$ and is the lowest energy state.

$$E_n = \left(\frac{h^2}{8mL^2}\right)n^2 \quad n = 1, 2, 3, \ldots \quad (41.14)$$

The **time-independent Schrödinger equation for a bound system** (with energy E) allows, in principle, the determination of the wave functions and energies of the allowed states if the potential energy function U is known.

$$-\frac{\hbar^2}{2m}\frac{d^2\psi}{dx^2} + U\psi = E\psi \quad (41.15)$$

The textbook describes the **application of the Schrödinger equation** to systems with three different potential energy functions: (i) infinite square well, (ii) potential energy well of finite height, and (iii) harmonic oscillator.

For an **infinite square well** (see figure at right), the potential energy U is zero inside the well and infinite outside; the Schrödinger equation can be stated as in Equation 41.16. *The wave function and quantized energy values are as those found for a particle in a box*. See Equations 41.12 and 41.14.

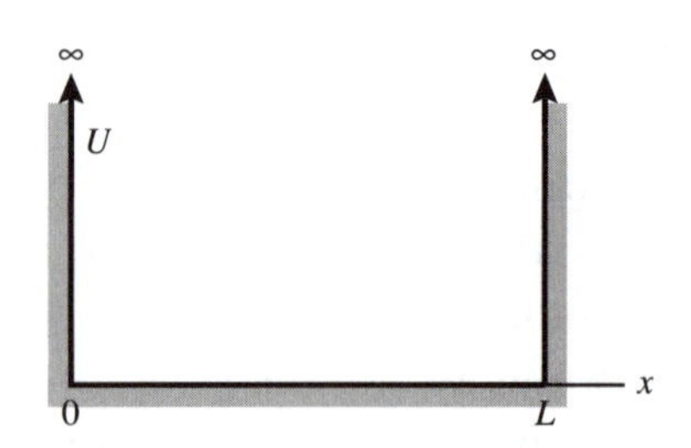

$$\frac{d^2\psi}{dx^2} = -\frac{2mE}{\hbar^2}\psi = -k^2\psi \quad (41.16)$$

where $k = \dfrac{\sqrt{2mE}}{\hbar}$

For a **potential-energy well of finite height** (see figure at right), the potential U is zero inside the well and has a value greater than E outside.

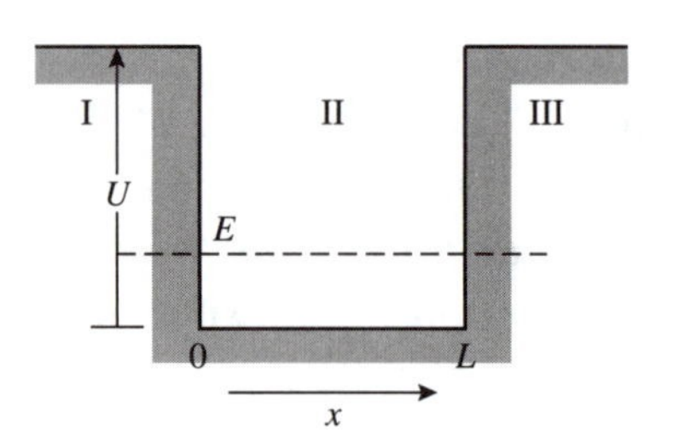

In **Regions I and III** (outside the potential well) the Schrödinger equation can be written as in Equation 41.20. In these regions outside the potential well, the wave functions decay exponentially with distance.

$$\frac{d^2\psi}{dx^2} = C^2\psi \quad (41.20)$$

where $C^2 = \dfrac{2m(U - E)}{\hbar^2}$

$\psi_{\text{I}} = Ae^{Cx} \quad (x < 0)$

$\psi_{\text{III}} = Be^{-Cx} \quad (x > L)$

In **Region II** (inside the potential well) the wave functions are sinusoidal. *The values of the constants A, B, F, and G can be determined from the boundary conditions and normalization.*

$\psi_{\text{II}} = F\sin kx + G\cos kx$

The **transmission coefficient** T represents the probability that a particle will tunnel or penetrate a barrier of width L where the potential energy is greater than the energy of the particle.

$$T \approx e^{-2CL} \tag{41.22}$$

$$\text{where} \quad C = \frac{\sqrt{2m(U - E)}}{\hbar} \tag{41.23}$$

The **potential energy function and energy level diagram of a simple harmonic oscillator** with frequency ω is shown in the figure at right. Note that the energy levels are equally spaced.

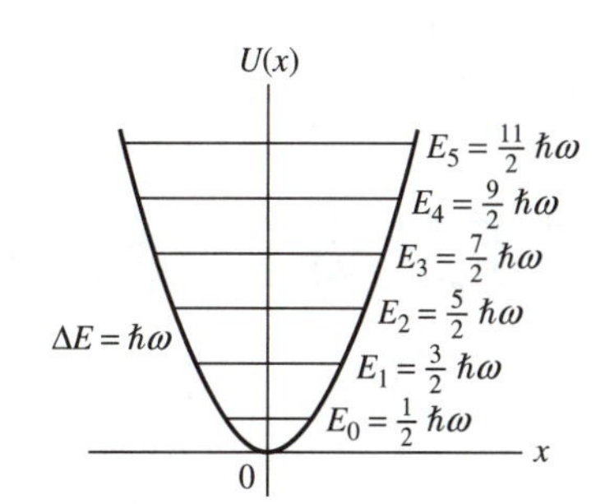

The **ground state wave function for the harmonic oscillator** contains a constant B that must be determined from the normalization condition. *Wave functions for all excited states include an exponential factor.*

$$\psi = Be^{-(m\omega/2\hbar)x^2} \tag{41.26}$$

The **energy levels of the harmonic oscillator** are quantized. *The ground state corresponds to $n = 0$.*

$$E = (n + \tfrac{1}{2})\hbar\omega \qquad n = 0, 1, 2, \ldots \tag{41.27}$$

REVIEW CHECKLIST

You should be able to:

- Describe the concept of a wave function for the representation of matter waves and state in equation form the normalization condition and expectation value of the coordinate. (Section 41.1)

- Given the wave function for a particle in one-dimensional motion, calculate the wave length, momentum, energy, and probability of finding the particle within a specified distance interval. (Section 41.1)

- State the wave function and calculate wavelength, energy levels, and quantum numbers for a particle in a box. Calculate the probability of finding a particle within a finite range of coordinate values. (Section 41.2)

- Demonstrate that a given wave function is a solution to the Schrödinger equation, Equation 41.15. Sketch the probability density for a particle in an infinitely deep square well. (Sections 41.3 and 41.4)

- Find and sketch the potential energy as a function of position given the wave function for a particle in an infinite square well potential. (Sections 41.3 and 41.4)

- State the boundary conditions on the wave functions for a particle in a finite height potential barrier. Sketch the wave function and probability density for the particle. Calculate the transmission coefficient for tunneling or penetration by an electron through a rectangular barrier. (Sections 41.5 and 41.6)

- Determine the amplitude constant B in the wave function of a harmonic oscillator by normalization. Find the total energy of an oscillator given the wave function. (Section 41.7)

ANSWER TO AN OBJECTIVE QUESTION

1. The probability of finding a certain quantum particle in the section of the x axis between $x = 4$ nm and $x = 7$ nm is 48%. The particle's wave function $\psi(x)$ is constant over this range. What numerical value can be attributed to $\psi(x)$, in units of $\text{nm}^{-1/2}$? (a) 0.48 (b) 0.16 (c) 0.12 (d) 0.69 (e) 0.40

Answer The square of the wave function at point x gives the probability per unit length of finding the particle at point x. The average value of $|\psi|^2$ multiplied by the width between two points gives the probability of finding the particle in that range. For ψ with a constant value, we have $\psi^2(3\text{ nm}) = 0.48$ for the probability of finding the particle in the range stated. Then solving for ψ gives $\psi = \sqrt{0.16/\text{nm}} = 0.4/\sqrt{\text{nm}}$. This is answer (e).

□ □ □ □

ANSWER TO A CONCEPTUAL QUESTION

2. Discuss the relationship between ground-state energy and the uncertainty principle.

Answer Consider a particle bound to a restricted region of space. If its minimum energy were zero, then the particle could have zero momentum and zero uncertainty in its momentum. At the same time, the uncertainty in its position would not be infinite, but equal to the width of the region. In such a case, the uncertainty product $\Delta x \Delta p_x$ would be zero, violating the uncertainty principle. This contradiction proves that the minimum energy of the particle is not zero.

□ □ □ □

SOLUTIONS TO SELECTED PROBLEMS

1. A free electron has a wave function

$$\psi(x) = Ae^{i\left(5.00 \times 10^{10} x\right)}$$

where x is in meters. Find its (a) de Broglie wavelength, (b) momentum, and (c) kinetic energy in electron volts.

Solution

Conceptualize: We cannot draw a simple graph of Ae^{ikx}, because it has both real and imaginary parts. The statement of the problem does not include the time variation of the wave function, but it would be as in $\psi(x,t) = Ae^{ikx-i\omega t} = A\cos(kx - \omega t) + Ai\sin(kx - \omega t)$. In a word, this wave function is for the wave motion of a quantum particle moving toward the right with a particular wave number.

Categorize: The wave function represents the state of the electron, so it contains information about its wavelength, momentum, and energy that we can read out.

Analyze:

(a) The wave function, $\psi(x) = Ae^{i(5\times10^{10}x)} = A\cos(5\times10^{10}x) + iA\sin(5\times10^{10}x)$, will go through one full cycle between $x_1 = 0$ and $(5.00\times10^{10})x_2 = 2\pi$. The wavelength is then

$$\lambda = x_2 - x_1 = \frac{2\pi}{5.00\times10^{10}\ \text{m}^{-1}} = 1.26\times10^{-10}\ \text{m} \qquad \blacksquare$$

To say the same thing, we can inspect $Ae^{i(5\times10^{10}x)}$ to see that the wave number is $k = 5.00\times10^{10}\ \text{m}^{-1} = 2\pi/\lambda$.

(b) Since $\lambda = h/p$, the momentum is

$$p = \frac{h}{\lambda} = \frac{6.63\times10^{-34}\ \text{J}\cdot\text{s}}{1.26\times10^{-10}\ \text{m}} = 5.28\times10^{-24}\ \text{kg}\cdot\text{m/s} \qquad \blacksquare$$

(c) The electron's kinetic energy is

$$K = \tfrac{1}{2}mu^2 = \frac{p^2}{2m}: \quad K = \frac{\left(5.28\times10^{-24}\ \text{kg}\cdot\text{m/s}\right)^2}{2\left(9.11\times10^{-31}\ \text{kg}\right)}\left(\frac{1\,\text{eV}}{1.60\times10^{-19}\ \text{J}}\right) = 95.5\,\text{eV} \qquad \blacksquare$$

[We use u to represent the speed of a particle with mass in chapters 39, 40, and 41.]

Finalize: Its relativistic total energy is 511 keV + 95.5 eV. This electron could be coasting along in a field-free vacuum tube after being fired from a 95.5-volt electron gun. Its wavelength is suitable for being diffracted by a crystal in a Davisson-Germer experiment.

---

7. An electron is contained in a one-dimensional box of length 0.100 nm. (a) Draw an energy-level diagram for the electron for levels up to $n = 4$. (b) Photons are emitted by the electron making downward transitions that could eventually carry it from the $n = 4$ state to the $n = 1$ state. Find the wavelengths of all such photons.

Solution

Conceptualize: This problem is about a one-dimensional model for an atom or for a quantum dot. We will see whether the photon wavelengths are around the visible part of the electromagnetic spectrum.

Categorize: In part (a), we can draw a diagram that parallels our treatment of mechanical waves under boundary conditions. In each standing-wave state, we measure the distance d from one node to the next (N to N), and base our solution upon that. In part (b) we identify the loss of energy by the electron with the Planck energy of an emitted photon.

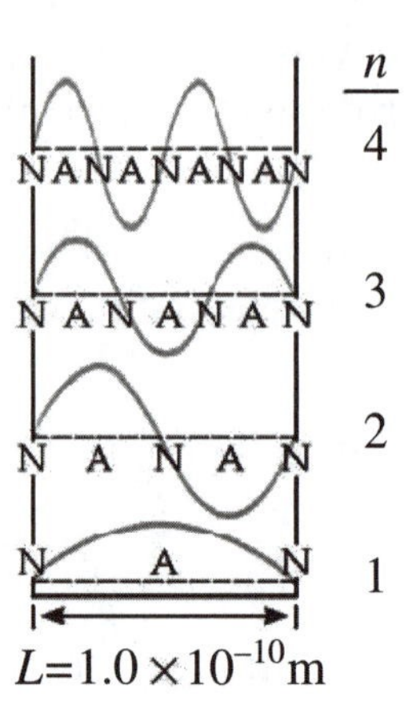

Analyze:

(a) Since $d_{\text{N to N}} = d = \frac{\lambda}{2}$ and $\lambda = \frac{h}{p}$

the electron's momentum in each state is $p = \frac{h}{\lambda} = \frac{h}{2d}$

The electron's energy is $K = \frac{p^2}{2m_e} = \frac{h^2}{8m_e d^2} = \frac{1}{d^2}\frac{(6.63\times10^{-34}\ \text{J}\cdot\text{s})^2}{8(9.11\times10^{-31}\ \text{kg})}$

Evaluating, $K = \frac{6.03\times10^{-38}\ \text{J}\cdot\text{m}^2}{d^2} = \frac{3.77\times10^{-19}\ \text{eV}\cdot\text{m}^2}{d^2}$

In state 1, $d = 1.00\times10^{-10}$ m $K_1 = 37.7$ eV

In state 2, $d = 5.00\times10^{-11}$ m $K_2 = 151$ eV

In state 3, $d = 3.33\times10^{-11}$ m $K_3 = 339$ eV

In state 4, $d = 2.50\times10^{-11}$ m $K_4 = 603$ eV

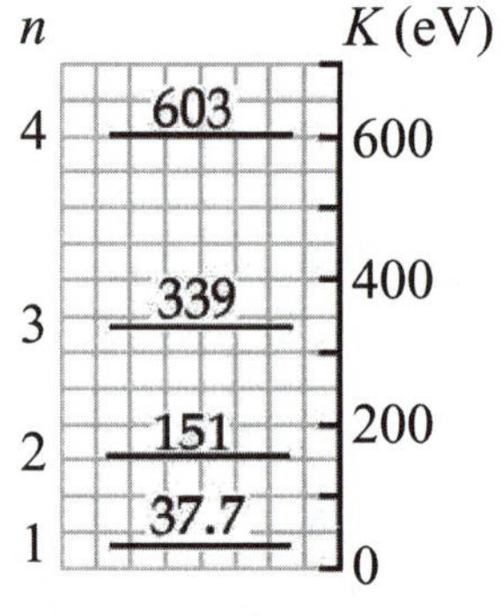

These energy levels are shown in the diagram. ■

(b) When the charged, massive electron inside the box makes a downward transition from one energy level to another, a chargeless, massless photon comes out of the box, carrying the difference in energy ΔE. Its wavelength is

$$\lambda = \frac{c}{f} = \frac{hc}{\Delta E} = \frac{(6.63\times10^{-34}\ \text{J}\cdot\text{s})(3.00\times10^{8}\ \text{m/s})}{\Delta E(1.602\times10^{-19}\ \text{J/eV})} = \frac{1.24\times10^{-6}\ \text{eV}\cdot\text{m}}{\Delta E}$$

For the downward tumble from state 4 to state 2, for example, the photon energy is (603 – 151) eV = 452 eV and the photon wavelength is

$$\lambda = \frac{hc}{\Delta E} = \frac{1.24\times10^{-6}\ \text{eV}\cdot\text{m}}{452\ \text{eV}} = 2.74\ \text{nm}$$

The wavelengths of light emitted in each transition are given in the table. ■

Transition	4 →3	4 →2	4 →1	3 →2	3 →1	2 →1
ΔE (eV) Wavelength (nm)	264 4.71	452 2.74	565 2.20	188 6.59	302 4.12	113 11.0

Finalize: This is a good problem to understand thoroughly. Note which equations for wavelength and energy apply to the electron and which equations for wavelength and energy apply to the photon. The size of the box is too small to model the space available to an outer electron in a real atom, so the photon wavelengths are in the ultraviolet or x-ray part of the electromagnetic spectrum. We could say the box models the space available to an inner electron in a multi-electron atom.

---

11. The nuclear potential energy that binds protons and neutrons in a nucleus is often approximated by a square well. Imagine a proton confined in an infinitely high square well of length 10.0 fm, a typical nuclear diameter. Assuming the proton makes a transition from the $n = 2$ state to the ground state, calculate (a) the energy and (b) the wavelength of the emitted photon. (c) Identify the region of the electromagnetic spectrum to which this wavelength belongs.

Solution

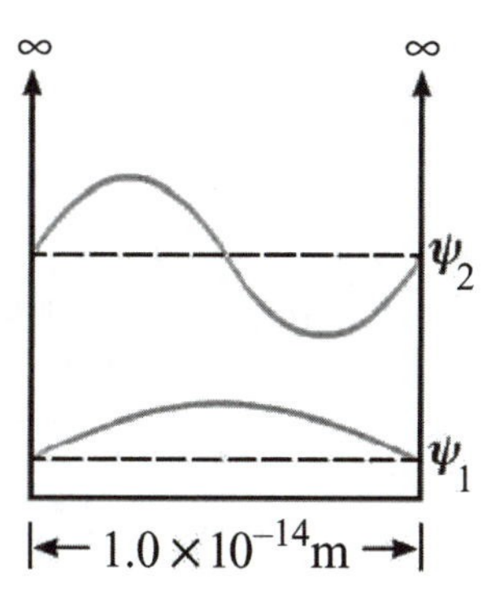

Conceptualize: Nuclear radiation from nucleon transitions is usually in the form of gamma rays with high energies and short wavelengths.

Categorize: The energy of the massive particle can be obtained from the wavelength of the standing wave corresponding to each level. The transition between energy levels will result in the emission of a photon with this energy difference.

Analyze: In energy level 1, the node-to-node distance of the standing wave is 1.00×10^{-14} m, so the wavelength is twice this distance: $h/p = 2.00 \times 10^{-14}$ m. The proton's kinetic energy is

$$K = \tfrac{1}{2}mu^2 = \frac{p^2}{2m} = \frac{h^2}{2m\lambda^2} = \frac{\left(6.63 \times 10^{-34}\ \text{J}\cdot\text{s}\right)^2}{2\left(1.67 \times 10^{-27}\ \text{kg}\right)\left(2.00 \times 10^{-14}\ \text{m}\right)^2}$$

$$K = \frac{3.29 \times 10^{-13}\ \text{J}}{1.60 \times 10^{-19}\ \text{J/eV}} = 2.06 \times 10^{6}\ \text{eV} = 2.06\ \text{MeV}$$

In the first excited state, level 2, the node-to-node distance is half as large as in state 1. The momentum is two times larger and the energy is four times larger: $K = 8.23$ MeV

The pRoton has mass, has charge, moves slowly compared to light in a standing-wave state, and stays inside the nucleus. When it falls from level 2 to level 1, its energy change is 2.06 MeV – 8.23 MeV = –6.17 MeV. Therefore, we know that a pHoton (a traveling wave with no mass and no charge) is emitted at the speed of light, and (a) that it has an energy of +6.17 MeV. ■

(b) Its frequency is

$$f = \frac{E}{h} = \frac{(6.17 \times 10^6 \text{ eV})(1.60 \times 10^{-19} \text{ J/eV})}{6.63 \times 10^{-34} \text{ J} \cdot \text{s}} = 1.49 \times 10^{21} \text{ Hz}$$

and its wavelength is

$$\lambda = \frac{c}{f} = \frac{3.00 \times 10^8 \text{ m/s}}{1.49 \times 10^{21} \text{ s}^{-1}} = 2.02 \times 10^{-13} \text{ m}$$ ■

(c) This is a gamma ray, according to the electromagnetic spectrum chart shown in Figure 34.13 of the text and in Problem 34.45 in this student manual. ■

Finalize: The radiated photons are energetic gamma rays as we expected for a nuclear transition. In the above calculations, we assumed that the proton was not relativistic ($u < 0.1c$), but we should check this assumption for the highest energy state we examined ($n = 2$):

$$u = \sqrt{\frac{2K}{m}} = \sqrt{\frac{2(8.23 \times 10^6 \text{ eV})(1.60 \times 10^{-19} \text{ J/eV})}{1.67 \times 10^{-27} \text{ kg}}} = 3.97 \times 10^7 \text{ m/s} = 0.132c$$

This appears to be a borderline case where we could better use relativistic equations, but our classical treatment should give reasonable results, to better than $(0.132)^2 = 2\%$ accuracy.

---

13. (a) Use the quantum-particle-in-a-box model to calculate the first three energy levels of a neutron trapped in an atomic nucleus of diameter 20.0 fm. (b) Explain whether the energy-level differences have a realistic order of magnitude.

Solution

Conceptualize: Compare this problem to Problem 11 just above. The proton and neutron have nearly the same mass. They both move as standing waves. In this problem the box is twice as long, so the wavelength in each state should be twice as large, the momentum $p = h/\lambda$ half as large, and the kinetic energy one-fourth as large.

Categorize: In both Problems 7 and 11, just preceding, we drew pictures of standing waves and reasoned from the pictures about node-to-node distance, wavelength, momentum, and kinetic energy. For variety, here we will reason from the energy equation derived in the chapter for a particle in a linear box.

Analyze:

(a) The energy of a quantum particle confined to a line segment is

$$E_n = \frac{h^2 n^2}{8mL^2} \quad \text{Here we have for the ground state}$$

$$E_1 = \frac{\left(6.626\times10^{-34}\ \text{J}\cdot\text{s}\right)^2 (1)^2}{8\left(1.67\times10^{-27}\text{kg}\right)\left(2.00\times10^{-14}\text{m}\right)^2} = 8.22\times10^{-14}\text{J} = 0.513\ \text{MeV} \quad \blacksquare$$

and for the first and second excited states, which are states 2 and 3,

$$E_2 = 4E_1 = 2.05\ \text{MeV} \qquad \text{and} \qquad E_3 = 9E_1 = 4.62\ \text{MeV} \quad \blacksquare$$

(b) Yes, the energy differences are on the order of 1 MeV, which is a typical energy for a γ-ray photon, as emitted by a nucleus in an excited state. ■

Stated differently: Scattering experiments show that an atomic nucleus is a three-dimensional object always less than 15 fm in diameter. This one-dimensional box 20 fm long is a good model in energy terms.

Finalize: The energies of the two lowest quantum states in this problem really are one-fourth of the energies of the lowest two energy levels in problem 11. In common language or as the title of an old TV show, a "quantum leap" means a very large change. In many real cases it is a change by a large factor, but by a tiny fraction of a joule in energy.

---

21. A quantum particle in an infinitely deep square well has a wave function

$$\psi_1(x) = \sqrt{\frac{2}{L}}\ \sin\left(\frac{\pi x}{L}\right)$$

for $0 \le x \le L$ and is zero otherwise. (a) Determine the probability of finding the particle between $x = 0$ and $x = L/3$. (b) Use the result of this calculation and a symmetry argument to find the probability of finding the particle between $x = L/3$ and $x = 2L/3$. Do not re-evaluate the integral.

Solution

Conceptualize: The textbook's Active Figure 41.4, at the bottom of column (b), shows the probability density $|\psi_1|^2$ for finding the particle at different points along the length of the well. We reproduce this graph here. Consider the area under the curve between $x = 0$ and $x = L/3$. The size of this area represents the probability of finding the particle in this section of its range. Because the curve is low near the boundary and high at the center, the probability is much less than one-third; we might estimate about 0.2.

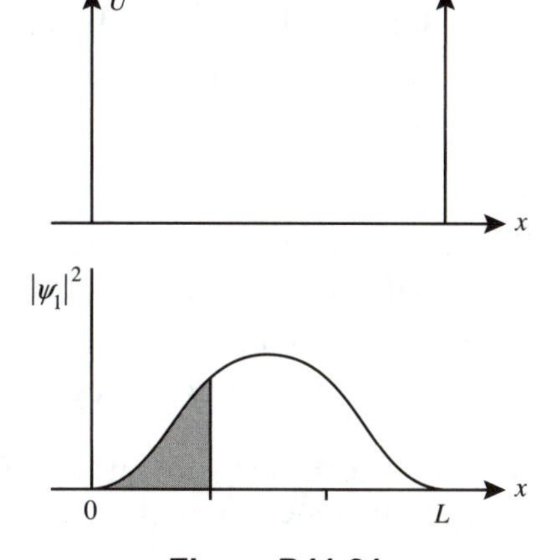

Figure P41.21

Categorize: We must square the given wave function and integrate it over x between the limits 0 and $L/3$ to find the probability. That is all we have to do, because the coefficient $\sqrt{2/L}$ has been chosen "for normalization" to make the whole area under the graph line equal to 1, representing the 100% probability of finding the particle anywhere.

Analyze:

(a) The probability is

$$P=\int_0^{L/3}|\psi|^2\,dx=\int_0^{L/3}\frac{2}{L}\sin^2\left(\frac{\pi x}{L}\right)dx=\frac{2}{L}\int_0^{L/3}\left(\frac{1}{2}-\frac{1}{2}\cos\frac{2\pi x}{L}\right)dx$$

$$=\frac{1}{L}\int_0^{L/3}dx-\frac{1}{L}\int_0^{L/3}\frac{L}{2\pi}\left(\cos\frac{2\pi x}{L}\right)\frac{2\pi dx}{L}$$

$$P=\left(\frac{x}{L}-\frac{1}{2\pi}\sin\frac{2\pi x}{L}\right)\Bigg|_0^{L/3}=\left(\frac{1}{3}-\frac{1}{2\pi}\sin\frac{2\pi}{3}\right)=\left(\frac{1}{3}-\frac{\sqrt{3}}{4\pi}\right)=0.196 \qquad \blacksquare$$

(b) The probability density is symmetric about $x=\frac{L}{2}$. Thus, the probability of finding the particle between $x=\frac{2L}{3}$ and $x=L$ is the same 0.196. Therefore, the probability of finding it in the range $\frac{L}{3}\le x\le\frac{2L}{3}$ is $P=1.00-2(0.196)=0.609$ $\blacksquare$

Finalize: Our 20% estimate for part (a) was right on. Classically, the particle would move back and forth with constant speed between the walls, and the probability of finding the particle is the same for all points between the walls. Thus, the classical probability of finding the particle in any range equal to one-third of the available space is $P_{\text{classical}}=\frac{1}{3}$. The result of part (a) is significantly smaller and the result of (b) is much larger, because of the curvature of the graph of the probability density. We could say, because the quantum particle moves as a wave.

---

24. Show that the wave function $\psi=Ae^{i(kx-\omega t)}$ is a solution to the Schrödinger equation, Equation 41.15, where $k=2\pi/\lambda$ and $U=0$.

Solution

Conceptualize: Just as the motion of a classical particle must be in accord with Newton's laws, a *wave function* for a quantum particle with mass must be a solution to the *wave equation*. This wave function describes a traveling wave. The condition $U=0$ can be realized by letting it move through a region free of fields or other particles.

Categorize: We substitute the wave function given into the Schrödinger equation and show that it gives equality for all x and t. Because the Schrödinger equation contains a second derivative, we must evaluate the first and second derivatives of the suggested function.

Analyze: From $\psi = Ae^{i(kx - \omega t)}$ **[1]**

we evaluate $\dfrac{d\psi}{dx} = ikAe^{i(kx - \omega t)}$

and $\dfrac{d^2\psi}{dx^2} = -k^2 Ae^{i(kx - \omega t)}$ **[2]**

We substitute Equations [1] and [2] into the Schrödinger equation, so that Equation 41.15,

$$-\frac{\hbar^2}{2m}\frac{d^2\psi}{dx^2} + U\psi = E\psi$$

becomes the test equation $\left(-\dfrac{\hbar^2}{2m}\right)\left(-k^2 Ae^{i(kx - \omega t)}\right) + 0 = EAe^{i(kx - \omega t)}$ **[3]**

The wave function $\psi = Ae^{i(kx - \omega t)}$ is a solution to the Schrödinger equation if Equation [3] is true. Both sides depend on A, x, and t in the same way, so we can cancel several factors, and determine that we have a solution if $\dfrac{\hbar^2 k^2}{2m} = E$

But this is true for a nonrelativistic particle with mass in a region where the potential energy is zero, since

$$\frac{\hbar^2 k^2}{2m} = \frac{1}{2m}\left(\frac{h}{2\pi}\right)^2\left(\frac{2\pi}{\lambda}\right)^2 = \underbrace{\frac{(h/\lambda)^2}{2m} = \frac{p^2}{2m}}_{\text{using de Broglie's equation}} = \frac{m^2u^2}{2m} = \tfrac{1}{2}mu^2 = \underbrace{K = K + U}_{\text{recall } U=0} = E$$

where K is the kinetic energy. Therefore, the given wave function does satisfy Equation 41.15. ■

Finalize: The terms in Schrödinger's equation represent the kinetic energy of the particle times its wave function, the potential energy of the particle times its wave function, and the total energy of the particle times its wave function. When the potential energy is some particular function of position, the wave function must have some particular form.

---

28. Suppose a quantum particle is in its ground state in a box that has infinitely high walls (see Active Fig. 41.4a). Now suppose the left-hand wall is suddenly lowered to a finite height and width. (a) Qualitatively sketch the wave function for the particle a short time later. (b) If the box has a length L, what is the wavelength of the wave that penetrates the left-hand wall?

Solution

Conceptualize: If a classical particle is rolling around inside a shoebox on a table and you cut open one wall, the particle will come rolling out with its energy unchanged. Possibly it will bounce a few times from the other shoebox walls before its exit. The quantum particle also has no reason to change its energy.

Categorize: The state of the quantum particle is not having a definite position and a definite velocity, but having a wave function, of which we can sketch the real part. It will look similar to a standing wave with two nodes inside the box, an exponential decay inside the wall, and a traveling wave outside the box.

Analyze: Before the wall is lowered, the wave function looks like the wave in the first frame of the diagram.

We assume that the wall is lowered to have a height in energy terms that is greater than the energy of the particle. We assume the potential energy outside the wall is equal to that inside the box. We assume the wall is so thick that the particle has a small tunneling probability.

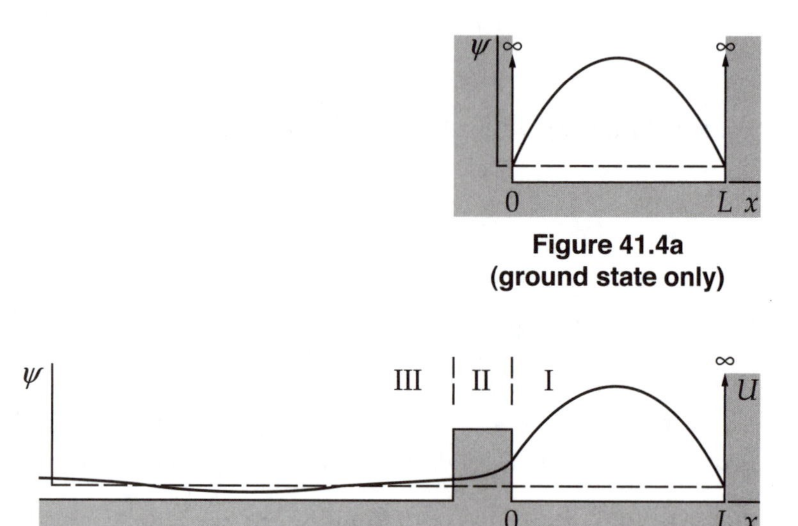

Figure 41.4a (ground state only)

(a) The wave function will look much the same in region I. The wave function will show exponential decay in region II. This means that in region I the wave function must change slightly, to be slightly greater than zero at the left wall. A small-amplitude traveling wave in region III represents the tunneled probability leaving the scene. This is illustrated in the diagram. ■

(b) Since the wave function in region I has not changed much, it still has about the same energy. If the particle tunnels through the wall, its energy remains constant. With the same potential energy on both side of the wall, the particle will have the same kinetic energy after tunneling. Thus the momentum will remain the same and the wavelength in region III will be the same as before, namely $2d_{NN} = 2L$. ■

Finalize: An experimental investigation could show that the particle is detected outside the box at some particular moment after the wall is lowered. The wave function does not predict this moment or represent the change from still-mostly-inside to outside-forevermore. This seeming incompleteness is the subject of Schrödinger's-cat discussions.

---

30. An electron with kinetic energy $E = 5.00$ eV is incident on a barrier of width $L = 0.200$ nm and height $U = 10.0$ eV (Fig. P41.30). What is the probability that the electron (a) tunnels through the barrier? (b) Is reflected?

Energy

Electron

L

U

E

O

x

Figure P41.30

Solution

Conceptualize: Since the barrier energy is higher than the kinetic energy of the electron, transmission is not likely, but should be possible since the barrier is not infinitely high or wide.

Categorize: The probability of transmission is found from Equations 41.22 and 41.23.

Analyze: The decay constant for the wave function inside the barrier is:

$$C = \frac{\sqrt{2m(U-E)}}{\hbar} = \frac{\sqrt{2(9.11\times10^{-31}\text{ kg})(10.0\text{ eV}-5.00\text{ eV})(1.60\times10^{-19}\text{ J/eV})}}{6.63\times10^{-34}\text{ J}\cdot\text{s}/2\pi}$$

$$C = 1.14\times10^{10}\text{ m}^{-1}$$

(a) The approximate probability of transmission is

$$T \approx e^{-2CL} = e^{-2(1.14\times10^{10}\text{ m}^{-1})(2.00\times10^{-10}\text{ m})} = 0.010\,3 \qquad \blacksquare$$

(b) If the electron does not tunnel, it is reflected, with probability

$$1 - 0.010\,3 = 0.990 \qquad \blacksquare$$

Finalize: Our expectation was correct; there is only a 1% chance that the electron will penetrate the barrier. This tunneling probability would be greater if the barrier were narrower, shorter, or if the kinetic energy of the electron were greater.

Related Comment: A typical scanning tunneling electron microscope (STM) can be built for less than $5 000, and can be tied directly to a home computer. The STM essentially consists of a needle (1) that is mounted on a few piezoelectric crystals (2). When a voltage is applied across the piezoelectric crystals, the crystals change their shape, and the needle moves.

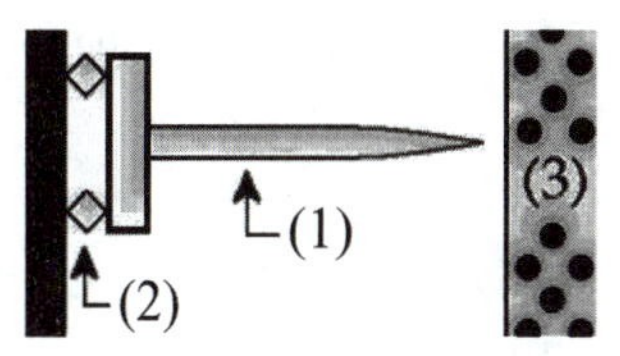

The STM charges the needle, and moves the tip of the needle to within a few tenths of a nanometer of the test sample (3); the remaining gap provides the energy barrier that is required for tunneling. When the electron cloud of one of the needle's atoms is close to the electron cloud of one of the sample's atoms, a relatively large number of electrons tunnel from one to the other, constituting a current. The current is measured by a computer. As the needle scans across the surface of the sample, an image of the atoms is generated.

---

34. The design criterion for a typical scanning tunneling microscope (STM) specifies that it must be able to detect, on the sample below its tip, surface features that differ in height by only 0.002 00 nm. If the electron transmission coefficient is e^{-2CL} with $C = 10.0\text{ nm}^{-1}$, what percentage change in electron transmission must the electronics of the STM be able to detect to achieve this resolution?

Solution

Conceptualize: The exquisite sensitivity of the scanning tunneling microscope is basically due to the strong dependence of the tunneling current on the size of the gap between the tip and the surface. But two thousandths of a nanometer is so very small that the electronics may have to detect a quite small change in current, like a fraction of one percent.

Categorize: We have to find a percentage difference between two values of e^{-2CL}, with L differing by 0.002 nm.

Analyze: With transmission coefficient e^{-2CL}, the fractional change in transmission is

$$\frac{e^{-2(10.0/\text{nm})L}-e^{-2(10.0/\text{nm})(L+0.002\,00\text{ nm})}}{e^{-2(10.0/\text{nm})L}}=1-e^{-20.0(0.002\,00)}=0.039\,2=3.92\%$$ ■

Finalize: It should be straightforward to measure a change in current of this size, larger than we anticipated. How does a scanning tunneling microscope scan? How are very small motions of the tip controlled? A piezoelectric crystal is one where a distortion in shape produces an electrical potential difference. Such materials are not too uncommon—large voltages can be produced by small distortions of quartz crystals in flint, used with steel to strike sparks, and by crushing wintergreen Life Savers candies, as children do in the dark. And the effect is reversible. Applying a voltage to a piezoelectric crystal makes it distort by a small amount, reversibly and controllably. In Figure 41.11 in the textbook, the three rectangular silver blocks represent a convenient arrangement of three piezoelectric crystals to control the tip position in three-dimensional space with three variable power supplies.

---

36. A one-dimensional harmonic oscillator wave function is

$$\psi = Axe^{-bx^2}$$

(a) Show that ψ satisfies Equation 41.24. (b) Find b and the total energy E. (c) Is this wave function for the ground state or for the first excited state?

Solution

Conceptualize: A wave function for a particular quantum state must be a solution satisfying the Schrödinger equation. Here is a problem with the same logical structure, that belongs back in Chapter 5 on motion of a classical particle:

> A 0.5-kg particle moves with its position in meters given by $x = 4 - 5t + ct^2 + 7t^3$ where t is in seconds. (a) Show that its motion satisfies Newton's second law if the total force on the particle increases with time as $\Sigma F = 21t$. (b) Find the required value of c.
>
> **Solution:** Newton's second law is about the acceleration, so we compute from the function $x(t)$ first the velocity and then $a(t)$. The first derivative is $v = dx/dt = 0 - 5 + 2ct + 21t^2$. The acceleration is $a = dv/dt = 2c + 42t$. Now we test whether the motion satisfies Newton's second law by substituting total force, mass, and acceleration into the equation expressing the law: We need the equation $\Sigma F = ma$ to be true. That is, $21t = (0.5)(2c + 42t)$ where the units are newtons on both sides. We test the truth of the equation $21t = c + 21t$. This equation is (a) true for all t, (b) provided $c = 0$. The proof is complete.

Categorize: The wave function to be tested represents one particular state of the quantum particle. For it to "move in this way"—for it to be in this state—the wave function must satisfy the Schrödinger equation, which represents the law of motion for all nonrelativistic quantum particles with mass. When we prove by substitution that the given wave function satisfies the Schrödinger equation with the harmonic-oscillator potential-energy function, we will find requirements that must be satisfied by the parameter b controlling the width of the wave function, and by the energy of the state. Good practice for the chain of logic here is Problem 42 in Chapter 16.

Analyze: From $\psi = Axe^{-bx^2}$ we have $\dfrac{d\psi}{dx} = Ae^{-bx^2} - 2bx^2Ae^{-bx^2}$

and $\dfrac{d^2\psi}{dx^2} = -2bxAe^{-bx^2} - 4bxAe^{-bx^2} + 4b^2x^3Ae^{-bx^2} = -6b\psi + 4b^2x^2\psi$

(a) The Schrödinger equation for a quantum particle in a harmonic-oscillator potential is

$$-\frac{\hbar^2}{2m}\frac{d^2\psi}{dt^2} + \tfrac{1}{2}m\omega^2x^2\psi = E\psi$$

or

$$\frac{d^2\psi}{dt^2} = -\frac{2mE}{\hbar^2}\psi - \frac{m^2\omega^2}{\hbar^2}x^2\psi$$

Substituting into it, we have

$$-6b\psi + 4b^2x^2\psi = -\left(\frac{2mE}{\hbar^2}\right)\psi + \left(\frac{m\omega}{\hbar}\right)^2 x^2\psi$$

For this to be true as an identity, it must be true for all values of x. So we must have both $-6b = -\dfrac{2mE}{\hbar^2}$, from matching the coefficients of the ψ terms, and $4b^2 = \left(\dfrac{m\omega}{\hbar}\right)^2$, from matching the coefficients of the $x^2\psi$ terms. With the required values of b and E, we do have a solution of the Schrödinger equation. ■

(b) Therefore $b = \dfrac{m\omega}{2\hbar}$ and $E = \dfrac{3b\hbar^2}{m} = \dfrac{3}{2}\hbar\omega$ ■

(c) The wave function is that of the first excited state. We can tell because the energy satisfies $E_n = (n + ½)\hbar\omega$ with $n = 1$. ■

Finalize: Observe that the Schrödinger equation puts no restriction on the value of A. An analogy: light waves of all amplitudes can move according to their wave equation $c^2\partial^2E/\partial x^2 = \partial^2E/\partial t^2$ where E now stands for electric field magnitude. The amplitude of the wave function ψ is determined instead by normalization. Another analogy: a photon is a quantum particle with zero mass. It moves as a wave, with its wave function $E(x,t)$ required to satisfy the wave equation just quoted. (And the magnetic field must satisfy a similar wave equation.) The intensity of a light beam is proportional to E^2, and is measured as a rate of photon bombardment of a detector, whether or not individual photons can be distinguished. The probability density of a beam of electrons is $|\psi|^2$, and can be measured as a rate of bombardment of a detector.

---

51. For a quantum particle described by a wave function $\psi(x)$, the expectation value of a physical quantity $f(x)$ associated with the particle is defined by

$$\langle f(x)\rangle \equiv \int_{-\infty}^{\infty} \psi^* f(x)\psi\, dx$$

For a particle in an infinitely deep one-dimensional box extending from $x = 0$ to $x = L$, show that

$$\langle x^2 \rangle = \frac{L^2}{3} - \frac{L^2}{2n^2\pi^2}$$

Solution

Conceptualize: An example in the chapter text proved that the average experimental value of position x for a particle in this state is $L/2$. The average value of x^2 over many experimental trials will not be just $L^2/4$, but something larger than that, because of the uncertainty spread in the particle's position.

Categorize: We note that the wave function for a quantum particle in a box is

$$\psi_n(x) = A\sin\left(\frac{n\pi x}{L}\right) \quad \text{where} \quad A = \sqrt{2/L}$$

Substituting x^2 for $f(x)$ and noting that $\psi^*\psi = |\psi|^2$, we are to evaluate $\langle x^2 \rangle = \int_{-\infty}^{\infty} x^2 |\psi|^2 \, dx$.

Analyze: Performing the integral,

$$\begin{aligned}
\langle x^2 \rangle &= \left(\frac{2}{L}\right)\int_0^L x^2 \sin^2\left(\frac{n\pi x}{L}\right)dx = \left(\frac{1}{L}\right)\int_0^L x^2\left[1 - \cos\left(\frac{2n\pi x}{L}\right)\right]dx \\
&= \left(\frac{1}{L}\right)\int_0^L x^2\,dx - \left(\frac{1}{L}\right)\int_0^L x^2\cos\left(\frac{2n\pi x}{L}\right)dx \\
&= \left(\frac{1}{L}\right)\frac{x^3}{3}\bigg|_0^L - \left(\frac{1}{L}\right)\left[x^2\frac{L}{2n\pi}\sin\left(\frac{2n\pi x}{L}\right)\right]_0^L + \left(\frac{1}{L}\right)\int_0^L 2x\frac{L}{2n\pi}\sin\left(\frac{2n\pi x}{L}\right)dx \\
&= \frac{L^2}{3} - 0 + \frac{1}{n\pi}\left[-x\frac{L}{2n\pi}\cos\left(\frac{2n\pi x}{L}\right)\right]_0^L + \frac{1}{n\pi}\int_0^L \frac{L}{2n\pi}\cos\left(\frac{2n\pi x}{L}\right)dx \\
&= \frac{L^2}{3} + \frac{1}{n\pi}\left[-L\frac{L}{2n\pi}1\right] + 0 = \frac{L^2}{3} - \frac{L^2}{2n^2\pi^2}
\end{aligned}$$

■

Finalize: In the ground state $n = 1$ we have

$$\langle x^2 \rangle = \frac{L^2}{3} - \frac{L^2}{2\pi^2} = 0.283L^2$$

In higher and higher states as $n \to \infty$ the expectation value approaches $0.333\ L^2$. An elementary theorem in statistics says that the root-mean-square uncertainty in x is given by $\Delta x = \sqrt{\langle x^2 \rangle - \langle x \rangle^2}$. This quantity is smallest in the ground state, where it is $0.181\ L$.

55. A quantum particle has a wave function

$$\psi(x) = \begin{cases} \sqrt{\dfrac{2}{a}}e^{-x/a} & \text{for } x > 0 \\ 0 & \text{for } x < 0 \end{cases}$$

(a) Find and sketch the probability density. (b) Find the probability that the particle will be at any point where $x < 0$. (c) Show that ψ is normalized and then (d) find the probability of finding the particle between $x = 0$ and $x = a$.

Solution

Conceptualize: Only an infinite value of potential energy could produce the discontinuity in the derivative of the wave function at $x = 0$. Nothing at all could produce the discontinuity in the wave function itself. Aside from those features, $\psi(x)$ could be a physically reasonable wave function, modeling an electron tied to a strong, very tiny attractor at $x = 0$, right next to an impenetrable wall.

Categorize: Think of a as some known distance such as 2 nm. The sketch will emphasize that the particle's state is spread out in space. The particle will never be found at $x < 0$. Normalization will require a 100% probability that it be somewhere in $0 < x < \infty$. The probability of its being between $x = 0$ and $x = a$ will be notably less than 1.

Analyze:

(a) The probability density is

$$|\psi|^2 = \begin{cases} \dfrac{2}{a}e^{-2x/a} & \text{for } x > 0 \\ 0 & \text{for } x < 0 \end{cases}$$

Its graph shows that the particle is restricted to the positive side of the x axis and it is less and less likely to be found at larger and larger values of x.

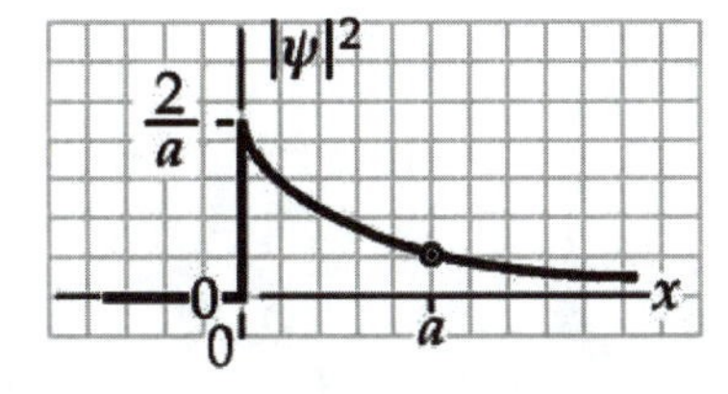

■

(b) The particle has zero probability of being at any point where $x < 0$. ■

(c) For normalization, $\int_{\text{all } x} |\psi|^2 dx = 1$. This condition becomes

$$0 + \int_0^\infty \frac{2}{a} e^{-2x/a} dx = -\int_0^\infty e^{-2x/a}(-2dx/a) = -e^{-2x/a}\Big|_0^\infty = -[0-1] = 1$$ ■

Thus, $\sqrt{2/a}$ was the right normalization coefficient for ψ in the first place.

(d) The probability of finding the particle in $0 < x < a$ is

$$\int_0^a \frac{2}{a} e^{-2x/a} dx = -e^{-2x/a}\Big|_0^a = -\left[e^{-2} - 1\right] = 0.865$$ ■

Finalize: On the graph the whole area under the line is finite and is in fact the number 1. The area lying to the left of the tic shown for $x = a$ is 86.5% of the total area.

---

58. An electron is represented by the time-independent wave function

$$\psi(x) = \begin{cases} Ae^{-\alpha x} & \text{for } x > 0 \\ Ae^{+\alpha x} & \text{for } x < 0 \end{cases}$$

(a) Sketch the wave function as a function of x. (b) Sketch the probability density representing the likelihood that the electron is found between x and $x + dx$. (c) Only an infinite value of potential energy could produce the discontinuity in the derivative of the wave function at $x = 0$. Aside from this feature, argue that $\psi(x)$ can be a physically reasonable wave function. (d) Normalize the wave function. (e) Determine the probability of finding the electron somewhere in the range

$$-\frac{1}{2\alpha} \le x \le \frac{1}{2\alpha}$$

Solution

Conceptualize: This could be a model for an electron tied to the center of a very tiny attractor, forming a one-dimensional quantum dot.

Categorize: Think of α as some known number such as 5/nm. The sketches emphasize that the electron state is spread out in space. The probability in part (e) will be notably less than 1. The probability in part (d) will be 1 because we think of the probability of finding the electron anywhere at all, from $-\infty$ to $+\infty$ on the x axis.

Analyze:

(a) The diagram shows the wave function.

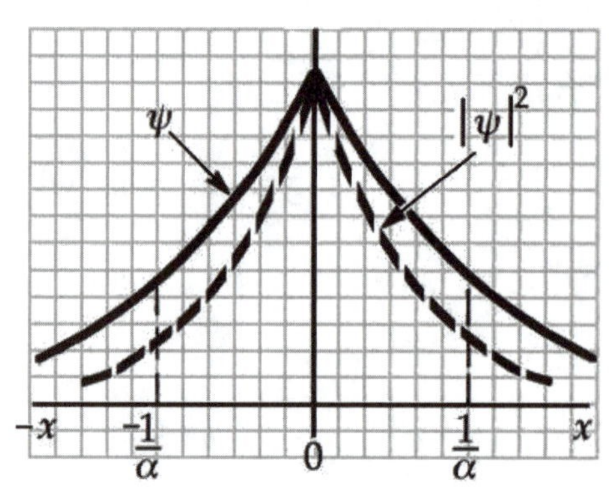

■

(b) For $x > 0$, the probability of finding the particle between coordinate x and coordinate $x + dx$ is

$$P_{x \text{ to } x+dx} = P(x)dx = |\psi|^2 dx = A^2 e^{-2\alpha x} dx \qquad \blacksquare$$

For $x < 0$ this probability is

$$P_{x \text{ to } x+dx} = P(x)dx = |\psi|^2 dx = A^2 e^{+2\alpha x} dx \qquad \blacksquare$$

The same diagram shows the probability distribution function as $|\psi|^2$. It shows exponential decays steeper than those of ψ, because it has 2α instead of α in the exponent.

(c) This can be reasonable since (1) ψ is continuous, (2) $\psi \to 0$ as $x \to \pm\infty$, and (3) the waveform can represent an electron bound by an infinitely deep, infinitely narrow potential well at $x = 0$. The wave function's derivative has no single value at the origin, but the wave function need not be differentiable at a point where the potential is infinite. (4) The wave function is integrable and can be normalized, as we show in part (d). ■

(d) Because ψ is symmetric, $\int_{-\infty}^{\infty} |\psi|^2 dx = 2\int_0^{\infty} |\psi|^2 dx = 1$

or the normalization requirement is $2A^2 \int_0^{\infty} e^{-2\alpha x} dx = 1$

Integrating,

$$\frac{2A^2}{-2\alpha}\int_0^{\infty} e^{-2\alpha x}(-2\alpha dx) = \frac{2A^2}{-2\alpha} e^{-2\alpha x}\Big|_0^{\infty} = \left[\frac{2A^2}{-2\alpha}\right]\left[e^{-\infty} - e^0\right] = A^2/\alpha = 1$$

which gives $A = \sqrt{\alpha}$ ■

(e) The probability of finding the particle between $-1/2\alpha$ and $+1/2\alpha$ is

$$P_{-1/2\alpha \text{ to } 1/2\alpha} = 2P_{0 \text{ to } 1/2\alpha} = 2\int_0^{1/2\alpha} |\psi|^2 dx = 2\int_0^{1/2\alpha} \left(\sqrt{\alpha}\right)^2 e^{-2\alpha x} dx$$

$$= \left[\frac{2\alpha}{-2\alpha}\right]\left[e^{-2\alpha/2\alpha} - 1\right] = \left[1 - e^{-1}\right] = 0.632 \qquad \blacksquare$$

Finalize: This can be another good problem to understand thoroughly. The wave function contains all the information about the particle's state, so it is possible to find lots of things from the wave function. The problem may be more tangible to you if you assume $\alpha = 5$/nm from the start so that definite answers show up as definite numbers. The graph of ψ would then start from $2.24/\sqrt{\text{nm}}$ at $x = 0$ and the graph of $|\psi|^2$ would start from 5/nm. Extra practice: Taking $\alpha = 5$/nm, substitute the wave function into the Schrödinger equation, with also the assumption that $U = 4.00$ eV for $x > 0$ and for $x < 0$. Prove that it satisfies the Schrödinger equation at all points other than $x = 0$, and evaluate the particle's energy. The answer is $E = 3.05$ eV.

42

Atomic Physics

EQUATIONS AND CONCEPTS

Balmer series wavelengths for four visible emission lines in the spectrum of hydrogen can be calculated using the empirical Equation 42.1. R_H is called the Rydberg constant. The figure shows the lines in the visible region of the spectrum and the short wavelength limit of the Balmer series.

λ (nm)
486.1 656.3
364.6 410.2 434.1

$$\frac{1}{\lambda} = R_H\left(\frac{1}{2^2} - \frac{1}{n^2}\right) \quad n = 3, 4, 5, \ldots \quad (42.1)$$

$$R_H = \frac{k_e e^2}{2a_0 hc} = 1.097\,373\,2 \times 10^7 \text{ m}^{-1}$$

Basic **postulates of Bohr's model** for the hydrogen atom:

- The **electron moves in circular orbits** around the proton.
- The **electron in stationary (stable) orbits** does not emit energy as radiation.
- **A photon is emitted** (carrying away energy) when the atomic electron undergoes a transition from a higher-energy stationary state to a lower-energy stationary state. *The frequency of the emitted photon is proportional to the difference between energies of the initial and final states.*

$$E_i - E_f = hf \quad (42.5)$$

- The **angular momentum** of the electron about the nucleus must be quantized in units of $n\hbar$. *This condition determines the radii of the allowed electron orbits.*

$$m_e vr = n\hbar \quad n = 1, 2, 3, \ldots \quad (42.6)$$

The **total energy of the hydrogen atom** (electron-proton bound system) is negative.

$$E = -\frac{k_e e^2}{2r} \quad (42.9)$$

Radii of the allowed orbits have discrete (quantized) values that can be expressed in terms of the Bohr radius, a_0. The **Bohr radius** corresponds to $n = 1$ (see the figure below).

$$r_n = \frac{n^2\hbar^2}{m_e k_e e^2} \quad n = 1, 2, 3, \ldots \qquad (42.10)$$

$$a_0 = \frac{\hbar^2}{m_e k_e e^2} = 0.052\ 9\ \text{nm} \qquad (42.11)$$

$$r_n = n^2 a_0 = n^2(0.052\ 9\ \text{nm}) \qquad (42.12)$$

The **first three circular orbits** predicted by the Bohr model of the hydrogen atom are shown in the figure at right (not to scale).

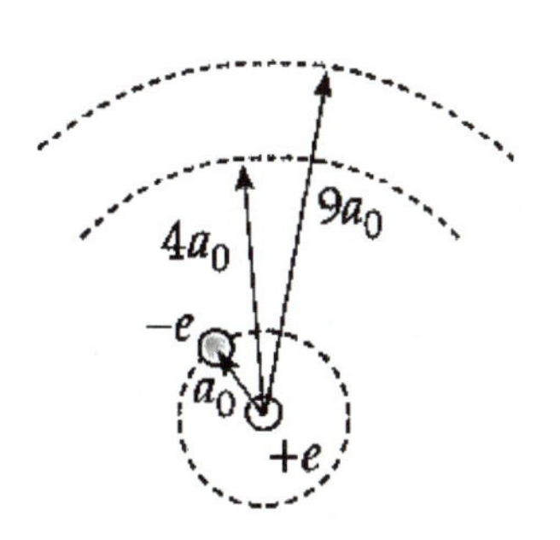

Quantized energy level values can be expressed in units of electron volts (eV). *The lowest allowed energy state or ground state corresponds to the principal quantum number n = 1. The absolute value of the ground state energy is equal to the ionization energy of the atom. The energy level approaches E = 0 as r approaches infinity.*

$$E_n = -\frac{13.606\ \text{eV}}{n^2} \quad n = 1, 2, 3, \ldots \qquad (42.14)$$

A **photon of frequency** f (and corresponding wavelength λ) is emitted when an electron undergoes a transition from an outer orbit (higher-energy level) to an inner orbit (lower-energy level).

$$f = \frac{E_i - E_f}{h} \qquad (42.15)$$

Photon wavelengths can be calculated using Equation 42.17. *Wavelength, rather than frequency, is the experimentally measured quantity.*

$$\frac{1}{\lambda} = R_{\text{H}}\left(\frac{1}{n_f^{\,2}} - \frac{1}{n_i^{\,2}}\right) \qquad (42.17)$$

The **lines in the hydrogen spectrum** can be arranged into several series. Each series of lines corresponds to an assigned value of the quantum number of the final energy level, n_f. Individual lines within a given series result from transitions from successively higher energy levels to the n_f level. Within each series, the difference in energy between adjacent energy levels

Substituting into Equation (42.17):

Lyman series: $n_f = 1$; $n_i = 2, 3, 4, \ldots$

Balmer series: $n_f = 2$; $n_i = 3, 4, 5, \ldots$

Paschen series: $n_f = 3$; $n_i = 4, 5, 6, \ldots$

Brackett series: $n_f = 4$; $n_i = 5, 6, 7, \ldots$

becomes smaller as n_i increases. Transitions in the Lyman, Balmer, and Paschen series are shown in the figure. The minimum wavelength in each series corresponds to $n_i \to \infty$. The **correspondence principle** states that for large quantum numbers (n), quantum physics agrees with classical physics.

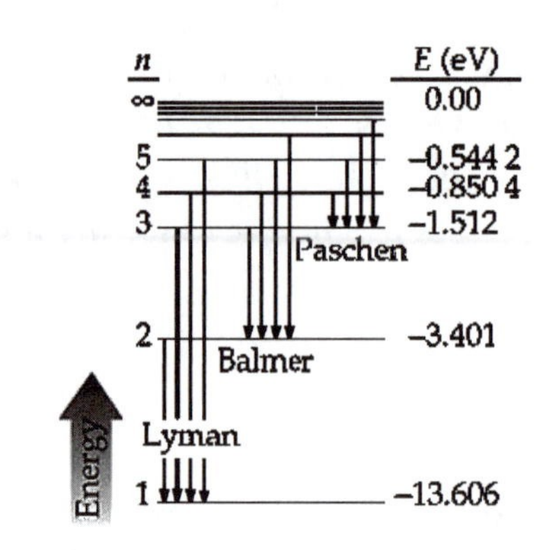

The **potential energy function** for the hydrogen atom depends on the radius of the allowed orbit of the electron. *There is no dependence on angular coordinates.*

$$U(r) = -k_e \frac{e^2}{r} \tag{42.20}$$

The **energy level values for hydrogen** obtained from the quantum model are the same as the results of the Bohr model. Compare Equations 42.14 and 42.21. The parameter a_0 is the Bohr radius.

$$E_n = -\left(\frac{k_e e^2}{2a_0}\right)\frac{1}{n^2} = -\frac{13.606\text{ eV}}{n^2} \tag{42.21}$$

$$n = 1, 2, 3, \ldots$$

The **1*s* state wave function for hydrogen** depends only on the radial distance r. *All s states have spherical symmetry.* The 1*s* state is the ground state of the atom.

$$\psi_{1s}(r) = \frac{1}{\sqrt{\pi a_0^{\,3}}} e^{-r/a_0} \tag{42.22}$$

The **radial probability density** for the 1*s* state of hydrogen is defined as the probability of finding the electron in a spherical shell of radius r and thickness dr. As shown in the figure, the largest value of $P_{1s}(r)$ corresponds to $r = a_0$.

$$P_{1s}(r) = \left(\frac{4r^2}{a_0^{\,3}}\right) e^{-2r/a_0} \tag{42.25}$$

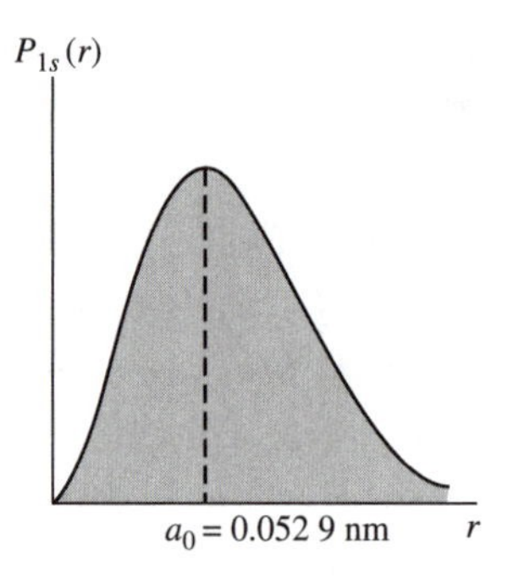

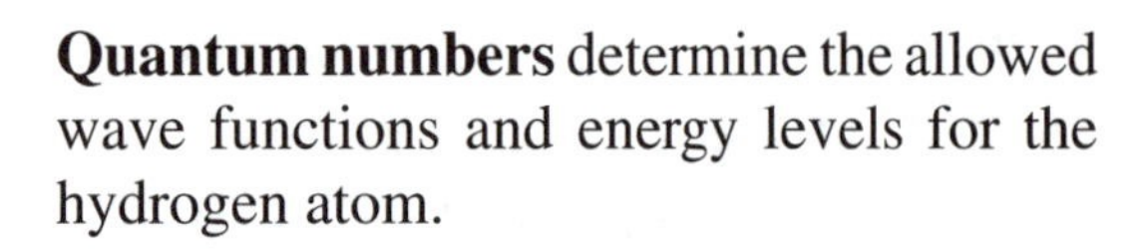

Quantum numbers determine the allowed wave functions and energy levels for the hydrogen atom.

The **principal quantum number** n is associated with allowed orbits and determines the energy states of the atom. *All states having the same principal quantum number form a shell.*

The **orbital quantum number** ℓ determines discrete values of the magnitude of the angular momentum, L. *There are n different possible values of ℓ. All states having the same values of n and ℓ form a subshell.*

$$L = \sqrt{\ell(\ell+1)}\,\hbar$$
$$\ell = 0, 1, 2, \ldots, n-1 \tag{42.27}$$

The **orbital magnetic quantum number** m_ℓ specifies the allowed values of the z component of the orbital angular momentum vector ($\vec{\mathbf{L}}$). *When an atom is placed in an external magnetic field, the vector $\vec{\mathbf{L}}$ lies on the surface of a cone and precesses about the z axis.*

$$L_z = m_\ell \hbar \tag{42.28}$$
$$m_\ell = -\ell, -\ell+1, -\ell+2, \ldots, 0, 1, 2, \ldots, \ell$$

The **Zeeman effect** ("splitting" of spectral lines) is due to the quantization of the angle θ that $\vec{\mathbf{L}}$ makes with the direction of an external magnetic field. The allowed values of θ are given by Equation 42.29. The figure illustrates the case when $n = 2$ and $\ell = 1$. In the absence of an external magnetic field a single line corresponding to the first excited state in hydrogen is observed. When an external magnetic field is present, this single line is split into three lines representing slightly different energy levels corresponding to the $(2\ell + 1)$ or 3 possible values of m_ℓ.

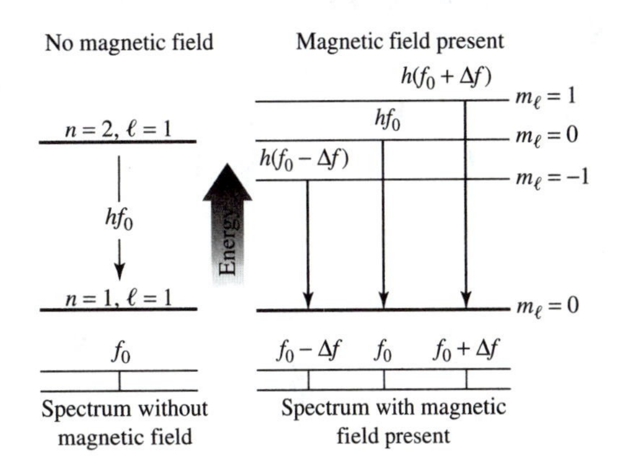

$$\cos\theta = \frac{L_z}{L} = \frac{m_\ell}{\sqrt{\ell(\ell+1)}} \tag{42.29}$$

The **spin quantum number** s has only one value. This quantum number is not derived from the Schrödinger equation; it originates in the relativistic properties of the electron and it determines the magnitude of the spin angular momentum of an electron (S).

$$S = \sqrt{s(s+1)}\,\hbar = \frac{\sqrt{3}}{2}\hbar \tag{42.30}$$
$$s = \frac{1}{2}$$

The **spin magnetic quantum number** m_s specifies the space orientation, relative to a z axis, of the spin angular momentum vector.

$$S_z = m_s \hbar = \pm\frac{1}{2}\hbar \tag{42.31}$$
$$m_s = \pm\frac{1}{2}$$

The **exclusion principle** dictates that no two electrons in the same atom can have the exact same set of quantum numbers.

Selection rules for allowed atomic transitions are expressed as allowed changes in the orbital quantum number and the orbital magnetic quantum number when an electron undergoes a transition between stationary states.

$$\Delta\ell = \pm 1 \quad and \quad \Delta m_\ell = 0, \pm 1 \qquad (42.34)$$

Energy levels of multielectron atoms are calculated by taking into account the shielding effect of the nuclear charge by inner-core electrons. *The atomic number is replaced by an effective atomic number* Z_{eff}, *which depends on the value of n.*

$$E_n = -\frac{13.6\,Z_{eff}^2}{n^2}\ \text{eV} \qquad (42.36)$$

SUGGESTIONS, SKILLS, AND STRATEGIES

Quantum numbers for the hydrogen atom are listed below. *The Pauli exclusion principle states that no two electrons in an atom can have the same set of quantum number values.*

Quantum Number	Name	Allowed Values	Allowed States
n	Principal quantum number	1, 2, 3, . . .	any integer
ℓ	Orbital quantum number	0,1, 2, . . ., $(n-1)$	n
m_ℓ	Orbital magnetic quantum number	$-\ell, -\ell+1, -\ell+2, \ldots, 0, 1, 2, \ldots, \ell$	$2\ell+1$
m_s	Spin magnetic quantum number	1/2, −1/2	2

All energy states with the same principal quantum number, n, form a shell. These shells are identified by the spectroscopic notation K, L, M, . . . corresponding to $n = 1, 2, 3, \ldots$.

All energy states having the same values of n and ℓ form a subshell. The letter designations $s, p, d, f \ldots$ correspond to values of $\ell = 0, 1, 2, \ldots$.

The orbital magnetic quantum number m_ℓ determines the possible orientations of the electron's orbital angular momentum vector in the presence of an external magnetic field.

The two possible values of the spin magnetic quantum number m_s correspond to the two possible directions of the electron's intrinsic spin.

REVIEW CHECKLIST

You should be able to:

- Calculate the wavelength and principal quantum number for lines in the hydrogen spectrum. (Section 42.1)

- For an electron in an orbit with a given value of n, calculate: orbital radius, linear momentum, angular momentum, and kinetic energy. Construct an energy level diagram for the hydrogen atom and calculate the ionization energy. (Section 42.3)
- State the wave function equation and sketch the radial probability density function for the 1s state of hydrogen. Calculate the probability of finding an electron in the range between r and $r + dr$. Show that a given wave function is a solution to the Schrödinger equation. (Section 42.5)
- Describe qualitatively what each of the quantum numbers n, ℓ, m_ℓ, s, and m_s implies concerning atomic structure. State the allowed values which may be assigned to each quantum number. Associate the customary shell and subshell notations with allowed combinations of quantum numbers n and ℓ. (Section 42.6)
- State the possible sets of quantum numbers (ℓ, m_ℓ, m_s) associated with a given value of n (e.g., $n = 2$ as shown in Table 42.4 of your text). Calculate the possible values of the orbital angular momentum, L, corresponding to a given value of the principal quantum number. Find the allowed values for L_z (the component of the angular momentum vector along the direction of an external magnetic field) for a given value of L. (Section 42.6)
- State the Pauli exclusion principle and describe its relevance to the periodic table of the elements. Show how the exclusion principle leads to the known electronic ground state configuration of the light elements. (Section 42.7)
- Calculate the x-ray energy and wavelength associated with electron transitions between the L and K shells within an element of known Z number. (Section 42.8)

ANSWER TO AN OBJECTIVE QUESTION

9. Which of the following electronic configurations are *not* allowed for an atom? Choose all correct answers. (a) $2s^2 2p^6$ (b) $3s^2 3p^7$ (c) $3d^7 4s^2$ (d) $3d^{10} 4s^2 4p^6$ (e) $1s^2 2s^2 2d^1$

Answer In (b) it is impossible to have seven electrons in a p subshell. With $\ell = 1$, the possibilities for m_ℓ are -1, 0, and $+1$, and the possibilities for m_s are $+1/2$ and $-1/2$. So only six electrons can fit into a p subshell. In (e) it is impossible for a 2d state to exist. With $n = 2$, the possible values for ℓ are 0 and 1. The shell with $n = 2$ has just a 2s and a 2p subshell.

□ □ □ □

ANSWERS TO SELECTED CONCEPTUAL QUESTIONS

5. Could the Stern–Gerlach experiment be performed with ions rather than neutral atoms? Explain.

Answer Practically speaking, no. Since ions have a net charge, the magnetic force $q\vec{\mathbf{v}} \times \vec{\mathbf{B}}$ would deflect the beam, making it very difficult to separate ions with different magnetic-moment orientations.

□ □ □ □

7. Discuss some consequences of the exclusion principle.

Answer If the Pauli exclusion principle were not valid, the elements and their chemical behavior would be grossly different because every electron would end up in the lowest energy level of the atom. All substances would be nearly alike in their chemistry. Atoms would be only weakly bound in molecules, and complicated molecules might not exist. Most materials would have a much higher density, and the spectra of atoms would be very simple. There would be much less color in the world.

□ □ □ □

9. Why do lithium, potassium, and sodium exhibit similar chemical properties?

Answer The three elements have similar electronic configurations, with filled inner shells plus a single outer electron in an *s* orbital. Since atoms typically interact through their unfilled outer shells, and the outer shell of each of these atoms is similar, the chemical interactions of the three atoms are also similar.

□ □ □ □

SOLUTIONS TO SELECTED PROBLEMS

5. (a) What value of n_i is associated with the 94.96-nm spectral line in the Lyman series of hydrogen? (b) **What If?** Could this wavelength be associated with the Paschen series? (c) Could this wavelength be associated with the Balmer series?

Solution

Conceptualize: Its wavelength makes this an ultraviolet spectral line, rather far to the left of the visible hydrogen spectrum in the text's Figure 42.1. The electron must tumble down into $n = 1$ as its final state, but we cannot guess the initial state. In Active Figure 42.8, the photon energies in the Balmer and Paschen series are much smaller than for ultraviolet light.

Categorize: We can use the empirical Rydberg equation for wavelengths of hydrogen spectral lines.

Analyze: Our equation is $\frac{1}{\lambda} = R_\text{H}\left(\frac{1}{n_f^{\ 2}} - \frac{1}{n_i^{\ 2}}\right)$ where $R_\text{H} = 1.097 \times 10^7 \text{ m}^{-1}$.

With our notation we have identified what Rydberg did not know, that the integers are the principal quantum numbers of the original and final atomic states in the photon emission process. For the Lyman series, we have

$$n_f = 1 \quad \text{and} \quad n_i = 2, 3, 4 \ldots$$

We solve for the quantum number of the original state

$$\frac{1}{n_i^2} = \frac{1}{n_f^2} - \frac{1}{R_H\lambda} \qquad\qquad n_i = \left(\frac{1}{n_f^2} - \frac{1}{R_H\lambda}\right)^{-1/2}$$

(a) and substitute the given values.

$$n_i = \left(\frac{1}{1^2} - \frac{1}{94.96 \times 10^{-9}\,\text{m} \times 1.097 \times 10^7\,\text{m}^{-1}} \right)^{-1/2} = 5$$ ■

The electron makes a transition from energy level 5 to the ground state to emit light in this spectral line.

(b) and (c) By Figure 42.8, spectral lines in the Balmer and Paschen series all have much longer wavelengths, since much smaller energy losses put the atom into energy levels 2 or 3. The expressions

$$n_i = \left(\frac{1}{2^2} - \frac{1}{94.96 \times 10^{-9}\,\text{m} \times 1.097 \times 10^7\,\text{m}^{-1}} \right)^{-1/2}$$

and $$n_i = \left(\frac{1}{3^2} - \frac{1}{94.96 \times 10^{-9}\,\text{m} \times 1.097 \times 10^7\,\text{m}^{-1}} \right)^{-1/2}$$

are imaginary quantities, not real positive integers. The Lyman-delta wavelength given cannot be part of the Balmer or the Paschen series. ■

Finalize: We could have solved this problem by computing the energy of a photon of the given wavelength and using that energy to measure up from the ground state on the hydrogen energy-level diagram to see how high the atom must have been initially.

Another way to compare the different spectral series of hydrogen with one another is to say that the spacing of energy levels varies from very wide between 1 and 2 to narrower and narrower going up. Only transitions with level 1 at the bottom end have enough energy to be associated with the emission or absorption of ultraviolet light. Only transitions with level 2 at the lower end can have the right energy for visible light. All other transitions radiate or absorb infrared photons.

---

6. According to classical physics, a charge e moving with an acceleration a radiates energy at a rate

$$\frac{dE}{dt} = -\frac{1}{6\pi \epsilon_0} \frac{e^2 a^2}{c^3}$$

(a) Show that an electron in a classical hydrogen atom (see Fig. 42.5) spirals into the nucleus at a rate

$$\frac{dr}{dt} = -\frac{e^4}{12\pi^2 \epsilon_0^2 m_e^2 c^3} \frac{1}{r^2}$$

(b) Find the time interval over which the electron reaches $r = 0$, starting from $r_0 = 2.00 \times 10^{-10}$ m.

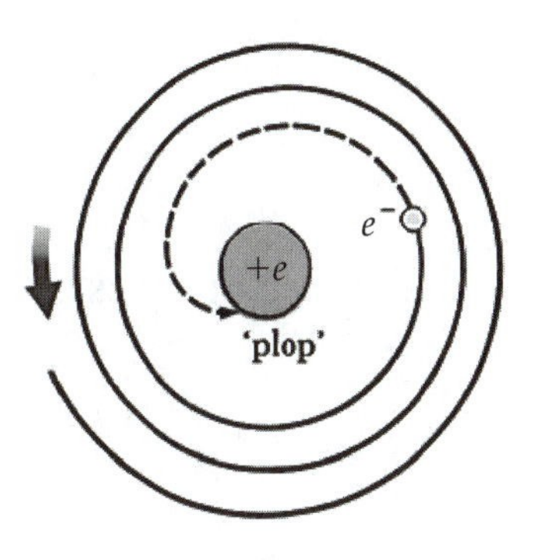

Figure 42.5

Solution

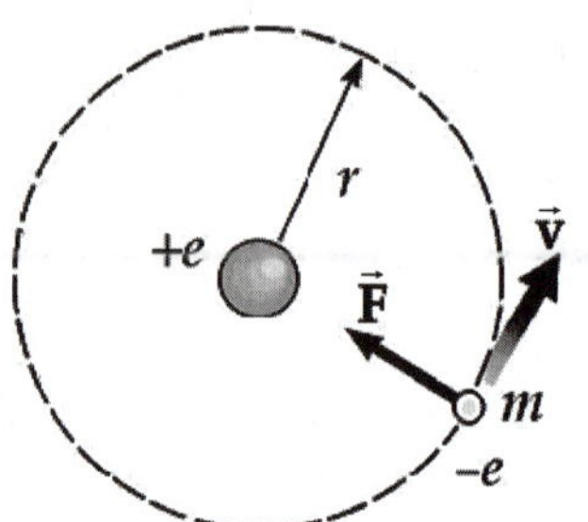

Conceptualize: This problem emphasizes that a theory has a limited range of applicability. Classical mechanics does not describe an electron.

Categorize: We imagine that Newton's second law describes a single electron moving around a proton. As its energy decreases according to the given radiation equation, its radius must decrease.

Analyze: According to a classical model, the electron moving as a particle in uniform circular motion about the proton in the hydrogen atom experiences a force $k_e e^2/r^2$; and from Newton's second law, $F = ma$, its acceleration is $k_e e^2/m_e r^2$.

(a) Using the fact that the Coulomb constant is $k_e = \dfrac{1}{4\pi\in_0}$

we have the electron's acceleration $a = \dfrac{v^2}{r} = \dfrac{k_e e^2}{m_e r^2} = \dfrac{e^2}{4\pi\in_0 m_e r^2}$ **[1]**

From the Bohr structural model of the atom, we can write the total energy of the atom as

$$E = -\frac{k_e e^2}{2r} = -\frac{e^2}{8\pi\in_0 r} \quad \text{so} \quad \frac{dE}{dt} = \frac{e^2}{8\pi\in_0 r^2}\frac{dr}{dt} = -\frac{1}{6\pi\in_0}\frac{e^2 a^2}{c^3} \qquad \textbf{[2]}$$

Substituting [1] into [2] for a, solving for dr/dt, and simplifying gives

$$\frac{dr}{dt} = -\frac{4r^2}{3c^3}\left(\frac{e^2}{4\pi\in_0 m_e r^2}\right)^2 = -\frac{e^4}{12\pi^2\in_0{}^2 r^2 m_e{}^2 c^3} \quad \blacksquare$$

(b) We can evaluate the constant to express dr/dt in a simpler form:

$$\frac{dr}{dt} = -\frac{A}{r^2} = -\frac{3.15\times10^{-21}}{r^2}\frac{\text{m}^3}{\text{s}}$$

Thus the time required to spiral in to the proton is described by

$$-\int_{2.00\times10^{-10}\text{ m}}^{0} r^2 dr = 3.15\times10^{-21}\int_0^T dt$$

Integration gives

$$T = \left(3.17\times10^{20}\frac{\text{s}}{\text{m}^3}\right)\frac{r^3}{3}\Bigg|_0^{2.00\times10^{-10}\text{ m}} = 8.46\times10^{-10}\text{ s} = 0.846\text{ ns} \quad \blacksquare$$

Finalize: We know that atoms last much longer than eight tenths of a nanosecond. Thus, classical physics does not hold (fortunately) for atomic systems.

14. A hydrogen atom is in its first excited state ($n = 2$). Using the Bohr model of the atom, calculate (a) the radius of the orbit, (b) the linear momentum of the electron, (c) the angular momentum of the electron, (d) the kinetic energy of the electron, (e) the potential energy of the system, and (f) the total energy of the system.

Solution

Conceptualize: The radius will be $4a_0$. The angular momentum will be $2\hbar$ so the linear momentum must be $2\hbar/4a_0$. The total energy is -13.6 eV/4. The kinetic energy is $+13.6$ eV/4 and the potential energy is $2(-13.6$ eV/4).

Categorize: We use the Bohr model of the hydrogen atom.

Analyze: We note, during our calculations, that the nominal velocity of the electron is less than 1% of the speed of light; therefore, we do not need to use relativistic equations.

(a) By Bohr's theory, $r_n = n^2(a_0) = 2^2(0.052\,9 \text{ nm}) = 2.12 \times 10^{-10} \text{ m}$ ■

(b) Since $m_e vr = n\hbar$,

$$p = m_e v = \frac{n\hbar}{r} = \frac{2(1.054\,6 \times 10^{-34} \text{ J}\cdot\text{s})}{2.12 \times 10^{-10} \text{ m}} = 9.97 \times 10^{-25} \text{ kg}\cdot\text{m/s}$$ ■

(c) $\vec{\mathbf{L}} = \vec{\mathbf{r}} \times \vec{\mathbf{p}}$ becomes $L = rp = n\hbar = 2.11 \times 10^{-34} \text{ J}\cdot\text{s}$ ■

(d) Next, the speed is $v = \dfrac{p}{m_e} = \dfrac{9.97 \times 10^{-25} \text{ kg}\cdot\text{m/s}}{9.11 \times 10^{-31} \text{ kg}} = 1.09 \times 10^6 \text{ m/s}$

So the kinetic energy is $K = \frac{1}{2} m_e v^2$

$$K = \frac{(9.11 \times 10^{-31} \text{ kg})(1.09 \times 10^6 \text{m/s})^2}{2} = \frac{5.45 \times 10^{-19} \text{ J}}{1.602 \times 10^{-19} \text{ J/eV}} = 3.40 \text{ eV}$$ ■

(e) From Chapter 25, the electric potential energy is $U = qV$:

$$U = -\frac{k_e e^2}{r} = -\frac{(8.99 \times 10^9 \text{ N}\cdot\text{m}^2/\text{C}^2)(1.602 \times 10^{-19} \text{ C})^2}{2.12 \times 10^{-10} \text{ m}}$$

and $U = -1.09 \times 10^{-18} \text{ J} = -6.80 \text{ eV}$ ■

(f) Thus the total energy is $E = K + U = -5.45 \times 10^{-19} \text{ J} = -3.40 \text{ J}$ ■

Finalize: The total energy ought to be equal to the negative of the kinetic energy. Our answer confirms this, but we needed to use more than 3 significant figures for some constants. For an accurate answer from data with fewer significant figures, we could calculate the total energy first:

$$E = -\frac{k_e e^2}{2r} = -\frac{\left(8.99\times10^9\ \text{N}\cdot\text{m}^2/\text{C}^2\right)\left(1.602\times10^{-19}\ \text{C}\right)^2}{2\left(2.12\times10^{-10}\ \text{m}\right)} = -5.45\times10^{-19}\ \text{J} \quad \blacksquare$$

$$K = E - U = 5.45\times10^{-19}\ \text{J} = 3.40\ \text{J} \quad \blacksquare$$

---

17. (a) Construct an energy-level diagram for the He^+ ion, for which $Z = 2$, using the Bohr model. (b) What is the ionization energy for He^+?

Solution

Conceptualize: The helium ion has a nucleus with charge $+2e$. It attracts the single orbiting electron twice as strongly as a hydrogen nucleus does, so the absolute values of the energies will be larger than for hydrogen.

Categorize: We will work from the Bohr theory, generalized for any value of atomic number Z.

Analyze: From Equation 42.19,

$$E_n = -\frac{k_e e^2 Z^2}{2a_0 n^2} = \frac{(-13.6\ \text{eV})(2)^2}{n^2}$$

For the lowest energy level, $n = 1$ and $E_1 = -54.4$ eV

For the first excited state, $n = 2$ and $E_2 = -13.6$ eV

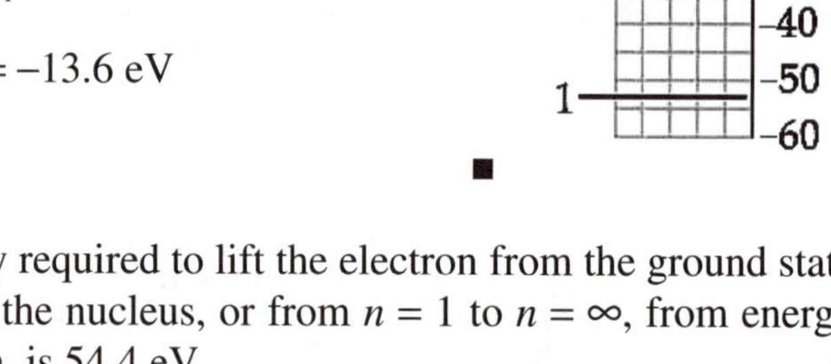

(a) The diagram is shown. ■

(b) The ionization energy is the energy required to lift the electron from the ground state to the point of breaking free from the nucleus, or from $n = 1$ to $n = \infty$, from energy -54.4 eV to 0 eV. The answer, then, is 54.4 eV. ■

Finalize: The Sun and many other stars show spectral lines of ionized helium. Helium was in this way discovered on the Sun before it was isolated on Earth.

---

18. A general expression for the energy levels of one-electron atoms and ions is

$$E_n = -\frac{\mu k_e^2 q_1^2 q_2^2}{2\hbar^2 n^2}$$

Here μ is the reduced mass of the atom, given by $\mu = m_1 m_2/(m_1 + m_2)$, where m_1 is the mass of the electron and m_2 is the mass of the nucleus; k_e is the Coulomb constant; and q_1 and

q_2 are the charges of the electron and the nucleus, respectively. The wavelength for the $n = 3$ to $n = 2$ transition of the hydrogen atom is 656.3 nm (visible red light). What are the wavelengths for this same transition in (a) positronium, which consists of an electron and a positron, and (b) singly ionized helium? *Note:* A positron is a positively charged electron.

Solution

Conceptualize: The reduced mass of positronium is **less** than hydrogen, so the photon energy will be **less** for positronium than for hydrogen. This means that the wavelength of the emitted photon will be **longer** than 656.3 nm. On the other hand, helium has about the same reduced mass but more charge than hydrogen, so its transition energy will be **larger**, corresponding to a wavelength **shorter** than 656.3 nm.

Categorize: All the factors in the above equation are constant for this problem except for the reduced mass and the nuclear charge. Therefore, the wavelength corresponding to the energy difference for the transition can be found simply from the ratio of mass and charge variables.

Analyze: For hydrogen, $\mu = \dfrac{m_p m_e}{m_p + m_e} \approx m_e$

The photon energy is $\Delta E = E_3 - E_2$

Its wavelength is $\lambda = 656.3$ nm, where $\lambda = \dfrac{c}{f} = \dfrac{hc}{\Delta E}$

(a) For positronium, $\mu = \dfrac{m_e m_e}{m_e + m_e} = \dfrac{m_e}{2}$

so the energy of each level is one half as large as in hydrogen, for which we could invent the name "protonium." The photon energy is inversely proportional to its wavelength, so for positronium,

$$\lambda_{32} = 2(656.3 \text{ nm}) = 1.31\ \mu\text{m (in the infrared region)}$$ ■

(b) For He^+, $\mu \approx m_e$, $q_1 = e$, and $q_2 = 2e$, so the transition energy is $2^2 = 4$ times as large as in hydrogen.

Then, $\lambda_{32} = \left(\dfrac{656}{4}\right)$ nm $= 164$ nm (in the ultraviolet region) ■

Finalize: As expected, the wavelengths for positronium and helium are respectively larger and smaller than for hydrogen. Other energy transitions should have wavelength shifts consistent with this pattern. It is important to remember that the reduced mass is not the total mass, but is generally close in magnitude to the smaller mass of the system (hence the name **reduced** mass).

---

20. An electron of momentum p is at a distance r from a stationary proton. The electron has kinetic energy $K = p^2/2m_e$. The atom has potential energy $U = -k_e e^2/r$ and total energy $E = K + U$. If the electron is bound to the proton to form a hydrogen atom, its average position is at the proton but the uncertainty in its position is approximately equal to the radius r of its orbit. The electron's average vector momentum is zero, but its average squared momentum is approximately equal to the squared uncertainty in its momentum as given by the uncertainty principle. Treating the atom as a one-dimensional system, (a) estimate the uncertainty in the electron's momentum in terms of r. Estimate the electron's (b) kinetic energy and (c) total energy in terms of r. The actual value of r is the one that *minimizes the total energy*, resulting in a stable atom. Find (d) that value of r and (e) the resulting total energy. (f) State how your answers compare with the predictions of the Bohr theory.

Solution

Conceptualize: Bohr's model makes an arbitrary and incorrect assumption that the angular momentum of a hydrogen atom in its ground state is $1\hbar$ instead of zero. This problem shows how important the Heisenberg uncertainty principle is, by using it instead of Bohr's quantization of angular momentum to determine the size of a hydrogen atom in its ground state, and the energy of the atom.

Categorize: We follow the step-by-step directions in the problem.

Analyze:

(a) The uncertainty principle is represented by $\Delta x \Delta p \geq \dfrac{\hbar}{2}$

Thus, if $\Delta x = r$, $\Delta p \geq \dfrac{\hbar}{2r}$ ■

(b) The minimum uncertainty would be attained only if the wave function had a particular (gaussian) waveform. We assume that the momentum uncertainty is just twice as large as its minimum possible value: $\Delta p = \dfrac{\hbar}{r}$. Then the kinetic energy is $K = \dfrac{p^2}{2m_e} = \dfrac{\hbar^2}{2m_e r^2}$. ■

(c) The electric potential energy is $U = -\dfrac{k_e e^2}{r}$ so the total energy is $E = \dfrac{\hbar^2}{2m_e r^2} - \dfrac{k_e e^2}{r}$ ■

(d) To minimize E as a function of r, we require $\dfrac{dE}{dr} = -\dfrac{\hbar^2}{m_e r^3} + \dfrac{k_e e^2}{r^2} = 0$

This implies $r = \dfrac{\hbar^2}{m_e k_e e^2}$ = Bohr radius ■

(e) Then the energy is

$$E = \frac{\hbar^2}{2m_e}\left(\frac{m_e k_e e^2}{\hbar^2}\right)^2 - k_e e^2\left(\frac{m_e k_e e^2}{\hbar^2}\right) = -\frac{m_e k_e^2 e^4}{2\hbar^2} = -13.6 \text{ eV}$$ ■

(f) The radius and this energy have the same values as those predicted by the Bohr theory for the ground state of hydrogen. ■

Finalize: In the Schrödinger theory the angular momentum of a hydrogen atom in its ground state is in fact zero, and not $1\hbar$ as in the Bohr theory; and the zero value is confirmed by experiment. The argument given in this problem comes out so neatly partly because we assumed $\Delta p = \frac{\hbar}{r}$ rather than $\Delta p = \frac{\hbar}{2r}$ or some other multiple of the minimum value required by the uncertainty principle. The argument given here cannot be generalized to excited states of the atom.

---

24. For a spherically symmetric state of a hydrogen atom, the Schrödinger equation in spherical coordinates is

$$-\frac{\hbar^2}{2m_e}\left(\frac{d^2\psi}{dr^2}+\frac{2}{r}\frac{d\psi}{dr}\right)-\frac{k_e e^2}{r}\psi = E\psi$$

(a) Show that the $1s$ wave function for an electron in hydrogen,

$$\psi_{1s}(r)=\frac{1}{\sqrt{\pi a_0{}^3}}e^{-r/a_0}$$

satisfies the Schrödinger equation. (b) What is the energy of the atom for this state?

Solution

Conceptualize: The potential energy function $k_e e^2/r$ makes this Schrödinger equation describe an electron in hydrogen. With this particular functional form of the potential energy term, the wave function must have another particular functional form, which we confirm.

Categorize: We use the quantum particle under boundary conditions model. We substitute the wave function and its derivatives into the Schrödinger equation, and then simplify the resulting equation. If we find that the resulting equation is true, then we know that the Schrödinger equation is satisfied.

Analyze: We evaluate the derivatives of the proposed wave function solution:

$$\frac{d\psi}{dr}=\frac{1}{\sqrt{\pi a_0{}^3}}\frac{d}{dr}\left(e^{-r/a_0}\right)=-\frac{1}{\sqrt{\pi a_0{}^3}}\left(\frac{1}{a_0}\right)e^{-r/a_0}=-\frac{\psi}{a_0} \qquad \textbf{[1]}$$

Differentiating again, $$\frac{d^2\psi}{dr^2}=-\frac{1}{\sqrt{\pi a_0{}^5}}\frac{d}{dr}\left(e^{-r/a_0}\right)=\frac{1}{\sqrt{\pi a_0{}^7}}e^{-r/a_0}=\frac{1}{a_0{}^2}\psi \qquad \textbf{[2]}$$

Substituting [1] and [2] into the Schrödinger equation, and noting that m is the electron's mass m_e, we have

$$-\frac{\hbar^2}{2m_e}\left(\frac{1}{a_0{}^2}-\frac{2}{a_0 r}\right)\psi-\frac{k_e e^2}{r}\psi = E\psi \qquad \textbf{[3]}$$

Substituting $\hbar^2 = m_e k_e e^2 a_0$ and canceling m_e and ψ we have

$$-\frac{k_e e^2 a_0}{2}\left(\frac{1}{a_0^2} - \frac{2}{a_0 r}\right) - \frac{k_e e^2}{r} = E$$

We find that the second and third terms add to zero, leaving the equation $E = -\frac{k_e e^2}{2a_0}$. This is a true statement of the ground state energy of hydrogen, so (a) the Schrödinger equation is satisfied, with (b) the value of energy just stated. ■

Finalize: The equation we were testing had functions of r on both sides. An important step in the proof was showing that they amounted to the same function of r, when we divided out the ψ.

---

30. How many sets of quantum numbers are possible for a hydrogen atom for which (a) $n = 1$, (b) $n = 2$, (c) $n = 3$, (d) $n = 4$, and (e) $n = 5$?

Solution

Conceptualize: Several state functions, with different angular momentum values, different z components of angular momentum, or different orientations of electron spin, have the same energy.

Categorize: We follow the rules about possible values for ℓ, for m_ℓ, and for m_s to tabulate the states for each value of the principal quantum number n. We will check our results to show that they agree with the general rule that the number of sets of quantum numbers for a shell is equal to $2n^2$.

Analyze:

(a) For $n = 1$, we have $\ell = 0$, $m_\ell = 0$, $m_s = \pm\frac{1}{2}$

n	ℓ	m_ℓ	m_s
1	0	0	−1/2
1	0	0	+1/2

This yields $2n^2 = 2(1)^2 = 2$ sets. ■

(b) For $n = 2$, we have

n	ℓ	m_ℓ	m_s
2	0	0	±1/2
2	1	−1	±1/2
2	1	0	±1/2
2	1	+1	±1/2

This yields $2n^2 = 2(2)^2 = 8$ sets. ■

Note that the number is twice the number of m_ℓ values. Also, for each ℓ there are $(2\ell + 1)$ different m_ℓ values. Finally, ℓ can take on values ranging from 0 to $n - 1$.

So the general expression is $\text{number} = \sum_{\ell=0}^{n-1} 2(2\ell + 1)$

The series is an arithmetic progression like 2 + 6 + 10 + 14.

The sum is $\sum_{0}^{n-1} 4\ell + \sum_{0}^{n-1} 2 = 4\left[\dfrac{n^2 - n}{2}\right] + 2n = 2n^2$

(c) $n = 3$, $2(1) + 2(3) + 2(5) = 2 + 6 + 10 = 18$: $2n^2 = 2(3)^2 = 18$ ■

(d) $n = 4$, $2(1) + 2(3) + 2(5) + 2(7) = 32$: $2n^2 = 2(4)^2 = 32$ ■

(e) $n = 5$, $32 + 2(9) = 32 + 18 = 50$: $2n^2 = 2(5)^2 = 50$ ■

Finalize: We have computed the number of states in two different ways for each of the assigned n values, and have shown that the results agree.

---

33. The ρ^- meson has a charge of $-e$, a spin quantum number of 1, and a mass 1 507 times that of the electron. The possible values for its spin magnetic quantum number are -1, 0, and 1. **What If?** Imagine that the electrons in atoms are replaced by ρ^- mesons. List the possible sets of quantum numbers for ρ^- mesons in the 3d subshell.

Solution

Conceptualize: There will be more possible states than for an electron because the meson can have three spin states instead of only two.

Categorize: We follow the rule that m_ℓ can range from $-\ell$ to $+\ell$ in integer steps, to tabulate the required values for n and ℓ and the possible values for m_ℓ and m_s.

Analyze: The 3d subshell has $n = 3$ and $\ell = 2$. Also, we have $s = 1$. Altogether we can have $n = 3$, $\ell = 2$, $m_\ell = -2, -1, 0, 1, 2$, $s = 1$, and $m_s = -1, 0, 1$, leading to the following table:

n	3	3	3	3	3	3	3	3	3	3	3	3	3	3	3
ℓ	2	2	2	2	2	2	2	2	2	2	2	2	2	2	2
m_ℓ	−2	−2	−2	−1	−1	−1	0	0	0	1	1	1	2	2	2
s	1	1	1	1	1	1	1	1	1	1	1	1	1	1	1
m_s	−1	0	1	−1	0	1	−1	0	1	−1	0	1	−1	0	1

■

Finalize: There are 15 possible states altogether. With electrons there would be only 10 possibilities.

---

40. Scanning through Figure 42.19 in order of increasing atomic number, notice that the electrons usually fill the subshells in such a way that those subshells with the lowest values of $n + \ell$ are filled first. If two subshells have the same value of $n + \ell$, the one with the lower value of n is generally filled first. Using these two rules, write the order in which the subshells are filled through $n + \ell = 7$.

Solution

Conceptualize: There is a pattern to the ground-state electron configurations of the elements. Writing down the whole configuration for the heaviest element, number 110, and then crossing out the appropriate number of electrons at the end, will give a good estimate of the configuration of any other element.

Categorize: We follow the directions given in the problem, remembering that the possible values for ℓ range from zero to $n - 1$.

Analyze:

$n + \ell$	1	2	3	4	5	6	7
subshell	1*s*	2*s*	2*p*, 3*s*	3*p*, 4*s*	3*d*, 4*p*, 5*s*	4*d*, 5*p*, 6*s*	4*f*, 5*d*, 6*p*, 7*s*

■

Finalize: Checking against the periodic table in Figure 42.19 shows that the prediction made from the two rules given in the problem works well for the ground-state electron configurations of atoms of most elements. But for one exception, consider silver (Ag). For it the rules suggest a configuration ending in . . . $5s^2 4d^9$. But this atom happens to have lower energy in the configuration . . . $5s^1 4d^{10}$. The energy advantage for having the full *d* subshell must be greater than for having a full *s* subshell.

---

49. Use the method illustrated in Example 42.5 to calculate the wavelength of the x-ray emitted from a molybdenum target ($Z = 42$) when an electron moves from the L shell ($n = 2$) to the K shell ($n = 1$).

Solution

Conceptualize: Most electrons in a multielectron atom are feeling forces due to one another as well as due to the nucleus, and their wave functions are quite complex. "Shells" interpenetrate in space. But an inner electron can be modeled as moving in the field of just the nucleus and few or no electrons below it, and in fact the Bohr theory gives a fair estimate for its energy.

Categorize: Following Example 42.5, we suppose the electron is originally in the L shell with just one other electron in the K shell between it and the nucleus . . .

Analyze: . . . so it moves in a field of effective charge $(42 - 1)e$. Its energy is then $E_L = -(42 - 1)^2(13.6 \text{ eV}/4) = -5.72 \text{ keV}$.

In the electron's final state we choose to estimate the charge holding it in orbit as $42e$, so its energy is $E_K = -(42)^2(13.6\ \text{eV}) = -24.0\ \text{keV}$.

The photon energy emitted is the absolute value of the difference,

$$E_\gamma = 24.0\ \text{keV} - 5.72\ \text{keV} = 18.3\ \text{keV} = 2.92 \times 10^{-15}\ \text{J}$$

Then $f = \dfrac{E}{h} = 4.41 \times 10^{18}\ \text{Hz}$ and $\lambda = \dfrac{c}{f} = 68\ \text{pm}$ ■

Finalize: In effect, we visualize the shells as like hollow nesting dolls. We imagine this particular electron as making a transition from the inner surface of the larger shell where it starts, to the inner surface of the smaller shell where it ends up. It is a little remarkable that this method works as well as it does, giving an estimate good to within a few percent.

---

55. A ruby laser delivers a 10.0-ns pulse of 1.00 MW average power. If the photons have a wavelength of 694.3 nm, how many are contained in the pulse?

Solution

Conceptualize: Lasers generally produce concentrated beams that are bright (except for IR or UV lasers that produce invisible beams). Since our eyes can detect light levels as low as a few photons, there are probably at least 1 000 photons in each pulse.

Categorize: From the pulse width and average power, we can find the energy delivered by each pulse. The number of photons can then be found by dividing the pulse energy by the energy of each photon, which is determined from the photon wavelength.

Analyze: The energy in each pulse is

$$E = P\Delta t = \left(1.00 \times 10^{6}\ \text{W}\right)\left(1.00 \times 10^{-8}\ \text{s}\right) = 1.00 \times 10^{-2}\ \text{J}$$

The energy of each photon is

$$E_\gamma = hf = \frac{hc}{\lambda} = \frac{\left(6.626 \times 10^{-34}\ \text{J}\cdot\text{s}\right)\left(3.00 \times 10^{8}\ \text{m/s}\right)}{694.3 \times 10^{-9}\ \text{m}} = 2.86 \times 10^{-19}\ \text{J}$$

So the number of photons in the pulse is

$$N = \frac{E}{E_\gamma} = \frac{1.00 \times 10^{-2}\ \text{J}}{2.86 \times 10^{-19}\ \text{J/photon}} = 3.49 \times 10^{16}\ \text{photons}$$ ■

Finalize: With 10^{16} photons/pulse, this laser beam should produce a bright red spot when the light reflects from a surface, even though the time between pulses is generally much longer than the width of each pulse. For comparison, this laser produces more photons in a single ten-nanosecond pulse than a typical 5 mW helium-neon laser produces over a full second (about 1.6×10^{16} photons/second). Human eyes require many more than 1 000 photons for something to appear bright.

---

61. Suppose a hydrogen atom is in the 2*s* state, with its wave function given by Equation 42.26. Taking $r = a_0$, calculate values for (a) $\psi_{2s}(a_0)$, (b) $|\psi_{2s}(a_0)|^2$, and (c) $P_{2s}(a_0)$.

Solution

Conceptualize: Schrödinger's theory is a quantitative theory. The wave function tells probabilities of finding the particle in particular places.

Categorize: We evaluate the wave function as stated in the problem.

Analyze: The wave function for the 2*s* state is given by

$$\psi_{2s}(r) = \frac{1}{4\sqrt{2\pi}}\left(\frac{1}{a_0}\right)^{3/2}\left(2 - \frac{r}{a_0}\right)e^{-r/2a_0}$$

(a) Substituting $r = a_0 = 0.529 \times 10^{-10}$ m, we find

$$\psi_{2s}(a_0) = \frac{1}{4\sqrt{2\pi}}\left(\frac{1}{0.529 \times 10^{-10}\ \text{m}}\right)^{3/2}(2-1)e^{-1/2} = 1.57 \times 10^{14}\ \text{m}^{-3/2} \quad \blacksquare$$

(b) $|\psi_{2s}(a_0)|^2 = \left(1.57 \times 10^{14}\ \text{m}^{-3/2}\right)^2 = 2.47 \times 10^{28}\ \text{m}^{-3}$ ■

(c) Using Equation 42.24 and the results to (b) gives

$$P_{2s}(a_0) = 4\pi a_0^{\,2}|\psi_{2s}(a_0)|^2 = 8.69 \times 10^{8}\ \text{m}^{-1} \quad \blacksquare$$

Finalize: The numbers are large because we are describing the probability *per volume* of finding the electron, and the electron really is in a small parcel of volume around the atomic nucleus. Scattering experiments can show the "shape" of an atomic electron cloud, and give results agreeing with the theory.

---

62. An electron in chromium moves from the $n = 2$ state to the $n = 1$ state without emitting a photon. Instead, the excess energy is transferred to an outer electron (one in the $n = 4$ state), which is then ejected by the atom. In this Auger (pronounced "ohjay") process, the ejected electron is referred to as an Auger electron. Use the Bohr theory to find the kinetic energy of the Auger electron.

Solution

Conceptualize: The chromium atom with nuclear charge $Z = 24$ starts with one vacancy in the $n = 1$ shell, perhaps produced by the absorption of an x-ray, which ionized the atom. An electron from the $n = 2$ shell tumbles down to fill the vacancy.

Categorize: We use the same ideas as for estimating a wavelength in the x-ray spectrum.

Analyze: We suppose that the electron that makes the transition is shielded from the electric field of the full nuclear charge by the one K-shell electron originally below it. With $Z = 24$, its original energy is

$$E = -(Z - 1)^2(13.6\text{ eV})\left(\frac{1}{2^2}\right) = -1.80\text{ keV}$$

Its final energy is $$E = -Z^2(13.6\text{ eV})\left(\frac{1}{1^2}\right) = -7.83\text{ keV}$$

The magnitude of the electron's energy loss is $7.83\text{ keV} - 1.80\text{ keV} = 6.04\text{ keV}$

Then, instead of coming out as an x-ray photon, this +6.04 keV can be transferred to the single 4*s* electron. Suppose that it is shielded by the 22 electrons in the K, L, and M shells. To break the outermost electron out of the atom, producing a Cr^{2+} ion, requires an energy investment of

$$E_{\text{ionize}} = \frac{(Z - 22)^2(13.6\text{ eV})}{4^2} = \frac{2^2(13.6\text{ eV})}{16} = 3.40\text{ eV}$$

Then the remaining energy that can appear as kinetic energy is

$$K = |\Delta E| - E_{\text{ionize}} = 6035\text{ eV} - 3.4\text{ eV} = 6.03\text{ keV}$$ ■

Because of conservation of momentum for the ion-electron system and the tiny mass of the electron compared to that of the Cr^{2+} ion, almost all of this kinetic energy will belong to the electron.

Finalize: As evidence that the relatively tiny 3.40 eV can be the right order of magnitude for the energy to remove a second electron from ionized chromium, note that the (first) ionization energy for neutral chromium is tabulated as 6.76 eV.

---

63. In the technique known as electron spin resonance (ESR), a sample containing unpaired electrons is placed in a magnetic field. Consider a situation in which a single electron (*not* contained in an atom) is immersed in a magnetic field. In this simple situation only two energy states are possible, corresponding to $m_s = \pm\frac{1}{2}$. In ESR, the absorption of a photon causes the electron's spin magnetic moment to flip from the lower energy state to the higher energy state. According to Section 29.5, the change in energy is $2\mu_B B$. (The lower energy state corresponds to the case in which the *z* component of the magnetic moment $\vec{\mu}_{\text{spin}}$ is aligned with the magnetic field, and the higher energy state corresponds to the case in which the *z* component of $\vec{\mu}_{\text{spin}}$ is aligned opposite to the field.) What is the photon frequency required to excite an ESR transition in a 0.350-T magnetic field?

Solution

Conceptualize: Application of the magnetic field splits an atomic energy level into two. The separation between the levels can be adjusted by changing the field. This action will change the frequency of the photons that are absorbed as they boost the atom from the lower to the higher state, and the photons that are emitted as the atom tumbles back down again. In this problem we model the experimental situation by ignoring the rest of the atom.

Categorize: As in Section 29.5, the magnetic moment feels torque $\vec{\tau} = \vec{\mu} \times \vec{\mathbf{B}}$ in an external field. The electron spin vector cannot be geometrically parallel or antiparallel to the field. But its z component must be either parallel or antiparallel to any applied field.

Analyze: To evaluate the energy difference, we imagine the z component of the electron's magnetic moment as continuously variable. In turning it from alignment with the field to the opposite direction, the field does work according to Equation 10.22,

$$W = \int dW = \int_0^{180^\circ} \tau \, d\theta = \int_0^{\pi} \mu B \sin\theta \, d\theta = -\mu B \cos\theta \Big|_0^{\pi} = 2\mu B$$

To make the electron flip, the photon must carry energy $\Delta E = 2\mu_B B = hf$.

Therefore,

$$f = \frac{2\mu_B B}{h} = \frac{2(9.27 \times 10^{-24} \text{ J/T})(0.350 \text{ T})}{6.63 \times 10^{-34} \text{ J} \cdot \text{s}} = 9.79 \times 10^9 \text{ Hz} \quad \blacksquare$$

Finalize: This is a microwave photon. Experimentally we identify a resonance phenomenon, because the sample will absorb at just this frequency, and will not absorb microwaves of slightly lower or higher frequency.

---

71. We wish to show that the most probable radial position for an electron in the 2s state of hydrogen is $r = 5.236a_0$. (a) Use Equations 42.24 and 42.26 to find the radial probability density for the 2s state of hydrogen. (b) Calculate the derivative of the radial probability density with respect to r. (c) Set the derivative in (b) equal to zero and identify three values of r that represent minima in the function. (d) Find two values of r that represent maxima in the function. (e) Identify which of the values in part (d) represents the highest probability.

Solution

Conceptualize: An example in the chapter found the most probable radial position for an electron in the ground state of hydrogen. Here we expect a larger modal position for an excited state. The textbook Active Figure 42.12 shows the radial probability distribution.

Categorize: We are using the quantum particle model. We differentiate to find the most probable radius.

Analyze: We use Equation 42.26:

$$\psi_{2s}(r) = \frac{1}{4\sqrt{2\pi}} \left(\frac{1}{a_0}\right)^{3/2} \left[2 - \frac{r}{a_0}\right] e^{-r/2a_0}$$

(a) By Equation 42.24, the radial probability distribution function is

$$P(r) = 4\pi r^2\psi^2 = \tfrac{1}{8}\left(\frac{r^2}{a_0^{\,3}}\right)\left(2 - \frac{r}{a_0}\right)^2 e^{-r/a_0}$$ ■

(b) Its extremes are given by

$$\frac{dP(r)}{dr} = \tfrac{1}{8}\left[\frac{2r}{a_0^{\,3}}\left(2 - \frac{r}{a_0}\right)^2 - \frac{2r^2}{a_0^{\,3}}\left(\frac{1}{a_0}\right)\left(2 - \frac{r}{a_0}\right) - \frac{r^2}{a_0^{\,3}}\left(2 - \frac{r}{a_0}\right)^2\left(\frac{1}{a_0}\right)\right]e^{-r/a_0} = 0$$

We factor in the following manner:

$$\tfrac{1}{8}\left(\frac{r}{a_0^{\,3}}\right)\left(2 - \frac{r}{a_0}\right)\left[2\left(2 - \frac{r}{a_0}\right) - \frac{2r}{a_0} - \frac{r}{a_0}\left(2 - \frac{r}{a_0}\right)\right]e^{-r/a_0} = 0$$

The left-hand side of this equation is the derivative requested. ■

(c) The roots of $dP/dr = 0$ at $r = 0$, $r = 2a_0$, and $r = \infty$ are minima, because $P(r) = 0$, as shown in textbook Figure 42.12. ■

Therefore, we focus instead on the roots given by

$$2\left(2 - \frac{r}{a_0}\right) - 2\frac{r}{a_0} - \frac{r}{a_0}\left(2 - \frac{r}{a_0}\right) = 4 - \frac{6r}{a_0} + \left(\frac{r}{a_0}\right)^2 = 0$$

(d) This equation has solutions $r = \left(3 \pm \sqrt{5}\right)a_0$. ■

We substitute two roots into $P(r)$:

When $r = \left(3 - \sqrt{5}\right)a_0 = 0.764a_0$ then $P(r) = \dfrac{0.0519}{a_0}$. This is the smaller peak in the textbook graph.

(e) When $r = \left(3 + \sqrt{5}\right)a_0 = 5.24a_0$ then $P(r) = \dfrac{0.191}{a_0}$. This is the higher peak in the textbook graph. The most probable value of r is $\left(3 + \sqrt{5}\right)a_0 = 5.24a_0$. ■

Finalize: Finding the most probable radial position told us the least probable radii as well. In particular, and as textbook Figure 42.12 shows, in this state the electron is exactly never at $r = 2a_0$. So then, how can it be sometimes in closer to the nucleus and most of the time farther away? The answer: The electron moves as a wave. It is in a standing-wave state with a nodal surface at this radius. Its motion is not that of a classical particle, moving through a succession of points one after another. In particular, the electron does not speed up to infinite speed to jump across the forbidden radius.

---

77. For hydrogen in the $1s$ state, what is the probability of finding the electron farther than $2.50a_0$ from the nucleus?

Solution

Conceptualize: From the red graph shown in textbook Figure 42.12, it appears that the probability of finding the electron beyond 2.5 a_0 may be about 20%.

Categorize: The precise probability can be found by integrating the 1*s* radial probability distribution function from $r = 2.50\ a_0$ to ∞.

Analyze: The general radial probability distribution function is

$P(r) = 4\pi r^2 |\psi|^2$. With $\psi_{1s} = \left(\pi a_0^{\ 3}\right)^{-1/2} e^{-r/a_0}$, it is $P(r) = 4r^2 a_0^{\ -3} e^{-2r/a_0}$

The required probability is then $P = \int_{2.50a_0}^{\infty} P(r)dr = \int_{2.50a_0}^{\infty} \frac{4r^2}{a_0^{\ 3}} e^{-2r/a_0} dr$

Let $z = 2r/a_0$ and $dz = 2dr/a_0$. Then we want $P = \frac{1}{2}\int_{5.00}^{\infty} z^2 e^{-z} dz$.

Performing this integration by parts, $P = -\frac{1}{2}\left(z^2 + 2z + 2\right)e^{-z}\Big|_{5.00}^{\infty}$

$$P = -\tfrac{1}{2}(0) + \tfrac{1}{2}(25.0 + 10.0 + 2.00)e^{-5.00} = \left(\tfrac{37}{2}\right)(0.006\ 74) = 0.125 \qquad \blacksquare$$

Finalize: The probability of 12.5% is less than the 20% we estimated, but close enough to be a reasonable result. As textbook Figure 42.12 suggests, the 1*s* probability density function peaks more sharply and closer to the origin than other states.

---

79. The positron is the antiparticle to the electron. It has the same mass and a positive electric charge of the same magnitude as that of the electron. Positronium is a hydrogen-like atom consisting of a positron and an electron revolving around each other. Using the Bohr model, find (a) the allowed distances between the two particles and (b) the allowed energies of the system.

Solution

Conceptualize: Since we are told that positronium is like hydrogen, we might expect the allowed radii and energy levels to be about the same as for hydrogen:

$$r = a_0 n^2 = \left(5.29 \times 10^{-11}\ \text{m}\right)n^2 \text{ and } E_n = (-13.6\ \text{eV})/n^2$$

Categorize: Imitating the textbook calculations for hydrogen, we can use the quantization of angular momentum of positronium to find the allowed radii and energy levels.

Analyze: Let r represent the distance between the electron and the positron. The two move in a circle of radius $r/2$ around their center of mass with opposite velocities. The total angular momentum is quantized according to $L_n = \frac{mvr}{2} + \frac{mvr}{2} = n\hbar$ where $n = 1, 2, 3, \ldots$.

For each particle, $\sum F = ma$ expands to $\dfrac{k_e e^2}{r^2} = \dfrac{mv^2}{r/2}$

We can eliminate $v = \dfrac{n\hbar}{mr}$ to find $\dfrac{k_e e^2}{r^2} = \dfrac{2mn^2\hbar^2}{m^2 r^3}$

(a) So the separation distances are

$$r = \frac{2n^2\hbar^2}{mk_e e^2} = 2a_0 n^2 = \left(1.06 \times 10^{-10}\ \text{m}\right)n^2 \quad \blacksquare$$

The orbital radii are $r/2 = a_0 n^2$, the same as for the electron in hydrogen.

(b) The energy can be calculated from $E = K + U = \frac{1}{2}mv^2 + \frac{1}{2}mv^2 - \dfrac{k_e e^2}{r}$

Since $mv^2 = \dfrac{k_e e^2}{2r}$, we have $E = \dfrac{k_e e^2}{2r} - \dfrac{k_e e^2}{r} = -\dfrac{k_e e^2}{2r} = \dfrac{-k_e e^2}{4a_0 n^2} = -\dfrac{6.80\ \text{eV}}{n^2}$ ■

Finalize: The allowed separations for positronium are twice as large as for hydrogen, while the energy levels are half as big in absolute value. One way to explain this is that, in a hydrogen atom, the proton is much more massive than the electron, so the proton remains nearly stationary with essentially no kinetic energy. On the other hand, in positronium the positron and electron have the same mass and therefore both have kinetic energy. Their motion separates them from each other and reduces the magnitude of their total energy compared with hydrogen.

---

81. (a) Use Bohr's model of the hydrogen atom to show that when the electron moves from the state n to the state $n - 1$, the frequency of the emitted light is

$$f = \left(\frac{2\pi^2 m_e k_e^{\,2} e^4}{h^3}\right)\frac{2n - 1}{n^2(n - 1)^2}$$

(b) Bohr's correspondence principle claims that quantum results should reduce to classical results in the limit of large quantum numbers. Show that as $n \to \infty$, this expression varies as $1/n^3$ and reduces to the classical frequency one expects the atom to emit. *Suggestion:* To calculate the classical frequency, note that the frequency of revolution is $v/2\pi r$, where v is the speed of the electron and r is given by Equation 42.10.

Solution

Conceptualize: "Correspondence" refers to a prediction of quantum theory agreeing with an expectation from classical physics, as we can expect if the quantum number is very large.

Categorize: We follow the directions given in the problem.

Analyze:

(a) Combining Equations 42.11 and 42.13, the energy of a state of the hydrogen atom is

$$E_n = -\frac{m_e k_e^2 e^4}{2n^2\hbar^2}$$

We use $\hbar = h/2\pi$ and identify the photon energy emitted in a transition between adjacent states as

$$hf = \Delta E = \frac{4\pi^2 m_e k_e^2 e^4}{2h^2}\left(\frac{1}{(n-1)^2} - \frac{1}{n^2}\right)$$

which reduces to

$$f = \frac{2\pi^2 m_e k_e^2 e^4}{h^3}\left(\frac{2n-1}{(n-1)^2 n^2}\right) \quad \blacksquare$$

(b) As $n \to \infty$, the "–1" terms lose importance and drop out, leaving the right-hand factor equal to

$$\lim_{n\to\infty}\left(\frac{2n-1}{(n-1)^2 n^2}\right) = \frac{2n}{n^4} = \frac{2}{n^3}$$

Thus the emission frequency becomes for n large $\quad f = \frac{4\pi^2 m_e k_e^2 e^4}{h^3}\left(\frac{1}{n^3}\right) \quad \blacksquare$

But in the Bohr model, $v^2 = \frac{k_e e^2}{m_e r}$ and $r = \frac{n^2 h^2}{4\pi^2 m_e k_e e^2}$

so the classical frequency is $\quad f = \frac{v}{2\pi r} = \frac{1}{2\pi r^{3/2}}\sqrt{\frac{k_e e^2}{m_e}}$

or $\quad f = \frac{1}{2\pi}\left(\sqrt{\frac{k_e e^2}{m_e}}\right)\left(\frac{4\pi^2 m_e k_e e^2}{n^2 h^2}\right)^{3/2}$

And this is identical to $\quad f = \frac{4\pi^2 m_e k_e^2 e^4}{h^3}\left(\frac{1}{n^3}\right) \quad \blacksquare$

Finalize: Experiments on macroscopic objects showed that an oscillating charge radiates electromagnetic waves at its own frequency. We have shown in the case of an electron orbiting a proton that this idea agrees with the quantum description of large orbits.

43

Molecules and Solids

EQUATIONS AND CONCEPTS

A **potential-energy function to model a molecule** must account for a force of repulsion between the atoms of the molecule at small distances of separation and attractive forces at larger separations. In Equation 43.1, the parameters A and B are associated with attractive and repulsive forces, respectively.

$$U(r) = -\frac{A}{r^n} + \frac{B}{r^m} \tag{43.1}$$

The **total energy of a diatomic molecule** is due to the combined effects of electronic, translational, rotational, and vibrational energy. Chapter 43 of your textbook considers energy levels related to rotational and vibrational motion; these are important in the interpretation of molecular spectra.

$$E = E_{el} + E_{trans} + E_{rot} + E_{vib}$$

A **diatomic molecule orientated along the *x* axis** as shown in the figure has two degrees of rotational freedom corresponding to rotation about the y and z axes.

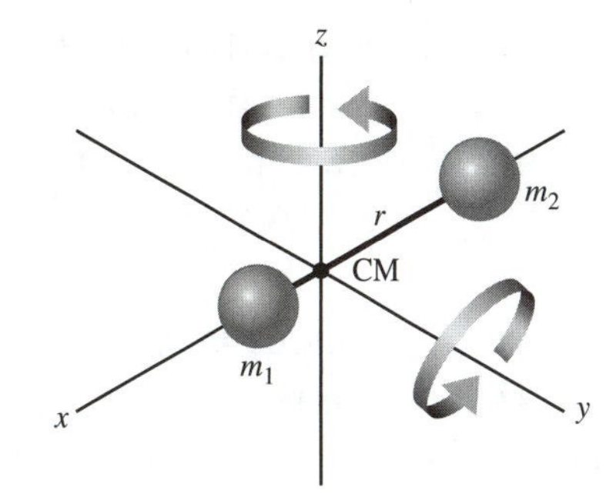

The **moment of inertia** (I) for a diatomic molecule (about an axis perpendicular to the axis of the molecule) can be written in terms of the reduced mass (μ) and the atomic separation distance (r).

$$I = \mu r^2 \tag{43.3}$$

$$\mu = \frac{m_1 m_2}{m_1 + m_2} \tag{43.4}$$

Allowed values of rotational angular momentum (and corresponding rotational frequencies) are determined by the rotational quantum number (J).

$$L = \sqrt{J(J+1)}\hbar \qquad J = 0, 1, 2 \ldots \tag{43.5}$$

The **rotational energy** of a molecule is quantized and depends on the value of the moment of inertia I.

$$E_{\text{rot}} = \frac{\hbar^2}{2I}J(J+1) \qquad (43.6)$$

$$J = 0, 1, 2 \ldots$$

The **energy of emitted and absorbed photons** is related to the energy separation between rotational levels and is governed by the selection rule for the rotational quantum number. In Equation 43.7, J refers to the *rotational quantum number of the higher energy state.* The selection rule for allowed rotational transitions is given by $\Delta J = \pm 1$.

$$E_{\text{photon}} = \frac{\hbar^2}{I}J = \frac{h^2}{4\pi^2 I}J \qquad (43.7)$$

$$J = 1, 2, 3, \ldots$$

A **diatomic molecule modeled as a simple harmonic oscillator**, as shown in the figure, has one vibrational degree of freedom. The atoms have mass values of m_1 and m_2; k is the effective spring constant.

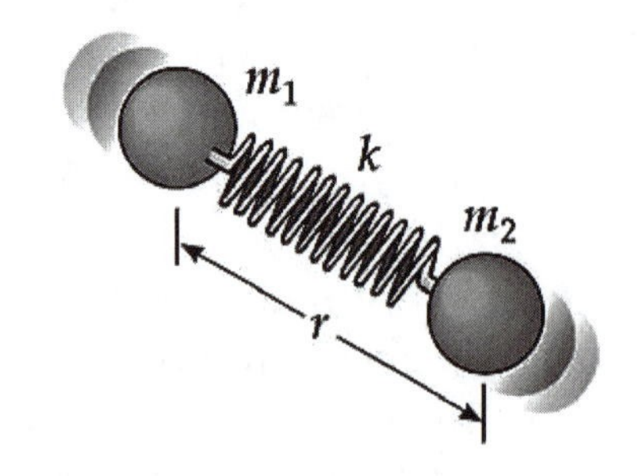

The **vibrational energy for a diatomic molecule** is quantized and is characterized by the vibrational quantum number (v). The selection rule for allowed vibrational transitions is given by $\Delta v = \pm 1$.

$$E_{\text{vib}} = \left(v + \tfrac{1}{2}\right)\frac{h}{2\pi}\sqrt{\frac{k}{\mu}} \qquad (43.10)$$

$$v = 0, 1, 2, \ldots$$

The **energy difference between successive vibrational levels** is hf, where f is the frequency of vibration. *The energy separation between successive vibrational levels is much larger than the energy difference between successive rotational levels.*

$$E_{\text{photon}} = \Delta E_{\text{vib}} = \frac{h}{2\pi}\sqrt{\frac{k}{\mu}} \qquad (43.11)$$

A **molecular absorption spectrum consists of two sets of lines**. When photon absorption by a molecule increases the vibrational quantum number v by one unit, the rotational quantum number J *can either increase or decrease by one unit.* In Equations 43.13 and 43.14, J refers to the quantum number of the *initial state.*

when $\Delta v = +1$ and $\Delta J = +1$ (43.13)

$$E_{\text{photon}} = hf + \frac{\hbar^2}{I}(J+1) \qquad J = 0, 1, 2, \ldots$$

when $\Delta v = +1$ and $\Delta J = -1$ (43.14)

$$E_{\text{photon}} = hf - \frac{\hbar^2}{I}J \qquad J = 1, 2, 3, \ldots$$

Refer to Figure 43.8 of your textbook.

The **ionic cohesive energy** U_0 of a solid is the minimum potential energy per ion pair of the system and occurs when the separation distance between ions (r) equals the equilibrium separation (r_0).

U_0 is the energy per ion pair necessary to separate the solid into a collection of positive and negative ions. The Madelung constant, α, is a dimensionless quantity characteristic of a specific crystal structure. The parameter m is a small integer.

$$U_0 = -\alpha k_e \frac{e^2}{r_0}\left(1 - \frac{1}{m}\right) \quad (43.18)$$

Ionic crystals:

- form stable, hard crystals
- are poor electrical conductors
- have high melting points
- are transparent to visible radiation
- absorb strongly in the infrared
- are soluble in polar liquids

The **Fermi-Dirac distribution function** $f(E)$ gives the probability of finding an electron in a particular energy state, E. The quantity E_F is called the Fermi energy (see Equation 43.25).

$$f(E) = \frac{1}{e^{(E-E_F)/k_BT} + 1} \quad (43.19)$$

Electron energy levels in metals are characterized by three quantum numbers. *In the ground state*, $n_x = n_y = n_z = 1$.

$$E = \frac{\hbar^2\pi^2}{2m_eL^2}\left(n_x^2 + n_y^2 + n_z^2\right) \quad (43.20)$$

The **number of allowed states per unit volume** that have energies between E and $E + dE$ is given by Equation 43.21. The function $g(E)$ is the **density-of-states function**.

$$g(E)dE = \frac{8\sqrt{2}\pi m_e^{3/2}}{h^3}E^{1/2}dE \quad (43.21)$$

The **number of electrons per unit volume that have energy between** E **and** $E + dE$ is the product of the number of allowed states per unit volume and the probability that a state is occupied.

$$N(E)dE = g(E)f(E)\,dE$$

$N(E)dE$ = number of electrons per unit volume with energy between E and $E + dE$.

$g(E)dE$ = number of allowed states per unit volume.

$f(E)$ = probability that a state is occupied.

The **total number of electrons per unit volume** n_e is the integral of $N(E)dE$ over all possible states.

$$n_e = \int_0^\infty N(E)dE$$

$$n_e = \frac{8\sqrt{2}\pi m_e^{3/2}}{h^3}\int_0^\infty \frac{E^{1/2}dE}{e^{(E-E_F)/k_BT} + 1} \quad (43.23)$$

The **Fermi energy**, E_F in the Fermi-Dirac distribution function in Equation 43.19, at 0 K is a function of the total number of electrons per unit volume (n_e).

$$E_F(0) = \frac{h^2}{2m_e}\left(\frac{3n_e}{8\pi}\right)^{2/3} \quad (43.25)$$

$$E_{\text{avg}} = \frac{3}{5}E_F \quad (43.26)$$

(average energy at 0 K)

The **current-voltage relationship for an ideal diode** depends on the direction of the applied voltage. For forward bias (when the p side of the junction is positive) the current increases exponentially with voltage; a reverse bias (the n side of the junction is positive) results in a small reverse current that reaches saturation value I_0. In the exponent of Equation 43.27, e is the electron charge and T is the absolute temperature. An example I vs. ΔV curve for a p-n junction is shown in the figure.

$$I = I_0\left(e^{e\Delta V/k_BT} - 1\right) \quad (43.27)$$

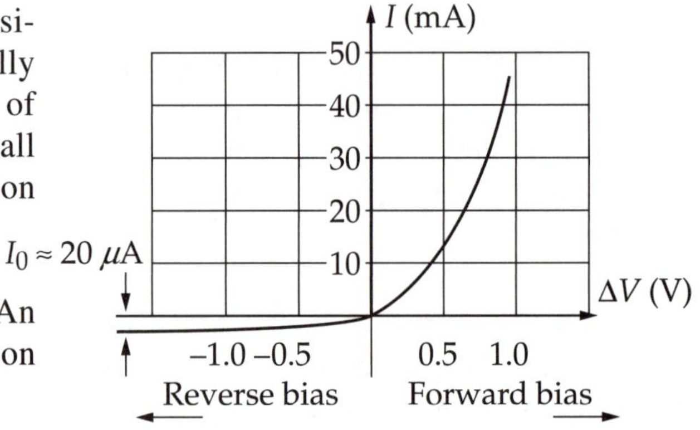

REVIEW CHECKLIST

You should be able to:

- Describe the essential bonding mechanisms involved in ionic, covalent, hydrogen, and van der Waals bonding. Given the potential energy function for a molecule, determine the value of r for which the energy is minimum and the value of E_0. (Section 43.1)
- Calculate the reduced mass and moment of inertia of a diatomic molecule about an axis through the center of mass. (Section 43.2)
- Calculate allowed values of rotational and vibrational energy levels in a diatomic molecule and determine energy differences between ground and excited states. (Section 43.2)
- Use the selection rules for rotational quantum number J and vibrational quantum number v to identify the possible absorptive transitions between $v = 0$ and $v = 1$. (Section 43.2)
- Describe the general properties of ionic crystals. Calculate the ionic cohesive energy of a diatomic molecule. (Section 43.3)
- Calculate the Fermi energy of metals at 0° C and the average energy of a free electron in a metal at 0° C. (Section 43.4)
- Use the band theory of solids as a basis for a qualitative discussion of the mechanisms for conduction in metals, insulators, and semiconductors. (Sections 43.5 and 43.6)

- Describe a *p-n* junction and the diffusion of electrons and holes through the junction. Discuss the fabrication and function of a junction diode and junction transistor. (Sections 43.6 and 43.7)

ANSWERS TO SELECTED OBJECTIVE QUESTIONS

1. Consider a typical material composed of covalently bonded diatomic molecules. Rank the following energies from the largest in magnitude to the smallest in magnitude. (a) the latent heat of fusion per molecule (b) the molecular binding energy (c) the energy of the first excited state of molecular rotation (d) the energy of the first excited state of molecular vibration.

Answer b > d > c > a. If you start with a solid sample and raise its temperature, it will typically melt first, then start emitting lots of far infrared light, then emit light with a spectrum peaking in the near infrared, and later have its molecules dissociate into atoms. Rotation of a diatomic molecule involves less energy than vibration. Absorption and emission of microwave photons, of frequency $\sim 10^{11}$ Hz, accompany excitation and de-excitation of rotational motion, while infrared photons, of frequency $\sim 10^{13}$ Hz, accompany changes in the vibration state of typical simple molecules.

□ □ □ □

5. As discussed in Chapter 27, the conductivity of metals decreases with increasing temperature due to electron collisions with vibrating atoms. In contrast, the conductivity of semiconductors increases with increasing temperature. What property of a semiconductor is responsible for this behavior? (a) Atomic vibrations decrease as temperature increases. (b) The number of conduction electrons and the number of holes increase steeply with increasing temperature. (c) The energy gap decreases with increasing temperature. (d) Electrons do not collide with atoms in a semiconductor.

Answer (b). First consider electric conduction in a metal. The number of conduction electrons is essentially fixed. They conduct electricity by having drift motion in an applied electric field superposed on their random thermal motion. At higher temperature, the ion cores vibrate more and are more efficient at scattering the conduction electrons flying among them. The mean time between collisions is reduced. The electrons have time to develop only a lower drift speed. The electric current is reduced, so we see the resistivity increasing with temperature.

Now consider an intrinsic semiconductor. At absolute zero its valence band is full and its conduction band is empty. It is an insulator, with very high resistivity. As the temperature increases, more electrons are promoted to the conduction band, leaving holes in the valence band. Then both electrons and holes move in response to an applied electric field. Thus we see the resistivity decreasing as temperature goes up.

□ □ □ □

ANSWER TO A CONCEPTUAL QUESTION

9. The energies of photons of visible light range between the approximate values 1.8 and 3.1 eV. Explain why silicon, with an energy gap of 1.14 eV at room temperature (see Table 43.3), appears opaque, whereas diamond, with an energy gap of 5.47 eV, appears transparent.

Answer Silicon absorbs photons of visible light, because the photons contain more than enough energy to promote an electron across the energy gap between the valence band and the conduction band. Diamond cannot absorb photons of visible light, because the photons do not have enough energy to raise the energy of electrons in diamond to make them cross the energy gap.

SOLUTIONS TO SELECTED PROBLEMS

2. Review. A K^+ ion and a Cl^- ion are separated by a distance of 5.00×10^{-10} m. Assuming the two ions act like charged particles, determine (a) the force each ion exerts on the other and (b) the potential energy of the two-ion system in electron volts.

Solution

Conceptualize: The force will be a tiny fraction of a newton; we might estimate it as a few eV divided by 0.5 nm, which is several 10^{-10} N. The energy will be a few eV.

Categorize: This problem applies the ideas of Chapters 23 and 25 to a molecular system.

Analyze:

(a) The force on each ion is directed towards the other:

$$F = \frac{k_e|q_1||q_2|}{r^2} = \frac{\left(8.99 \times 10^9 \text{ N}\cdot\text{m}^2/\text{C}^2\right)\left(1.60 \times 10^{-19} \text{ C}\right)^2}{\left(5.00 \times 10^{-10} \text{ m}\right)^2} = 9.21 \times 10^{-10} \text{ N} \quad \blacksquare$$

(b) The potential energy of the ion pair is

$$U = \frac{k_e q_1 q_2}{r} = \frac{\left(8.99 \times 10^9 \text{ N}\cdot\text{m}^2/\text{C}^2\right)\left(1.60 \times 10^{-19} \text{ C}\right)\left(-1.60 \times 10^{-19} \text{ C}\right)}{5.00 \times 10^{-10} \text{ m}}$$

$$U = -4.60 \times 10^{-19} \text{ J} = -2.88 \text{ eV} \quad \blacksquare$$

Finalize: Our estimates were good. Their energy is negative, which means that positive work must be put in to pull the ions apart to infinite separation.

9. An HCl molecule is excited to its second rotational energy level, corresponding to $J = 2$. If the distance between its nuclei is 0.127 5 nm, what is the angular speed of the molecule about its center of mass?

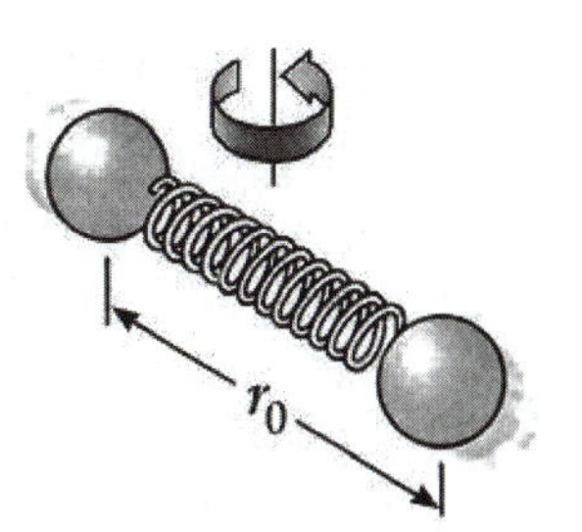

Solution

Conceptualize: For a system as small as a molecule, we can expect the angular speed to be some huge number of rad/s, much faster than typical everyday objects we encounter.

Categorize: The rotational energy is given by the angular momentum quantum number J. The angular speed can be calculated from the kinetic rotational energy and the moment of inertia of this one-dimensional molecule.

Analyze: For the HCl molecule in the $J = 2$ rotational energy level, we are given $r_0 =$ 0.127 5 nm. The rotational kinetic energy is

$$E_{rot} = \frac{\hbar^2}{2I}J(J+1)$$

With $J = 2$, $E_{rot} = \frac{3\hbar^2}{I} = \frac{1}{2}I\omega^2$ and $\omega = \sqrt{\frac{6\hbar^2}{I^2}} = \sqrt{6}\frac{\hbar}{I}$

Common isotopes of chlorine have mass numbers of 35 and 37. Here we will use the average weighted according to abundance. The moment of inertia of the molecule is given by

$$I = \mu r_0^2 = \frac{m_1 m_2}{m_1 + m_2}r_0^2 = \frac{(1.01\text{ u})(35.5\text{ u})}{1.01\text{ u} + 35.5\text{ u}}r_0^2 = (0.982\text{ u})r_0^2$$

$$I = (0.982\text{ u})(1.66 \times 10^{-27}\text{ kg/u})(1.275 \times 10^{-10}\text{ m})^2 = 2.65 \times 10^{-47}\text{ kg} \cdot \text{m}^2$$

Therefore, $\omega = \sqrt{6}\frac{\hbar}{I} = \sqrt{6}\left(\frac{1.055 \times 10^{-34}\text{ J} \cdot \text{s}}{2.65 \times 10^{-47}\text{ kg} \cdot \text{m}^2}\right) = 9.75 \times 10^{12}\text{ rad/s}$ ■

Finalize: This angular speed is hundreds of billions of times faster than the spin rate of a music CD (about 200 to 500 revolutions per minute, or $\omega \approx 20$ rad/s to 50 rad/s).

---

13. The effective spring constant describing the potential energy of the HI molecule is 320 N/m, and that for the HF molecule is 970 N/m. Calculate the minimum amplitude of vibration for (a) the HI molecule and (b) the HF molecule.

Solution

Conceptualize: A classical block on a spring can have any energy and any amplitude. A quantum oscillator's energy is quantized, with the steps give by Planck's constant times the frequency. If the frequency is on the order of 10^{13} Hz, the energy of the lowest state will be on the order of 0.1 eV. The oscillation amplitude in the lowest-energy state may be on the order of a picometer. The hydrogen iodide molecule has a weaker spring and greater mass, so its frequency should be significantly less than that of the hydrogen fluoride. We may then expect its ground-state energy and amplitude to be smaller.

Categorize: We will use classical ideas, expressed by equations such as $f = \frac{1}{2\pi}\sqrt{\frac{k}{m}}$ for a simple harmonic oscillator's frequency, which is the same for all quantum states, and $E = \frac{1}{2}kA^2$ for the energy. And we will neatly combine them with the quantum idea of energy quantization, expressed for the oscillator as $E = \left(v + \frac{1}{2}\right)hf$ with $v = 0$ for the ground state, which has minimum vibration amplitude.

Analyze: The mass to consider is the molecule's reduced mass. Iodine has atomic mass 126.90 u and a hydrogen atom is 1.007 9 u, so the reduced mass of HI is

$$\mu = m_1 m_2/(m_1 + m_2) = 127.90\text{ u}/127.91 = 0.999\,96\text{ u}$$

Now for the energy of the ground state we have

$$E = \tfrac{1}{2}kA^2 = \left(0 + \tfrac{1}{2}\right)hf = \frac{1}{2}\frac{h}{2\pi}\sqrt{\frac{k}{\mu}}$$

So the amplitude is $A = \sqrt{\frac{h}{2\pi k}\sqrt{\frac{k}{\mu}}} = \left(\frac{h}{2\pi}\right)^{1/2}\left(\frac{1}{k\mu}\right)^{1/4}$

(a) For HI we have

$$A = \left(\frac{6.626\times10^{-34}\text{ J}\cdot\text{s}}{2\pi}\right)^{1/2}\left(\frac{1}{(320\text{ N/m})(0.999\,96)(1.66\times10^{-27}\text{kg})}\right)^{1/4}$$
$$= 1.20\times10^{-11}\text{m}$$ ■

(b) For HF, $\mu = m_1 m_2/(m_1 + m_2) = 19.148\text{ u}/20.006 = 0.957\,12\text{ u}$

and $$A = \left(\frac{6.626\times10^{-34}\text{ J}\cdot\text{s}}{2\pi}\right)^{1/2}\left(\frac{1}{(970\text{ N/m})(0.957\,12)(1.66\times10^{-27}\text{kg})}\right)^{1/4}$$
$$= 9.22\times10^{-12}\text{m}$$ ■

Finalize: The amplitudes are on the order of 10 pm, not 1 pm. Hydrogen fluoride has the smaller minimum amplitude, but the k in $E = \frac{1}{2}kA^2$ makes its minimum energy larger than for hydrogen iodide, as we expected. An alternative chain of logic is to say that the proportionality of A to $k^{-1/4}$ was unexpected in detail, but makes A smaller when the stiffness constant gets larger. The combination of units looks unusual, so it is worthwhile to work out:

$$\left(\frac{\text{kg}\cdot\text{m}^2\cdot\text{s}}{\text{s}^2}\right)^{1/2}\left(\frac{1}{(\text{kg}\cdot\text{m/s}^2\cdot\text{m})(\text{kg})}\right)^{1/4} = \frac{\text{kg}^{1/2}\text{m}}{\text{s}^{1/2}}\frac{\text{s}^{2/4}}{\text{kg}^{2/4}} = \text{m}$$

26. Consider a one-dimensional chain of alternating positive and negative ions. Show that the potential energy associated with one of the ions and its interactions with the rest of this hypothetical crystal is

$$U(r) = -k_e \alpha \frac{e^2}{r}$$

where the Madelung constant is $\alpha = 2\ln 2$ and r is the distance between ions. *Suggestion:* Use the series expansion for $\ln(1 + x)$.

Solution

Conceptualize: A crystal has so many atoms that the environment of each atom can be modeled as going off to infinity. But the potential energy of one atom is only finite. A finite amount of energy is required to remove the atom from a crystal. We calculate the finite sum of such an infinite series in this problem.

Categorize: We use $U = qV = k_e q_1 q_2 / r$ for the energy associated with one ion pair, and add up the energy of one ion by including its interactions with all the others.

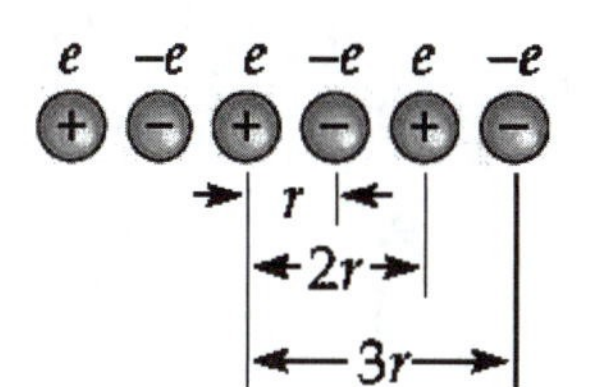

Analyze: We assume the ions are all singly ionized. The total potential energy is obtained by summing over all interactions of our ion with others:

$$U = \sum_{i \neq j} k_e \frac{q_i q_j}{r_{ij}} = -k_e \left[\frac{e^2}{r} + \frac{e^2}{r} - \frac{e^2}{2r} - \frac{e^2}{2r} + \frac{e^2}{3r} + \frac{e^2}{3r} - \frac{e^2}{4r} - \frac{e^2}{4r} + \cdots \right]$$

$$U = -2k_e \frac{e^2}{r} \left[1 - \tfrac{1}{2} + \tfrac{1}{3} - \tfrac{1}{4} + \cdots \right]$$

But from Appendix B.5,

$$\ln(1 + x) = x - \frac{x^2}{2} + \frac{x^3}{3} - \frac{x^4}{4} + \cdots$$

Our series follows this pattern with $x = 1$, so the potential energy of one ion due to its interactions with all the others is

$$U = (-2\ln 2) k_e \frac{e^2}{r} = -\alpha k_e \frac{e^2}{r} \qquad \blacksquare$$

Finalize: Here the Madelung constant is $2\ln 2 = 1.39$. A three-dimensional crystal structure with positive and negative ions stacked in some particular lattice would have a Madelung constant with some different value, but the energy of each ion would still be proportional to $k_e e^2/r$, where r is a distance describing the spacing of lattice sites.

---

27. Sodium is a monovalent metal having a density of 0.971 g/cm^3 and a molar mass of 23.0 g/mol. Use this information to calculate (a) the density of charge carriers and (b) the Fermi energy of sodium.

Solution

Conceptualize: "Monovalent" means that each atom contributes one electron to the conduction band. Then the density of charge carriers in space will be the same as the number density of atoms, rather larger than 10^{23} per cubic meter, from Avogadro's number. The Fermi energy should be a few electronvolts.

Categorize: The density tells us the mass of one cubic meter of sodium. Then the molar mass can tell us the number of moles in a cubic meter, and Avogadro's number reveals the number of atoms. Because each sodium atom contributes just one electron to the conduction band, this same number will be the density of charge carriers. Then Equation 43.25 will reveal the Fermi energy at temperature 0 K. The Fermi energy at room temperature is similar.

Analyze: (a) The mass per cubic meter is the density. The number of moles per cubic meter is ρ/M. The number of atoms per cubic meter is $\rho N_A/M$ and the density of conduction electrons is

$$n_e = \frac{\rho N_A}{M} = \frac{(0.971\text{ g/cm}^3)(6.02\times 10^{23}\text{ molecules/mole})}{23.0\text{ g/mole}}\frac{1\text{ atom}}{1\text{ molecule}}\frac{1\text{ conduction electron}}{1\text{ atom}}$$

$$= 2.54\times 10^{22}\,e/\text{cm}^3 = 2.54\times 10^{28}\,e/\text{m}^3 \quad \blacksquare$$

(b) $$E_F = \left(\frac{h^2}{2m_e}\right)\left(\frac{3n_e}{8\pi}\right)^{2/3} = \frac{(6.626\times 10^{-34}\text{ J}\cdot\text{s})^2}{2(9.11\times 10^{-31}\text{ kg})}\left(\frac{3(2.54\times 10^{28}\text{ / m}^3)}{8\pi}\right)^{2/3}$$

$$= 5.05\times 10^{-19}\text{ J}\frac{1\text{ eV}}{1.60\times 10^{-19}\text{ J}} = 3.15\text{ eV} \quad \blacksquare$$

Finalize: Notice that the Fermi energy is much larger than the energy k_BT characteristic of random thermal vibration. At room temperature (around 300 K), k_BT is about one-fortieth of one electronvolt. You can roughly think of the buzzing swarm of numerous, energetic conduction electrons as responsible for high electrical conductivity, high thermal conductivity, optical reflectivity, and other typical properties of metals.

---

33. Calculate the energy of a conduction electron in silver at 800 K, assuming the probability of finding an electron in that state is 0.950. The Fermi energy of silver is 5.48 eV at this temperature.

Solution

Conceptualize: Since there is a 95% probability of finding the electron in this state, its energy should be slightly less than the Fermi energy, as indicated by the graph in Figure 43.15.

Categorize: The electron energy can be found from the Fermi-Dirac distribution function.

Analyze: Taking $E_F = 5.48$ eV for silver at 800 K, and given $f(E) = 0.950$, we find

$$f(E) = \frac{1}{e^{(E-E_F)/k_BT} + 1} \qquad \text{so} \qquad e^{(E-E_F)/k_BT} = \frac{1}{f(E)} - 1$$

Then $$\frac{E - E_F}{k_BT} = \ln\left(\frac{1}{f(E)} - 1\right) \qquad \text{and} \qquad E = E_F + k_BT\ln\left(\frac{1}{f(E)} - 1\right)$$

$$E = 5.48 \text{ eV} + \left(1.38\times10^{-23}\,\frac{\text{J}}{\text{K}}\,800 \text{ K}\right)\ln\left(\frac{1}{0.950} - 1\right) = 5.28 \text{ eV}$$ ■

Finalize: As expected, the energy of the electron is slightly less than the Fermi energy, which is around 5 eV for most metals. There is just a 50% probability of an electron occupying a state at the Fermi energy, and a very low probability of finding an electron in an energy state well above the Fermi energy in a metal.

---

34. (a) Consider a system of electrons confined to a three-dimensional box. Calculate the ratio of the number of allowed energy levels at 8.50 eV to the number at 7.05 eV. (b) **What If?** Copper has a Fermi energy of 7.05 eV at 300 K. Calculate the ratio of the number of occupied levels in copper at an energy of 8.50 eV to the number at the Fermi energy. (c) How does your answer to part (b) compare with that obtained in part (a)?

Solution

Conceptualize: The gas of electrons we consider is the model used by the free-electron theory of metals. That theory suggests that considerably more electron states exist at the higher energy than at the Fermi energy, but that only a few of these states are occupied by electrons at room temperature.

Categorize: We must evaluate ratios based on the density of states function and the Fermi-Dirac distribution function, given in the text.

Analyze: The density of states at the energy E is $g(E) = CE^{1/2}$

(a) Hence, the required ratio is

$$R_{states} = \frac{g(8.50 \text{ eV})}{g(7.05 \text{ eV})} = \frac{C(8.50)^{1/2}}{C(7.05)^{1/2}} = 1.10$$ ■

(b) From Equation (43.22) we see that the number of occupied states between energy E and energy $E + dE$ is

$$N(E)dE = \frac{CE^{1/2}}{e^{(E-E_F)/k_BT} + 1}dE$$

Hence, the required ratio is

$$R_{occupied\ states} = \frac{N(8.50 \text{ eV})}{N(7.05 \text{ eV})} = \sqrt{\frac{8.50}{7.05}}\left[\frac{e^{(7.05-7.05)/k_BT} + 1}{e^{(8.50-7.05)/k_BT} + 1}\right]$$

At $T = 300$ K, we compute

$$k_B T = \left(1.380\ 65\times10^{-23}\ \frac{\text{J}}{\text{K}}\right)300.000\ \text{K}\left(\frac{1\ \text{eV}}{1.602\ 18\times10^{-19}\text{J}}\right) = 0.025\ 852\ 0\ \text{eV}$$

so $$R_{occ.st} = \left(\frac{8.50}{7.05}\right)^{1/2}\left(\frac{2}{e^{1.45/0.025\ 852\ 0}+1}\right) = 9.61\times10^{-25}$$ ■

With an exponent of 56.1, the derivative of the exponential function is so large that none of the digits in 9.61 is really significant. Different-looking answers would result from different choices of how precisely to represent the input data.

(c) The answer to part (b) is vastly smaller than the answer to (a). Very few states well above the Fermi energy are occupied at room temperature. ■

Finalize: In pictorial terms, a continuation of the sideways parabola in the text's Figure 43.16a shows the somewhat (not a lot) larger number of available electron states at the higher energy. The red line in Figure 43.16b, exponentially tailing down to very low values at high energy, shows how little temperature has affected the distribution of electrons among the states.

---

39. Show that the average kinetic energy of a conduction electron in a metal at 0 K is $E_{\text{avg}} = \frac{3}{5}E_{\text{F}}$. *Suggestion:* In general, the average kinetic energy is

$$E_{\text{avg}} = \frac{1}{n_e}\int EN(E)dE$$

where n_e is the density of particles, $N(E)\,dE$ is given by Equation 43.22, and the integral is over all possible values of the energy.

Solution

Conceptualize: In the textbook's Figure 43.16a, showing the distribution function as a function of energy, we are finding the horizontal coordinate of the center of mass. Because $N(E)$ slopes upward, the average energy is greater than half the Fermi energy.

Categorize: The problem tells us just what to do.

Analyze: We are to compute $E_{\text{av}} = \dfrac{1}{n_e}\displaystyle\int_0^\infty E\ N(E)\,dE$

where from Equation 43.22 $N(E) = \dfrac{CE^{1/2}}{e^{(E-E_F)/k_BT}+1} = Cf(E)\,E^{1/2}$

with $C = \dfrac{8\sqrt{2}\pi m_e^{3/2}}{h^3}$

But at $T = 0$ the Fermi-Dirac distribution function is $f(E) = 0$ for $E > E_F$

And $f(E) = 1$ for $E < E_F$

So we can take $N(E) = CE^{1/2}$ just for energies up to the Fermi energy. The average we want is then

$$E_{\text{av}} = \frac{1}{n_e}\int_0^{E_F} CE^{3/2}dE = \frac{2C}{5n_e}E_F^{\ 5/2}$$

But from Equation 43.24, $\dfrac{C}{n_e} = \frac{3}{2}E_F^{\ -3/2}$ so $E_{\text{av}} = \left(\frac{2}{5}\right)\left(\frac{3}{2}\right)\left(E_F^{\ -3/2}\right)E_F^{\ 5/2} = \frac{3}{5}E_F$ ■

Finalize: We have carried through the proof cited in the text at Equation 43.26. The average energy of a conduction electron in a solid metal is determined to be a few electron volts, with the value set by the spatial density of the conduction electrons.

---

44. Most solar radiation has a wavelength of 1 μm or less. (a) What energy gap should the material in a solar cell have if it is to absorb this radiation? (b) Is silicon an appropriate solar cell material (see Table 43.3)? Explain your answer.

Solution

Conceptualize: Since most photovoltaic solar cells are made of silicon, we can expect to demonstrate that this semiconductor is an appropriate material for these devices.

Categorize: For the material to absorb the longest-wavelength photons, its energy gap should be no larger than the photon energy.

Analyze: (a) The minimum energy of most of the photons in sunlight is

$$hf = \frac{hc}{\lambda} = \frac{\left(6.63\times10^{-34}\ \text{J}\cdot\text{s}\right)\left(3.00\times10^{8}\ \text{m/s}\right)}{10^{-6}\ \text{m}}\left(\frac{1\ \text{eV}}{1.60\times10^{-19}\ \text{J}}\right) = 1.24\ \text{eV}$$

Therefore, the energy gap in the absorbing material should be smaller than or equal to 1.24 eV. ■

(b) Silicon's energy gap of 1.14 eV means that it can absorb energy from nearly all of the photons in sunlight, and is an appropriate material for a solar energy collector. ■

Finalize: For heating, absorbing the light is enough. For generating electricity, silicon has low efficiency if it uses only the first 1.14 eV of the energy of each photon. Photovoltaic cells with layers of different materials to absorb different "colors," notably including infrared and near ultraviolet, are currently being developed. ■

---

48. A diode is at room temperature so that $k_BT = 0.025\,0$ eV. Taking the applied voltages across the diode to be +1.00 V (under forward bias) and −1.00 V (under reverse bias), calculate the ratio of the forward current to the reverse current if the diode is described by Equation 43.27.

Solution

Conceptualize: A diode can be thought of roughly as a one-way valve. We expect the forward current to be much larger than the backward current, so the ratio should be some large number. For the diode to be very useful in a circuit, the ratio might be 10^9 or larger, to be comparable to the ratio of the current in a copper wire to the current in a similarly-shaped insulator.

Categorize: The problem tells us the equation describing the current in an ideal diode. We need not calculate two explicit values for current to take the ratio.

Analyze: We are told to use the equation $I = I_0(e^{e\Delta V/k_BT} - 1)$ for each of the currents, caused by positive and negative potential differences of equal size. The ratio we are to evaluate is

$$\frac{I_0(e^{e(1\text{ V})/0.025\text{ eV}} - 1)}{I_0(e^{e(-1\text{ V})/0.025\text{ eV}} - 1)} = \frac{e^{40} - 1}{e^{-40} - 1} = \frac{2.35\times10^{17} - 1}{4.25\times10^{-18} - 1} = -2.35\times10^{17} \quad \blacksquare$$

The saturation current I_0 divides out; it would depend on the length and diameter of the diode, among other factors. The e in $e\Delta V$ means the same thing as the e in the unit eV, as both represent the magnitude of the electron charge.

Finalize: The negative sign in the ratio means, of course, that the reverse current is opposite the forward current in direction. The huge size of the ratio means that the diode works very well at stopping current in the reverse direction and allowing it in the forward direction.

---

51. A superconducting ring of niobium metal 2.00 cm in diameter is immersed in a uniform 0.020 0-T magnetic field directed perpendicular to the ring and carries no current. Determine the current generated in the ring when the magnetic field is suddenly decreased to zero. The inductance of the ring is 3.10×10^{-8} H.

Solution

Conceptualize: The resistance of a superconductor is zero, so the current is limited only by the change in magnetic flux and self-inductance. Therefore, unusually large currents (greater than 100 A) are possible.

Categorize: The change in magnetic field through the ring will induce an emf according to Faraday's law of induction. Since we do not know how fast the magnetic field is changing, we must use the ring's inductance and the geometry of the ring to calculate the magnetic flux, which can then be used to find the current.

Analyze: From Faraday's law (Eq. 31.1), we have

$$|\mathcal{E}| = \frac{|\Delta\Phi_B|}{\Delta t} = A\frac{|\Delta B|}{\Delta t} = L\frac{|\Delta I|}{\Delta t} \quad \text{or} \quad |\Delta I| = \frac{A|\Delta B|}{L} = \frac{\pi(0.010\,0\text{ m})^2(0.020\,0\text{ T})}{3.10\times10^{-8}\text{ H}} = 203\text{ A} \quad \blacksquare$$

The current is directed so as to produce its own magnetic field in the direction of the original field.

Finalize: This induced current should remain constant as long as the ring is superconducting. If the ring failed to be a superconductor (e.g., if it warmed above the critical temperature), the metal would have a non-zero resistance, and the current would quickly drop to zero. It is interesting to note that we were able to calculate the current in the ring without knowing the emf. In order to calculate the emf, we would need to know how quickly the magnetic field goes to zero.

---

59. Starting with Equation 43.17, show that the ionic cohesive energy of an ionically bonded solid is given by Equation 43.18.

Solution

Conceptualize: Equation 43.17 shows a model potential energy function for an ion in a crystal. It includes the effect of attraction to oppositely-charged neighbors, partially offset by repulsion of ions with like charges that are on the average more distant. And it includes the effect of strong short-range repulsion that keeps ions from interpenetrating with one another. The electrostatic term, proportional to r^{-1}, is long-range and shows no saturation—that is, the attraction or repulsion of a particular pair of ions is not affected by the presence of other charges around or between them. The exclusion-principle repulsion force is short-range and does show saturation. Its contribution to the energy is modeled as proportional to r^{-m} with $m > 1$. This force tends to keep the electron cloud of one ion from interpenetrating with its immediate neighbors. It is reasonable to expect that a particular equilibrium separation distance will exist.

Categorize: The total potential energy is given by Equation 43.17:

$$U_{\text{total}} = -\alpha k_e \frac{e^2}{r} + \frac{B}{r^m}$$

The potential energy will have its minimum value U_0 when $r = r_0$, where r_0 is the equilibrium spacing. We can find U_0 by finding r_0, knowing that…

Analyze: … at this point, the slope of the curve U versus r is zero.

That is, we require $\left.\dfrac{dU}{dr}\right|_{r=r_0} = 0$

with $$\frac{dU}{dr} = \frac{d}{dr}\left(-\alpha k_e \frac{e^2}{r} + \frac{B}{r^m}\right) = \alpha k_e \frac{e^2}{r^2} - \frac{mB}{r^{m+1}}$$

Setting this equal to zero with $r = r_0$, we have $$\alpha k_e \frac{e^2}{r_0^{\,2}} - \frac{mB}{r_0^{\,m+1}} = 0$$

Then we can evaluate $$B = \alpha \frac{k_e e^2}{m} r_0^{\,m-1}$$

Substituting this expression for B into U_{total} gives

$$U_0 = -\alpha k_e \frac{e^2}{r_0} + \alpha \frac{k_e e^2}{m} r_0^{m-1} \left(\frac{1}{r_0^m} \right) = -\alpha \frac{k_e e^2}{r_0} \left(1 - \frac{1}{m} \right)$$ ■

Finalize: We have successfully derived Equation 43.18. This ionic cohesive energy can be measured by seeing how much energy is required to pull an ion out of the crystal. The interionic distance r_0 is known from the density. The Madelung constant α can be computed for the crystal structure. Then the theory can be tested. We can see if m has a reasonable and constant value.

44

Nuclear Structure

EQUATIONS AND CONCEPTS

The **nuclear radius** is proportional to the cube root of the mass number (the total number of nucleons). *All nuclei have nearly the same density and can be modeled as tightly packed spheres.* The Fermi (fm) is a convenient unit in which to express nuclear dimensions.

$$r = aA^{1/3}$$
$$a = 1.2\ \text{fm} \quad (44.1)$$
$$1\ \text{fm} = 10^{-15}\ \text{m}$$

The **binding energy** of a nucleus equals the combined energy of the separated nucleons (protons and neutrons) minus the total energy of the bound system. *The total mass of a composite nucleus is less than the sum of the masses of the individual nucleons.*

$$E_b(\text{MeV}) = \left[ZM(\text{H}) + Nm_n - M(^A_Z\text{X})\right] \times 931.494\ \text{MeV/u} \quad (44.2)$$

m_n = neutron mass

$M(\text{H})$ = mass of neutral hydrogen atom

$M(^A_Z\text{X})$ = atomic mass of atom

The **semiempirical binding-energy formula** is based on the liquid drop model of the nucleus. C_1, C_2, C_3, and C_4 are constants related to volume, surface, Coulomb repulsion, and symmetry effects; they are adjusted to fit experimental data.

$$E_b = C_1 A - C_2 A^{2/3} - C_3 \frac{Z(Z-1)}{A^{1/3}} - C_4 \frac{(N-Z)^2}{A} \quad (44.3)$$

The **shell model** of the nucleus accounts for: (i) stability of nuclei with even values of A and (ii) peaks in the binding energy curve when Z or N equals one of the magic numbers.

Magic numbers of the shell model:

Z or N = 2, 8, 20, 28, 50, 82 (44.4)

The **exponential decay** law is illustrated in the graph of N versus t shown at right. The decay constant λ is characteristic of a specific isotope and is the probability of decay per nucleus per second. N_0 is the number of radioactive nuclei present at time $t = 0$. Time is shown in units of half-life ($T_{1/2}$). See Equation 44.8 below.

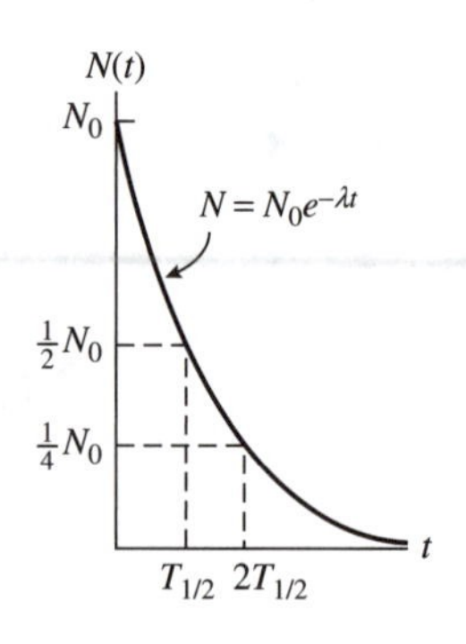

$$N = N_0 e^{-\lambda t} \tag{44.6}$$

The **decay rate** R (or **activity**) of a sample of radioactive nuclei is defined as the number of decays per second. R_0 is the decay rate at $t = 0$. *Note that both R and N decrease exponentially with time.*

$$R = \left|\frac{dN}{dt}\right| = \lambda N = \lambda N_0 e^{-\lambda t} = R_0 e^{-\lambda t} \tag{44.7}$$

Units of activity are the becquerel (Bq) and the curie (Ci). *The becquerel is the SI unit and the curie is defined as the activity of* 1 *gram of radium.*

$$1 \text{ Ci} \equiv 3.7 \times 10^{10} \text{ decays/s}$$

$$1 \text{ Bq} \equiv 1 \text{ decay/s}$$

The **half-life** of a radioactive substance ($T_{1/2}$) is the time interval required for half of a given number of radioactive nuclei in a sample of the substance to decay. *After an elapsed time of n half-lives,* where n can be an integer or a noninteger, *the number of radioactive nuclei remaining* can be found by using Equation 44.9.

$$T_{1/2} = \frac{\ln 2}{\lambda} = \frac{0.693}{\lambda} \tag{44.8}$$

$$N = N_0 \left(\tfrac{1}{2}\right)^n \tag{44.9}$$

Spontaneous decay of radioactive nuclei proceeds by one of the following process: alpha decay, beta decay, electron capture, or gamma decay. The radioactive decay process can be represented in the form of an equation. See Summary of Characteristics of Alpha, Beta, and Gamma Decay in *Suggestions, Skills, and Strategies. For any decay process, the sum of the mass numbers on each side of the equation must be equal and the sum of the atomic numbers on each side of the equation must also be equal.*

$$\underset{\text{Parent nucleus}}{\text{X}} \rightarrow \underset{\text{Daughter nucleus}}{\text{Y}} + \text{Emitted Radiation}$$

Alpha Decay

When a nucleus decays by alpha emission, the parent nucleus loses two neutrons and two protons. For alpha emission to occur, the mass of the parent nucleus $\left({}^{A}_{Z}\mathrm{X}\right)$ must be greater than the combined masses of the daughter nucleus $\left({}^{A-4}_{Z-2}\mathrm{Y}\right)$ and the emitted alpha particle. Two examples of alpha decay are shown.

$$ {}^{A}_{Z}\mathrm{X} \rightarrow {}^{A-4}_{Z-2}\mathrm{Y} + {}^{4}_{2}\mathrm{He} \qquad (44.10) $$

$$ {}^{238}_{92}\mathrm{U} \rightarrow {}^{234}_{90}\mathrm{Th} + {}^{4}_{2}\mathrm{He} \qquad (44.11) $$

$$ {}^{226}_{88}\mathrm{Ra} \rightarrow {}^{222}_{86}\mathrm{Rn} + {}^{4}_{2}\mathrm{He} \qquad (44.12) $$

The **disintegration energy** (Q) of the system (parent nucleus, daughter nucleus, and alpha particle) is required in order to conserve energy during spontaneous decay of the isolated parent nucleus. Q appears in the form of kinetic energy of the daughter nucleus and the alpha particle. *In Equation 44.14, mass values are in atomic mass units.*

$$ Q = \left(M_{\mathrm{X}} - M_{\mathrm{Y}} - M_{\alpha}\right) \times 931.494 \text{ MeV/u} \qquad (44.14) $$

Beta Decay

When a radioactive nucleus undergoes beta decay, the process is accompanied by a third particle required to conserve energy and momentum. Two modes of beta decay are: (1) Electron emission (e^-) with antineutrino $(\bar{\nu})$ and (2) positron emission (e^+) with neutrino (ν). For each mode of decay, the equations at right show: (i) The equation representing the process, (ii) an example of the decay mode, and (iii) an equation describing the origin of the electron or positron.

Electron Emission

(i) $$ {}^{A}_{Z}\mathrm{X} \rightarrow {}^{A}_{Z+1}\mathrm{Y} + \mathrm{e}^- + \bar{\nu} \qquad (44.19) $$

(ii) $$ {}^{14}_{6}\mathrm{C} \rightarrow {}^{14}_{7}\mathrm{N} + \mathrm{e}^- + \bar{\nu} \qquad (44.21) $$

(iii) $$ \mathrm{n} \rightarrow \mathrm{p} + \mathrm{e}^- + \bar{\nu} \qquad (44.23) $$

Positron Emission

(i) $$ {}^{A}_{Z}\mathrm{X} \rightarrow {}^{A}_{Z-1}\mathrm{Y} + \mathrm{e}^+ + \nu \qquad (44.20) $$

(ii) $$ {}^{12}_{7}\mathrm{N} \rightarrow {}^{12}_{6}\mathrm{C} + \mathrm{e}^+ + \nu \qquad (44.22) $$

(iii) $$ \mathrm{p} \rightarrow \mathrm{n} + \mathrm{e}^+ + \nu $$

Electron Capture

This is a decay process that competes with positron emission and occurs when a parent nucleus captures an orbital electron and emits a neutrino. *It is typically a K shell electron that is captured by the parent nucleus.*

$$ {}^{A}_{Z}\mathrm{X} + {}^{0}_{-1}\mathrm{e} \rightarrow {}^{A}_{Z-1}\mathrm{Y} + \nu \qquad (44.24) $$

Gamma Decay

Nuclei which undergo alpha or beta decay are often left in an excited energy state (indicated by the * symbol). The nucleus returns to the ground state by emission of one or more photons. The excited state of $^{12}C^*$ is an example of a state that decays by gamma emission. *Gamma decay results in no change in mass number or atomic number.*

$$^{A}_{Z}X^* \rightarrow {}^{A}_{Z}X + \gamma \qquad (44.25)$$

$$^{12}_{6}C^* \rightarrow {}^{12}_{6}C + \gamma \qquad (44.27)$$

Nuclear reactions can occur when target nuclei (X) are bombarded with energetic particles (a), resulting in a daughter or product nucleus (Y) and an outgoing particle (b). A nuclear reaction can be represented as shown in Equation 44.28 or in the corresponding compact form. *In these reactions the structure, identity, or properties of the target nuclei are changed.*

$$a + X \rightarrow Y + b \qquad (44.28)$$

$$X(a,b)Y \quad \text{(compact form)}$$

The **reaction energy** (Q) associated with a nuclear reaction is the difference between the initial and final energy resulting from the reaction.

$$Q = (M_a + M_X - M_Y - M_b)c^2 \qquad (44.29)$$

$Q > 0$ exothermic reaction

$Q < 0$ endothermic reaction

The **threshold energy** (E_{th}) is the minimum kinetic energy of the incident particle in Equation 44.28 in order for an endothermic reaction to occur.

$$E_{th} = -Q\left(1 + \frac{M_a}{M_x}\right)$$

The **nuclear magneton** (μ_n) is a unit of magnetic moment in which the spin magnetic moment of a nucleus is measured.

$$\mu_n \equiv \frac{e\hbar}{2m_p} = 5.05 \times 10^{-27}\ \text{J/T} \qquad (44.31)$$

SUGGESTIONS, SKILLS, AND STRATEGIES

The rest energy of a particle is given by $E_R = mc^2$. It is therefore often convenient to express the unified mass unit in terms of its equivalent energy: $1\ \text{u} = 1.660\,540 \times 10^{-27}$ kg or $1\ \text{u} = 931.494\ \text{MeV}/c^2$. When masses are expressed in units of u, energy values are then $E_R = m(931.494\ \text{MeV/u})$.

Equation 44.6 can be solved for the particular time t after which the number of remaining nuclei will be some specified fraction of the original number N_0. This can be done by taking the natural log of each side of Equation 44.6 to find $t = \frac{1}{\lambda}\ln\left(\frac{N_0}{N}\right)$.

Summary of Characteristics of Alpha, Beta, and Gamma Decay

Decay Mode	Pathway	Process
Alpha decay	${}^{A}_{Z}\mathrm{X} \rightarrow {}^{A-4}_{Z-2}\mathrm{X} + {}^{4}_{2}\mathrm{He}$	The parent nucleus emits an alpha particle and loses two protons and two neutrons.
Beta decay	${}^{A}_{Z}\mathrm{X} \rightarrow {}_{Z+1}^{\ \ A}\mathrm{Y} + \mathrm{e}^{-} + \bar{\nu}$ (electron emission) ${}^{A}_{Z}\mathrm{X} \rightarrow {}_{Z-1}^{\ \ A}\mathrm{Y} + \mathrm{e}^{+} + \nu$ (positron emission)	A nucleus can undergo beta decay in two ways. The parent nucleus can emit an electron (e^-) and antineutrino ($\bar{\nu}$) or emit a positron (e^+) and a neutrino (ν).
Gamma decay	${}^{A}_{Z}\mathrm{X}^{*} \rightarrow {}^{A}_{Z}\mathrm{X} + \gamma$ * parent nucleus in an excited state	A nucleus in an excited state decays to a lower energy state (often the ground state) and emits a gamma ray.

REVIEW CHECKLIST

You should be able to:

- Use appropriate nomenclature in describing the static properties of nuclei. Calculate nuclear radii and densities. (Section 44.1)
- Use the binding-energy curve to estimate the energy released by fission and fusion. Calculate values of nuclear binding energy per nucleon. (Section 44.2)
- Explain the four major effects which influence the binding energy according to the liquid drop model of the nucleus. (Section 44.3)
- Calculate the time interval required for the number of radioactive nuclei in a sample to decrease to a given fraction of the initial value. (Section 44.4)
- Identify each of the components of radiation that are emitted by the nucleus through natural radioactive decay and describe the basic properties of each. Write out typical equations to illustrate the processes of transmutation by alpha decay, beta decay,

gamma decay, and electron capture. Explain why the neutrino must be considered in the analysis of beta decay. (Sections 44.5 and 44.6)

- Calculate the Q value of given nuclear reactions and determine the threshold energy of endothermic reactions. Balance equations representing specific nuclear reactions. (Section 44.7)

ANSWER TO AN OBJECTIVE QUESTION

5. Two samples of the same radioactive nuclide are prepared. Sample G has twice the initial activity of sample H. **(i)** How does the half-life of G compare with the half-life of H? (a) It is two times larger. (b) It is the same. (c) It is half as large. **(ii)** After each has passed through five half-lives, how do their activities compare? (a) G has more than twice the activity of H. (b) G has twice the activity of H. (c) G and H have the same activity. (d) G has lower activity than H.

Answer **(i)** (b) Because they are the same nuclide (samples of the same isotope), they have the same half-life. Nothing in the physical or chemical environment of the sample can alter a nucleus's probability of decay, and certainly the sample size cannot alter it.

(ii) (b) The activities of G and H decrease in time together, but G will always have twice as many remaining parent nuclei, and will always have twice the activity of H. After five half-lives, each has 1/32 as many parent nuclei as it started with, and the ratio of their activities is $(2/32)/(1/32) = 2/1$, as at the start. We ignore statistical fluctuations that would only show up over short periods of observation, and the possibility that the daughter nucleus is also radioactive.

□ □ □ □

ANSWERS TO SELECTED CONCEPTUAL QUESTIONS

2. Explain why nuclei that are well off the line of stability in Figure 44.4 tend to be unstable.

Answer The nuclear force favors the formation of neutron-proton pairs, so a stable nucleus cannot be too far away from having equal numbers of protons and neutrons. This effect sets the upper boundary of the zone of stability on the neutron-proton diagram. All of the protons repel one another electrically, so a stable nucleus cannot have too many protons. This effect sets the lower boundary of the zone of stability, and also the upper end.

□ □ □ □

8. Why do nearly all the naturally occurring isotopes lie above the $N = Z$ line in Figure 44.4?

Answer As Z increases, extra neutrons are required to overcome the increasing electrostatic repulsion of the protons. The extra neutrons attract their neighbors, including protons, with the strong nuclear force.

□ □ □ □

15. If a nucleus such as ^{226}Ra initially at rest undergoes alpha decay, which has more kinetic energy after the decay, the alpha particle or the daughter nucleus? Explain your answer.

Answer The alpha particle and the daughter nucleus carry momenta of equal magnitudes in opposite directions. Since kinetic energy can be written as $p^2/2m$, the less massive alpha particle has much more of the decay energy than the recoiling nucleus.

□ □ □ □

16. Suppose it could be shown that the cosmic-ray intensity at the Earth's surface was much greater 10 000 years ago. How would this difference affect what we accept as valid carbon-dated values of the age of ancient samples of once-living matter? Explain your answer.

Answer If the cosmic ray intensity at the Earth's surface was much greater 10 000 years ago, the fraction of the Earth's carbon dioxide with the radioactive nuclide ^{14}C would also be greater at that time than now. Thus, there would initially be a greater fraction of ^{14}C in the organic artifacts, and we would believe that the artifact was more recent than it actually is.

For example, suppose that that actual ratio of atmospheric ^{14}C to ^{12}C, two half-lives (11 460 years) ago, was 2.6×10^{-12}. The current ratio of isotopes in a sample from that time would be $\left(\frac{1}{2}\right)\left(\frac{1}{2}\right)\left(2.6\times10^{-12}\right) = 0.65\times10^{-12}$. We would measure this ratio today; but, believing the initial ratio to be equal to the present value of 1.3×10^{-12}, we would think that the creature had died only one half-life (5 730 years) ago. We would calculate $\left(\frac{1}{2}\right)\left(1.3\times10^{-12}\right) = 0.65\times10^{-12}$.

□ □ □ □

SOLUTIONS TO SELECTED PROBLEMS

3. (a) Use energy methods to calculate the distance of closest approach for a head-on collision between an alpha particle having an initial energy of 0.500 MeV and a gold nucleus (^{197}Au) at rest. Assume the gold nucleus remains at rest during the collision. (b) What minimum initial speed must the alpha particle have to approach as close as 300 fm to the gold nucleus?

Solution

Conceptualize: The positively charged alpha particle ($q = +2e$) will be repelled by the positive gold nucleus ($Q = +79e$), so that the particles probably will not touch each other in the electrostatic "collision" of part (a). The closest the alpha particle can get to the gold nucleus would be if the two nuclei did touch, in which case the distance between their centers would be about 9 fm (using $r = aA^{1/3}$ for the radius of each nucleus). To get this close or even within 300 fm, the alpha particle must be traveling very fast, probably close to the speed of light (but of course v must be less than c).

Categorize: The initial kinetic energy is equal to the electric potential energy of the alpha-particle-gold-nucleus system at the distance of closest approach, r_{min}.

Analyze:

(a) $K_\alpha + 0 = 0 + U = k_e \dfrac{qQ}{r_{\text{min}}}$

$$r_{\text{min}} = k_e \frac{qQ}{K_\alpha} = \frac{\left(8.99 \times 10^9 \text{ N}\cdot\text{m}^2/\text{C}^2\right)(2)(79)\left(1.60 \times 10^{-19} \text{ C}\right)^2}{\left(0.500 \text{ MeV}\right)\left(1.60 \times 10^{-13} \text{ J/MeV}\right)} = 455 \text{ fm} \qquad \blacksquare$$

(b) Since $K_\alpha = \frac{1}{2}mv^2 = k_e \dfrac{qQ}{r_{\text{min}}}$, we find that $v = \sqrt{\dfrac{2k_e qQ}{mr_{\text{min}}}}$

$$v = \sqrt{\frac{2\left(8.99 \times 10^9 \text{ N}\cdot\text{m}^2/\text{C}^2\right)(2)(79)\left(1.60 \times 10^{-19} \text{ C}\right)^2}{4\left(1.66 \times 10^{-27} \text{ kg}\right)\left(3.00 \times 10^{-13} \text{ m}\right)}} = 6.04 \times 10^6 \text{ m/s} \qquad \blacksquare$$

Finalize: The minimum distance in part (a) is on the order of 100 times greater than the combined radii of the particles. For part (b), the alpha particle must have more than 0.5 MeV of energy since it gets closer to the nucleus than the 455 fm found in part (a). Even so, the speed of the alpha particle in part (b) is only about 2% of the speed of light so we are justified in not using a relativistic approach. In solving this problem, we ignored the effect of the electrons around the gold nucleus that tend to "screen" the nucleus so that the alpha particle sees a reduced positive charge. If this screening effect were considered, the potential energy would be slightly reduced and, for the same initial energy, the alpha particle could get closer to the gold nucleus.

---

7. A star ending its life with a mass of four to eight times the Sun's mass is expected to collapse and then undergo a supernova event. In the remnant that is not carried away by the supernova explosion, protons and electrons combine to form a neutron star with approximately twice the mass of the Sun. Such a star can be thought of as a gigantic atomic nucleus. Assume $r = aA^{1/3}$ (Equation 44.1). If a star of mass 3.98×10^{30} kg is composed entirely of neutrons ($m_n = 1.67 \times 10^{-27}$ kg), what would its radius be?

Solution

Conceptualize: The neutron star is really held together by gravity, but it is a good approximation to think of its density as that of nuclear matter. Do you expect a radius like that of the Earth?

Categorize: We find the number of neutrons and then apply the given radius equation to find the size of the object.

Analyze: The number of nucleons in the star is

$$A = \frac{2(1.99 \times 10^{30}\ \text{kg})}{1.67 \times 10^{-27}\ \text{kg}} = 2.38 \times 10^{57}$$

Therefore, $r = aA^{1/3} = (1.2 \times 10^{-15}\ \text{m})(2.38 \times 10^{57})^{1/3} = 16\ \text{km}$ ∎

Finalize: It is not the size of the Earth at all! The Sun itself is expected to become a *white dwarf star* after using up all its nuclear fuel in reactions fusing hydrogen to helium to carbon. Still having most of its present mass, it will then be a roughly Earth-size object made of atoms (carbon) that you can visualize as having no space between them. Its density will be in tons per teaspoonful. A star with mass greater than the Chandrasekhar limit of 1.4 times the mass of the Sun has enough gravity to crush its atoms at the end of its life. It becomes the neutron star considered here, with diameter the size of a city and a density of billions of tons per teaspoonful.

---

17. A pair of nuclei for which $Z_1 = N_2$ and $Z_2 = N_1$ are called *mirror isobars* (the atomic and neutron numbers are interchanged). Binding-energy measurements on these nuclei can be used to obtain evidence of the charge independence of nuclear forces (that is, proton–proton, proton–neutron, and neutron–neutron nuclear forces are equal). Calculate the difference in binding energy for the two mirror isobars ${}^{15}_{8}\text{O}$ and ${}^{15}_{7}\text{N}$. The electric repulsion among eight protons rather than seven accounts for the difference.

Solution

Conceptualize: The oxygen nucleus will be less stable. It will have a smaller value of binding energy.

Categorize: We imagine a reaction of assembling each nucleus from isolated protons and neutrons. Really, to use the measurements as tabulated, we imagine a reaction of assembling each whole atom from isolated hydrogen atoms and neutrons.

Analyze: For ${}^{15}_{8}\text{O}$ we have, using Equation 44.2, a binding energy of

$$\begin{aligned} E_b &= [8(1.007\,825)\ \text{u} + 7(1.008\,665)\ \text{u} - 15.003\,066\ \text{u}](931.494\ \text{MeV/u}) \\ &= 111.96\ \text{MeV} \end{aligned}$$

For ${}^{15}_{7}\text{N}$ we have similarly

$$\begin{aligned} E_b &= [7(1.007\,825)\ \text{u} + 8(1.008\,665)\ \text{u} - 15.000\,109\ \text{u}](931.494\ \text{MeV/u}) \\ &= 115.49\ \text{MeV} \end{aligned}$$

Therefore, the difference in the two binding energies is $\Delta E_b = 3.54$ MeV. ∎

Finalize: The lower mass of the nitrogen-15 nucleus is direct evidence that it is more stable. As an analogy, a roller skate is more stable at the bottom of a staircase than at the top. As the problem says, the electric repulsion among eight protons rather than seven accounts for the difference in binding energy.

---

19. Nuclei having the same mass numbers are called *isobars.* The isotope $^{139}_{57}\text{La}$ is stable. A radioactive isobar, $^{139}_{59}\text{Pr}$, is located below the line of stable nuclei in Figure 44.19 and decays by e^+ emission. Another radioactive isobar of $^{139}_{57}\text{La}$, $^{139}_{55}\text{Cs}$, decays by e^- emission and is located above the line of stable nuclei in Figure 44.19. (a) Which of these three isobars has the highest neutron-to-proton ratio? (b) Which has the greatest binding energy per nucleon? (c) Which do you expect to be heavier, $^{139}_{57}\text{Pr}$ or $^{139}_{57}\text{Cs}$?

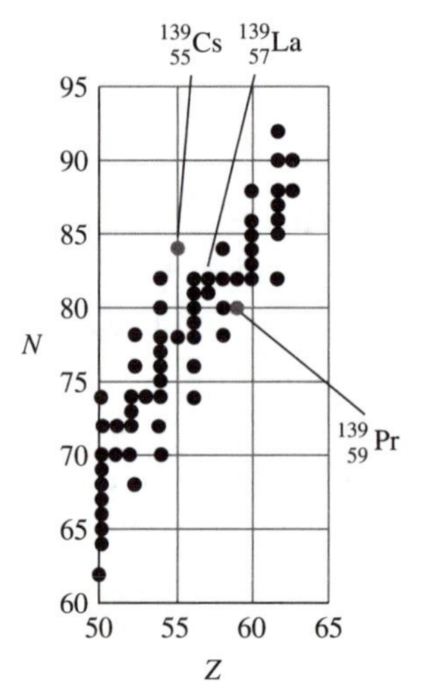

Figure P44.19

Solution

Conceptualize: This problem is about nuclear stability. An unstable nucleus is radioactive. Among unstable nuclei, a less stable nucleus has a shorter half-life.

Categorize: The neutron–proton plot of stable nuclei, Figure 44.4, suggests that special conditions have to be met for a nucleus to be stable. Figure P44.19 is an enlarged section of the overall plot.

Analyze:

(a) For $^{139}_{59}\text{Pr}$ the neutron number is $139 - 59 = 80$. For $^{139}_{55}\text{Cs}$ the neutron number is 84, so the Cs isotope has the greatest neutron-to-proton ratio. ■

(b) Binding energy per nucleon measures stability so it is greatest for the stable nucleus, the lanthanum isotope. Note also that it has a magic number of neutrons, 82. ■

(c) Cs-139 has 55 protons and 84 neutrons. Pr-139 has 59 protons and 80 neutrons. Because both are unstable, they are not shown on Figure 44.4 in the textbook. When we locate them on Figure P44.19 here, we see that cesium is a little farther away from the center of the zone of stable nuclei. Being less stable goes with being able to lose more energy in decay, and with having more mass. We therefore expect cesium to be heavier. ■

Finalize: For nuclear stability there needs to be a "just right" mix of neutrons and protons. Too many protons will tear the nucleus apart with their electrostatic repulsion. Too many neutrons will leave too many unpaired with protons. We do not have a precise predictive theory on what "just right" means. We guess bits of the recipe Nature is following from the nuclei Nature stews up.

---

22. Using the graph in Figure 44.5, estimate how much energy is released when a nucleus of mass number 200 fissions into two nuclei each of mass number 100.

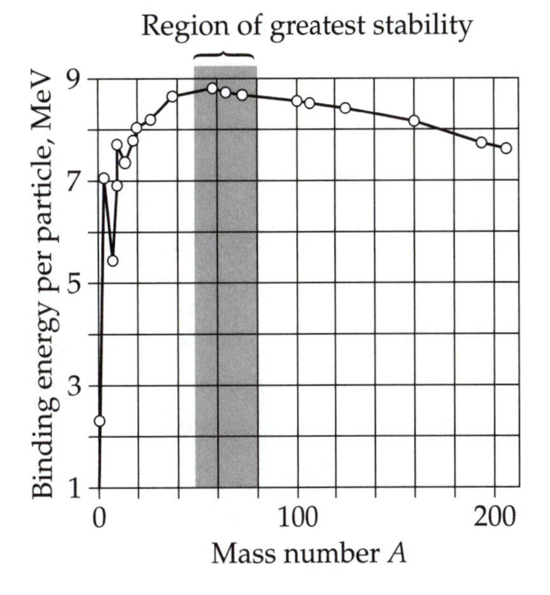

Figure 44.5

Solution

Conceptualize: A good way to think of the curve of binding energy, Figure 44.5, is to turn it upside down and think of it as the profile of a potentially slippery bathtub. Then it is the graph of total energy per nucleon versus mass number. Going downhill anywhere on the curve, toward the bathtub drain (which is the region of greatest stability) represents a way to get energy out of a nuclear reaction.

Categorize: We read the graph to think of the energy of a nucleus with $A = 200$, compared with the energy of two nuclei of $A = 100$, its potential fission products.

Analyze: The curve of binding energy shows that a heavy nucleus of mass number $A = 200$ has binding energy about

$$\left(7.8\frac{\text{MeV}}{\text{nucleon}}\right)(200\text{ nucleons}) \approx 1.56\text{ GeV}$$

Thus, it is less stable than its potential fission products, two middleweight nuclei of $A = 100$, together having binding energy

$$2(8.7\text{ MeV/nucleon})(100\text{ nucleons}) \approx 1.74\text{ GeV}$$

Fission then releases about $1.74\text{ GeV} - 1.56\text{ GeV} \approx 200\text{ MeV}$ ■

Finalize: Fission is the energy source of uranium bombs and of nuclear electric-generating plants. Still more energy could be released by asymmetric fission of an $A = 200$ nucleus into a fragment closer to the region of greatest stability and another heavier fragment.

---

25. A sample of radioactive material contains 1.00×10^{15} atoms and has an activity of 6.00×10^{11} Bq. What is its half-life?

Solution

Conceptualize: If the material maintained constant activity, the nuclei would last for 10^{15} nuclei/$(6 \times 10^{11}$ decays/s$) \approx 2\,000$ s. This can be a very rough estimate for the half-life.

Categorize: We identify the decay constant, which is the probability of decay within the next second, and use it to compute the half-life.

Analyze: The decay law is $dN/dt = -\lambda N$

Then the decay constant is

$$\lambda = -\frac{1}{N}\left(\frac{dN}{dt}\right) = -\left(\frac{1}{1.00\times10^{15}\text{ nuclei}}\right)\left(\frac{-6.00\times10^{11}\text{ nuclei}}{\text{s}}\right) = 6.00\times10^{-4}\text{ s}^{-1}$$

$$T_{1/2} = \frac{\ln 2}{\lambda} = 1\,160\text{ s} \qquad \text{(This is 19.3 minutes.)}$$ ■

Finalize: Any particular radioactive isotope has a characteristic half-life. Different nuclei can have half-lives over an infinitely wide range. In Table 44.2 you can find entries from fractions of a second to quadrillions of years.

A useful tip: Find the decay constant λ as early as possible in any problem about radioactive decay. It appears in lots of the equations.

---

26. A freshly prepared sample of a certain radioactive isotope has an activity of 10.0 mCi. After 4.00 h, its activity is 8.00 mCi. Find (a) the decay constant and (b) the half-life. (c) How many atoms of the isotope were contained in the freshly prepared sample? (d) What is the sample's activity 30.0 h after it is prepared?

Solution

Conceptualize: More tangibly, "freshly prepared" could mean "newly isolated" or "freshly purified." Over the course of 4 hours, this isotope lost 20% of its activity, so its half-life appears to be around 10 hours, which means that its activity after 30 hours (~3 half-lives) will be about 1 mCi. The decay constant and number of atoms are not so easy to estimate.

Categorize: From the rate equation, $R = R_0 e^{-\lambda t}$, we can find the decay constant λ, which can then be used to find the half life, the original number of atoms, and the activity at any other time, t.

Analyze:

(a) The decay constant is

$$\lambda = \frac{1}{t}\ln\left(\frac{R_0}{R}\right) = \frac{1}{4.00\text{ h}(3\,600\text{ s/h})}\ln\left(\frac{10.0\text{ mCi}}{8.00\text{ mCi}}\right) = 1.55\times10^{-5}\text{ s}^{-1} = 0.0558\text{ h}^{-1}$$ ■

(b) The half-life is $T_{1/2} = \dfrac{\ln 2}{\lambda} = \dfrac{0.693}{0.0558\text{ h}^{-1}} = 12.4\text{ h}$ ■

(c) The number of original atoms can be found if we convert the initial activity from curies into becquerels (decays per second): 1 Ci ≡ 3.7×10^{10} Bq.

$$R_0 = 10.0\text{ mCi} = (10.0\times10^{-3}\text{ Ci})(3.70\times10^{10}\text{ Bq/Ci}) = 3.70\times10^{8}\text{ Bq}$$

Since $R_0 = \lambda N_0$, the original number of nuclei is

$$N_0 = \frac{R_0}{\lambda} = \frac{3.70\times10^{8}\text{ decays/s}}{1.55\times10^{-5}\text{ s}} = 2.39\times10^{13}\text{ atoms}$$ ■

(d) The decay rate after thirty hours is

$$R = R_0 e^{-\lambda t} = (10.0\text{ mCi})e^{-(5.58\times10^{-2}\text{ h}^{-1})(30.0\text{ h})} = 1.88\text{ mCi}$$ ■

Finalize: Our estimate of the half-life was about 20% short because we did not account for the non-linearity of the decay rate. Consequently, our estimate of the final activity also fell short, but both of these calculated results are close enough to be reasonable.

The number of atoms is much less than one mole, so this appears to be a very small sample. To get a sense of how small, we can assume that the molar mass is on the order of 100 g/mol, so the sample has a mass of only

$$m \approx \frac{(2.4\times10^{13}\text{ atoms})(100\text{ g/mol})}{6.02\times10^{23}\text{ atoms/mol}} \approx 0.004\ \mu\text{g}$$

This sample is so small it cannot be measured by a commercial mass balance!

The problem states that this sample was "freshly prepared," from which we assumed that **all** the atoms within the sample are initially radioactive. Generally this is not the case, so that N_0 only accounts for the formerly radioactive atoms, and does not include additional atoms in the sample that were not radioactive. Realistically, then, the sample mass should be significantly greater than our estimate above.

---

29. The radioactive isotope ^{198}Au has a half-life of 64.8 h. A sample containing this isotope has an initial activity ($t = 0$) of 40.0 μCi. Calculate the number of nuclei that decay in the time interval between $t_1 = 10.0$ h and $t_2 = 12.0$ h.

Solution

Conceptualize: One microcurie is 3.7×10^4 decays per second, so the initial decay rate of this sample is about 1.5 million decays per second or 5 billion decays per hour. The answer to the problem will be somewhat less than 10 billion decays in two hours, because the activity is always decreasing, and will be less ten or twelve hours from now.

Categorize: We will find the number of nuclei remaining at the 10-h mark and the number at the 12-h instant. Then subtraction, not specified by any equation in the textbook, will tell us the number decaying in the time interval between.

Analyze: First, let us find λ and N_0 from the given information. The decay constant is

$$\lambda = \frac{\ln 2}{T_{1/2}} = \frac{0.693}{64.8\text{ h}} = 0.0107\text{ h}^{-1} = 2.97\times10^{-6}\text{ s}^{-1}$$

The original sample size is

$$N_0 = \frac{R_0}{\lambda} = \left(\frac{40.0\times10^{-6}\text{ Ci}}{2.97\times10^{-6}\text{ s}^{-1}}\right)\left(\frac{3.70\times10^{10}\text{ decays/s}}{1\text{ Ci}}\right) = 4.98\times10^{11}\text{ nuclei}$$

Since $N = N_0 e^{-\lambda t}$, the number of nuclei which decay between times t_2 and t_2 is

$$N_1 - N_2 = N_0\left(e^{-\lambda t_1} - e^{-\lambda t_2}\right)$$

$$N_1 - N_2 = \left(4.98\times10^{11}\right)\left[e^{-\left(0.0107\text{ h}^{-1}\right)\left(10.0\text{ h}\right)} - e^{-\left(0.0107\text{ h}^{-1}\right)\left(12.0\text{ h}\right)}\right]$$

$$N_1 - N_2 = 9.47\times10^{9}\text{ nuclei}$$ ■

Finalize: Our estimate was good. Notice that 0.01 hours is 36 seconds, but the decay probability per time is $0.0107\text{ h}^{-1} = 2.97\times10^{-6}\text{ s}^{-1}$. This sort of conversion is routinely necessary. Make sure you do not get it upside down.

---

37. Find the energy released in the alpha decay ${}^{238}_{92}\text{U} \rightarrow {}^{234}_{90}\text{Th} + {}^{4}_{2}\text{He}$.

Solution

Conceptualize: We expect on the order of 1 MeV.

Categorize: We apply to the decay a relativistic energy version of the isolated system model. Some of the rest energy of the parent nucleus is converted into kinetic energy of the decay products according to $Q = |\Delta m|c^2$.

Analyze: Table 44.2 contains the following values:

$$M\left({}^{238}_{92}\text{U}\right) = 238.050\,788\text{ u}$$

$$M\left({}^{234}_{90}\text{Th}\right) = 234.043\,601\text{ u}$$

$$M\left({}^{4}_{2}\text{He}\right) = 4.002\,603\text{ u}$$

We calculate the energy released by the reaction, its Q-value, as

$$Q = (M_{\text{U}} - M_{\text{Th}} - M_{\text{He}})(931.5\text{ MeV/u})$$

$$Q = (238.050\,788 - 234.043\,601 - 4.002\,603)(931.5) = 4.27\text{ MeV}$$ ■

Finalize: It is more than we guessed. A chemical reaction releases a few eV per molecule. Energy is millions of times more concentrated in the nucleus of an atom than in the electron cloud.

---

39. The nucleus ${}^{15}_{8}\text{O}$ decays by electron capture. The nuclear reaction is written ${}^{15}_{8}\text{O} + \text{e}^- \rightarrow {}^{15}_{7}\text{N} + \nu$. (a) Write the process going on for a single particle within the nucleus. (b) Disregarding the daughter's recoil, determine the energy of the neutrino.

Solution

Conceptualize: The oxygen nucleus swallows one of the inner electrons. The process is a decay in the sense that the system is dropping down into a lower energy state.

Categorize: We will find the energy by comparing masses before and after.

Analyze:

(a) The reaction for one particle is $e^- + p \rightarrow n + \nu$ ■

(b) Add 7 protons, 7 neutrons, and 7 electrons to each side to give

$$^{15}\text{O atom} \rightarrow {}^{15}\text{N atom} + \nu$$

It is the masses of neutral atoms that we can look up in a table.

From Table 44.2, $M(^{15}\text{O}) = M(^{15}\text{N}) + Q/c^2$

$$|\Delta M| = 15.003\ 066 - 15.000\ 109 = 0.002\ 957\ \text{u}$$

$$Q = (931.5\ \text{MeV/u})(0.002\ 957\ \text{u}) = 2.75\ \text{MeV}$$ ■

Finalize: One proton in the parent nucleus combines with the captured electron to become a neutron. Note that charge is conserved in this process, as in any process for an isolated system. We need a reaction stated in terms of neutral atoms to use the information listed in Table 44.2.

---

41. Enter the correct nuclide symbol in each open rectangle in Figure P44.41, which shows the sequences of decays in the natural radioactive series starting with the long-lived isotope uranium-235 and ending with the stable nucleus lead-207.

Solution

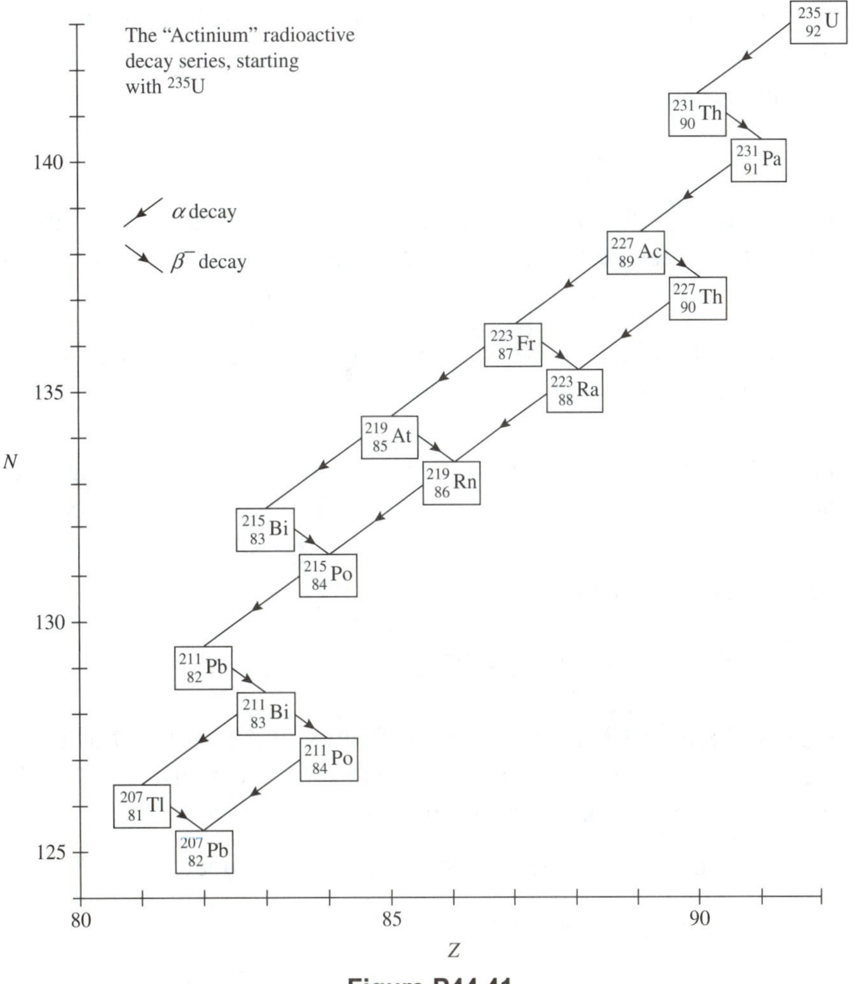

■

Figure P44.41

Conceptualize: You can think of a nucleus as a roiling mass of nucleons continuously colliding, pairing, and unpairing. It would be colossally improbable for 18 neutrons and 10 protons within a uranium-235 nucleus to bind to each other, come unstuck from their other neighbors, be at the surface, be moving fast, and be headed outward instead of inward. So the nucleus actually decays in a series of small steps.

Categorize: We use conservation laws for charge and for mass number. (In Chapter 46 we will identify the census number of nucleons A as the baryon number.) Whenever an $\alpha = {}^4_2\text{He}$ is emitted, Z drops by 2 and A by 4. Whenever an $e^- = {}^{0}_{-1}e$ is emitted, Z increases by 1 and A is unchanged.

Analyze: We find the chemical name by looking up Z in a periodic table. The values in the shaded boxes (${}^{235}\text{U}$ and ${}^{207}\text{Pb}$) were given; all others have been filled in as part of the solution.

Finalize: The mass number A is an integer equal to the total number of protons and neutrons. It is exactly conserved in each reaction. The mass M decreases a bit with each decay as the radioactive nuclei lose energy and become more nearly stable. Note that some nuclei can decay by more than one process. Our chart does not show the gamma rays that are emitted when daughter nuclei are left in excited states.

---

48. Natural gold has only one isotope, ${}^{197}_{79}\text{Au}$. If natural gold is irradiated by a flux of slow neutrons, electrons are emitted. (a) Write the reaction equation. (b) Calculate the maximum energy of the emitted electrons.

Solution

Conceptualize: The ${}^{197}_{79}\text{Au}$ will absorb a neutron. The conservation laws for charge and mass number (more generally called baryon number) will let us identify the product nucleus . . .

Categorize: . . . and the daughter after beta decay. Comparing masses will tell us how much energy is liberated in the overall reaction.

Analyze: The ${}^{197}_{79}\text{Au}$ absorbs a neutron ${}^1_0\text{n}$ to become ${}^{198}_{79}\text{Au}$, which emits an $e^- = {}^{0}_{-1}e$ to become ${}^{198}_{80}\text{Hg}$.

(a) For nuclei the net reaction is:

$${}^{197}_{79}\text{Au nucleus} + {}^1_0\text{n} \rightarrow {}^{198}_{80}\text{Hg nucleus} + {}^{0}_{-1}e + \bar{\nu} \qquad \blacksquare$$

Note the conservation of baryon number (which you can think of as nucleon census number and call mass number in this chapter) in the superscripts: $197 + 1 = 198 + 0$. Note the conservation of charge in the subscripts: $79 + 0 = 80 - 1$.

(b) Adding 79 electrons to both sides gives the reaction for neutral atoms:

$${}^{197}_{79}\text{Au atom} + {}^1_0\text{n} \rightarrow {}^{198}_{80}\text{Hg atom} + \bar{\nu}$$

From Table 44.2, 196.966 569 + 1.008 665 = 197.966 769 + 0 + Q/c^2. The energy released in the reaction is $Q = |\Delta m| c^2$

$$Q = (0.008\,465\text{ u})(931.5\text{ MeV/u}) = 7.89\text{ MeV}$$ ■

Finalize: The ancient alchemists could not transmute other metals into gold by mechanical or chemical means. This nuclear reaction transmutes gold into base metal. For starting with a stable nucleus, the energy released may seem surprisingly large.

It is our choice of units—unified atomic mass units—that makes the masses of atoms so nearly numerically equal to the integer values for the mass numbers *A* of the atomic nuclei. But the mass numbers *A* and atomic numbers *Z* go into reasoning about *what* reaction occurs. The masses *M* go into different reasoning that lets us figure out the reaction energy.

---

49. The following reactions are observed:

$$^{9}_{4}\text{Be} + \text{n} \rightarrow {}^{10}_{4}\text{Be} + \gamma \qquad Q = 6.812\text{ MeV}$$
$$^{9}_{4}\text{Be} + \gamma \rightarrow {}^{8}_{4}\text{Be} + \text{n} \qquad Q = -1.665\text{ MeV}$$

Calculate the masses of ^{8}Be and ^{10}Be in unified mass units to four decimal places from these data.

Solution

Conceptualize: The mass of each isotope in atomic mass units will be approximately the number of nucleons (8 or 10), also called the mass number. The electrons are much less massive and contribute only about 0.03% to the total mass.

Categorize: In addition to summing the mass of the subatomic particles, the net mass of the isotopes must account for the binding energy that holds the atom together.

Analyze: Table 44.2 give the mass of beryllium-9. We compute the second answer first. We can write the first reaction as

$$^{9}\text{Be} + \text{n} \rightarrow {}^{10}\text{Be} + \gamma + 6.812\text{ MeV}$$

In terms of masses, we write

$$m\left({}^{10}\text{Be}\right) = m\left({}^{9}\text{Be}\right) + m_n - \frac{6.812\text{ MeV}}{931.5\text{ MeV/u}}$$

(The gamma ray has no rest energy.)

Then $m\left({}^{10}\text{Be}\right) = 9.012\,182 + 1.008\,665 - 0.007\,313 = 10.013\,5\text{ u}$ ■

Similarly we write the second reaction as

$$^{9}\text{Be} + \gamma \rightarrow {}^{8}\text{Be} + \text{n} - 1.665\text{ MeV}$$

Therefore $m\left({}^{8}\text{Be}\right) = m\left({}^{9}\text{Be}\right) - m_n + \dfrac{1.665\text{ MeV}}{931.5\text{ MeV/u}}$

or $m\left({}^{8}\text{Be}\right) = 9.012182 - 1.008\,665 + 0.001\,787 = 8.005\,3\text{ u}$ ■

Finalize: As expected, both isotopes have masses slightly greater than their mass numbers. We were asked to calculate the masses to four decimal places, but with the available data, the results could be reported accurately to as many as six decimal places.

---

50. Construct a diagram like that of Figure 44.19 for the cases when I equals (a) $\frac{5}{2}$ and (b) 4.

Solution

Conceptualize: Each diagram will show the allowed value for the angular momentum of a particular nucleus, and the set of possible values for the component of angular momentum that can be measured along a particular direction, as by applying a magnetic field.

Categorize: It is the quantum particle under boundary conditions model that is behind the general rules: With angular momentum quantum number I, the magnitude of the angular momentum must be $\sqrt{I(I+1)}\,\hbar$. Whether I is an integer or a half-integer, the allowed values for one component of angular momentum being measured range from $+I\hbar$ to $+(I-1)\hbar$ to … to $-I\hbar$. Conditions that the wave function for a quantum particle must satisfy, for self-consistency under rotations in three-dimensional space, impose these requirements. We call a component being measured the z component. It can be measured more directly, as in a nuclear magnetic resonance experiment, or less directly, as from the way the angular momentum influences the intrinsic energy levels of a system and the number of available states within an energy level.

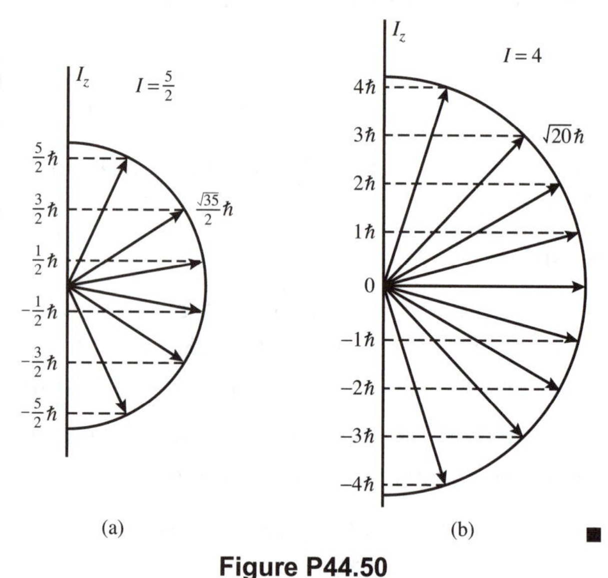

Figure P44.50

Analyze:

(a) With $I = 5/2$, the magnitude of the angular momentum is

$$\sqrt{I(I+1)}\,\hbar = \sqrt{\tfrac{5}{2}(\tfrac{5}{2}+1)}\,\hbar = \sqrt{35}\,\hbar/2$$

$$= 2.958\,04(6.626\times10^{-34}\text{ J}\cdot\text{s})/2\pi = 3.119\times10^{-34}\text{ kg}\cdot\text{m}^2/\text{s}$$

The z component can take the values $+5\hbar/2$, $+3\hbar/2$, $+\hbar/2$, $-\hbar/2$, $-3\hbar/2$, and $-5\hbar/2$. These identifications are shown in part (a) of the diagram.

(b) Similarly, with $I = 4$, the magnitude of the angular momentum of a nucleus is $\sqrt{I(I+1)}\,\hbar = \sqrt{4(4+1)}\,\hbar = \sqrt{20}\,\hbar$ and its z component must have one of the nine values $+4\hbar$, $+3\hbar$, $+2\hbar$, $+\hbar$, 0, $-\hbar$, $-2\hbar$, $-3\hbar$, $-4\hbar$, as shown in part (b) of the diagram.

Finalize: As was mentioned in the chapter on atomic physics, the name *spatial quantization* is applied to the way the angular momentum vector must make one of a set of exact angles with the direction of any particular magnetic field we apply. The other two components of angular momentum cannot be individually measured at the same time—having such a measured value is forbidden by the uncertainty principle. Spatial quantization is a thoroughly counterintuitive and nonclassical effect. If we apply a southward magnetic field and then suddenly switch to a westward one, a particle's value of L_z in the first field has no correlation with its value in the second field. It can sometimes happen, as by particle-emission processes, that two electrons have opposite spins adding to zero, with one electron spin-up and the other spin-down. If their spin components are both measured in southward magnetic fields, one will be south and the other north. If instead you choose to measure the spin component of one electron in a westward magnetic field and find that it is east, then the other electron must immediately have a westward magnetic field component, even if there is no possibility for a signal to be passed to it from the first electron. You could keep the result of the first electron's spin measurement hidden in a sealed envelope until after the second is measured. This is one case of *quantum entanglement,* which Einstein called a "spooky action at a distance."

69. Free neutrons have a characteristic half-life of 10.4 min. What fraction of a group of free neutrons with kinetic energy 0.040 0 eV decays before traveling a distance of 10.0 km?

Solution

Conceptualize: We guess that these neutrons are moving pretty fast, so they are not in flight for very long, and perhaps one percent may decay.

Categorize: The fraction that will remain is given by the ratio N/N_0 where $N/N_0 = e^{-\lambda\Delta t}$ and Δt is the time interval it takes the neutron to travel a distance of d = 10.0 km. Further, we use the particle under constant speed model.

Analyze: The time of flight is given by $\Delta t = d/v$.

Since $K = \frac{1}{2}mv^2$, $$\Delta t = \frac{d}{\sqrt{\frac{2K}{m}}} = \frac{10.0\times10^3\text{ m}}{\sqrt{\frac{2(0.040\,0\text{ eV})\left(1.60\times10^{-19}\text{ J/eV}\right)}{1.67\times10^{-27}\text{ kg}}}} = 3.61\text{ s}$$

The decay constant is

$$\lambda = \frac{0.693}{T_{1/2}} = \frac{0.693}{(10.4\text{ min})(60\text{ s/min})} = 1.11\times10^{-3}\text{ s}^{-1}$$

Therefore we have

$$\lambda\Delta t = \left(1.11\times10^{-3}\text{ s}^{-1}\right)(3.61\text{ s}) = 4.01\times10^{-3} = 0.004\,01$$

And the fraction remaining is $$\frac{N}{N_0} = e^{-\lambda\Delta t} = e^{-0.004\,01} = 0.996\,0$$

Hence, the fraction that has decayed in this time interval is

$$1 - \frac{N}{N_0} = 0.004\,00 \quad \text{or} \quad 0.400\%$$ ■

Finalize: We made a good guess! In designing fission reactors, to be considered in Chapter 45, neutron loss by decay is a minor concern.

45

Applications of Nuclear Physics

EQUATIONS AND CONCEPTS

The **fission of a uranium nucleus** by bombardment with a low-energy neutron results in the production of fission fragments and neutrons. *Equation 45.3 is typical of many possible combinations of fission fragments* X *and* Y. U* *is a short-lived intermediate state.*

$$ {}_0^1\text{n} + {}_{92}^{235}\text{U} \rightarrow {}_{92}^{236}\text{U}^* \rightarrow \text{X} + \text{Y} + \text{neutrons} \qquad (45.2)$$

$$ {}_0^1\text{n} + {}_{92}^{235}\text{U} \rightarrow {}_{56}^{141}\text{Ba} + {}_{36}^{92}\text{Kr} + 3\left({}_0^1\text{n}\right) \qquad (45.3)$$

The **fusion reactions** shown here are those most likely to be used as the basis of the design and operation of a fusion power reactor. The Q values refer to the energy released in each reaction.

$$ {}_1^2\text{H} + {}_1^2\text{H} \rightarrow {}_2^3\text{He} + {}_0^1\text{n} \quad (Q = 3.27\text{ MeV})$$

$$ {}_1^2\text{H} + {}_1^2\text{H} \rightarrow {}_1^3\text{H} + {}_1^1\text{H} \quad (Q = 4.03\text{ MeV})$$

$$ {}_1^2\text{H} + {}_1^3\text{H} \rightarrow {}_2^4\text{He} + {}_0^1\text{n} \quad (Q = 17.59\text{ MeV}) \qquad (45.4)$$

Lawson's criterion states the conditions under which a net power output of a fusion reactor is possible. In these expressions, n is the **plasma density** (number of ions per cubic cm) and τ is the plasma **confinement time** (the time during which the interacting ions are maintained at a temperature equal to or greater than that required for the reaction to proceed).

$$ n\tau \geq 10^{14}\text{ s/cm}^3 \quad \text{(D-T reaction)}$$
$$ n\tau \geq 10^{16}\text{ s/cm}^3 \quad \text{(D-D reaction)} \qquad (45.5)$$

Several different **radiation units** are used to quantify the amount of radiation interacting with a material:

- One **roentgen** (R) is the amount of ionizing radiation that produces, by ionization, an electric charge of 3.33×10^{-10} C in one cm^3 of air at STP.

- One **rad** (radiation absorbed dose) is the amount of radiation that increases the energy of 1 kg of absorbing material by 1×10^{-2} J.

- **RBE** (relative biological effectiveness) is a factor defined as the number of rads of X-radiation or gamma radiation that produces the same biological effect as 1 rad of the radiation being used.

- The radiation **dose in rem** (radiation equivalent man) is the product of the dose in rad and the appropriate relative biological effectiveness factor. Current SI units of radiation exposure and dose are the gray (Gy) and the sievert (Sv)

$$\text{Dose in rem} \equiv \text{dose in rad} \times \text{RBE} \quad (45.6)$$

$$1 \text{ sievert (Sv)} = 100 \text{ rem}$$

$$1 \text{ gray (Gy)} = 100 \text{ rad}$$

REVIEW CHECKLIST

You should be able to:

- Write several possible equations for the fission of ^{235}U and describe the sequence of events that occurs during the fission process. Calculate the mass of ^{235}U required to release a given quantity of energy by fission. (Section 45.2)

- Describe the basic design features and control mechanisms in a fission reactor including the functions of the moderator, control rods, and heat exchange system. Identify some major safety considerations and potential environmental hazards in the operation of a fission reactor. (Section 45.3)

- Describe the basis of energy release in fusion and write several nuclear reactions which might be used in a fusion-powered reactor. Explain the three basic parameters required of a successful thermonuclear power reactor. Calculate the rate of fuel consumption required to maintain a given power output by the D-T reaction. (Section 45.4)

- Define the units of radiation exposure or dose and explain the significance of the RBE factor. Calculate the thickness of shielding required to reduce the intensity of gamma rays to a given fraction of the intensity of the incident beam. (Section 45.5)

- Describe the basic principle of operation of the cloud chamber, ion chamber, Geiger counter, semiconductor diode detector, scintillation detector, and spark chamber. (Section 45.6)

ANSWER TO AN OBJECTIVE QUESTION

7. In the operation of a Geiger counter, is the amplitude of the current pulse (a) proportional to the kinetic energy of the particle producing the pulse, (b) proportional to the number of particles entering the tube to produce the pulse, (c) proportional to the RBE factor of the type of particle producing the pulse, or (d) independent of all these factors?

Answer (d). As long as the particle triggering the avalanche discharge has enough energy to ionize one atom of the gas inside the tube, the amount of its kinetic energy does not significantly affect the size of the pulse, so (a) is untrue. A single particle produces a single pulse, so (b) is untrue. If there is a high flux of particles, the tube will fire again very soon after its latency period, so the counting rate will be high, but the size of each pulse is controlled by the tube power supply. Different kinds of particles may have different efficiencies in ionizing gas atoms, but there is no necessary correlation between that effect and the relative biological effectiveness of the kind of particle. Thus (c) is untrue.

□ □ □ □

ANSWERS TO SELECTED CONCEPTUAL QUESTIONS

1. Why is water a better shield against neutrons than lead or steel?

Answer The hydrogen nuclei in water molecules have mass similar to that of a neutron, so that they can efficiently rob a fast-moving neutron of kinetic energy as they scatter it. Once the neutron is slowed down, a hydrogen nucleus can absorb it in the reaction ${}_0^1\text{n} + {}_1^1\text{H} \rightarrow {}_1^2\text{H} + \gamma$.

□ □ □ □

5. Discuss the similarities and differences between fusion and fission.

Answer Fusion of light nuclei to a heavier nucleus releases energy. Fission of a heavy nucleus to lighter nuclei releases energy. Both processes are steps towards greater stability on the curve of binding energy, Figure 44.5. The energy release per nucleon is typically greater for fusion, and this process is harder to control.

□ □ □ □

SOLUTIONS TO SELECTED PROBLEMS

5. List the nuclear reactions required to produce ${}^{233}\text{U}$ from ${}^{232}\text{Th}$ under fast neutron bombardment.

Solution

Conceptualize: Thorium-232 is the only isotope of thorium found in nature. It is not hard to mine from the Earth's crust. A reaction such as this might be used for breeding nuclear fuel.

Categorize: We use conservation laws for Z and A to figure out what thorium becomes when it absorbs a neutron, and then what it will spontaneously decay to.

Analyze: First, the thorium is bombarded:

$$ {}^{1}_{0}\mathrm{n} + {}^{232}_{90}\mathrm{Th} \rightarrow {}^{233}_{90}\mathrm{Th} \qquad \blacksquare $$

Then, the thorium decays by beta emission: $\quad {}^{233}_{90}\mathrm{Th} \rightarrow {}^{233}_{91}\mathrm{Pa} + {}^{0}_{-1}\mathrm{e} + \bar{\nu} \qquad \blacksquare$

Protactinium-233 has more neutrons than the more stable protactinium-231, so it too decays by beta emission:

$$ {}^{233}_{91}\mathrm{Pa} \rightarrow {}^{233}_{92}\mathrm{U} + {}^{0}_{-1}\mathrm{e} + \bar{\nu} \qquad \blacksquare $$

Finalize: We represent the electron as ${}^{0}_{-1}\mathrm{e}$. The electron is not an atom, so we might not call the −1 subscript its atomic number; but this is the charge of the electron in units of the elementary charge. Writing it there lets us explicitly see conservation of charge across each reaction. The electron does not have zero mass, so we might not call the 0 superscript its atomic mass number, but it is the electron's baryon number. You can think of this as a conserved census number of protons and neutrons together. Writing it as a superscript lets us explicitly see conservation of baryon number across each reaction.

---

9. Review. Suppose seawater exerts an average frictional drag force of 1.00×10^5 N on a nuclear-powered ship. The fuel consists of enriched uranium containing 3.40% of the fissionable isotope ${}^{235}_{92}\mathrm{U}$ and the ship's reactor has an efficiency of 20.0%. If 200 MeV is released per fission event, how far can the ship travel per kilogram of fuel?

Solution

Conceptualize: Nuclear fission is much more efficient for converting mass to energy than burning fossil fuels. However, without knowing the rate of diesel fuel consumption for a comparable ship, it is difficult to estimate the nuclear fuel use rate. It seems plausible that a ship could cross the Atlantic Ocean with only a few kilograms of nuclear fuel, so a reasonable range of uranium fuel consumption might be 100 km/kg to 10 000 km/kg.

Categorize: The fuel consumption rate can be found from the energy released by the nuclear fuel and the work required to push the ship through the water. We use the particle in equilibrium model. The thrust force P exerted on the propeller by the water around it must be equal in magnitude to the backward friction (drag) force on the hull.

Analyze: One kg of enriched uranium contains 3.40% ${}^{235}_{92}\mathrm{U}$, so the mass of uranium-235 is $m_{235} = 0.034\,0(1\,000\text{ g}) = 34.0\text{ g}$

In terms of number of nuclei, this is equivalent to

$$ N_{235} = (34.0\text{ g})\left(\frac{1}{235\text{ g/mol}}\right)\left(6.02 \times 10^{23}\text{ atoms/mol}\right) = 8.71 \times 10^{22}\text{ nuclei} $$

If all these nuclei fission, the energy released is equal to

$$\left(8.71\times10^{22}\text{ nuclei}\right)\left(200\times10^{6}\text{ eV/nucleus}\right)\left(1.602\times10^{-19}\text{ J/eV}\right)=2.79\times10^{12}\text{ J}$$

Now, for the engine,

$$\text{efficiency}=\frac{\text{work output}}{\text{heat input}}\quad\text{or}\quad e=\frac{P\Delta r\cos\theta}{Q_h}$$

So the distance the ship can travel per kilogram of uranium fuel is

$$\Delta r=\frac{eQ_h}{P\cos(0°)}=\frac{0.200\left(2.79\times10^{12}\text{ J}\right)}{1.00\times10^{5}\text{ N}}=5.58\times10^{6}\text{ m}\quad\blacksquare$$

Finalize: The ship can travel 5 580 km/kg of uranium fuel, which is on the high end of our prediction range. The distance between New York and Paris is 5 851 km, so this ship could cross the Atlantic Ocean on just about one kilogram of uranium fuel.

---

17. According to one estimate, there are 4.40×10^6 metric tons of world uranium reserves extractable at \$130/kg or less. We wish to determine if these reserves are sufficient to supply all of the world's energy needs. About 0.700% of naturally occurring uranium is the fissionable isotope ^{235}U. (a) Calculate the mass of ^{235}U in the reserve in grams. (b) Find the number of moles of ^{235}U in the reserve. (c) Find the number of ^{235}U nuclei in the reserve. (d) Assuming 200 MeV is obtained from each fission reaction and all this energy is captured, calculate the total energy in joules that can be extracted from the reserve. (e) Assuming the rate of world power consumption remains constant at 1.5×10^{13} J/s, how many years could the uranium reserve provide for all the world's energy needs? (f) What conclusion can be drawn?

Solution

Conceptualize: A million tons of coal would be burned quite quickly in providing energy for worldwide human use. A nucleus can release ten million times more energy than the chemical energy associated with atomic electrons, but we should not expect a million tons of uranium to keep greedy people happy for centuries.

Categorize: We will follow the directions in the problem step by step. Some of the steps have the form of conversions, involving Avogadro's number at one step. The last calculation will use the definition of power.

Analyze:

(a) Do not think of the "reserve" as being held in reserve. We are depleting it as fast as we choose. The remaining current balance of irreplaceable ^{235}U is 0.7% of the whole mass of uranium:

$$0.007\,00\times4.40\times10^{6}\text{ tonnes}\frac{10^{3}\text{kg}}{1\text{ tonne}}\frac{10^{3}\text{g}}{1\text{ kg}}=3.08\times10^{10}\text{g}\quad\blacksquare$$

(b) $n = \dfrac{m}{M} = \dfrac{3.08 \times 10^{10} \text{ g}}{235 \text{ g/mole}} = 1.31 \times 10^{8} \text{ mole}$ ■

(c) $N = nN_{\text{A}} = 1.31 \times 10^{8} \text{ mole} \dfrac{6.02 \times 10^{23} \text{ atom}}{1 \text{ mole}} \dfrac{1 \text{ nucleus}}{1 \text{ atom}} = 7.89 \times 10^{31} \text{ nuclei}$ ■

(d) We imagine each nucleus as fissioning, to release

$$7.89 \times 10^{31} \text{ fissions} \frac{200 \text{ MeV}}{1 \text{ fission}} \left(\frac{10^{6}}{\text{M}} \right) \frac{1.60 \times 10^{-19} \text{ C}}{1 \text{ e}} \left(\frac{1 \text{ J/C}}{1 \text{ V}} \right) = 2.52 \times 10^{21} \text{ J}$$ ■

(e) The definition of power is represented by P = (energy converted)/Δt so we have

$$\Delta t = \frac{\text{energy}}{P} = \frac{2.52 \times 10^{21} \text{ J}}{1.5 \times 10^{13} \text{ J/s}} = 1.68 \times 10^{8} \text{ s} \frac{1 \text{ yr}}{3.156 \times 10^{7} \text{ s}} = 5.33 \text{ yr}$$ ■

(f) So far from really managing the resource or from planning responsibly for seven generations, we could destroy the reserve before many of you get out of college. The uranium available at \$130/kg is not sufficient to supply world energy use for a significant interval of time. ■

Finalize: Somewhat more uranium is available at higher prices, but it is best to think of uranium as we do of fossil fuels. We can use some as one of several energy sources, while we turn to a future of renewable energy (or maybe fusion) for a sustainable population.

---

25. To understand why plasma containment is necessary, consider the rate at which an unconfined plasma would be lost. (a) Estimate the rms speed of deuterons in a plasma at a temperature of 4.00×10^{8} K. (b) **What If?** Estimate the order of magnitude of the time interval during which such a plasma would remain in a 10-cm cube if no steps were taken to contain it.

Solution

Conceptualize: At room temperature, molecules in the air move faster than sound. With small mass and a very high temperature, these deuterons may be moving at millions of meters per second, to escape from a reaction zone in nanoseconds.

Categorize: The average kinetic energy per particle $\frac{1}{2}m\overline{v^2}$ must equal the thermal energy $\frac{3}{2}k_{\text{B}}T$. We will also use the particle under constant speed model.

Analyze:

(a) Taking $m \approx 2m_p$ for deuterons, we have $\frac{1}{2}m\overline{v^2} = \frac{3}{2}k_{\text{B}}T$

The root-mean-square speed is

$$v_{rms} = \sqrt{\frac{3k_{\text{B}}T}{2m_p}} = \sqrt{\frac{3(1.38 \times 10^{-23} \text{ J/K})(4.00 \times 10^{8} \text{ K})}{2(1.67 \times 10^{-27} \text{ kg})}} = 2.23 \times 10^{6} \text{ m/s}$$ ■

(b) $\Delta t = \dfrac{x}{v} = \dfrac{0.100 \text{ m}}{2.23 \times 10^{6} \text{ m/s}} \sim 10^{-7} \text{ s}$ ■

Finalize: As predicted, it is all gone in less than a microsecond. Just moving fast is one good way to escape, but fusing plasma can also escape or stop fusing for other reasons. Containing it has been compared to keeping water in an inverted pitcher.

---

28. Another series of nuclear reactions that can produce energy in the interior of stars is the carbon cycle first proposed by Hans Bethe in 1939, leading to his Nobel Prize in Physics in 1967. This cycle is most efficient when the central temperature in a star is above 1.6×10^7 K. Because the temperature at the center of the Sun is only 1.5×10^7 K, the following cycle produces less than 10% of the Sun's energy. (a) A high-energy proton is absorbed by ^{12}C. Another nucleus, *A*, is produced in the reaction, along with a gamma ray. Identify nucleus *A*. (b) Nucleus *A* decays through positron emission to form nucleus *B*. Identify nucleus *B*. (c) Nucleus *B* absorbs a proton to produce nucleus *C* and a gamma ray. Identify nucleus *C*. (d) Nucleus *C* absorbs a proton to produce nucleus *D* and a gamma ray. Identify nucleus *D*. (e) Nucleus *D* decays through positron emission to produce nucleus *E*. Identify nucleus *E*. (f) Nucleus *E* absorbs a proton to produce nucleus *F* plus an alpha particle. Identify nucleus *F*. (g) What is the significance of the final nucleus in the last step of the cycle outlined in part (f)?

Solution

Conceptualize: The Sun is generally described as getting its energy from the fusion of hydrogen into helium, so this set of reactions, beginning with carbon, will be interesting.

Categorize: We can just use conservation of charge and conservation of the integer total number of protons and neutrons (conservation of baryon number) to identify the *Z* and *A* values of each product nucleus. From the periodic table we can identify the chemical element with each *Z* value.

Analyze:

(a) By adding $1 + 6 = 7$ and $1 + 12 = 13$, we have ${}^{1}_{1}\text{H} + {}^{12}_{6}\text{C} \rightarrow {}^{13}_{7}\text{N} + \gamma$ so nucleus *A* is ^{13}N. ■

(b) Now $13 - 0 = 13$ and $7 - 1 = 6$, so the positron decay is ${}^{13}_{7}\text{N} \rightarrow {}^{13}_{6}\text{C} + {}^{0}_{1}\text{e} + \nu$ and nucleus *B* is ^{13}C. ■

(c) Similarly, we have ${}^{1}_{1}\text{H} + {}^{13}_{6}\text{C} \rightarrow {}^{14}_{7}\text{N} + \gamma$ and nucleus *C* is ^{14}N. ■

(d) The hydrogen nuclei keep piling on like rugby players after a tackle. We have ${}^{1}_{1}\text{H} + {}^{14}_{7}\text{N} \rightarrow {}^{15}_{8}\text{O} + \gamma$ and nucleus *D* is ^{15}O. ■

(e) Now ${}^{15}_{8}\text{O} \rightarrow {}^{15}_{7}\text{N} + {}^{0}_{1}\text{e} + \nu$, so nucleus *E* is ^{15}N. ■

(f) We calculate $15 + 1 - 4 = 12$ and $7 + 1 - 2 = 6$ to identify ${}^{1}_{1}\text{H} + {}^{15}_{7}\text{N} \rightarrow {}^{12}_{6}\text{C} + {}^{4}_{2}\text{He}$ and nucleus *F* is ^{12}C. ■

(g) The original carbon-12 nucleus is returned. One carbon nucleus can participate in the fusions of colossal numbers of hydrogen nuclei, four after four. Carbon is a catalyst. ■

The two positrons immediately annihilate with electrons according to ${}^{0}_{1}\text{e} + {}^{0}_{-1}\text{e} \rightarrow 2\gamma$. The overall reaction, obtained by adding all eight reactions, can be represented as

$${}^{1}_{1}\text{H} + {}^{12}_{6}\text{C} + {}^{1}_{1}\text{H} + {}^{1}_{1}\text{H} + {}^{1}_{1}\text{H} + 2\,{}^{0}_{-1}\text{e} \rightarrow {}^{4}_{2}\text{He} + {}^{12}_{6}\text{C} + 7\gamma + 2\nu$$

This simplifies to $4\left({}^{1}_{1}\text{H}\right) + 2\,{}^{0}_{-1}\text{e} \rightarrow {}^{4}_{2}\text{He} + 2\nu$. The net reaction is identical to the net reaction in the proton–proton cycle which predominates in the Sun. In energy terms the reaction can be considered as $4\left({}^{1}_{1}\text{H atom}\right) \rightarrow {}^{4}_{2}\text{He atom} + 26.7\text{ MeV}$, where the Q value of energy output was computed in Chapter 39, Problem 57 and again in Problem 59 in this Chapter 45, both included in this manual.

Finalize: A higher temperature is required for the carbon cycle than for the proton–proton cycle, because of the stronger electrical repulsion of the carbon and nitrogen nuclei, with 6 or 7 protons, for the extra protons colliding with them. Bethe unexpectedly won the Nobel Prize for this work twenty-eight years after doing it. He remarked that he had previously thought that physics would be carried forward by a cadre of people smarter than he was, but then had to admit that progress requires everyone to do their best.

---

33. Review. The danger to the body from a high dose of gamma rays is not due to the amount of energy absorbed; rather, it is due to the ionizing nature of the radiation. As an illustration, calculate the rise in body temperature that results if a "lethal" dose of 1 000 rad is absorbed strictly as internal energy. Take the specific heat of living tissue as 4 186 J/kg · °C.

Solution

Conceptualize: A thousand units of radiation-absorbed-dose sounds like a lot. When they quickly warm things up in a microwave oven, some people say they "nuke" them. A "radiation burn" is a real injury. So it will be interesting to calculate the temperature increase in this definite case.

Categorize: We will look up the definition of a rad, and use the relationship between an increase in internal energy and an increase in temperature. We imagine that the body is surrounded by ideal thermal insulation so that none of the absorbed energy escapes.

Analyze: By definition, one rad increases the energy of one kilogram of the absorbing material by 1.00×10^{-2} J. The energy starts as energy carried by electromagnetic radiation, and turns entirely into internal energy. The 1 000 rad or 10.0 gray = 10.0 Gy will then put 10.0 J/kg into the body, to raise its temperature by the same amount as 10.0 J/kg of energy input by heat from a higher-temperature energy source. In $Q = mc\Delta T$ we have $Q/m = 10.0$ J/kg and

$$\Delta T = \frac{Q}{m}\frac{1}{c} = 10.0\,\frac{\text{J}}{\text{kg}}\,\frac{1}{4\ 186\ \text{J/kg}\cdot{}^{\circ}\text{C}} = 2.39 \times 10^{-3}\ {}^{\circ}\text{C}$$ ■

Finalize: This millidegree temperature change is quite harmless. Your average body temperature undergoes a much larger decrease when you drink a can of cold soda pop. Still the thousand-rad dose of gamma radiation does terrible damage to living tissue on a molecular level.

---

35. A small building has become accidentally contaminated with radioactivity. The longest-lived material in the building is strontium-90. ($^{90}_{38}$Sr has an atomic mass 89.907 7 u, and its half-life is 29.1 yr. It is particularly dangerous because it substitutes for calcium in bones.) Assume the building initially contained 5.00 kg of this substance uniformly distributed throughout the building and the safe level is defined as less than 10.0 decays/min (which is small compared to background radiation). How long will the building be unsafe?

Solution

Conceptualize: Two fission bombs have been used in war and many have been tested. They create a mix of radioactive products, such as strontium-90, dispersed into the atmosphere. A localized blob containing five kilograms of a radioactive isotope with a short half-life will have a very high decay rate originally, but you can contain it, keep people away, and wait for the activity to go down to a low value. With five kilograms of an isotope with a very long half-life, the decay rate will be very low. We and our human and non-human ancestors have always lived with such radioactive materials in the Earth's crust. It is the medium-half-life materials, from strontium-90 to plutonium-239 (24 120 yr), that give us the most trouble. We estimate centuries.

Categorize: We can find the sample size as a number of nuclei, and then the original decay rate. Then the law of radioactive decay will tell us the time interval required for the activity to get down.

Analyze: The number of nuclei in the original sample is

$$N_0 = \frac{\text{mass present}}{\text{mass of nucleus}} = \frac{5.00 \text{ kg}}{(89.907\,7 \text{ u})(1.66 \times 10^{-27} \text{ kg/u})} = 3.35 \times 10^{25} \text{ nuclei}$$

The decay constant is $\lambda = \frac{\ln 2}{T_{1/2}} = \frac{0.693}{29.1 \text{ yr}} = 2.38 \times 10^{-2} \text{ yr}^{-1} = 4.53 \times 10^{-8} \text{ min}^{-1}$

The original activity is

$$R_0 = \lambda N_0 = (4.53 \times 10^{-8} \text{ min}^{-1})(3.35 \times 10^{25} \text{ nuclei}) = 1.52 \times 10^{18} \text{ decays/min}$$

The law of decay then gives us

$$\frac{R}{R_0} = \frac{10.0 \text{ decays/min}}{1.52 \times 10^{18} \text{ decays/min}} = 6.59 \times 10^{-18} = e^{-\lambda t}$$

and the time interval is

$$t = \frac{-\ln(R/R_0)}{\lambda} = \frac{-\ln(6.59 \times 10^{-18})}{2.38 \times 10^{-2} \text{ yr}^{-1}} = 1\,660 \text{ yr}$$ ■

Finalize: What sort of containment can exclude all curiosity seekers, pranksters, criminals, and irresponsible politicians for millennia? The people working on this question do not have any big successes to report.

39. In a Geiger tube, the voltage between the electrodes is typically 1.00 kV and the current pulse discharges a 5.00-pF capacitor. (a) What is the energy amplification of this device for a 0.500-MeV electron? (b) How many electrons participate in the avalanche caused by the single initial electron?

Solution

Conceptualize: A million electron volts sounds like a lot of energy, but it is small compared to the energy in the discharge that the particle triggers. We expect an amplification factor in the millions, and a really large number of electrons to participate in the pulse.

Categorize: We will find the electric potential energy stored in the capacitor. Its ratio to 0.5 MeV is the amplification factor. We can find the charge on the capacitor as it is originally charged, and write it as a number of electrons.

Analyze:

(a) The energy amplification is

$$\frac{E}{E_0} = \frac{\frac{1}{2}C\Delta V^2}{0.500\text{ MeV}} = \frac{\frac{1}{2}(5.00 \times 10^{-12}\text{ F})(1.00 \times 10^3\text{ V})^2}{(0.500\text{ MeV})(1.60 \times 10^{-13}\text{ J/MeV})} = 3.12 \times 10^7 \quad \blacksquare$$

(b) The number of electrons is

$$N = \frac{Q}{e} = \frac{C\Delta V}{e} = \frac{(5.00 \times 10^{-12}\text{ F})(1.00 \times 10^3\text{ V})}{1.60 \times 10^{-19}\text{ C}} = 3.12 \times 10^{10}\text{ electrons} \quad \blacksquare$$

Finalize: The energy of the discharge is $3.12 \times 10^7 \times 0.5 \times 10^6$ eV = 1.56×10^{13} eV. The average energy of one of the electrons participating in the discharge is 1.56×10^{13} eV/(3.12×10^{10}) = 500 eV. There is a voltage-reduction factor of 1000, along with the 31-million-fold energy amplification factor, so that one particle can dependably trigger an avalanche of billions.

---

41. When gamma rays are incident on matter, the intensity of the gamma rays passing through the material varies with depth x as $I(x) = I_0 e^{-\mu x}$, where I_0 is the intensity of the radiation at the surface of the material (at $x = 0$) and μ is the linear absorption coefficient. For low-energy gamma rays in steel, take the absorption coefficient to be 0.720 mm^{-1}. (a) Determine the "half-thickness" for steel, that is, the thickness of steel that would absorb half the incident gamma rays. (b) In a steel mill, the thickness of sheet steel passing into a roller is measured by monitoring the intensity of gamma radiation reaching a detector below the rapidly moving metal from a small source immediately above the metal. If the thickness of the sheet changes from 0.800 mm to 0.700 mm, by what percentage does the gamma-ray intensity change?

Solution

Conceptualize: You can think of the absorption of gamma rays by steel as basically like the absorption of visible light by a filter, such as the lenses of sunglasses. Energy comes in carried by electromagnetic radiation. The portion that does not make it out the other side turns into internal energy, slightly raising the temperature of the material.

Categorize: The exponential decay in space described by the equation in the problem parallels the exponential decay in time characteristic of radioactive decay. So a certain thickness of the absorbing material will reduce the radiation intensity by half. Twice that thickness will reduce it to one-quarter of the incident intensity, and so on. The steel-mill thickness gauge described can monitor the sheet thickness in real time so that the roller pressure can be adjusted to make the thickness of the roller output more uniform.

Analyze:

(a) Let x_h represent the unknown half-thickness, according to

$$\tfrac{1}{2}I_0 = I_0 e^{-\mu x_h} \qquad \text{Then} \qquad 2 = e^{+\mu x_h} \qquad \text{and} \qquad \mu x_h = \ln 2$$

$$\text{so } x_h = \frac{\ln 2}{\mu} = \frac{\ln 2}{0.720\,/\,\text{mm}} = 0.963 \text{ mm}$$ ■

(b) The intensity reaching the detector through $x_1 = 0.8$ mm of steel is $I_1 = I_0 e^{-\mu x_1}$. That transmitted by thickness $x_2 = 0.7$ mm is $I_2 = I_0 e^{-\mu x_2}$. The fractional change is

$$\frac{I_2 - I_1}{I_1} = \frac{I_0 e^{-\mu x_2} - I_0 e^{-\mu x_1}}{I_0 e^{-\mu x_1}} = e^{\mu(x_1 - x_2)} - 1 = e^{(0.720/\text{mm})(0.100\text{ mm})} - 1$$

$$= e^{0.0720} - 1 = +0.074\,7 = 7.47\%$$

As the thickness decreases, the intensity increases by 7.47%. ■

Finalize: The fractional change in thickness has magnitude 12.5%, but the smaller change in gamma ray intensity should be easily measurable, say from a counting rate in a Geiger tube or from a current in an ionization chamber. The gamma radiation passing through the steel does no harm; the photons stopping inside the steel do not damage it, but produce only a small temperature increase. Dragging a radiation source through a pipeline can be a good way to check for cracks and faulty welds.

Note that it is impractical or impossible to set up shielding that will block *all* of the radiation from some source. The best that can be done is to reduce its intensity to be small compared to background radiation. For citizen education and in emergency situations, scientists must be careful to inform the public clearly and accurately about radiation hazards and their mitigation.

---

50. The half-life of tritium is 12.3 yr. If the TFTR fusion reactor contained 50.0 m^3 of tritium at a density equal to 2.00×10^{14} ions/cm^3, how many curies of tritium were in the plasma? State how this value compares with a fission inventory (the estimated supply of fissionable material) of 4×10^{10} Ci.

Solution

Conceptualize: It is difficult to estimate the activity of the tritium in the fusion reactor without actually calculating it; however, we might expect it to be a small fraction of the fission (not fusion) inventory.

Categorize: The decay rate (activity) can be found by multiplying the decay constant by the number of ${}^3_1\text{H}$ particles. The decay constant can be found from the half-life of tritium, and the number of particles from the density and volume of the plasma.

Analyze: The number of hydrogen-3 nuclei is

$$N = \left(50.0\ \text{m}^3\right)\left(2.00 \times 10^{14}\ \frac{\text{particles}}{\text{cm}^3}\right)\left(100\ \text{cm/m}\right)^3 = 1.00 \times 10^{22}\ \text{particles}$$

The decay constant is $\lambda = \dfrac{\ln 2}{T_{1/2}} = \left(\dfrac{0.693}{12.3\ \text{yr}}\right)\left(\dfrac{1\ \text{yr}}{3.16 \times 10^7\ \text{s}}\right) = 1.78 \times 10^{-9}\ \text{s}^{-1}$

The activity is then

$$R = \lambda N = \left(1.78 \times 10^{-9}\ \text{s}^{-1}\right)\left(1.00 \times 10^{22}\ \text{nuclei}\right) = 1.78 \times 10^{13}\ \text{Bq}$$

In curies this is $R = \left(1.78 \times 10^{13}\ \text{Bq}\right)\left(\dfrac{1\ \text{Ci}}{3.70 \times 10^{10}\ \text{Bq}}\right) = 482\ \text{Ci}$ ■

Finalize: Even though 482 Ci is a large amount of radioactivity, it is smaller than 4.00×10^{10} Ci by about a hundred million times. Loss of containment is a smaller hazard for a fusion power reactor than for a fission reactor. ■

---

55. Carbon detonations are powerful nuclear reactions that temporarily tear apart the cores inside massive stars late in their lives. These blasts are produced by carbon fusion, which requires a temperature of approximately 6×10^8 K to overcome the strong Coulomb repulsion between carbon nuclei. (a) Estimate the repulsive energy barrier to fusion, using the temperature required for carbon fusion. (In other words, what is the average kinetic energy of a carbon nucleus at 6×10^8 K?) (b) Calculate the energy (in MeV) released in each of these "carbon-burning" reactions:

$$^{12}\text{C} + {}^{12}\text{C} \rightarrow {}^{20}\text{Ne} + {}^{4}\text{He}$$

$$^{12}\text{C} + {}^{12}\text{C} \rightarrow {}^{24}\text{Mg} + \gamma$$

(c) Calculate the energy in kilowatt-hours given off when 2.00 kg of carbon completely fuse according to the first reaction.

Solution

Conceptualize: The core of a massive star goes through stages late in the life of the star, getting hotter and hotter as it fuses heavier and heavier nuclei until it gets to iron. We expect less than 1 MeV as the thermal energy of the colliding nuclei and many MeV's as the energy output, amounting to a huge number of kilowatt-hours per kilogram.

Categorize: We review the energy of a particle in an ideal gas. We use Einstein's equation about loss of mass appearing as energy output.

Analyze:

(a) At 6×10^8 K, the carbon nuclei have an average energy of

$$\tfrac{3}{2}k_B T = 1.5\left(8.62 \times 10^{-5}\ \text{eV/K}\right)\left(6 \times 10^8\ \text{K}\right) = 8 \times 10^4\ \text{eV} \qquad \blacksquare$$

(b) The energy released in the first reaction is

$$Q = E = \left[2m\left({}^{12}\text{C}\right) - m\left({}^{20}\text{Ne}\right) - m\left({}^{4}\text{He}\right)\right]c^2$$

$$E = (24.000\ 000 - 19.992\ 440 - 4.002\ 603)(931.5)\ \text{MeV} = 4.62\ \text{MeV} \qquad \blacksquare$$

In the second case, the energy released is

$$E = \left[2m\left({}^{12}\text{C}\right) - m\left({}^{24}\text{Mg}\right)\right](931.5\ \text{MeV/u})$$

$$E = (24.000\ 000 - 23.985\ 042)(931.5)\ \text{MeV} = 13.9\ \text{MeV} \qquad \blacksquare$$

(c) The energy released equals the energy of reaction of the number of carbon nuclei in a 2.00-kg sample, which corresponds to

$$\Delta E = \left(\frac{2\ 000\ \text{g}}{12\ \text{g/mol}}\right)\left(\frac{6.02 \times 10^{23}\ \text{atoms}}{1\ \text{mol}}\right)\left(\frac{1\ \text{fusion}}{2\ \text{atoms}}\right)\left(\frac{4.62\ \text{MeV}}{\text{fusion}}\right)\left(\frac{1\ \text{kWh}}{2.25 \times 10^{19}\ \text{MeV}}\right)$$

$$\Delta E = 10.3 \times 10^6\ \text{kWh} \qquad \blacksquare$$

Finalize: To prove that 1 kWh is 2.25×10^{19} MeV, just review facts you should definitely know: k = 1 000, W = J/s, h = 3 600 s, M = 1 000 000, $e = 1.60 \times 10^{-19}$ C, and the really important V = J/C.

The carbon fusion reactions involve 24 nucleons, but put out much less than 24 times the energy of simple hydrogen fusion. The aging star burns through its last resources in a small fraction of the time it spent as a normal star. Tapping into carbon fusion to run a power plant is beyond the potential of the human race.

---

59. Consider the two nuclear reactions

$$\text{(I)} \quad \text{A} + \text{B} \rightarrow \text{C} + \text{E}$$

$$\text{(II)} \quad \text{C} + \text{D} \rightarrow \text{F} + \text{G}$$

(a) Show that the net disintegration energy for these two reactions ($Q_{\text{net}} = Q_{\text{I}} + Q_{\text{II}}$) is identical to the disintegration energy for the net reaction

$$\text{A} + \text{B} + \text{D} \rightarrow \text{E} + \text{F} + \text{G}$$

(b) One chain of reactions in the proton–proton cycle in the Sun's core is

$$^1_1\text{H} + ^1_1\text{H} \rightarrow ^2_1\text{H} + ^0_1\text{e} + \nu$$

$$^0_1\text{e} + ^0_{-1}\text{e} \rightarrow 2\gamma$$

$$^1_1\text{H} + ^2_1\text{H} \rightarrow ^3_2\text{He} + \gamma$$

$$^1_1\text{H} + ^3_2\text{He} \rightarrow ^4_2\text{He} + ^0_1\text{e} + \nu$$

$$^0_1\text{e} + ^0_{-1}\text{e} \rightarrow 2\gamma$$

Based on part (a), what is Q_{net} for this sequence?

Solution

Conceptualize: It seems natural that we should be able to add up reactions and use only the net reaction to determine the net energy output. We will prove this fact. Adding up the reactions in part (b) will give the net reaction 4 hydrogen-1 atoms → helium-4 atom. An example in the textbook shows that the energy output is 26.7 MeV, so we should get this value again.

Categorize: It is all based on the Einstein mass-energy relation.

Analyze: In the symbols given in the problem, the energy output of the first reaction is

$$Q_{\text{I}} = (M_A + M_B - M_C - M_E)c^2$$

and that of the second reaction is $Q_{\text{II}} = (M_C + M_D - M_F - M_G)c^2$

The net energy output is then

$$\begin{aligned} Q_{\text{net}} &= (M_A + M_B - M_C - M_E + M_C + M_D - M_F - M_G)c^2 \\ &= (M_A + M_B + M_D - M_E - M_F - M_G)c^2 \end{aligned}$$

(a) This value is identical to Q for the reaction A + B + D → E + F + G. Thus, any product (like "C") that is a reactant in a subsequent reaction disappears from the energy balance. ■

(b) Adding all five reactions, we have $4\left(^1_1\text{H}\right) + 2\left(^0_{-1}\text{e}\right) \rightarrow ^4_2\text{He} + 2\nu$.

Here the symbol ^1_1H represents a proton, the nucleus of a hydrogen-1 atom. So that we may use the tabulated masses of neutral atoms for calculation, we add two electrons to each side of the reaction. The four electrons on the initial side are enough to make four hydrogen atoms and the two electrons on the final side constitute, with the alpha particle, a neutral helium atom. The atomic-electronic binding energies are negligible compared to Q_{net}.

We use the arbitrary symbol "${}^{1}_{1}$H atom" to represent a neutral atom. Then we have the reaction $4\left({}^{1}_{1}\text{H atom}\right) \rightarrow {}^{4}_{2}\text{He atom} + 2\nu$

and the corresponding mass-energy equation

$$4(1.007\ 825\ \text{u}) = 4.002\ 603\ \text{u} + Q_{\text{net}}/c^2$$

The energy released is

$$Q_{\text{net}} = [4(1.007\ 825\ \text{u}) - 4.002\ 603\ \text{u}](931.5\ \text{MeV/u}) = 26.7\ \text{MeV}$$ ■

Finalize: Both parts worked out just right. The neutrinos created in the fusion reaction have negligible mass, but carry off some energy, so the star has to stay hot on somewhat less than 26.7 MeV per reaction.

---

63. Assume a deuteron and a triton are at rest when they fuse according to the reaction

$${}^{2}_{1}\text{H} + {}^{3}_{1}\text{H} \rightarrow {}^{4}_{2}\text{He} + {}^{1}_{0}\text{n}$$

Determine the kinetic energy acquired by the neutron.

Solution

Conceptualize: The products of this nuclear reaction are an alpha particle and a neutron, with total kinetic energy of 17.6 MeV, as stated in Equation 45.4. In order to conserve momentum, the lighter neutron will have a larger velocity than the more massive alpha particle (which consists of two protons and two neutrons). Since the kinetic energy of the particles is proportional to the square of their velocities but only linearly proportional to their mass, the neutron should have the larger kinetic energy, somewhere between 8.8 and 17.6 MeV.

Categorize: Conservation of system linear momentum and energy can be applied to find the kinetic energy of the neutron. We first suppose the particles are moving nonrelativistically.

Analyze: The momentum of the alpha particle and that of the neutron must add to zero, so their velocities must be in opposite directions with magnitudes related by

$$m_n\vec{\mathbf{v}}_n + m_\alpha\vec{\mathbf{v}}_\alpha = 0 \quad \text{or} \quad (1.008\ 7\ \text{u})v_n = (4.002\ 6\ \text{u})v_\alpha$$

At the same time, their kinetic energies must add to 17.6 MeV:

$$E = \tfrac{1}{2}m_n v_n^{\ 2} + \tfrac{1}{2}m_\alpha v_\alpha^{\ 2} = \tfrac{1}{2}(1.008\ 7\ \text{u})v_n^{\ 2} + \tfrac{1}{2}(4.002\ 6\ \text{u})v_\alpha^{\ 2} = 17.6\ \text{MeV}$$

Substitute $v_\alpha = 0.252\ 0\ v_n$

to obtain $E = (0.504\ 35\ \text{u})v_n^2 + (0.127\ 10\ \text{u})v_n^2 = 17.6\ \text{MeV}\left(\dfrac{1\ \text{u}}{931.494\ \text{MeV}/c^2}\right)$

$$v_n = \sqrt{\frac{0.018\ 9c^2}{0.631\ 45}} = 0.173c = 5.19 \times 10^7\ \text{m/s}$$

Since this speed is not too much greater than $0.1c$, we can get a reasonable estimate of the kinetic energy of the neutron from the classical equation,

$$K = \tfrac{1}{2}mv^2 = \tfrac{1}{2}(1.008\ 7\ \text{u})(0.173c)^2\left(\frac{931.494\ \text{MeV}/c^2}{\text{u}}\right) = 14.1\ \text{MeV}$$ ■

Finalize: The kinetic energy of the neutron is within the range we predicted. For a more accurate calculation of the kinetic energy, we should use relativistic expressions. Conservation of system momentum gives

$$\gamma_n m_n \vec{\mathbf{v}}_n + \gamma_\alpha m_\alpha \vec{\mathbf{v}}_\alpha = 0 \qquad \text{or} \qquad 1.008\ 7\frac{v_n}{\sqrt{1 - v_n^2/c^2}} = 4.002\ 6\frac{v_\alpha}{\sqrt{1 - v_\alpha^2/c^2}}$$

yielding $\dfrac{v_\alpha^2}{c^2} = \dfrac{v_n^2}{15.746c^2 - 14.746v_n^2}$

Then $(\gamma_n - 1)m_n c^2 + (\gamma_\alpha - 1)m_\alpha c^2 = 17.6\ \text{MeV}$

and $v_n = 0.171c$, implying that $(\gamma_n - 1)m_n c^2 = 14.0\ \text{MeV}$ ■

---

64. (a) Calculate the energy (in kilowatt-hours) released if 1.00 kg of ^{239}Pu undergoes complete fission and the energy released per fission event is 200 MeV. (b) Calculate the energy (in electron volts) released in the deuterium–tritium fusion reaction

$$^2_1\text{H} + {}^3_1\text{H} \to {}^4_2\text{He} + {}^1_0\text{n}$$

(c) Calculate the energy (in kilowatt-hours) released if 1.00 kg of deuterium undergoes fusion according to this reaction. (d) **What If?** Calculate the energy (in kilowatt-hours) released by the combustion of 1.00 kg of coal if each $\text{C} + \text{O}_2 \to \text{CO}_2$ reaction yields 4.20 eV. (e) List advantages and disadvantages of each of these methods of energy generation.

Solution

Conceptualize: The problem compares one kilogram of very different fuels. We estimate that the energy from fission (in part (a)) is more than a million times the chemical energy (part (d)) and that the fusion energy (part (c)) will be more than ten times larger still.

Categorize: In parts (a), (c), and (d) we can find the correspondences by multiplying the quantity of fuel by continued conversion factors. In part (b) we will compare masses before and after the reaction to find the Q value.

Analyze:

(a) The energy release of the plutonium is

$$E = (1\ 000\ \text{g})\left(\frac{6.02 \times 10^{23}\ \text{nuclei}}{239\ \text{g}}\right)\left(\frac{200 \times 10^6\ \text{eV}}{1\ \text{nucleus}}\right)\left(\frac{1.60 \times 10^{-19}\ \text{J/eV}}{3.60 \times 10^6\ \text{J/kWh}}\right)$$

$$E = 2.24 \times 10^7\ \text{kWh}$$ ■

(b) We obtain the masses of the particles from Table 44.2. The total mass of reactants is $m_{\text{before}} = 2.014\,102\text{ u} + 3.016\,049\text{ u} = 5.030\,151\text{ u}$

The mass of the products is $m_{\text{after}} = 4.002\,603\text{ u} + 1.008\,665\text{ u} = 5.011\,268\text{ u}$

For the change in mass we have $|\Delta m| = m_{\text{before}} - m_{\text{after}} = 0.018\,883\text{ u}$

The energy released is

$$\Delta E = |\Delta m|c^2 = (0.018\,883\text{ u})(931.494\text{ MeV/u}) = 17.6\text{ MeV}$$ ■

(c) One kilogram of deuterium can then release energy

$$E = \left(1\,000\text{ g }{}_1^2\text{H}\right)\left(\frac{6.02\times10^{23}\text{ deuterons}}{2.014\text{ g }{}_1^2\text{H}}\right)\left(\frac{17.6\text{ MeV}}{{}_1^2\text{H fusion}}\right)\left(\frac{1.60\times10^{-13}\text{ J/MeV}}{3.60\times10^{6}\text{ J/kWh}}\right)$$

$$E = 2.34\times10^{8}\text{ kWh}$$ ■

(d) Coal is essentially pure carbon. Assuming complete combustion, the energy output of one kilogram is

$$\left(1\,000\text{ g C}\right)\left(\frac{6.02\times10^{23}\text{ C atoms}}{12.0\text{ g C}}\right)\left(\frac{4.20\text{ eV}}{\text{C atom}}\right)\left(\frac{1.60\times10^{-19}\text{ J/eV}}{3.60\times10^{6}\text{ J/kWh}}\right) = 9.36\text{ kWh}$$ ■

(e) You likely pay the electric company to burn coal, because coal is cheap at this moment in human history. The limit on supply will inevitably drive up the price of fossil fuels. Worldwide, electrically transmitted energy from fission is an immediate option. Fission may offer advantages of reduced carbon dioxide emissions, reduced overall pollution, and even a reduction in the radiation released into the atmosphere, compared to mining and burning coal. However, the Chernobyl explosion and other accidents demonstrate that there is a continuing significant risk of catastrophic disaster. For a fair comparison to a coal-fired plant, we should think of a nuclear generating station paying for long-term disposal of its waste and paying for its own insurance, without government subsidies. In rich countries we urgently need to reduce total energy use and to employ renewable energy sources. We hope that fusion reactors will become practical in the future, because fusion might be both safer and less expensive than either coal or fission. ■

Finalize: Our answer to part (b) agrees with the statement in the chapter text. For each kilogram of fuel, the energy output of fission is on the order of ten million times larger than that of chemical burning, and the energy output of fusion ten times larger still.

46

Particle Physics and Cosmology

EQUATIONS AND CONCEPTS

Pions are present in three varieties, corresponding to three charge states: π^+, π^-, and π^0. Pions are very unstable; the equations at right show a decay mode for each type.

$$\pi^+ \rightarrow \pi^0 + e^+ + \nu_e$$

$$\pi^- \rightarrow \mu^- + \bar{\nu}_\mu$$

$$\pi^0 \rightarrow \gamma + \gamma$$

Muons come in two varieties: μ^- and μ^+ (the antiparticle). Example decay processes are shown.

$$\mu^+ \rightarrow e^+ + \nu_e + \bar{\nu}_\mu$$

$$\mu^- \rightarrow e^- + \nu + \bar{\nu} \qquad (46.1)$$

We model the universe as composed of field particles and matter particles. The exchange of field particles mediates the interactions of the matter particles.

- **Gluons** are the field particles for the strong force.
- **Photons** are exchanged by charged particles in electromagnetic interactions.
- $\mathbf{W^+}$, $\mathbf{W^-}$, and $\mathbf{Z^0}$ **particles** mediate the weak force.
- **Gravitons** are hypothetical quantum particles of the gravitational field.

Matter particles fall into two broad classifications: hadrons and leptons.

Hadrons are particles that interact via all fundamental forces. They are composed of quarks.

Classes of hadrons:

- **Mesons** have spin = 0 or 1.
- **Baryons** have spin = 1/2 or 3/2.

Leptons have no structure and do not interact via the strong force. Six types of leptons exist: electron, muon, tau particle, and their three associated neutrinos.

Leptons have spin = 1/2.

All particles have a **corresponding antiparticle**. *A particle and its antiparticle have the same mass and spin, and have opposites of all other characteristics* (electric charge, baryon number, strangeness, etc.). For a particle composed of quarks, its antiparticle is composed of the corresponding antiquarks. There are a few neutral mesons that are their own antiparticle.

Conservation laws for elementary particles:

Relativistic total energy is conserved in all reactions; an apparent fluctuation in energy (ΔE) can occur in the creation of a virtual particle provided that the particle exists for a time no longer than $\Delta t = \hbar/2\Delta E$.

Electric charge (a scalar quantity) is conserved in all reactions.

Linear momentum and **angular momentum** (vector quantities) are conserved in all reactions.

Baryon number: $B = +1$ (baryons), -1 (antibaryons), 0 (all other particles) is conserved whenever a reaction or decay occurs. The sum of baryon numbers before the event must equal the sum following the event as illustrated in the example reaction at right.

Example reaction showing conservation of baryon number:

$$\mathrm{p} + \mathrm{n} \rightarrow \mathrm{p} + \mathrm{p} + \mathrm{n} + \overline{\mathrm{p}}$$
$$(1 + 1 = 1 + 1 + 1 - 1)$$

Electron lepton number: $L_e = +1$ (electron and electron neutrino), -1 (antileptons), 0 (all other particles) is conserved whenever a reaction or decay occurs. The sum of the electron lepton numbers before an event must equal the sum of the electron lepton numbers following the event. See the example reaction to the right.

Example reaction showing conservation of electron lepton number:

$$\mathrm{n} \rightarrow \mathrm{p} + \mathrm{e}^- + \overline{\nu}_e$$
$$(0 = 0 + 1 - 1)$$

Muon lepton number (L_μ) and **tau lepton number** (L_τ) are also conserved quantities.

Strangeness ($S = +1, -1$, or 0) is a conserved quantity in decays or reactions which occur via the strong force.

Hubble's law states a linear relationship between the velocity of a galaxy and its distance R from the Earth. The constant H is called the **Hubble constant**.

$$v = HR \quad (46.4)$$

$$H \approx 22 \times 10^{-3}\ \mathrm{m/s \cdot ly}$$

Quarks and antiquarks make up baryons and mesons; the identity of each particle is determined by the particular combination of quarks. *Each baryon consists of three quarks and each meson consists of one quark and one antiquark.* The table at right lists the charge and baryon number for six quarks and the corresponding antiquarks.

Quark, antiquark		Charge	Baryon number
Up:	$u, \bar{u}$	$+\frac{2}{3}e, -\frac{2}{3}e$	$\frac{1}{3}, -\frac{1}{3}$
Down:	$d, \bar{d}$	$-\frac{1}{3}e, +\frac{1}{3}e$	$\frac{1}{3}, -\frac{1}{3}$
Strange:	$s, \bar{s}$	$-\frac{1}{3}e, +\frac{1}{3}e$	$\frac{1}{3}, -\frac{1}{3}$
Charmed:	$c, \bar{c}$	$+\frac{2}{3}e, -\frac{2}{3}e$	$\frac{1}{3}, -\frac{1}{3}$
Bottom:	$b, \bar{b}$	$-\frac{1}{3}e, +\frac{1}{3}e$	$\frac{1}{3}, -\frac{1}{3}$
Top:	$t, \bar{t}$	$+\frac{2}{3}e, -\frac{2}{3}e$	$\frac{1}{3}, -\frac{1}{3}$

REVIEW CHECKLIST

You should be able to:

- Identify the four fundamental forces in nature and the corresponding field particles or quanta through which these forces are mediated. Calculate energy and mass values involved in pair production and pair annihilation. (Sections 46.1 and 46.2)
- Estimate ranges of forces based on the mass of the field particles. (Section 46.3)
- Outline the broad classification of particles and the characteristic properties of the classes (relative mass value, spin, decay mode). Complete a proposed reaction by identifying a particle in either the reactants or products. Calculate momentum and energy involved in pion decay. (Sections 46.3 and 46.4)
- Determine whether or not a suggested decay can occur based on the conservation of baryon number and the conservation of electron lepton number. Determine whether or not a predicted reaction/decay will occur based on the conservation of strangeness for the strong and electromagnetic interactions. (Sections 46.5 and 46.6)
- Calculate threshold energies for reactions of elementary particles. (Section 46.7)
- Show that the charge, baryon number, and strangeness of a specified particle equal the sums of the corresponding numbers for the constituent quarks. Identify particles corresponding to specified quark combinations. (Section 46.9)
- Make calculations based on Hubble's law. (Section 46.12)

ANSWER TO AN OBJECTIVE QUESTION

6. Define the average density of the solar system ρ_{SS} as the total mass of the Sun, planets, satellites, rings, asteroids, icy outliers, and comets, divided by the volume of a sphere around the Sun large enough to contain all these objects. The sphere extends about halfway to the nearest star, with a radius of approximately 2×10^{16} m, about two light-years. How does this average density of the solar system compare with the critical density ρ_c required for the Universe to stop its Hubble's-law expansion? (a) ρ_{SS} is much greater than ρ_c. (b) ρ_{SS} is approximately or precisely equal to ρ_c. (c) ρ_{SS} is much less than ρ_c. (d) It is impossible to determine.

Answer (a) The average density of the solar system, as defined in the question, can be considered a measure of the average density of the galaxy. It is much greater than the critical density that would make gravity a strong enough force to stop the expansion of the Universe. Recall that that critical density is in turn greater than the actual average density of the Universe. It is the great gulfs between galaxies, empty as far as we know, and especially the colossal spaces between galaxy clusters, including the voids that show up on the largest scales we can map, that bring the average concentration of matter in the Universe way down.

We can estimate ρ_{SS} numerically. Most of the mass of the solar system is that in the Sun, 1.99×10^{30} kg. The volume of the sphere defined in the question is $(4/3)\pi r^3 = (4/3)\pi(2 \times 10^{16}\text{ m})^3 = 3 \times 10^{49}\text{ m}^3$. Then we have $\rho_{SS} = m/V = 1.99 \times 10^{30}\text{ kg}/3 \times 10^{49}\text{ m}^3 = 6 \times 10^{-20}\text{ kg/m}^3$. This is ten million times larger than the critical density $3H^2/8\pi G$ computed in the chapter text. One could say that gravity in the solar system, and in the Milky Way galaxy, is such a strong force that it has stopped the Hubble expansion in our locality. Hubble's law as an observational result describes the expansion of distances between clusters of galaxies.

□ □ □ □

ANSWERS TO SELECTED CONCEPTUAL QUESTIONS

4. Describe the properties of baryons and mesons and the important differences between them.

Answer There are two types of hadrons, called baryons and mesons. Hadrons interact primarily through the strong force and are not elementary particles, being composed of either three quarks (baryons), or a quark and an antiquark (mesons). Baryons have a nonzero baryon number with a spin of either 1/2 or 3/2. Mesons have baryon number of zero, and a spin of either 0 or 1.

□ □ □ □

5. The Ξ^0 particle decays by the weak interaction according to the decay mode $\Xi^0 \rightarrow \Lambda^0 + \pi^0$. Would you expect this decay to be fast or slow? Explain.

Answer This decay should be slow, since decays which occur via the weak interaction typically take 10^{-10} s or longer to occur. This decay cannot occur by the strong interaction, because at the level of quarks it is uss → uds + $\bar{d}$d, changing the net strangeness quantum number.

□ □ □ □

9. How many quarks are in each of the following: (a) a baryon, (b) an antibaryon, (c) a meson, (d) an antimeson? (e) How do you explain that baryons have half-integral spins whereas mesons have spins of 0 or 1?

Answer (a) and (b) All baryons and antibaryons consist of three quarks. (c) and (d) All mesons and antimesons consist of two quarks. When one quantum particle with spin 1/2 combines with another particle to form a new particle, the spin of the new particle must be either 1/2 greater or 1/2 less than that of the original. Since quarks have spins of 1/2, it follows that all baryons (which consist of three quarks) must have half-integral spins, and all mesons (which consist of two quarks) must have spins of 0 or 1.

□ □ □ □

SOLUTIONS TO SELECTED PROBLEMS

5. A photon with an energy $E_\gamma = 2.09$ GeV creates a proton–antiproton pair in which the proton has a kinetic energy of 95.0 MeV. What is the kinetic energy of the antiproton? *Note:* $m_p c^2 = 938.3$ MeV.

Solution

Conceptualize: An antiproton has the same mass as a proton, so it seems reasonable to expect that both particles will have similar kinetic energies.

Categorize: The total energy of each particle is the sum of its rest energy and its kinetic energy. Conservation of system energy requires that the total energy before this pair production event equal the total energy after.

Analyze: $E_\gamma = \left(E_{Rp} + K_p\right) + \left(E_{R\bar{p}} + K_{\bar{p}}\right)$

The energy of the photon is given as $E_\gamma = 2.09 \text{ GeV} = 2.09 \times 10^3 \text{ MeV}$

From Table 46.2 or from the problem statement, we see that the rest energy of both the proton and the antiproton is

$$E_{Rp} = E_{R\bar{p}} = m_p c^2 = 938.3 \text{ MeV}$$

If the kinetic energy of the proton is observed to be 95.0 MeV, the kinetic energy of the antiproton is

$$K_{\bar{p}} = E_\gamma - E_{R\bar{p}} - E_{Rp} - K_p = 2.09 \times 10^3 \text{ MeV} - 2(938.3 \text{ MeV}) - 95.0 \text{ MeV}$$

$$K_{\bar{p}} = 118 \text{ MeV}$$ ■

Finalize: The kinetic energy of the antiproton is slightly (~20%) greater than the proton. The magnitude of the momentum of each product particle is less than 1/4 of the gamma ray's momentum. Thus, another particle must have been involved to satisfy system momentum conservation. It could be a pre-existing heavy nucleus with which the photon collided. This means that the nucleus must have also carried away some of the energy. For a heavy nucleus this could have been small, as little as 3 MeV for a favorable geometry. The actual value cannot be determined without further information. This extra particle also explains why the energy need not be shared equally between the proton and antiproton.

---

6. One mediator of the weak interaction is the Z^0 boson, with mass 91 GeV/c^2. Use this information to find the order of magnitude of the range of the weak interaction.

Solution

Conceptualize: Following Yukawa, we visualize the weak interaction as acting between two matter particles when they exchange a virtual Z^0 boson. The boson should not exist according to classical energy conservation, and . . .

Categorize: . . . it cannot exist for a time interval longer than that allowed by the uncertainty principle. This is the condition that tells us about the range of the force.

Analyze: The rest energy of the Z^0 boson is $E_0 = 91$ GeV. The maximum time a virtual Z^0 boson can exist is found from

$$\Delta E \Delta t \geq \tfrac{1}{2}\hbar \quad \text{or} \quad \Delta t = \frac{\hbar}{2\Delta E} = \frac{1.055 \times 10^{-34}\ \text{J}\cdot\text{s}}{2(91\ \text{GeV})\left(1.60 \times 10^{-10}\ \text{J/GeV}\right)} = 3.62 \times 10^{-27}\ \text{s}$$

The maximum distance it can travel in this time interval is

$$d = c\Delta t = \left(3.00 \times 10^{8}\ \text{m/s}\right)\left(3.62 \times 10^{-27}\ \text{s}\right) \sim 10^{-18}\ \text{m} \qquad \blacksquare$$

The distance d is an approximate value for the range of the weak interaction.

Finalize: Our answer agrees with the datum in Table 46.1.

---

9. A neutral pion at rest decays into two photons according to $\pi^0 \rightarrow \gamma + \gamma$. Find the (a) energy, (b) momentum, and (c) frequency of each photon.

Solution

Conceptualize: A pion is not a very massive particle, but we guess that it is heavy enough for these to be gamma-ray photons.

Categorize: We use the energy and momentum versions of the isolated system model.

Analyze: Since the pion is at rest and system momentum is conserved in the decay, the two gamma rays must have equal amounts of momentum in opposite directions. So they must share equally in the energy of the pion.

We have the original mass $m_{\pi^0} = 135.0\ \text{MeV}/c^2$ (Table 46.2)

(a) Therefore the energy of each photon is $E_\gamma = 67.5 \text{ MeV} = 1.08 \times 10^{-11} \text{ J}$ ■

(b) Its momentum is $p = \dfrac{E_\gamma}{c} = \dfrac{67.5 \text{ MeV}}{3.00 \times 10^8 \text{ m/s}} = 3.60 \times 10^{-20} \text{ kg} \cdot \text{m/s}$ ■

(c) and its frequency is $f = \dfrac{E_\gamma}{h} = 1.63 \times 10^{22} \text{ Hz}$ ■

Finalize: Gamma ray is right! This electromagnetic wave is at the top of the spectrum chart in the text's Figure 34.13. Currently [December 2009] the Fermi Space Telescope is searching for gamma rays at just this energy from starburst galaxies, as compelling evidence that supernova explosions and their aftermath produce cosmic rays. The logic: In a region where many stars are being born, numerous large-mass stars are dying as supernovas. Expanding shock waves around them accelerate charged particles to energies up to the TeV range. The charged particles themselves do not go off in straight lines—they get all mixed up in direction by magnetic fields. Those that we on Earth receive as cosmic rays do not reveal their source to us. But some of the particles collide or decay within the source galaxy to produce pions, which in turn decay into photons by the process considered here. We on Earth can sample the gamma rays and (we hope) trace them straight back to the source.

---

13. The following reactions or decays involve one or more neutrinos. In each case, supply the missing neutrino (ν_e, ν_μ, or ν_τ) or antineutrino.

(a) $\pi^- \to \mu^- + ?$ (b) $K^+ \to \mu^+ + ?$

(c) $? + p \to n + e^+$ (d) $? + n \to p + e^-$

(e) $? + n \to p + \mu^-$ (f) $\mu^- \to e^- + ? + ?$

Solution

Conceptualize: There are just six choices: an electron neutrino, a muon neutrino, a tau neutrino, or the antineutrino version of each one.

Categorize: By following the conservation of electron-lepton number, muon-lepton number, and tau-lepton number we will obtain each answer.

Analyze:

(a) The muon-lepton number in the reaction $\pi^- \to \mu^- + ?$ changes from 0 to 1 + ?, so the muon-lepton number of the final particle must be −1 and the reaction must be $\pi^- \to \mu^- + \bar{\nu}_\mu$ to give $L_\mu : 0 \to 1 - 1$ ■

(b) Following the same logic we have $K^+ \to \mu^+ + \nu_\mu$ so that the muon lepton number will be conserved according to $L_\mu : 0 \to -1 + 1$ ■

(c) $\bar{\nu}_e + p^+ \to n + e^+$ $L_e : -1 + 0 \to 0 - 1$ ■

(d) $\nu_e + \text{n} \rightarrow \text{p}^+ + \text{e}^-$ $\quad L_e: 1 + 0 \rightarrow 0 + 1$ ■

(e) $\nu_\mu + \text{n} \rightarrow \text{p}^+ + \mu^-$ $\quad L_\mu: 1 + 0 \rightarrow 0 + 1$ ■

(f) $\mu^- \rightarrow \text{e}^- + \bar{\nu}_e + \nu_\mu$ $\quad L_\mu: 1 \rightarrow 0 + 0 + 1$ and $L_e: 0 \rightarrow 1 - 1 + 0$ ■

Finalize: The problem asked us to identify neutrinos, but we could have used the same rules to determine whether an electron, muon, or tau participated in a reaction, if the problem had told us what kind of neutrino it was.

---

15. Determine which of the following reactions can occur. For those that cannot occur, determine the conservation law (or laws) violated.

(a) $\text{p} \rightarrow \pi^+ + \pi^0$ (b) $\text{p} + \text{p} \rightarrow \text{p} + \text{p} + \pi^0$

(c) $\text{p} + \text{p} \rightarrow \text{p} + \pi^+$ (d) $\pi^+ \rightarrow \mu^+ + \nu_\mu$

(e) $\text{n} \rightarrow \text{p} + \text{e}^- + \bar{\nu}_e$ (f) $\pi^+ \rightarrow \mu^+ + \text{n}$

Solution

Conceptualize: Conservation laws include those for a system's total energy, momentum, angular momentum, charge, baryon number, three lepton numbers, strangeness, charm, bottomness, and topness.

Categorize: The particles participating here all have zero strangeness, charm, bottomness, and topness. Their motion determines their energy, momentum, and angular momentum. So we have only conservation of charge, baryon number, and lepton numbers to check.

Analyze:

(a) $\text{p} \rightarrow \pi^+ + \pi^0$ Baryon number conservation is violated. If the reaction occurred, the baryon number would change according to $1 \rightarrow 0 + 0$ ■

(b) $\text{p} + \text{p} \rightarrow \text{p} + \text{p} + \pi^0$ This reaction can occur. ■

(c) $\text{p} + \text{p} \rightarrow \text{p} + \pi^+$ Baryon number is violated: $1 + 1 \rightarrow 1 + 0$ ■

(d) $\pi^+ \rightarrow \mu^+ + \nu_\mu$ This reaction can occur. ■

(e) $\text{n} \rightarrow \text{p} + \text{e}^- + \bar{\nu}_e$ This reaction can occur. ■

(f) $\pi^+ \rightarrow \mu^+ + \text{n}$ Violates baryon number: $0 \rightarrow 0 + 1$ and violates muon-lepton number: $0 \rightarrow -1 + 0$ Further, if the pion is not colliding with another particle, it does not have enough energy to undergo this decay. ■

Finalize: Note that reactions proceeding by the weak interaction can alter the strangeness, charm, topness, and bottomness of an isolated system.

---

18. A Λ^0 particle at rest decays into a proton and a π^- meson. (a) Use the data in Table 46.2 to find the Q value for this decay in MeV. (b) What is the total kinetic energy shared by the proton and the π^- meson after the decay? (c) What is the total momentum shared by the proton and the π^- meson? (d) The proton and π^- meson have momenta with the same magnitude after the decay. Do they have equal kinetic energies? Explain.

Solution

Conceptualize: This reaction, among others, is shown in the bubble-chamber photographs of the text's Figure 46.7 and 46.10. We expect the less massive pion to carry off the majority of the energy.

Categorize: We use the relativistic energy and relativistic momentum versions of the isolated-system model.

Analyze:

(a) To conserve charge, the decay reaction is $\Lambda^0 \to \mathrm{p} + \pi^-$.

We look up in the table the rest energy of each particle:

$$m_\Lambda c^2 = 1\,115.6 \text{ MeV} \qquad m_\mathrm{p} c^2 = 938.3 \text{ MeV} \qquad m_\pi c^2 = 139.6 \text{ MeV}$$

The Q value of the reaction, representing the energy output, is the difference between starting rest energy and final rest energy, and is the kinetic energy of the products:

$$Q = (1\,115.6 - 938.3 - 139.6) \text{ MeV} = 37.7 \text{ MeV} \quad \blacksquare$$

(b) The original kinetic energy is zero in the process considered here, so the whole Q becomes the kinetic energy of the products

$$K_\mathrm{p} + K_\pi = 37.7 \text{ MeV} \quad \blacksquare$$

(c) The lambda particle is at rest. Its momentum is zero. System momentum is conserved in the decay, so the total vector momentum of the proton and the pion must be zero. ■

(d) The proton and the pion move in precisely opposite directions with precisely equal momentum magnitudes. Because their masses are different, their kinetic energies are not the same. ■

The mass of the π-meson is much less than that of the proton, so it carries much more kinetic energy. We can find the energy of each. Let p represent the magnitude of the momentum of each. Then the total energy of each particle is given by $E^2 = (pc)^2 + (mc^2)^2$ and its kinetic energy is $K = E - mc^2$. For the total kinetic energy of the two particles we have

$$\sqrt{m_p^2c^4 + p^2c^2} - m_pc^2 + \sqrt{m_\pi^2c^4 + p^2c^2} - m_\pi c^2 = Q = m_\Lambda c^2 - m_pc^2 - m_\pi c^2$$

Proceeding to solve for pc, we find

$$m_p^2c^4 + p^2c^2 = m_\Lambda^2c^4 - 2m_\Lambda c^2\sqrt{m_\pi^2c^4 + p^2c^2} + m_\pi^2c^4 + p^2c^2$$

$$\sqrt{m_\pi^2c^4 + p^2c^2} = \frac{m_\Lambda^2c^4 - m_p^2c^4 + m_\pi^2c^4}{2m_\Lambda c^2} = \frac{1\,115.6^2 - 938.3^2 + 139.6^2}{2(1\,115.6)} \text{MeV} = 171.9 \text{ MeV}$$

$$pc = \sqrt{171.9^2 - 139.6^2} \text{ MeV} = 100.4 \text{ MeV}$$

Then the kinetic energies are

$$K_p = \sqrt{938.3^2 + 100.4^2} - 938.3 = 5.35 \text{ MeV}$$

and $K_\pi = \sqrt{139.6^2 + 100.4^2} - 139.6 = 32.3 \text{ MeV}$

Finalize: Observe that charge, baryon number, and three lepton numbers, as well as system energy and system momentum, are conserved in this decay. The decay occurs by the weak interaction and strangeness is not conserved. The proton has 6.7 times more rest energy compared to the pion and carries 6.0 times more kinetic energy. It is relativity that makes these two ratios different.

---

19. Determine whether or not strangeness is conserved in the following decays and reactions.

(a) $\Lambda^0 \to \text{p} + \pi^-$ (b) $\pi^- + \text{p} \to \Lambda^0 + \text{K}^0$

(c) $\bar{\text{p}} + \text{p} \to \bar{\Lambda}^0 + \Lambda^0$ (d) $\pi^- + \text{p} \to \pi^- + \Sigma^+$

(e) $\Xi^- \to \Lambda^0 + \pi^-$ (f) $\Xi^0 \to \text{p} + \pi^-$

Solution

Conceptualize: We could think about the quark composition of each particle, but it is simplest to compare the total strangeness quantum number before and after, based on . . .

Categorize: . . . the listings for the individual particles in Table 46.2.

Analyze:

(a) For $\Lambda^0 \to \text{p} + \pi^-$ we have strangeness: $-1 \to 0 + 0$. Since -1 does not equal 0, strangeness is not conserved. ■

(b) Following the same procedure, for $\pi^- + \text{p} \to \Lambda^0 + \text{K}^0$ we have strangeness: $0 + 0 \to -1 + 1$. This reduces to $0 = 0$ so strangeness is conserved. ■

(c) $\bar{\text{p}} + \text{p} \to \bar{\Lambda}^0 + \Lambda^0$ gives strangeness: $0 + 0 \to +1 - 1$. Since $0 = 0$, strangeness is conserved. ■

(d) $\pi^- + \text{p} \to \pi^- + \Sigma^+$ gives strangeness: $0 + 0 \to 0 - 1$. Since 0 does not equal -1, strangeness is not conserved. ■

(e) $\Xi^- \to \Lambda^0 + \pi^-$ gives strangeness: $-2 \to -1 + 0$. Since -2 does not equal -1, strangeness is not conserved. ■

(f) $\Xi^0 \to \text{p} + \pi^-$ gives strangeness: $-2 \to 0 + 0$. Since -2 does not equal 0, strangeness is not conserved. ■

Finalize: Note again that reactions proceeding by the weak interaction can alter the strangeness of an isolated system.

---

25. If a K_S^0 meson at rest decays in 0.900×10^{-10} s, how far does a K_S^0 meson travel if it is moving at $0.960c$?

Solution

Conceptualize: The motion of the K_S^0 particle is relativistic. Just like the astronaut who leaves for a distant star and returns to find his family long gone, the kaon appears to us to have a longer lifetime.

Categorize: We identify the effect of time dilation and use the particle under constant speed model.

Analyze: The time-dilated lifetime is

$$t = \gamma t_0 = \frac{0.900 \times 10^{-10}\ \text{s}}{\sqrt{1 - v^2/c^2}} = \frac{0.900 \times 10^{-10}\ \text{s}}{\sqrt{1 - (0.960)^2}} = 3.21 \times 10^{-10}\ \text{s}$$

During this time interval, we see the kaon travel at $0.960c$. It travels for a distance of

$$d = vt = (0.960)\left(3.00 \times 10^8\ \text{m/s}\right)\left(3.21 \times 10^{-10}\ \text{s}\right) = 0.092\,6\ \text{m} = 9.26\ \text{cm} \quad \blacksquare$$

Finalize: This distance can fit into a particle detector, such as the bubble-chamber photographs in the text's Figures 46.7 and 46.10.

---

32. Analyze each of the following reactions in terms of constituent quarks and show that each type of quark is conserved.

(a) $\pi^+ + \text{p} \rightarrow \text{K}^+ + \Sigma^+$

(b) $\text{K}^- + \text{p} \rightarrow \text{K}^+ + \text{K}^0 + \Omega^-$

(c) Determine the quarks in the final particle for this reaction:

$\text{p} + \text{p} \rightarrow \text{K}^0 + \text{p} + \pi^+ + ?$

(d) In the reaction in part (c), identify the mystery particle.

Solution

Conceptualize: We should see the same quarks differently arranged on the two sides of each reaction.

Categorize: We look up the quark constituents of the particles in Tables 46.4 and 46.5.

Analyze:

(a) The reaction in terms of quarks reads $\bar{\text{d}}\text{u} + \text{uud} \rightarrow \text{u}\bar{\text{s}} + \text{uus}$, which nets 3u, 0d, 0s before and after. $\blacksquare$

(b) Following the same procedure gives $\bar{\text{u}}\text{s} + \text{uud} \rightarrow \text{u}\bar{\text{s}} + \text{d}\bar{\text{s}} + \text{sss}$, which shows conservation at 1u, 1d, 1s before and after. $\blacksquare$

(c) $uud + uud \rightarrow d\bar{s} + uud + u\bar{d} + uds$ has 4u, 2d, 0s before and after. ■

(d) A uds is either a Λ^0 or a Σ^0. ■

Finalize: With larger mass, the Σ^0 can be thought of as an excited state of the Λ^0, with the quarks inside it moving around faster.

---

43. The first quasar to be identified and the brightest found to date, 3C 273 in the constellation Virgo, was observed to be moving away from the Earth at such high speed that the observed blue 434-nm H_γ line of hydrogen is Doppler-shifted to 510 nm, in the green portion of the spectrum (Fig. P46.43). (a) How fast is the quasar receding? (b) Edwin Hubble discovered that all objects outside the local group of galaxies are moving away from us, with speeds v proportional to their distances R. Hubble's law is expressed as $v = HR$, where Hubble's constant has the approximate value $H \approx 22 \times 10^{-3}$ m/s·ly. Determine the distance from the Earth to this quasar.

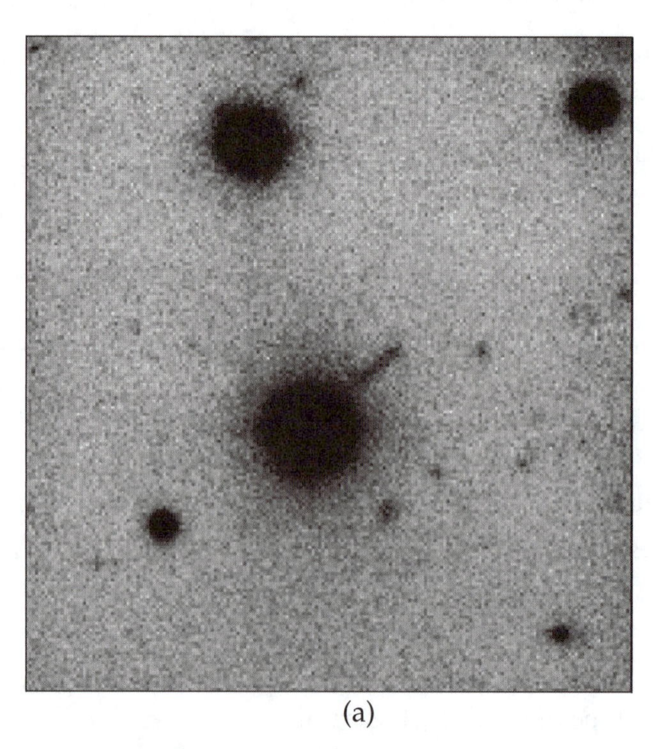

(a)

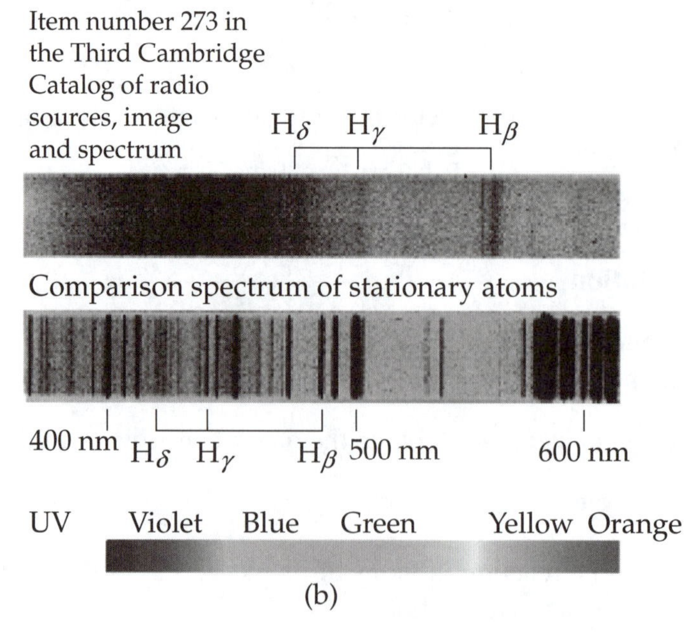

(b)

Figure P46.43

Solution

Conceptualize: The problem states that the quasar is moving very fast, and since there is a significant red shift of the light, the quasar must be moving away from Earth at a relativistic speed ($v > 0.1c$). Quasars are very distant astronomical objects, and since our universe is estimated to be about 14 billion years old, we should expect this quasar to be some billions of light-years away.

Categorize: We use the treatment of the relativistic Doppler effect in Chapter 39 to find the speed of the quasar from the Doppler red shift, and this speed can then be used to find the distance using Hubble's law.

Analyze:

(a) We use primed symbols to represent observed Doppler-shifted values and unprimed symbols to represent values as they would be measured by an observer stationary relative to the source. Doppler-shift equations from Chapter 17 do not apply to electromagnetic waves, because the speed of source or observer relative to some medium cannot be defined for these waves. Instead, we use Equation 39.10, expressing it as

$$f' = \frac{c}{\lambda'} = \sqrt{\frac{1+v/c}{1-v/c}}\,f = \sqrt{\frac{1+v/c}{1-v/c}}\,\frac{c}{\lambda}$$

where v is the velocity of mutual approach. Then we have

$$\frac{\lambda'}{\lambda} = \sqrt{\frac{1-v/c}{1+v/c}} \quad \text{and proceed to solve:} \quad \left(\frac{\lambda'}{\lambda}\right)^2 = \frac{1-v/c}{1+v/c}$$

$$\left(\frac{\lambda'}{\lambda}\right)^2 + \left(\frac{\lambda'}{\lambda}\right)^2 \frac{v}{c} = 1 - \frac{v}{c} \qquad \left(\frac{\lambda'}{\lambda}\right)^2 - 1 = -\frac{v}{c}\left[\left(\frac{\lambda'}{\lambda}\right)^2 + 1\right]$$

$$\frac{v}{c} = -\frac{(\lambda'/\lambda)^2 - 1}{(\lambda'/\lambda)^2 + 1} = -\frac{(510\text{ nm}/434\text{ nm})^2 - 1}{(510\text{ nm}/434\text{ nm})^2 + 1} = \frac{(1.18)^2 - 1}{(1.18)^2 + 1}$$

$$= -\frac{1.381 - 1}{1.381 + 1} = -0.160$$

The negative sign indicates that the quasar is moving away from us, or us from it. The speed of recession that the problem asks for is then

$v = 0.160c$ (or 16.0% of the speed of light) ■

(b) Hubble's law asserts that the universe is expanding at a constant rate so that the speeds of galaxies are proportional to their distance R from Earth, as described by $v = HR$.

So, $$R = \frac{v}{H} = \frac{0.160\left(3.00 \times 10^8\text{ m/s}\right)}{2.2 \times 10^{-2}\text{ m/s}\cdot\text{ly}} = 2.18 \times 10^9\text{ ly}$$ ■

Finalize: The speed and distance of this quasar are consistent with our predictions. It appears that this quasar is quite far from Earth but the universe was comparable to its present age when the quasar was radiating the light we now receive.

---

52. The energy flux carried by neutrinos from the Sun is estimated to be on the order of 0.400 W/m^2 at Earth's surface. Estimate the fractional mass loss of the Sun over 10^9 yr due to the emission of neutrinos. The mass of the Sun is 1.989×10^{30} kg. The Earth–Sun distance is 1.496×10^{11} m.

Solution

Conceptualize: Our Sun is estimated to have a life span of about 10 billion years, so in this problem, we are examining the radiation of neutrinos over a considerable fraction of the Sun's life. However, the mass carried away by the neutrinos is a very small fraction of the total mass involved in the Sun's nuclear fusion process, so even over this long time interval, the mass of the Sun may not change significantly (probably by less than 1%).

Categorize: The change in mass of the Sun can be found from the energy flux received by the Earth and Einstein's famous equation, $E = mc^2$. We will also use the inverse square law and the definition of power.

Analyze: Since the neutrino flux from the Sun reaching the Earth is 0.4 W/m², the total energy emitted per second by the Sun in neutrinos in all directions is that which would irradiate the surface of a great sphere around it, with the Earth's orbit as its equator.

$$(0.400\ \text{W/m}^2)(4\pi r^2) = (0.400\ \text{W/m}^2)\left[4\pi(1.496\times 10^{11}\ \text{m})^2\right] = 1.12\times 10^{23}\ \text{W}$$

In a period of 10^9 yr, the Sun emits a total energy of $E = P\Delta t$.

$$E = (1.12\times 10^{23}\ \text{J/s})(10^9\ \text{yr})(3.156\times 10^7\ \text{s/yr}) = 3.55\times 10^{39}\ \text{J}$$

carried by neutrinos. This energy corresponds to an annihilated mass according to

$$E = m_\nu c^2 = 3.55\times 10^{39}\ \text{J} \quad \text{so} \quad m_\nu = 3.94\times 10^{22}\ \text{kg}$$

Since the Sun has a mass of 1.989×10^{30} kg, this corresponds to a loss of only about 1 part in 50 400 000 of the Sun's mass over 10^9 yr in the form of neutrinos. ■

Finalize: It appears that the neutrino flux changes the mass of the Sun by so little that it would be difficult to measure the difference in mass, even over its lifetime!

---

56. A Σ^0 particle at rest decays according to $\Sigma^0 \to \Lambda^0 + \gamma$. Find the gamma-ray energy.

Solution

Conceptualize: The Λ^0 and the Σ^0 have the same quark composition. You can think of the more massive Σ^0 as an excited state of the Λ^0, with the u, d, and s quarks in it moving around more rapidly. From Table 46.2, the rest energies are $E_{0\Sigma} = m_\Sigma c^2 = 1\,192.5$ MeV and $E_{0\Lambda} = m_\Lambda c^2 = 1\,115.6$ MeV. The photon has zero rest energy. Subtracting gives the Q value of the reaction as 76.9 MeV. This energy must be shared between the lambda particle and the photon. The photon will have most of the decay energy but not all of it, because the lambda particle must move in the opposite direction to the photon with enough speed to have an equal amount of momentum.

Categorize: We use the energy and momentum versions of the isolated system model. The lambda particle will be moving at less than one-tenth of the speed of light, so we can use the classical relation $K = \frac{1}{2}mv^2 = p^2/2m$ for it. The only relativistic equations we need are the identification of the Q value as $m_\Sigma c^2 - m_\Lambda c^2$ and $E = pc$ for the photon.

Analyze: For $\Sigma^0 \to \Lambda^0 + \gamma$ conservation of system energy in the decay requires

$$E_{0\Sigma} = \left(E_{0\Lambda} + K_\Lambda\right) + E_\gamma \qquad \text{or} \qquad m_\Sigma c^2 = \left(m_\Lambda c^2 + \frac{p_\Lambda^2}{2m_\Lambda}\right) + E_\gamma$$

or $$m_\Sigma c^2 = \left(m_\Lambda c^2 + \frac{p_\Lambda^2 c^2}{2m_\Lambda c^2}\right) + E_\gamma$$

System momentum conservation gives $|p_\Lambda| = |p_\gamma|$, so the last result may be written as

$$m_\Sigma c^2 = m_\Lambda c^2 + \frac{p_\gamma^2 c^2}{2m_\Lambda c^2} + E_\gamma$$

Recognizing that $p_\gamma c = E_\gamma$ we now have $Q = \dfrac{E_\gamma^2}{2m_\Lambda c^2} + E_\gamma$

Solving this quadratic equation gives

$$E_\gamma = m_\Lambda c^2\left(\sqrt{1 + \frac{2Q}{m_\Lambda c^2}} - 1\right) = (1\,115.6 \text{ MeV})\left(\sqrt{1 + \frac{2(76.9)}{1\,115.6}} - 1\right)$$

$$E_\gamma = 74.4 \text{ MeV}$$ ■

Finalize: The kinetic energy of the lambda particle is only about $Q - E_\gamma = 2.48$ MeV, so its speed is

$$v = \sqrt{\frac{2K}{m}} = c\sqrt{\frac{2K}{mc^2}} = 3.00\times10^8 \ \frac{\text{m}}{\text{s}}\sqrt{\frac{2(2.48)}{1\,115.6}} = 0.066\ 7c = 2.00\times10^7 \ \frac{\text{m}}{\text{s}}$$

This is less than $0.1c$, confirming our assumption that it moves nonrelativistically.

---

67. Assume the average density of the Universe is equal to the critical density. (a) Prove that the age of the Universe is given by $2/(3H)$. (b) Calculate $2/(3H)$ and express it in years.

Solution

Conceptualize: The reciprocal of Hubble's constant is a time interval. If the Universe had a very low density of matter to generate very little gravitation, the Hubble's-law expansion would proceed at nearly constant speed. For two objects now a large distance d apart, that distance opened up over the age of the Universe T according to $d = vT = HdT$, so that $T = d/v = d/Hd = 1/H$ would be the time since the Big Bang. With significant density, as in this problem, gravitational attraction slows the expansion. The average expansion speed d/T is higher than the current speed Hd and the age of the Universe is less than $1/H$. We are here to prove that it is two-thirds as large, in the case of a Universe with the critical density.

Categorize: The critical density is defined in terms of the Big Bang's setting objects moving with sufficient kinetic energy to make the expansion continue forever, with speeds approaching zero. We will use conservation of system energy for a large section of the Universe, and also the definition of the Hubble constant.

Analyze:

(a) Consider a sphere around us of radius R large compared to the size of galaxy clusters. If the matter, of mass M, inside the sphere has the critical density, then a galaxy of mass m at the surface of the sphere is moving just at its escape speed, according to

$$K + U_g = 0 \qquad \text{or} \qquad \frac{1}{2}mv^2 - GMm/R = 0$$

Note that all the matter outside the sphere does not generate gravitational fields inside the sphere, just as, according to Gauss's law, symmetrically placed charges outside a shell do not produce an electric field inside the shell. The energy of the galaxy-sphere system is conserved, so this equation is true throughout the history of the Universe, from when $v = dR/dt$ was large and R was small soon after the Big Bang, to now when v is smaller and R vastly larger. Then we have $\left(\frac{dR}{dt}\right)^2 = \frac{2GM}{R}$ so $\frac{dR}{dt} = R^{-1/2}\sqrt{2GM}$

Integrating from the start of the Universe to the present epoch T, we have

$$\int_0^R R^{1/2}dR = \sqrt{2GM}\int_0^T dt \qquad \left.\frac{R^{3/2}}{3/2}\right|_0^R = \sqrt{2GM}\, t\big|_0^T$$

$$\frac{2}{3}R^{3/2} = \sqrt{2GM}\, T \qquad \text{and} \qquad T = \frac{2}{3}\frac{R^{3/2}}{\sqrt{2GM}} = \frac{2}{3}\frac{R}{\sqrt{2GM/R}}$$

From above, $v = \sqrt{\frac{2GM}{R}}$ so $T = \frac{2}{3}\frac{R}{v}$

Now Hubble's law summarizes observational results as

$$v = HR \qquad \text{so} \qquad T = \frac{2}{3}\frac{R}{HR} = \frac{2}{3H}$$ ■

(b) $T = \frac{2}{3H} = \frac{2}{3(22\times10^{-3}\,\text{m/s}\cdot\text{ly})}\frac{3.00\times10^{8}\,\text{m/s}}{1\,\text{ly/yr}} = 9.09\times10^{9}\,\text{yr}$ ■

Finalize: Knowing about how stars of different masses age and die lets astronomers measure the ages of clusters of stars, each comprising stars of different masses that were born at the same time. Some star clusters are older than this nine billion years, so the Universe must definitely be older still. Then the Universe must have less than the critical density. Whether humans find it congenial or even possible to think about, the Universe appears to be a one-shot deal, starting at a particular moment a finite time ago and continuing to expand and cool forever.

Our calculation can be thought of as including the effect of dark matter in slowing the expansion of the Universe, by including the gravitating mass of the dark matter in the factor M. But then our calculation does not include the effect of mysterious dark energy in speeding up the expansion. Perhaps the value of H that we used, characteristic of our current neighborhood and epoch, is too large to characterize the Universe throughout its history. Then our conclusions about the density and future of the Universe are tentative.